Small Air-Cooled Engine 1990 - 1994

SERVICE MANUAL ■ 1ST EDITION

An INTERTEC® Publication

P.O. Box 12901 ■ Overland Park, KS 66282-2901

Cover photo courtesy of Briggs & Stratton Corporation

Small Air-Cooled Engine *1990-1994*

SERVICE MANUAL ■ *1ST EDITION*

Small Air-Cooled Engine Manufacturers:

- Acme
- Briggs & Stratton
- Craftsman
- Homelite
- Honda

- Kawasaki
- Kohler
- Lawn Boy
- Shindaiwa
- Stihl

- Tanaka
- Tecumseh
- Toro
- Wisconsin Robin
- Yamaha

INTERTEC PUBLISHING

President and CEO Raymond E. Maloney

Group Vice President Dan Fink

EDITORIAL

Editorial Director
Randy Stephens

Editors
Mike Hall
Mark Jacobs

Technical Writers
Robert Mills
Michael Morlan
Edward Scott
Ron Wright

Product Planning and Support Manager
Terry Distin

Lead Editorial Assistant
Shirley Renicker

Editorial Assistants
Veronica Bollin
Elizabeth Couzens
Dylan Goodwin

Technical Illustrators
Steve Amos
Robert Caldwell
Mitzi McCarthy
Diana Kirkland

MARKETING

Marketing Services Manager
Katherine Hughes

Advertising Assistant
Hilary Lindsey

Art Director
Anita Blattner

SALES AND ADMINISTRATION

Director of Sales
Dutch Sadler

Accounts Manager
Ted Metzger

Sales Coordinator
Lynn Reynolds

Customer Service and Administration Manager
Joan Dickey

The following books and guides are published by Intertec Publishing

---

CLYMER SHOP MANUALS
Boat Motors and Drives
Motorcycles and ATVs
Snowmobiles
Personal Watercraft

ABOS/INTERTEC BLUE BOOKS AND TRADE-IN GUIDES
Recreational Vehicles
Outdoor Power Equipment
Agricultural Tractors
Lawn and Garden Tractors
Motorcycles and ATVs
Snowmobiles
Boats and Motors
Personal Watercraft

AIRCRAFT BLUEBOOK-PRICE DIGEST
Airplanes
Helicopters

I&T SHOP MANUALS
Tractors

INTERTEC SERVICE MANUALS
Snowmobiles
Outdoor Power Equipment
Personal Watercraft
Gasoline and Diesel Engines
Recreational Vehicles
Boat Motors and Drives
Motorcycles
Lawn and Garden Tractors

CONTENTS

GENERAL

ENGINE SERVICE SECTIONS

DUAL DIMENSIONS

This service manual provides specifications in both the Metric (SI) and U.S. Customary systems of measurement. The first specification is given in the measuring system used during manufacture, while the second specification (given in parenthesis) is the converted measurement. For instance, a specification of "0.28 mm (0.011 inch)" would indicate that the equipment was manufactured using the metric system of measurement and U.S. equivalent of 0.28 mm is 0.011 inch.

FUNDAMENTALS SECTION
ENGINE FUNDAMENTALS

OPERATING PRINCIPLES

The small engines used to power lawn mowers, garden tractors and many other items of power equipment in use today are basically similar. All are technically known as "Internal Combustion Reciprocating Engines."

The source of power is heat formed by the burning of a combustible mixture of petroleum products and air. In a reciprocating engine, this burning takes place in a closed cylinder containing a piston. Expansion resulting from the heat of combustion applies pressure on the piston to turn a shaft by means of a crank and connecting rod.

The fuel:air mixture may be ignited by means of an electric spark (Otto Cycle Engine) or by heat formed from compression of air in the engine cylinder (Diesel Cycle Engine). The complete series of events which must take place in order for the engine to run may occur in one revolution of the crankshaft (two strokes of the piston in cylinder) which is referred to as a "Two Stroke Cycle Engine," or in two revolutions of the crankshaft (four strokes of the piston in cylinder) which is referred to as a "Four-Stroke Cycle Engine."

OTTO CYCLE. In a spark ignited engine, a series of five events is required in order for the engine to provide power. This series of events is called the "Cycle" (or "Work Cycle") and is repeated in each cylinder of the engine as long as work is being done. This series of events which comprise the "Cycle" are as follows:

1. The mixture of fuel and air is pushed into the cylinder by atmospheric pressure when the pressure within the engine cylinder is reduced by the piston moving downward in the cylinder (or by applying pressure to the fuel:air mixture as by crankcase compression in the crankcase of a "Two-Stroke Cycle Engine" which is described in a later paragraph).
2. The mixture of fuel and air is compressed by the piston moving upward in the cylinder.
3. The compressed fuel:air mixture is ignited by a timed electric spark.
4. The burning fuel:air mixture expands, forcing the piston downward in the cylinder thus converting the

chemical energy generated by combustion into mechanical power.
5. The gaseous products formed by the burned fuel:air mixture are exhausted from the cylinder so that a new "Cycle" can begin.

The above described five events which comprise the work cycle of an engine are commonly referred to as (1), INTAKE; (2), COMPRESSION; (3), IGNITION; (4), EXPANSION (POWER); and (5), EXHAUST.

DIESEL CYCLE. The Diesel Cycle differs from the Otto Cycle in that air alone if drawn into the cylinder during the intake period. The air is heated from being compressed by the piston moving upward in the cylinder, then a finely atomized charge of fuel is injected into the cylinder where it mixes with the air and is ignited by the heat of the compressed air. In order to create sufficient heat to ignite the injected fuel, an engine operating on the Diesel Cycle must compress the air to a much greater degree than an engine operating on the Otto Cycle where the fuel:air mixture is ignited by an electric spark. The power and exhaust events of the Diesel Cycle are similar to the power and exhaust events of the Otto Cycle.

TWO-STROKE CYCLE ENGINES. Two-stroke cycle engines may be of the Otto Cycle (spark ignition) or Diesel Cycle (compression ignition) type. However, since the two-stroke cycle engines listed in the repair section of this manual are all of the Otto Cycle type, operation of two-stroke Diesel Cycle engines will not be discussed in this section.

In two-stroke cycle engines, the piston is used as a sliding valve for the cylinder intake and exhaust ports. See Fig. 1-1. The intake and exhaust ports are both open when the piston is at the bottom of its downward stroke (bottom dead center or "BDC"). The exhaust port is open to atmospheric pressure; therefore, the fuel:air mixture must be elevated to a higher than atmospheric pressure in order for the mixture to enter the cylinder. As the crankshaft is turned from BDC and the piston starts on its upward stroke, the intake and exhaust ports are closed and the fuel:air mixture in the cylinder is compressed.

When piston is at or near the top of its upward stroke (top dead center or "TDC"), an electric spark across the electrode gap of the spark plug ignites the fuel:air mixture. As the crankshaft turns past TDC and the piston starts on its downward stroke, the rapidly burning fuel:air mixture expands and forces the piston downward. As the piston nears bottom of its downward stroke, the cylinder exhaust port is opened and the burned gaseous products from combustion of the fuel:air mixture flows out the open port. Slightly further downward travel of the piston opens the cylinder intake port and a fresh charge of fuel:air mixture is forced into the cylinder. Since the exhaust port remains open, the incoming flow of fuel:air mixture helps clean (scavenge) any remaining burned gaseous products from the cylinder. As the crankshaft turns past BDC and the piston starts on its upward stroke, the cylinder intake and exhaust ports are closed and a new cycle begins.

Since the fuel:air mixture must be elevated to a higher than atmospheric pressure to enter the cylinder of a two-stroke cycle engine, a compressor pump must be used. Coincidentally, downward movement of the piston decreases the volume of the engine crankcase. Thus, a compressor pump is made available by sealing the engine crankcase and connecting the carburetor to a port in the crankcase. When the piston moves upward, volume of the crankcase is increased which lowers pressure within the crankcase to below atmospheric. Air will then be forced through the carburetor, where fuel is mixed with the air, and on into the engine crankcase. In order for downward movement of the piston to compress the fuel:air mixture in the crankcase, a valve must be provided to close the carburetor to crankcase port. Three different types of valves are used. In Fig. 1-1, a reed type inlet valve is shown in the schematic diagram of the two-stroke cycle engine. Spring steel reeds (R) are forced open by atmospheric pressure as shown in view "B" when the piston is on its upward stroke and pressure in the crankcase is below atmospheric. When piston reaches TDC, the reeds close as shown in view "A" and fuel:air mixture is trapped in the crankcase to be compressed by downward movement of the piston. In Fig 1-2, a schematic diagram of a two-stroke cycle engine is shown in which the piston is utilized as a sliding carburetor-crankcase port (third port) valve. In Fig. 1-3, a schematic diagram of a two-stroke cycle engine is shown in which a slotted disc (rotary valve) attached to the engine crankshaft opens the carburetor-crankcase port when the piston is on its upward stroke. In each of the three basic designs shown, a transfer port (TP—Fig. 1-2) connects the crankcase compression chamber to the cylinder; the transfer port is the cylinder intake port through which the compressed fuel:air mixture in the crankcase is transferred to the cylinder when the piston is at bottom of stroke as shown in view "A."

Due to rapid movement of the fuel:air mixture through the crankcase, the crankcase cannot be used as a lubricating oil sump because the oil would be carried into the cylinder. Lubrication is accomplished by mixing a small amount of oil with the fuel; thus, lubricating oil for the engine moving parts is carried into the crankcase with the fuel:air mixture. Normal lubricating oil to fuel mixture ratios vary from one part of oil mixed with 16 to 50 parts of fuel by volume. In all instances, manufacturer's recommendations for fuel-oil mixture ratio should be observed.

FOUR-STROKE CYCLE. In a four-stroke engine operating on the Otto Cycle (spark ignition), the five events of the cycle take place in four strokes of the piston, or in two revolutions of the engine crankshaft. Thus, a power stroke occurs only on alternate downward strokes of the piston.

In view "A" of Fig. 1-4, the piston is on the first downward stroke of the cycle. The mechanically operated intake valve has opened the intake port and, as the downward movement of the piston has reduced the air pressure in the cylinder to below atmospheric pressure, air is forced through the carburetor, where fuel is mixed with the air, and into the cylinder through the open intake port. The intake valve remains open and the fuel:air mixture continues

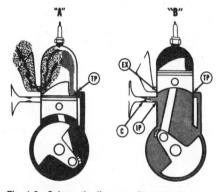

Fig. 1-2—Schematic diagram of two-stroke cycle engine operating on Otto Cycle. Engine differs from that shown in Fig. 1-1 in that piston is utilized as a sliding valve to open and close intake (carburetor to crankcase) port (IP) instead of using reed valve (R—Fig. 1-1).

C. Carburetor
EX. Exhaust port
IP. Intake port
    (carburetor to
    crankcase)
TP. Transfer port
    (crankcase to
    cylinder)

to flow into the cylinder until the piston reaches the bottom of its downward stroke. As the piston starts on its first upward stroke, the mechanically operated intake valve closes and, since the exhaust valve is closed, the fuel:air mixture is compressed as in view "B."

Just before the piston reaches the top of its first upward stroke, a spark at the spark plug electrodes ignites the compressed fuel:air mixture. As the engine crankshaft turns past top center, the burning fuel:air mixture expands rapidly and forces the piston downward on its power stroke as shown in view "C." As the piston reaches the bottom of the power stroke, the mechanically oper-

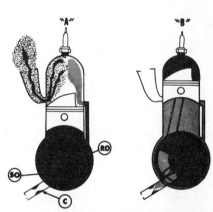

Fig. 1-3—Schematic diagram of two-stroke cycle engine similar to those shown in Figs. 1-1 and 1-2 except that a rotary carburetor to crankcase port valve is used. Disc driven by crankshaft has rotating opening (RO) which uncovers stationary opening (SO) in crankcase when piston is on upward stroke. Carburetor is (C).

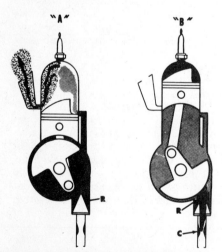

Fig. 1-1—Schematic diagram of a two-stroke engine operating on the Otto Cycle (spark ignition). View "B" shows piston near top of upward stroke and atmospheric pressure is forcing air through carburetor (C), where fuel is mixed with the air, and the fuel:air mixture enters crankcase through open reed valve (R). In view "A", piston is near bottom of downward stroke and has opened the cylinder exhaust and intake ports; fuel:air mixture in crankcase has been compressed by downward stroke of piston and flows into cylinder through open port. Incoming mixture helps clean burned exhaust gases from cylinder.

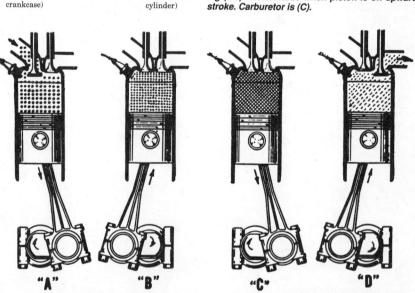

Fig. 1-4—Schematic diagram of four-stroke cycle engine operating on the Otto (spark ignition) cycle. In view "A", piston is on first downward (intake) stroke and atmospheric pressure is forcing fuel:air mixture from carburetor into cylinder through the open intake valve. In view "B", both valves are closed and piston is on its first upward stroke compressing the fuel:air mixture in cylinder. In view "C", spark across electrodes of spark plug has ignited fuel:air mixture and heat of combustion rapidly expands the burning gaseous mixture, forcing the piston on its second downward (expansion or power) stroke. In view "D", exhaust valve is open and piston on its second upward (exhaust) stroke forces the burned mixture from cylinder. A new cycle then starts as in view "A."

ated exhaust valve starts to open and as the pressure of the burned fuel:air mixture is higher than atmospheric pressure, it starts to flow out the open exhaust port. As the engine crankshaft turns past bottom center, the exhaust valve is almost completely open and remains open during the upward stroke of the piston as shown in view "D." Upward movement of the piston pushes the remaining burned fuel:air mixture out of the exhaust port. Just before the piston reaches the top of its second upward or exhaust stroke, the intake valve opens and the exhaust valve closes. The cycle is completed as the crankshaft turns past top center and a new cycle begins as the piston starts downward as shown in view "A."

In a four-stroke cycle engine operating on the Diesel Cycle, the sequence of events of the cycle is similar to that described for operation on the Otto Cycle, but with the following exceptions: On the intake stroke, air only is taken into the cylinder. On the compression stroke, the air is highly compressed which raises the temperature of the air. Just before the piston reaches top dead center, fuel is injected into the cylinder and is ignited by the heated, compressed air. The remainder of the cycle is similar to that of the Otto Cycle.

CARBURETOR FUNDAMENTALS

OPERATING PRINCIPLES

Function of the carburetor on a spark-ignition engine is to atomize the fuel and mix the atomized fuel in proper proportions with air flowing to the engine intake port or intake manifold. Carburetors used on engines that are to be operated at constant speeds and under even loads are of simple design since they only have to mix fuel and air in a relatively constant ratio. On engines operating at varying speeds and loads, the carburetors must be more complex because different fuel:air mixtures are required to meet the varying demands of the engine.

FUEL:AIR MIXTURE RATIO REQUIREMENTS. To meet the demands of an engine being operated at varying speeds and loads, the carburetor must mix fuel and air at different mixture ratios. Fuel:air mixture ratios required for different operating conditions are approximately as follows:

	Fuel	Air
Starting, cold weather ..	1 lb.	7 lbs.
Accelerating	1 lb.	9 lbs.
Idling (no-load)........	1 lb.	11 lbs.
Part open throttle......	1 lb.	15 lbs.
Full load, open throttle .	1 lb.	13 lbs.

BASIC DESIGN. Carburetor design is based on the venturi principle which simply means that a gas or liquid flowing through a necked-down section (venturi) in a passage undergoes an increase in velocity (speed) and a decrease in pressure as compared to the velocity and pressure in full size sections of the passage. The principle is illustrated in Fig. 2-1, which shows air passing through carburetor venturi. The figures given for air speeds and vacuum are approximate for a typical wide-open throttle operating condition. Due to low pressure (high vacuum) in the venturi, fuel is forced out through the fuel nozzle by the atmospheric pressure (0 vacuum) on the fuel; as fuel is emitted from the nozzle, it is atomized by the high velocity air flow and mixes with the air.

In Fig. 2-2, the carburetor choke plate and throttle plate are shown in relation in the venturi. Downward pointing arrows indicate air flow through carburetor.

At cranking speeds, air flows through the carburetor venturi at a slow speed; thus, the pressure in the venturi does not usually decrease to the extent that atmospheric pressure on the fuel will force fuel from the nozzle. If the choke plate is closed as shown by dotted line in Fig. 2-2, air cannot enter into the carburetor and pressure in the carburetor decreases greatly as the engine is turned at cranking speed. Fuel can then flow from the fuel nozzle. In manufacturing the carburetor choke plate or disc, a small hole or notch is cut in the plate so that some air can flow through the plate when it is in closed position to provide air for the starting fuel:air mixture. In some instances after starting a cold engine, it is advantageous to leave the choke plate in a partly closed position as the restriction of air flow will decrease the air pressure in carburetor venturi, thus causing more fuel to flow from the nozzle, resulting in a richer fuel:air mixture. The choke plate or disc should be in full open position for normal engine operation.

If, after the engine has been started the throttle plate is in the wide-open position as shown by the solid line in Fig. 2-2, the engine can obtain enough fuel and air to run at dangerously high speeds. Thus, the throttle plate or disc must be partly closed as shown by the dotted lines to control engine speed. At no load, the engine requires very little

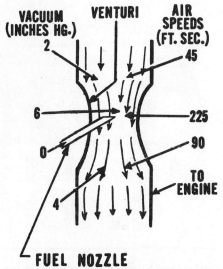

Fig. 2-1—Drawing illustrating the venturi principle upon which carburetor design is based. Figures at left are inches of mercury vacuum and those at right are air speeds in feet per second that are typical of conditions found in a carburetor operating at wide open throttle. Zero vacuum in fuel nozzle corresponds to atmospheric pressure.

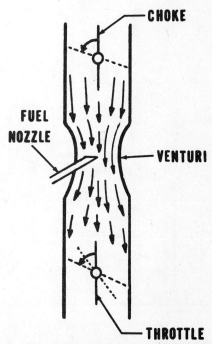

Fig. 2-2—Drawing showing basic carburetor design. Text explains operation of the choke and throttle valves. In some carburetors, a primer pump may be used instead of the choke valve to provide fuel for the starting fuel:air mixture.

air and fuel to run at its rated speed and the throttle must be moved on toward the closed position as shown by the dash lines. As more load is placed on the engine, more fuel and air are required for the engine to operate at its rated speed and the throttle must by moved closer to the wide open position as shown by the solid line. When the engine is required to develop maximum power or speed, the throttle must be in the wide open position.

Although some carburetors may be as simple as the basic design just described, most engines require more complex design features to provide variable fuel:air mixture ratios for different operating conditions. These design features will be described in the following paragraphs which outline the different carburetor types.

CARBURETOR TYPES

Carburetors used on small engines are usually classified by types as to method of delivery of fuel to the carburetor fuel nozzle. The following paragraphs describe the features and operating principles of the different type carburetors from the most simple suction lift type to the more complex float and diaphragm types.

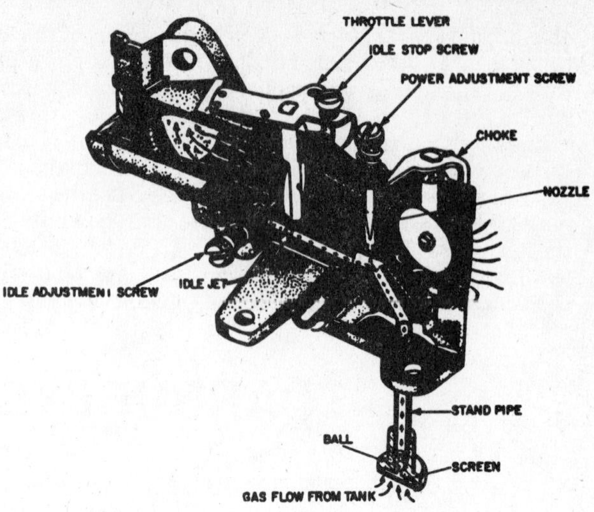

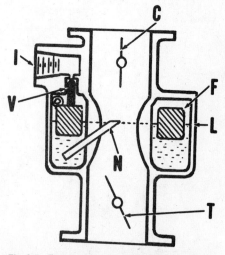

Fig. 2-4—Cut-away drawing of a suction lift carburetor used on one well known make engine. Ball in stand pipe prevents fuel in carburetor from flowing back into fuel tank.

SUCTION LIFT CARBURETOR. A cross-sectional drawing of a typical suction lift carburetor is shown in Fig. 2-3. Due to the low pressure at the orifice (O) of the fuel nozzle and to atmospheric pressure on the fuel in fuel supply tank, fuel is forced up through the fuel pipe and out of the nozzle into the carburetor venturi where it is mixed with the air flowing through the venturi. A check ball is located in the lower end of the fuel pipe to prevent pulsations of air pressure in the venturi from forcing fuel back down through the fuel pipe. The lower end of the fuel pipe has a fine mesh screen to prevent foreign material or dirt in fuel from entering the fuel nozzle. Fuel:air ratio can be adjusted by opening or closing the adjusting needle (N) slightly; turning the needle in will decrease flow of fuel out of nozzle orifice (O).

In Fig. 2-4, a cut-away view is shown of a suction type carburetor used on several models of a popular make small engine. This carburetor features an idle fuel passage, jet and adjustment screw. When carburetor throttle is nearly closed (engine is at low idle speed), air pressure is low (vacuum is high) at inner side of throttle plate. Therefore, atmospheric pressure in fuel tank will force fuel through the idle jet and adjusting screw orifice where it is emitted into the carburetor throat and mixes with air passing the throttle plate. The adjustment screw is turned in or out until an optimum fuel:air mixture is obtained and engine runs smoothly at idle speed. When the throttle is opened to increase engine speed, air velocity through the venturi increases, air pressure in the venturi decreases and fuel is emitted from the nozzle. Power adjustment screw (high speed fuel needle) is turned in or out to obtain proper fuel:air

mixture for engine running under operating speed and load.

FLOAT TYPE CARBURETOR. The principle of float type carburetor operation is illustrated in Fig. 2-5. Fuel is delivered at inlet (I) by gravity with fuel tank placed above carburetor, or by a fuel lift pump when tank is located below carburetor inlet. Fuel flows into the open inlet valve (V) until fuel level (L) in bowl lifts float against fuel valve needle and closes the valve. As fuel is emitted from the nozzle (N) when engine is running, fuel level will drop, lowering the float and allowing valve to open so that fuel will enter the carburetor to meet the requirements of the engine.

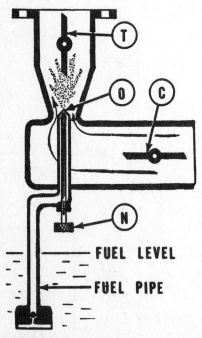

Fig. 2-3—Principle of suction lift carburetor is illustrated in above drawing. Atmospheric pressure on fuel forces fuel up through pipe and out nozzle orifice (O). Needle (N) is used to adjust amount of fuel flowing from nozzle to provide correct fuel:air mixture for engine operation. Choke (C) and throttle (T) valves are shown in wide open position.

Fig. 2-5—Drawing showing basic float type carburetor design. Fuel must be delivered under pressure either by gravity or by use of fuel pump, to the carburetor fuel inlet (I). Fuel level (L) operates float (F) to open and close inlet valve (V) to control amount of fuel entering carburetor. Also shown are the fuel nozzle (N), throttle (T) and choke (C).

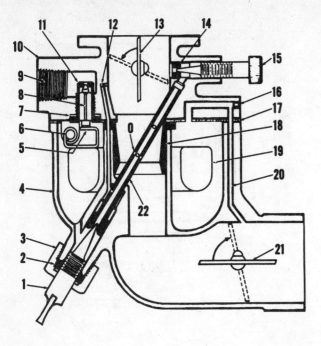

O. Orifice
1. Main fuel needle
2. Packing
3. Packing nut
4. Carburetor bowl
5. Float tang
6. Float hinge pin
7. Gasket
8. Inlet valve
9. Fuel inlet
10. Carburetor body
11. Inlet valve seat
12. Vent
13. Throttle plate
14. Idle orifice
15. Idle fuel needle
16. Plug
17. Gasket
18. Venturi
19. Float
20. Fuel bowl vent
21. Choke
22. Fuel nozzle

sure at venturi decreases and fuel will flow from openings (O) in nozzle instead of through orifice in idle seat (14). When engine is running at high speed, pressure in nozzle (22) is less than at vent (12) opening in carburetor throat above venturi. Thus, air will enter vent and travel down the vent into the nozzle and mix with the fuel in the nozzle. This is referred to as air bleeding and is illustrated in Fig. 2-7.

Many different designs of float type carburetors will be found when servicing the different makes and models of small engines. Reference should be made to the engine repair section of this manual for adjustment and overhaul specifications. Refer to carburetor servicing paragraphs in fundamentals sections for service hints.

DIAPHRAGM TYPE CARBURETOR. Refer to Fig. 2-8 for cross-sectional drawing showing basic design of a diaphragm type carburetor. Fuel is delivered to inlet (I) by gravity with fuel tank above carburetor, or under pressure from a fuel pump. Atmospheric pressure is maintained on lower side of diaphragm (D) through vent hole (V). When choke plate (C) is closed and engine is cranked, or when engine is running, pressure at orifice (O) is less than atmospheric pressure; this low pressure, or vacuum, is transmitted to fuel chamber (F) above diaphragm through nozzle channel (N). The higher (atmospheric) pressure at lower side of diaphragm will then push the diaphragm upward compressing spring (S) and allowing inlet valve (IV) to open and fuel will flow into the fuel chamber.

Some diaphragm type carburetors are equipped with an integral fuel pump. Although design of the pump

In Fig. 2-6, a cut-away view of a well known make of small engine float type carburetor is shown. Atmospheric pressure is maintained in fuel bowl through passage (20) which opens into carburetor air horn ahead of the choke plate (21). Fuel level is maintained at just below level of opening (O) in nozzle (22) by float (19) actuating inlet valve needle (8). Float height can be adjusted by bending float tang (5).

When starting a cold engine, it is necessary to close the choke plate (21) as shown by dotted lines so as to lower the air pressure in carburetor venturi (18) as engine is cranked. Then, fuel will flow up through nozzle (22) and will be emitted from openings (O) in nozzle. When an engine is hot, it will start on a

leaner fuel:air mixture than when cold and may start without the choke plate being closed.

When engine is running at slow idle speed (throttle plate nearly closed as indicated by dotted lines in Fig. 2-6), air pressure above the throttle plate is low and atmospheric pressure in fuel bowl forces fuel up through orifice in seat (14) where it mixes with air passing the throttle plate. The idle fuel mixture is adjustable by turning needle (15) in or out as required. Idle speed is adjustable by turning the throttle stop screw (not shown) in or out to control amount of air passing the throttle plate.

When throttle plate is opened to increase engine speed, velocity of air flow through venturi (18) increases, air pres-

Fig. 2-7—Illustration of air bleed principle explained in text.

C. Choke
D. Diaphragm
F. Fuel chamber
I. Fuel inlet
IV. Inlet valve needle
L. Lever
N. Nozzle
O. Orifice
P. Pivot pin
S. Spring
T. Throttle
V. Vent
VS. Valve seat

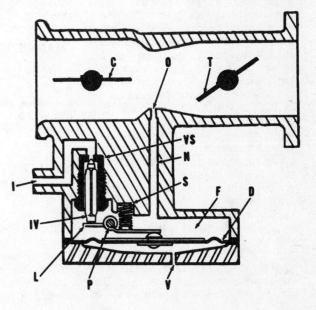

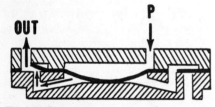

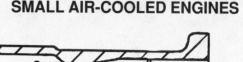

Fig. 2-10—Cross-sectional view of a popular make diaphragm type carburetor with integral fuel pump. Refer to Fig. 2-8 for view of basic diaphragm carburetor and to Fig. 2-9 for views showing operation of the fuel pump.

C. Choke
FI. Fuel inlet
IN. Idle fuel adjusting needle
IO. Idle orifice
MN. Main fuel adjusting needle
MO. Main orifice
P. Pulsation channel (fuel pump)
S. Screen
SO. Secondary orifice
T. Throttle
V. Vent (atmosphere to carburetor diaphragm)

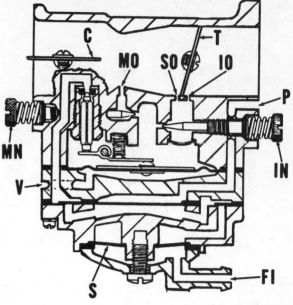

Fig. 2-9—Operating principle of diaphragm type fuel pump is illustrated in above drawings. Pump valves (A & B) are usually a part of diaphragm (D). Pump inlet is (I) and outlet is (O). Chamber above diaphragm is connected to engine crankcase by passage (C). When piston is on upward stroke, vacuum (V) at crankcase passage allows atmospheric pressure to force fuel into pump fuel chamber as shown in middle drawing. When piston is on downward stroke, pressure (P) expands diaphragm downward forcing fuel out of pump as shown in lower drawing.

may vary as to type of check valves, etc., all operate on the principle shown in Fig. 2-9. A channel (C) (or pulsation passage) connects one side of the diaphragm to the engine crankcase. When engine piston is on upward stroke, vacuum (V) (lower than atmospheric pressure) is present in channel; thus atmospheric pressure on fuel forces inlet valve (B) open and fuel flows into chamber below the diaphragm as shown in middle view. When piston is on down-ward stroke, pressure (P) (higher than atmospheric pressure) is present in channel (C); thus, the pressure forces the diaphragm downward closing the inlet valve (B) and causes the fuel to flow out by the outlet valve (A) as shown in lower view.

In Fig. 2-10, a cross-sectional view of a popular make diaphragm type carburetor, with integral diaphragm type pump, is shown.

IGNITION SYSTEM FUNDAMENTALS

The ignition system provides a properly timed surge of extremely high voltage electrical energy which flows across the spark plug electrode gap to create the ignition spark. Small engines may be equipped with either a magneto or battery ignition system. A magneto ignition system generates electrical energy, intensifies (transforms) this electrical energy to the extremely high voltage required and delivers this electrical energy at the proper time for the ignition spark. In a battery ignition system, a storage battery is used as a source of electrical energy and the system transforms the relatively low electrical voltage from the battery into the high voltage required and delivers the high voltage at proper time for the ignition spark. Thus, the function of the two systems is somewhat similar except for the basic source of electrical energy. The fundamental operating principles of ig-

nition systems are explained in the following paragraphs.

MAGNETISM AND ELECTRICITY

The fundamental principles upon which ignition systems are designed are presented in this section. As the study of magnetism and electricity is an entire scientific field, it is beyond the scope of this manual to fully explore these subjects. However, the following information will impart a working knowledge of basic principles which should be of value in servicing small engines.

MAGNETISM. The effects of magnetism can be shown easily while the theory of magnetism is too complex to be presented here. The effects of magnetism were discovered many years ago when fragments of iron ore were found

to attract each other and also attract other pieces of iron. Further, it was found that when suspended in air, one end of the iron ore fragment would always point in the direction of the North Star. The end of the iron ore fragment pointing north was called the "north pole" and the opposite end the "south pole." By stroking a piece of steel with a "natural magnet," as these iron ore fragments were called, it was found that the magnetic properties of the natural magnet could be transferred or "induced" into the steel.

Steel which will retain magnetic properties for an extended period of time after being subjected to a strong magnetic field are called "permanent magnets;" iron or steel that loses such magnetic properties soon after being subjected to a magnetic field are called "temporary magnets." Soft iron will lose magnetic properties almost immedi-

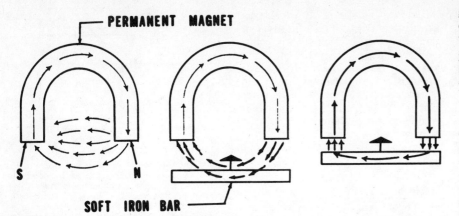

Fig. 3-1—In left view, field of force of permanent magnet is illustrated by arrows showing direction of magnetic force from north pole (N) to south pole (S). In center view, lines of magnetic force are being attracted by soft iron bar that is being moved into the magnetic field. In right view, the soft iron bar has been moved close to the magnet and the field of magnetic force is concentrated within the bar.

ately after being removed from a magnetic field, and so is used where this property is desirable.

The area affected by a magnet is called a "field of force." The extent of this field of force is related to the strength of the magnet and can be determined by use of a compass. In practice, it is common to illustrate the field of force surrounding a magnet by lines as shown in Fig. 3-1 and field of force is usually called "lines of force" or "flux." Actually, there are no "lines," however, this is a convenient method of illustrating the presence of the invisible magnetic forces and if a certain magnetic force is defined as a "line of force," then all magnetic forces may be measured by comparison. The number of "lines of force" making up a strong magnetic field is enormous.

Most materials when placed in a magnetic field are not attracted by the magnet, do not change the magnitude or direction of the magnetic field, and so are called "non-magnetic materials." Materials such as iron, cobalt, nickel or their alloys, when placed in a magnetic field will concentrate the field of force and hence are magnetic conductors or "magnetic materials." There are no materials known in which magnetic fields will not penetrate and magnetic lines of force can be deflected only by magnetic materials or by another magnetic field.

Alnico, an alloy containing aluminum, nickel and cobalt, retains magnetic properties for a very long period of time after being subjected to a strong magnetic field and is extensively used as a permanent magnet. Soft iron, which loses magnetic properties quickly, is used to concentrate magnetic fields as in Fig 3-1.

ELECTRICITY. Electricity, like magnetism, is an invisible physical force whose effects may be more readily

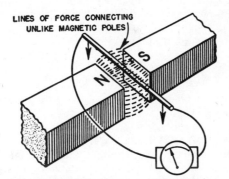

Fig. 3-2—When a conductor is moved through a magnetic field so as to cut across lines of force, a potential voltage will be induced in the conductor. If the conductor is a part of a completed electrical circuit, current will flow through the circuit as indicated by the gauge.

explained than the theory of what electricity consists of. All of us are familiar with the property of electricity to produce light, heat and mechanical power. What must be explained for the purpose of understanding ignition system operation is the inter-relationship of magnetism and electricity and how the ignition spark is produced.

Electrical current may be defined as a flow of energy in a conductor which, in some ways, may be compared to flow of water in a pipe. For electricity to flow, there must be a pressure (voltage) and a complete circuit (closed path) through which the electrical energy may return, a comparison being a water pump and a pipe that receives water from the outlet (pressure) side of the pump and returns the water to the inlet side of the pump. An electrical circuit may be completed by electricity flowing through the earth (ground), or through the metal framework of an engine or other equipment ("grounded" or "ground" connections). Usually, air is an insulator through which electrical energy will not flow. However, if the force (voltage) becomes

great, the resistance of air to the flow of electricity is broken down and a current will flow, releasing energy in the form of a spark. By high voltage electricity breaking down the resistance of the air gap between the spark plug electrodes, the ignition spark is formed.

ELECTROMAGNETIC INDUCTION. The principle of electro-magnetic induction is as follows: When a wire (conductor) is moved through a field of magnetic force so as to cut across the lines of force (flux), a potential voltage or electromotive force (emf) is induced in the wire. If the wire is part of a completed electrical circuit, current will flow through the circuit as illustrated in Fig. 3-2. It should be noted that the movement of the wire through the lines of magnetic force is a relative motion; that is, if the lines of force of a moving magnetic field cut across a wire, this will also induce an emf to the wire.

The direction of an induced current is related to the direction of magnetic force and also to the direction of movement of the wire through the lines of force, or flux. The voltage of an induced current is related to the strength, or concentration of lines of force, of the magnetic field and to the rate of speed at which the wire is moved through the flux. If a length of wire is wound into a coil and a section of the coil is moved through magnetic lines of force, the voltage induced will be proportional to the number of turns of wire in the coil.

ELECTRICAL MAGNETIC FIELDS. When current is flowing in a wire, a magnetic field is present around the wire as illustrated in Fig. 3-3. The direction of lines of force of this magnetic field is related to the direction of current in the wire. This is known as the left hand rule and is stated as follows: If a wire carrying a current is grasped in the left hand with thumb pointing in direction current is flowing, the curved fingers will point in the direction of lines of magnetic force (flux) encircling the wire.

If a current is flowing in a wire that is wound into a coil, the magnetic flux surrounding the wire converge to form

Fig. 3-3—A magnetic field surrounds a wire carrying an electrical current. The direction of magnetic force is indicated by the "left hand rule"; that is, if thumb of left hand points in direction that electrical current is flowing in conductor, fingers of left hand will indicate direction of magnetic force.

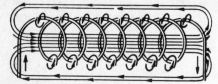

Fig. 3-4—When a wire is wound in a coil, the magnetic force created by a current in the wire will tend to converge in a single strong magnetic field as illustrated. If the loops of the coil are wound closely together, there is little tendency for lines of force to surround individual loops of the coil.

a stronger magnetic field as shown in Fig. 3-4. If the coils of wire are very close together, there is little tendency for magnetic flux to surround individual loops of the coil and a strong magnetic field will surround the entire coil. The strength of this field will vary with the current flowing through the coil.

STEP-UP TRANSFORMERS (IGNITION COILS). In both battery and magneto ignition systems, it is necessary to step-up, or transform, a relatively low primary voltage to the 15,000 to 20,000 volts required for the ignition spark. This is done by means of an ignition coil which utilizes the inter-relationship of magnetism and electricity as explained in preceding paragraphs.

Basic ignition coil design is shown in Fig. 3-5. The coil consists of two separate coils of wire which are called the primary coil winding and the secondary coil winding, or simply the primary winding and secondary winding. The primary winding as indicated by the heavy, black line is of larger diameter wire and has a smaller number of turns when compared to the secondary winding indicated by the light line.

A current passing through the primary winding creates a magnetic field (as indicated by the "lines of force") and this field, concentrated by the soft iron core, surrounds both the primary and secondary windings. If the primary winding current is suddenly interrupted, the magnetic field will collapse and the lines of force will cut through the coil windings. The resulting induced voltage in the secondary winding is greater than the voltage of the current that was flowing in the primary winding and is related to the number of turns of wire in each winding. Thus:

Induced secondary voltage = primary voltage ×

$$\frac{\text{No. of turns in secondary winding}}{\text{No. of turns in primary winding}}$$

For example, if the primary winding of an ignition coil contained 100 turns of wire and the secondary winding contained 10,000 turns of wire, a current having an emf of 200 volts flowing in the primary winding, when suddenly interrupted, would result in an emf of:

$$200 \text{ Volts} \times \frac{10,000 \text{ turns of wire}}{100 \text{ turns of wire}}$$

$$= 20,000 \text{ volts}$$

SELF-INDUCTANCE. It should be noted that the collapsing magnetic field resulting from the interrupted current in the primary winding will also induce a current in the primary winding. This effect is termed "self-inductance." This self-induced current is such as to oppose any interruption of current in the primary winding, slowing the collapse of the magnetic field and reducing the efficiency of the coil. The self-induced primary current flowing across the slightly open breaker switch, or contact points, will damage the contact surfaces due to the resulting spark.

To momentarily absorb, then stop the flow of current across the contact points, a capacitor or, as commonly called, a

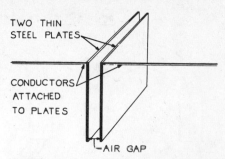

Fig. 3-6A—Drawing showing construction of a simple condenser. Capacity of such a condenser to absorb current is limited due to the relatively small surface area. Also, there is a tendency for current to arc across the air gap. Refer to Fig. 3-7 for construction of typical ignition system condenser.

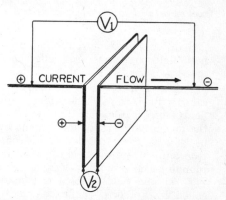

Fig. 3-6B—A condenser in an electrical circuit will absorb flow of current until an opposing voltage (V2) is built up across condenser plates which is equal to the voltage (V1) of the electrical current.

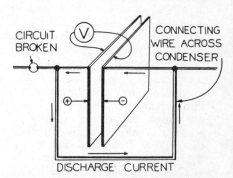

Fig. 3-6C—When flow of current is interrupted in circuit containing condenser (circuit broken), the condenser will retain a potential voltage (V). If a wire is connected across the condenser, a current will flow in reverse direction of charging current until condenser is discharged (voltage across condenser plates is zero).

condenser is connected in parallel with the contact points. A simple condenser is shown in Fig. 3-6A. Operating principles of a condenser are illustrated in Fig. 3-6B and Fig. 3-6C. The capacity of such a condenser to absorb current (capacitance) is limited by the small surface area of the plates. To increase capacity to absorb current, the condenser used in ignition systems is constructed as shown in Fig. 3-7.

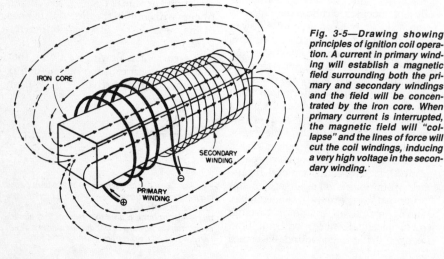

Fig. 3-5—Drawing showing principles of ignition coil operation. A current in primary winding will establish a magnetic field surrounding both the primary and secondary windings and the field will be concentrated by the iron core. When primary current is interrupted, the magnetic field will "collapse" and the lines of force will cut the coil windings, inducing a very high voltage in the secondary winding.

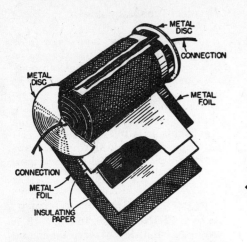

Fig. 3-7—Drawing showing construction of typical ignition system condenser. Two layers of metal foil, insulated from each other with paper, are rolled tightly together and a metal disc contacts each layer, or strip, of foil. Usually, one disc is grounded through the condenser shell.

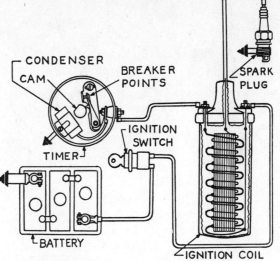

Fig. 3-9—Schematic diagram of typical battery ignition system. On unit shown, breaker points are actuated by timer cam; on some units, the points may be actuated by cam on engine camshaft. Refer to Fig. 3-10 for cutaway view of typical battery ignition coil. In view above, primary coil winding is shown as heavy black line (outside coil loops) and secondary winding is shown by lighter line (inside coil loops).

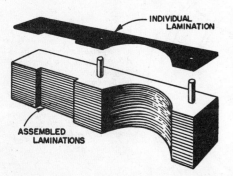

Fig. 3-8—To prevent formation of "eddy currents" within soft iron cores used to concentrate magnetic fields, core is assembled of plates or "laminations" that are insulated from each other. In a solid core, there is a tendency for counteracting magnetic forces to build up from stray currents induced in the core.

BATTERY IGNITION SYSTEMS

Some small engines are equipped with a battery ignition system. A schematic diagram of a typical battery ignition system for a single cylinder engine is shown in Fig. 3-9. Designs of battery ignition systems may vary, especially as to location of breaker points and method for actuating the points; however, all operate on the same basic principles.

BATTERY IGNITION SYSTEM PRINCIPLES. Refer to the schematic diagram on Fig. 3-9. When the timer cam is turned so that the contact points are closed, a current is established in the primary circuit by the emf of the battery. This current flowing through the primary winding of the ignition coil establishes a magnetic field concentrated in the core laminations and surrounding the windings. A cutaway view of a typical ignition coil is shown in Fig. 3-10. At the proper time for the ignition spark, the contact points are opened by the timer cam and the primary ignition circuit is interrupted. The condenser, wired in parallel with the breaker contact points between the timer terminal and ground, absorbs the self-induced current in the primary circuit for an instant and brings the flow of current to a quick, controlled stop. The magnetic field surrounding the coil rapidly cuts the primary and secondary windings creating an emf as high as 250 volts in the primary winding and up to 25,000 volts in the secondary winding. Current absorbed by the condenser is discharged as the cam closes the breaker points, grounding the condenser lead wire.

Due to resistance of the primary winding, a certain period of time is required for maximum primary current flow after the breaker contact points are

closed. At high engine speeds, the points remain closed for a smaller interval of time, hence the primary current does not build up to maximum and secondary voltage is somewhat less than at low engine speed. However, coil design is such that the minimum voltage available at high engine speed exceeds the

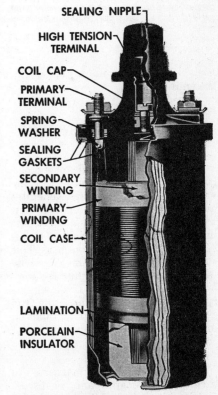

Fig. 3-10—Cutaway view of typical battery ignition system coil. Primary winding consists of approximately 200-250 turns (loops) of heavier wire; secondary winding consists of several thousand turns of fine wire. Laminations concentrate the magnetic lines of force and increase efficiency of the coil.

EDDY CURRENTS.

EDDY CURRENTS. It has been found that when a solid soft iron bar is used as a core for an ignition coil, stray electrical currents are formed in the core. These stray, or "eddy currents," create opposing magnetic forces causing the core to become hot and also decrease the efficiency of the coil. As a means of preventing excessive formation of eddy currents within the core, or other magnetic field carrying parts of a magneto, a laminated plate construction as shown in Fig. 3-8 is used instead of solid material. The plates, or laminations, are insulated from each other by a natural oxide coating formed on the plate surfaces or by coating the plates with varnish. The cores of some ignition coils are constructed of soft iron wire instead of plates and each wire is insulated by a varnish coating. This type construction serves the same purpose as laminated plates.

Fig. 3-11—Exploded view of a flywheel type magneto in which the breaker points (14) are actuated by a cam on engine camshaft. Push rod (9) rides against cam to open and close points. In this type unit, an ignition spark is produced only on alternate revolutions of the flywheel as the camshaft turns at one-half engine speed.

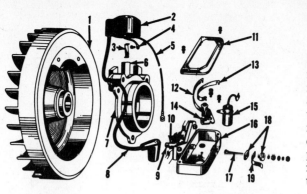

1. Flywheel
2. Ignition coil
3. Coil clamps
4. Coil ground lead
5. Breaker-point lead
6. Armature core (laminations)
7. Crankshaft bearing retainer
8. High tension lead
9. Push rod
10. Bushing
11. Breaker box cover
12. Point lead strap
13. Breaker-point spring
14. Breaker-point assy.
15. Condenser
16. Breaker box
17. Terminal bolt
18. Insulators
19. Ground (stop) spring

Flywheel Type Magnetos

The term "flywheel type magneto" is derived from the fact that the engine flywheel carries the permanent magnets and is the magneto rotor. See Fig. 3-11 for a typical flywheel magneto system. In some similar systems, the magneto rotor is mounted on the engine crankshaft as is the flywheel, but is a part separate from the flywheel. A drawing of a typical flywheel magneto ignition coil is shown in Fig. 3-12.

FLYWHEEL MAGNETO OPERATING PRINCIPLES. In Fig. 3-13, a cross-sectional view of a typical engine

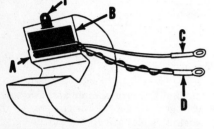

Fig. 3-12—Drawing showing construction of a typical flywheel magneto ignition coil. Primary winding (A) consists of about 200 turns of wire. Secondary winding (B) consists of several thousand turns of fine wire. Coil primary and secondary ground connection is (D); primary connection to breaker point and condenser terminal is (C); and coil secondary (high tension) terminal is (T).

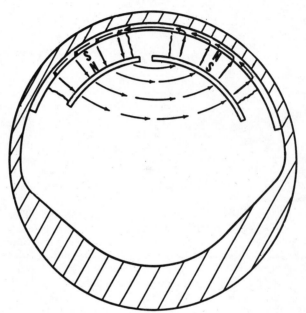

Fig. 3-13—Cutaway view of typical engine flywheel used with flywheel magneto type ignition system. The permanent magnets are usually cast into the flywheel. For flywheel type magnetos having the ignition coil and core mounted to outside of flywheel, magnets would be flush with outer diameter of flywheel.

normal maximum voltage required for the ignition spark.

MAGNETO IGNITION SYSTEMS

By utilizing the principles of magnetism and electricity as outlined in previous paragraphs, a magneto generates an electrical current of relatively low voltage, then transforms this voltage into the extremely high voltage necessary to produce the ignition spark. This surge of high voltage is timed to create the ignition spark and ignite the compressed fuel:air mixture in the engine cylinder at the proper time in the Otto cycle as described in the paragraphs on fundamentals of engine operation principles.

Two different types of magnetos are used on small engines and, for discussion in this section of the manual, will be classified as "flywheel type magnetos" and "self-contained unit type magnetos." The most common type of ignition system found on small engines is the flywheel type magneto.

Fig. 3-14—View showing flywheel turned to a position so that lines of force of the permanent magnets are concentrated in the left and center core legs and are interlocking the coil windings.

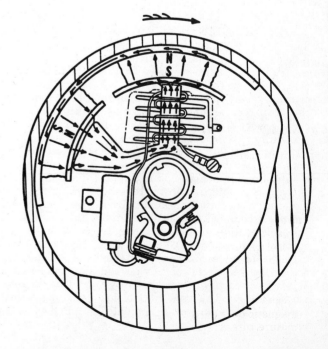

Fig. 3-15—View showing flywheel turned to a position so the lines of force of the permanent magnets are being withdrawn from the left and center core legs and are being attracted by the center and right core legs. While this event is happening, the lines of force are cutting up through the coil windings section between the left and center legs and are cutting down through the section between the right and center legs as indicated by the heavy black arrows. As the breaker points are now closed by the cam, a current is induced in the primary ignition circuit as the lines of force cut through the coil windings.

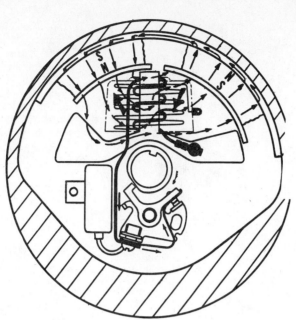

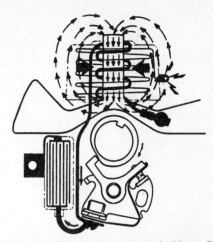

Fig. 3-17—View showing magneto ignition coil, condenser and breaker points at same instant as illustrated in Fig. 3-16; however, arrows shown above illustrate lines of force of the electro-magnetic field established by current in primary coil windings rather than the lines of force of the permanent magnets. As the current in the primary circuit ceases to flow, the electro-magnetic field collapses rapidly, cutting the coil windings as indicated by heavy arrows and inducing a very high voltage in the secondary coil winding resulting in the ignition spark.

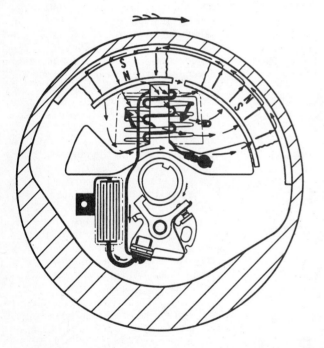

Fig. 3-16—The flywheel magnets have now turned slightly past the position shown in Fig. 3-15 and the rate of movement of lines of magnetic force cutting through the coil windings is at the maximum. At this instant, the breaker points are opened by the cam and flow of current in the primary circuit is being absorbed by the condenser, bringing the flow of current to a quick, controlled stop. Refer now to Fig. 3-17.

ting down through the coil windings section between the center and right legs. If the right hand rule, as explained in a previous paragraph, is applied to the lines of force cutting through the coil sections, it is seen that the resulting emf induced in the primary circuit will cause a current to flow through the primary coil windings and the breaker points which have now been closed by action of the cam.

At the instant the movement of the lines of force cutting through the coil winding sections is at the maximum rate, the maximum flow of current is obtained in the primary circuit. At this time, the cam opens the breaker points interrupting the primary circuit and for an instant, the flow of current is absorbed by the condenser as illustrated in Fig. 3-16. An emf is also induced in the secondary coil windings, but the voltage is not sufficient to cause current to flow across the spark plug gap.

The flow of current in the primary windings created a strong electromagnetic field surrounding the coil windings and up through the center leg of the armature core as shown in Fig. 3-17. As the breaker points were opened by the cam, interrupting the primary circuit, this magnetic field starts to collapse cutting the coil windings as indicated by the heavy black arrows. The emf induced in the primary circuit would be sufficient to cause a flow of current across the opening breaker points were it not for the condenser absorbing the flow of current and bringing it to a controlled stop. This allows the electro-

flywheel (magneto rotor) is shown. The arrows indicate lines of force (flux) of the permanent magnets carried by the flywheel. As indicated by the arrows, direction of force of the magnetic field is from the north pole (N) of the left magnet to the south pole (S) of the right magnet.

Figs. 3-14, 3-15, 3-16, and 3-17 illustrate the operational cycle of the flywheel type magneto. In Fig. 3-14, the flywheel magnets have moved to a position over the left and center legs of the armature (ignition coil) core. As the magnets moved into this position, their magnetic field was attracted by the armature core as illustrated in Fig. 3-1

and a potential voltage (emf) was induced in the coil windings. However, this emf was not sufficient to cause current to flow across the spark plug electrode gap in the high tension circuit and the points were open in the primary circuit.

In Fig. 3-15, the flywheel magnets have moved to a new position to where their magnetic field is being attracted by the center and right legs of the armature core, and is being withdrawn from the left and center legs. As indicated by the heavy black arrows, the lines of force are cutting up through the section of coil windings between the left and center legs of the armature and are cut-

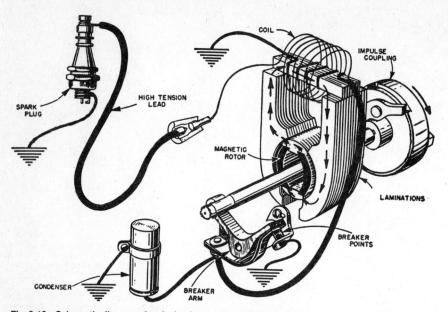

Fig. 3-18—Schematic diagram of typical unit type magneto for single cylinder engine. Refer to Figs. 3-19 and 3-20 for views showing construction of impulse couplings.

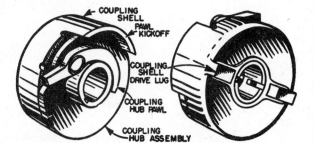

Fig. 3-19—Views of typical impulse coupling for magneto driven by engine shaft with slotted drive connection. Coupling drive spring is shown in Fig. 3-20.

magnetic field to collapse at such a rapid rate to induce a very high voltage in the coil high tension or secondary windings. This voltage, in the order of 15,000 to 25,000 volts, is sufficient to break down the resistance of the air gap between the spark plug electrodes and a current will flow across the gap. This creates the ignition spark which ignites the compressed fuel:air mixture in the engine cylinder.

Self-Contained Unit Type Magnetos

Some four-stroke cycle engines are equipped with a magneto which is a self-contained unit. This type magneto is driven from the engine timing gears via a gear or coupling. All components of the magneto are enclosed in one housing and the magneto can be removed from the engine as a unit.

UNIT TYPE MAGNETO OPERATING PRINCIPLES. In Fig. 3-18, a schematic diagram of a unit type magneto is shown. The magneto rotor is driven through an impulse coupling (shown at right side of illustration). The function of the impulse coupling is to

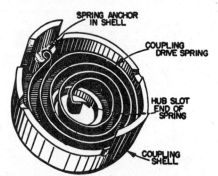

Fig. 3-20—View showing impulse coupling shell and drive spring removed from coupling hub assembly. Refer to Fig. 3-19 for views of assembled unit.

increase the rotating speed of the rotor, thereby increasing magneto efficiency, at engine cranking speeds.

A typical impulse coupling for a single cylinder engine magneto is shown in Fig. 3-19. When the engine is turned at cranking speed, the coupling hub pawl engages a stop pin in the magneto housing as the engine piston is coming up on compression stroke. This stops rotation of the coupling hub assembly and magneto rotor. A spring within the coupling

shell (see Fig. 3-20) connects the shell and coupling hub; as the engine continues to turn, the spring winds up until the pawl kickoff contacts the pawl and disengages it from the stop pin. This occurs at the time an ignition spark is required to ignite the compressed fuel:air mixture in the engine cylinder. As the pawl is released, the spring connecting the coupling shell and hub unwinds and rapidly spins the magneto rotor.

The magneto rotor (see Fig. 3-18) carries permanent magnets. As the rotor turns, alternating the position of the magnets, the lines of force of the magnets are attracted, then withdrawn from the laminations. In Fig. 3-18, arrows show the magnetic field concentrated within the laminations, or armature core. Slightly further rotation of the magnetic rotor will place the magnets to where the laminations will have greater attraction for opposite poles of the magnets. At this instant, the lines of force as indicated by the arrows will suddenly be withdrawn and an opposing field of force will be established in the laminations. Due to this rapid movement of the lines of force, a current will be induced in the primary magneto circuit as the coil windings are cut by the lines of force. At the instant the maximum current is induced in the primary windings, the breaker points are opened by a cam on the magnetic rotor shaft interrupting the primary circuit. The lines of magnetic force established by the primary current (refer to Fig. 3-5) will cut through the secondary windings at such a rapid rate to induce a very high voltage in the secondary (or high tension) circuit. This voltage will break down the resistance of the spark plug electrode gap and a spark across the electrodes will result.

At engine operating speeds, centrifugal force will hold the impulse coupling hub pawl (see Fig. 3-19) in a position so that it cannot engage the stop pin in magneto housing and the magnetic rotor will be driven through the spring (Fig. 3-20) connecting the coupling shell to coupling hub. The impulse coupling retards the ignition spark, at cranking speeds, as the engine piston travels closer to top dead center while the magnetic rotor is held stationary by the pawl and stop pin. The difference in degrees of impulse coupling shell rotation between the position of retarded spark and normal running spark is known as the impulse coupling lag angle.

SOLID-STATE IGNITION SYSTEM

BREAKERLESS MAGNETO SYSTEM. The solid-state (breakerless) magneto ignition system operates somewhat on the same basic principles as the conventional type flywheel magneto previously described. The main difference is that the breaker contact points are replaced by a solid-state electronic Gate Controlled Switch (GCS) which has no moving parts. Since, in a conventional system breaker points are closed over a longer period of crankshaft rotation than is the "GCS", a diode has been added to the circuit to provide the same characteristics as closed breaker points.

BREAKERLESS MAGNETO OPERATING PRINCIPLES. The same basic principles for electro-magnetic induction of electricity and formation of magnetic fields by electrical current as outlined for the conventional flywheel type magneto also apply to the solid state magneto. Therefore the principles of the different components (diode and GCS) will complete the operating principles of the solid-state magneto.

The diode is represented in wiring diagrams by the symbol shown in Fig. 3-21. The diode is an electronic device that will permit passage of electrical current in one direction only. In electrical schematic diagrams, current flow is opposite direction the arrow part of symbol is pointing.

The symbol shown in Fig. 3-22 is used to represent the gate controlled switch (GCS) in wiring diagrams. The GCS acts as a switch to permit passage of current from cathode (C) terminal to

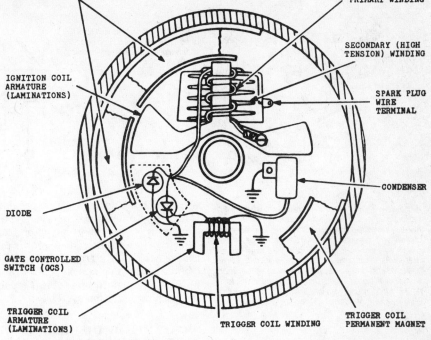

Fig. 3-23—Schematic diagram of typical breakerless magneto ignition system. Refer to Figs. 3-24, 3-25 and 3-26 for schematic views of operating cycle.

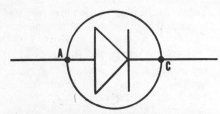

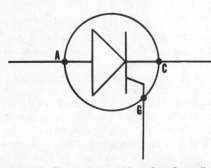

Fig. 3-21—In a diagram of an electrical circuit, the diode is represented by the symbol shown above. The diode will allow current to flow in one direction only, from cathode (C) to anode (A).

Fig. 3-22—The symbol used for a Gate Controlled Switch (GCS) in an electrical diagram is shown above. The GCS will permit current to flow from cathode (C) to anode (A) when "turned on" by a positive electrical charge at gate (G) terminal.

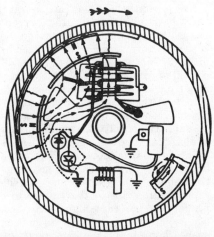

Fig. 3-24—View showing flywheel of breakerless magneto system at instant of rotation where lines of force of ignition coil magnets are being drawn into left and center legs of magneto armature. The diode (See Fig. 3-21) acts as a closed set of breaker points in completing the primary ignition circuit at this time.

anode (A) terminal when in "ON" state and will not permit electric current to flow when in "OFF" state. The GCS can be turned "ON" by a positive surge of electricity at the gate (G) terminal and will remain "ON" as long as current remains positive at the gate terminal or as long as current is flowing through the GCS from the cathode (C) terminal to anode (A) terminal.

The basic components and wiring diagram for the solid-state breakerless magneto are shown schematically in

Fig. 3-23. In Fig. 3-24, the magneto rotor (flywheel) is turning and the ignition coil magnets have just moved into position so that their lines of force are cutting the ignition coil windings and producing a negative surge of current in the primary windings. The diode allows current to flow opposite to the direction of diode symbol arrow and action is same as conventional magneto with breaker points closed. As rotor (flywheel) continues to turn as shown in Fig. 3-25, direction of magnetic flux lines will reverse in the armature center leg. Direction of current will change in the primary coil circuit and the previously conducting diode will be shut off. At this point, neither diode is conducting. As voltage begins to build up as rotor continues to turn, the condenser acts as a buffer to prevent excessive voltage build up at the GCS before it is triggered.

When the rotor reaches the approximate position shown in Fig. 3-26, maximum flux density has been achieved in the center leg of the armature. At this time the GCS is triggered. Triggering is accomplished by the triggering coil armature moving into the field of a permanent magnet which induces a positive voltage on the gate of the GCS. Primary coil current flow results in the formation of an electromagnetic field around primary coil which inducts a voltage of sufficient potential in the secondary coil windings to "fire" the spark plug.

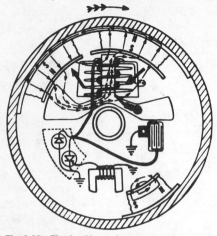

Fig. 3-25—Flywheel is turning to point where magnetic flux lines through armature center leg will reverse direction and current through primary coil circuit will reverse. As current reverses, diode which was previously conducting will shut off and there will be no current. When magnetic flux lines have reversed in armature center leg, voltage potential will again build up, but since GCS is in "OFF" state, no current will flow. To prevent excessive voltage build up, the condenser acts as a buffer.

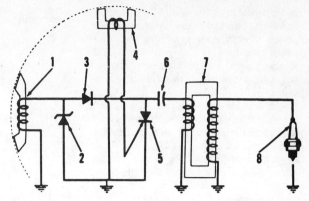

Fig. 3-27—Schematic diagram of a typical capacitor discharge ignition system.

1. Generating coil
2. Zener diode
3. Diode
4. Trigger coil
5. Gate controlled switch
6. Capacitor
7. Pulse transformer (coil)
8. Spark plug

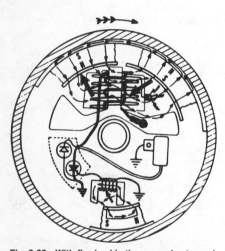

Fig. 3-26—With flywheel in the approximate position shown, maximum voltage potential is present in windings of primary coil. At this time the triggering coil armature has moved into the field of a permanent magnet and a positive voltage is induced on the gate of the GCS. The GCS is triggered and primary coil current flows resulting in the formation of an electromagnetic field around the primary coil which inducts a voltage of sufficient potential in the secondary windings to "fire" the spark plug.

When the rotor (flywheel) has moved the magnets past the armature, the GCS will cease to conduct and revert to the "OFF" state until it is triggered. The condenser will discharge during the time that the GCS was conducting.

CAPACITOR DISCHARGE SYSTEM. The capacitor discharge (CD) ignition system uses a permanent magnet rotor (flywheel) to induce a current in a coil, but unlike the conventional flywheel magneto and solid-state breakerless magneto described previously, the current is stored in a capacitor (condenser). Then the stored current is discharged through a transformer coil to create the ignition spark. Refer to Fig. 3-27 for a schematic of a typical capacitor discharge ignition system.

CAPACITOR DISCHARGE OPERATING PRINCIPLES. As the permanent flywheel magnets pass by the input generating coil (1—Fig. 3-27), the current produced charges capacitor (6). Only half of the generated current passes through diode (3) to charge the capacitor. Reverse current is blocked by diode (3), but passes through Zener diode (2) to complete the reverse circuit. Zener diode (2) also limits maximum voltage of the forward current. As the flywheel continues to turn and magnets pass the trigger coil (4), a small amount of electrical current is generated. This current opens the gate controlled switch (5), allowing the capacitor to discharge through the pulse transformer (7). The rapid voltage rise in the transformer primary coil induces a high voltage secondary current which forms the ignition spark when it jumps the spark plug gap.

THE SPARK PLUG

In any spark ignition engine, the spark plug (See Fig. 3-28) provides the means for igniting the compressed fuel:air mixture in the cylinder. Before an electric charge can move across an air gap, the intervening air must be charged with electricity, or ionized. If the spark plug is properly gapped and the system is not shorted, not more than 7,000 volts may be required to initiate a spark. Higher voltage is required as the spark plug warms up, or if compression pressures or the distance of the air gap is increased. Compression pressures are highest at full throttle and relatively slow engine speeds, therefore, high voltage requirements or a lack of available

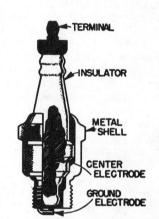

Fig. 3-28—Cross-sectional drawing of spark plug showing construction and nomenclature.

secondary voltage most often shows up as a miss during maximum acceleration from a slow engine speed.

There are many different types and sizes of spark plugs which are designed for a number of specific requirements.

THREAD SIZE. The threaded, shell portion of the spark plug and the attaching hole in the cylinder are manufactured to meet certain industry established standards. The diameter is referred to as "Thread Size." Those commonly used are: 10 mm, 14 mm, 18 mm, 7/8 inch and 1/2 inch pipe. The 14 mm plug is almost universal for small engine use.

REACH. The length of thread, and the thread depth in cylinder head or wall are also standardized throughout the industry. This dimension is measured from gasket seat of plug to cylinder end of thread. See Fig. 3-29. Four different reach plugs commonly used are 3/8 inch, 7/16 inch, 1/2 inch and 3/4 inch. The first two mentioned are the ones commonly used in small engines.

HEAT RANGE. During engine operation, part of the heat generated during combustion is transferred to the spark plug, and from the plug to the cylinder through the shell threads and gasket. The operating temperature of the spark plug plays an important part in engine operation. If too much heat is retained by the plug, the fuel:air mixture may be ignited by contact with heated surface before the ignition spark occurs. If not enough heat is retained, partially burned combustion products (soot, carbon and oil) may build up on the plug tip resulting in "fouling" or shorting out of the plug. If this happens, the secondary current is dissipated uselessly as it is generated instead of bridging the plug gap as a useful spark, and the engine will misfire.

The operating temperature of the plug tip can be controlled, within limits, by altering the length of the path the heat must follow to reach the threads and gasket of the plug. Thus, a plug with a short, stubby insulator around the center electrode will run cooler than one with a long, slim insulator. Refer to Fig. 3-30. Most plugs in the more popular sizes are available in a number of heat ranges which are interchangeable within the group. The proper heat range is determined by engine design and type of service. Refer to SPARK PLUG SERVICING FUNDAMENTALS for additional information on spark plug selection.

SPECIAL TYPES. Sometimes, engine design features or operating conditions call for special plug types designed for a particular purpose. Of special interest when dealing with two-cycle engines is the spark plug shown in the left hand view, Fig. 3-29. In the design of this plug, the ground electrode is shortened so that its end aligns with center of insulated electrode rather than com-

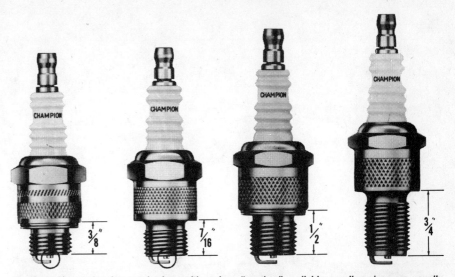

Fig. 3-29—Views showing spark plugs with various "reaches" available; small engines are usually equipped with a spark plug having a 3/8 inch reach. A 3/8 inch reach spark plug measures 3/8 inch from firing end of shell to gasket surface of shell. The two plugs at left side illustrate the difference in plugs normally used in two-stroke cycle and four-stroke cycle engines; refer to the circled electrodes. Spark plug at left has a shortened ground electrode and is specifically designed for two-stroke cycle engines. Second spark plug from left is normally used in four-stroke cycle engines although some two-stroke cycle engines may use this type plug.

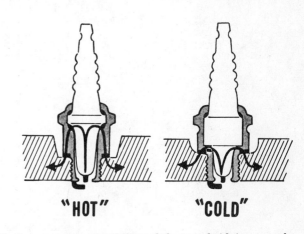

Fig. 3-30—Spark plug tip temperature is controlled by the length of the path heat must travel to reach cooling surface of the engine cylinder head.

pletely overlapping as with the conventional plug. This feature reduces the possibility of the gap bridging over by carbon formations.

ENGINE POWER AND TORQUE RATINGS

The following paragraphs discuss the terms used in expressing engine horsepower and torque ratings and explain the methods for determining the different ratings. Some small engine repair shops are now equipped with a dynamometer for measuring engine torque and/or horsepower and the mechanic should be familiar with terms, methods of measurement and how actual power developed by an engine can vary under different conditions.

GLOSSARY OF TERMS

FORCE. Force is an action against an object that tends to move the object from a state of rest, or to accelerate the movement of an object (Fig. 4-1). For use in calculating torque or horsepower, force is measured in pounds.

WORK. When a force moves an object from a state of rest, or accelerates the movement of an object, work is done (Fig. 4-2). Work is measured by multiplying the force applied by the distance the force moves the object, or:

$$\text{work} = \text{force} \times \text{distance}.$$

Thus, if a force of 50 pounds moved an object 50 feet, work done would equal 50 pounds times 50 feet, or 2500 pounds-feet (or as it is usually expressed, 2500 foot-pounds).

Fig. 4-1—A force, measured in pounds, is defined as an action tending to move an object or to accelerate movement of an object.

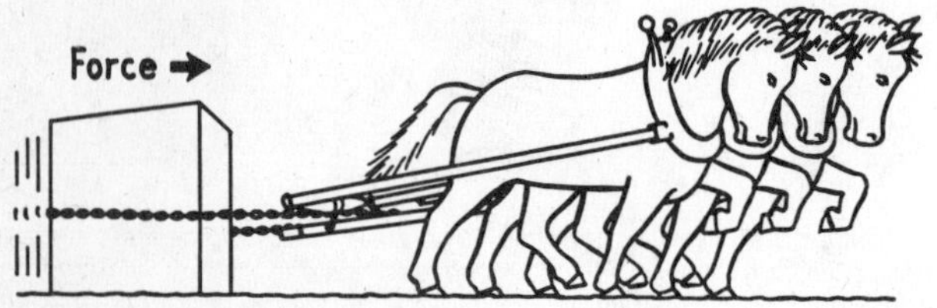

Fig. 4-2—If a force moves an object from a state of rest or accelerates movement of an object, then work is done.

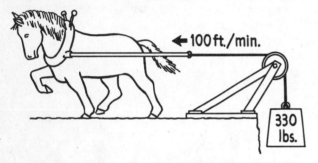

Fig. 4-3—This horse is doing 33,000 foot-pounds of work in one minute, or one horsepower.

RATED HORSEPOWER. An engine being operated under a load equal to the maximum horsepower available (brake horsepower) will not have reserve power for overloads and is subject to damage from overheating and rapid wear. Therefore, when an engine is being selected for a particular load, the engine's brake horsepower rating should be in excess of the expected normal operating load. Usually, it is recommended that the engine not be operated in excess of 80 percent of the engine maximum brake horsepower rating; thus, the "rated horsepower" of an engine is usually equal to 80 percent of maximum horsepower that the engine will develop.

TORQUE. In many engine specifications, a "torque rating" is given. Engine torque can be defined simply as the turning effort exerted by the engine output shaft when under load. Thus, it is possible to calculate engine horsepower being developed by measuring torque being developed and engine output speed. Refer to the following paragraphs.

MEASURING ENGINE TORQUE AND HORSEPOWER

THE PRONY BRAKE. The prony brake is the most simple means of testing engine performance. Refer to diagram in Fig. 4-4. A torque arm is attached to a brake on wheel mounted on engine output shaft. The torque arm, as the brake is applied, exerts a force (F) on scales. Engine torque is computed by multiplying the force (F) times the length of the torque arm radius (R), or:

$$\text{engine torque} = F \times R$$

If, for example, the torque arm radius (R) is 2 feet and the force (F) being exerted by the torque arm on the scales is 6 pounds, engine torque would be 2 feet × 6 pounds or 12 foot-pounds.

To calculate engine horsepower being developed by use of the prony brake, we

POWER. Power is the rate at which work is done; thus, if:

work = force × distance,

then:

$$\text{power} = \frac{\text{force} \times \text{distance}}{\text{time}}$$

From the above formula, it is seen that power must increase if the time in which work is done decreases.

HORSEPOWER. Horsepower is a unit of measurement of power. Many years ago, James Watt, a Scotsman noted as the inventor of the steam engine, evaluated one horsepower as being equal to doing 33,000 foot-pounds of work in one minute (Fig. 4-3). This evaluation has been universally accepted since that time. Thus, the formula for determining horsepower is:

$$\text{horsepower} = \frac{\text{pounds} \times \text{feet}}{33,000 \times \text{minutes}}$$

When referring to engine horsepower ratings, one usually finds the rating expressed as brake horsepower or rated horsepower, or sometimes as both.

BRAKE HORSEPOWER. Brake horsepower is the maximum horsepower available from an engine as determined by use of a dynamometer, and is usually stated as maximum observed brake horsepower or as corrected brake horsepower. As will be noted in a later paragraph, observed brake horsepower of a specific engine will vary under different conditions of temperature and atmospheric pressure. Corrected brake horsepower is a rating calculated from observed brake horsepower and is a means of comparing engines tested at varying conditions. The method for calculating corrected brake horsepower will be explained in a later paragraph.

Fig. 4-4—Diagram showing a prony brake on which the torque being developed by an engine can be measured. By also knowing the RPM of the engine output shaft, engine horsepower can be calculated.

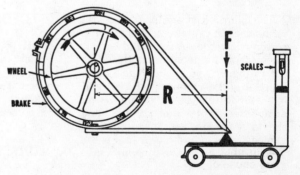

must also count revolutions of the engine output shaft for a specific length of time. In the formula for calculating horsepower:

$$\text{horsepower} = \frac{\text{feet} \times \text{pounds}}{33,000 \times \text{minutes}}$$

feet will equal the circumference transcribed by the torque arm radius multiplied by the number of engine output shaft revolutions. Thus:

$$\text{feet} = 2 \times 3.14 \times \text{radius} \times \text{revolutions.}$$

Pounds in the formula will equal the force (F) of the torque arm. If, for example, the force (F) is 6 pounds, torque arm radius is 2 feet and engine output shaft speed is 3300 revolutions per minute, then:

$$\text{horsepower} = \frac{2 \times 3.14 \times 2 \times 3300 \times 6}{33,000 \times 1}$$

or,

$$\text{horsepower} = 7.54$$

DYNAMOMETERS. Some commercial dynamometers for testing small engines are now available, although the cost may be prohibitive for all but the larger small engine repair shops. Usually, these dynamometers have a hydraulic loading device and scales indicating engine speed and load; horsepower is then calculated by use of a slide rule type instrument. For further information on commercial dynamometers, refer to manufacturers listed in special service tool section of this manual.

HOW ENGINE HORSEPOWER OUTPUT VARIES

Engine efficiency will vary with the amount of air taken into the cylinder on each intake stroke. Thus, air density has a considerable effect on the horsepower output of a specific engine. As air density varies with both temperature and atmospheric pressure, any change in air temperature, barometric pressure, or elevation will cause a variance in observed engine horsepower. As a general rule, engine horsepower will:

A. Decrease approximately 3 percent for each 1000 foot increase above 1000 ft. elevation;

B. Decrease approximately 3 percent for each 1 inch drop in barometric pressure; or,

C. Decrease approximately 1 percent for each 10° F rise in temperature.

Thus, to fairly compare observed horsepower readings, the observed readings should be corrected to standard temperature and atmospheric pressure conditions of 60° F, and 29.92 inches of mercury. The correction formula specified by the Society of Automotive Engineers is somewhat involved; however, for practical purposes, the general rules stated above can be used to approximate the corrected brake horsepower of an engine when the observed maximum brake horsepower is known.

For example, suppose the engine horsepower of 7.54 as found by use of the prony brake was observed at an altitude of 3000 feet and at a temperature of 100° F. At standard atmospheric pressure and temperature conditions, we could expect an increase of 4 percent due to temperature (100°-60° × 1% per 10°) and an increase of 6 percent due to altitude (3000 ft. - 1000 ft. × 3% per 1000 ft.) or a total increase of 10 percent. Thus, the corrected maximum horsepower from this engine would be approximately 7.54 + .75, or approximately 8.25 horsepower.

SERVICE SECTION
TROUBLESHOOTING

When servicing an engine to correct a specific complaint, such as engine will not start, is hard to start, etc., a logical step-by-step procedure should be followed to determine the cause of trouble before performing any service work. This procedure is "TROUBLESHOOTING" (Fig. 5-1).

The following procedures, as related to a specific complaint or trouble, have proven to be a satisfactory method for quickly determining the cause of trouble in a number of small engine repair shops.

NOTE: It is not suggested that the troubleshooting procedure as outlined in the following paragraphs be strictly adhered to at all times. In many instances, customer's comments on when trouble was encountered will indicate cause of trouble. Also, the mechanic will soon develop a diagnostic technique that can only come with experience. In addition to the general troubleshooting procedure, the reader should also refer to special notes following this section and to the information included in the engine, carburetor and magneto servicing fundamentals sections.

**If Engine Will Not Start—
Or Is Hard To Start**

1. If engine is equipped with a rope or crank starter, turn crankshaft slowly. As the engine piston is coming up on compression stroke, a definite resistance to turning should be felt on rope or crank. See Fig. 5-2. This resistance should be

Fig. 5-1—Diagnosing cause of trouble, or "troubleshooting" is an important factor in servicing small engines.

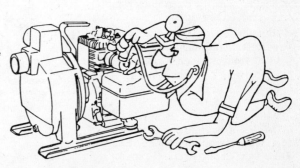

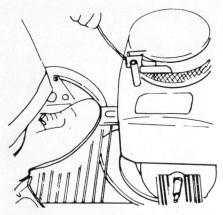

Fig. 5-2—Checking engine compression by slowly cranking engine; a definite resistance should be felt on starter rope each time piston comes up on compression stroke.

Fig. 5-3—Ignition spark can be checked using special test plug shown above.

Fig. 5-5—Condensation can cause water and rust to form in fuel tank even though only clean fuel has been poured into tank.

noted every other crankshaft revolution on a single cylinder four-stroke cycle engine and on every revolution of a two-stroke cycle engine crankshaft. If alternate hard and easy turning is noted, the engine compression can be considered as not the cause of trouble at this time.

NOTE: Compression gauges for small gasoline engines are available and are useful in troubleshooting engine service problems.

Where available from engine manufacturer, specifications will be given for engine compression pressure in the engine service sections of this manual. On engines having electric or impulse starters, remove spark plug and ground it against the engine. Install compression gauge in spark plug hole, crank the engine and check engine compression; if gauge is not available, hold thumb so that spark plug hole is partly covered. An alternating blowing and suction action should be noted as the engine is cranked.

If very little or no compression is noted, refer to appropriate engine repair section for repair of engine. Low or no compression is an indication of blown head gasket, valves not sealing, piston rings and/or cylinder worn, hole in piston or broken connecting rod. If check indicates engine is developing compression, proceed to step 2.

2. Remove spark plug wire and hold wire terminal about ⅛ inch (3 mm) away from cylinder. (On wires having rubber spark plug boot, insert a small screw or bolt in terminal.)

NOTE: If available, use of a spark test plug is recommended. See Fig. 5-3.

While cranking engine, a bright blue spark should snap across the ⅛ inch (3 mm) gap. If spark is weak or yellow, or if no spark occurs while cranking engine, refer to following IGNITION SYSTEM SERVICE section for information on appropriate type system.

NOTE: A test plug with ⅛ inch (3 mm) gap is available or a test plug can be made by adjusting the electrode gap of a spark plug to 0.125 inch (3 mm).

If spark is satisfactory, remove and inspect spark plug. Refer to SPARK PLUG SERVICING under following IGNITION SYSTEM SERVICE section. If in doubt about spark plug condition, install a new plug.

NOTE: Before installing plug, be sure to check electrode gap with proper gauge and, if necessary, adjust to value given in engine repair section of this manual. DO NOT guess or check gap with a "thin dime" or other inaccurate measuring device. A few thousandths variation from correct spark plug electrode gap can make an engine run unsatisfactorily, or under some conditions, not start at all. See Fig. 5-4.

If ignition spark is satisfactory and engine will not start with new plug, proceed with step 3.

3. If engine compression and ignition spark seem to be OK, trouble within the fuel system should be suspected.

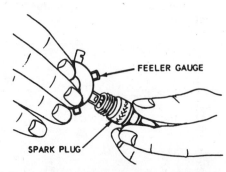

Fig. 5-4—Be sure to check spark plug electrode gap with proper size feeler gauge and adjust gap to specification recommended by manufacturer.

Remove and clean or renew air cleaner or cleaner element. Check fuel tank (Fig. 5-5) and be sure it is full of fresh gasoline (four-stroke cycle engines) or fresh gasoline and lubricating oil mixture (two-stroke cycle engines) as prescribed by engine manufacturer. Refer to LUBRICATION paragraph in each two-stroke cycle engine service (engine repair) section for proper fuel:oil mixture for each make and model. If equipped with a fuel shut-off valve, be sure valve is open.

If engine is equipped with remote throttle controls that also operate carburetor choke plate, check to be sure that when controls are placed in choke position, carburetor choke plate is fully closed, If not, adjust control linkage so that choke will fully close; then, try to start engine. If engine does not start after several turns, remove air cleaner assembly; carburetor throat should be wet with gasoline, If not, check for reason fuel is not getting to carburetor. On models with gravity feed from fuel tank to carburetor (fuel tank above carburetor), disconnect fuel line at carburetor to see that fuel is flowing through the line. If no fuel is flowing, remove and clean fuel tank, fuel line and any fuel filters or shut-off valve.

On models having a fuel pump separate from carburetor, remove fuel line at carburetor and crank engine through several turns; fuel should spurt from open line, If not, disconnect fuel line from tank to fuel pump at pump connection. If fuel will not run from open line, remove and clean the fuel tank, line and if so equipped, fuel filter and/or shut-off valve. If fuel runs from open line, remove and overhaul or renew the fuel pump.

After making sure that clean, fresh fuel is available at carburetor, again try to start engine. If engine will not start, refer to recommended initial adjustments for carburetor in appropriate engine repair section of this manual and adjust carburetor idle and/or main fuel needles.

If engine will not start when compression and ignition test OK and clean, fresh fuel is available to carburetor, remove and clean or overhaul carburetor as outlined in following CARBURETOR SERVICE section.

4. The preceding troubleshooting techniques are based on the fact that to run, an engine must develop compression, have an ignition spark and receive the proper fuel:air mixture. In some instances, there are other factors involved. Refer to the special notes following this section for service hints on finding common causes of engine trouble that may not be discovered in normal troubleshooting procedure.

If Engine Starts, Then Stops

This complaint is usually due to fuel starvation, but may be caused by a faulty ignition system. Recommended troubleshooting procedure is as follows:

1. Remove and inspect fuel tank cap; on all except a few early two-stroke cycle engines, fuel tank is vented through a breather in fuel tank cap so that air can enter the tank as fuel is used. If engine stops after running several minutes, a clogged breather should be suspected. On some engines, it is possible to let the engine run with fuel tank cap removed and if this permits engine to run without stopping, clean or renew the cap.

CAUTION: Be sure to observe safety precautions before attempting to run engine without fuel tank cap in place. If there is any danger of fuel being spilled on engine or spark entering open tank, DO NOT attempt to run engine without fuel tank cap in place. If in doubt, try a new cap.

2. If clogged breather in fuel tank cap is eliminated as cause of trouble, a partially clogged fuel filter or fuel line should be suspected. Remove and clean fuel tank and line and if so equipped, clean fuel shut-off valve and/or fuel tank filter. On some en-

gines, a screen or felt type fuel filter is located in the carburetor fuel inlet; refer to engine repair section for appropriate engine make and model for carburetor construction.

3. After cleaning fuel tank, line, filter, etc., if trouble is still encountered, a sticking or faulty carburetor inlet needle valve, float or diaphragm may be the cause of trouble. Remove, disassemble and clean carburetor using data in engine repair section and in following CARBURETOR SERVICE section as a guide.

4. If fuel system is eliminated as cause of trouble by performing procedure outlined in steps 1, 2 and 3, check magneto or battery ignition coil on tester if such equipment is available. If not, check for ignition spark immediately after engine stops. Renew coil, condenser and breaker points if no spark is noted. Also, on four-stroke cycle engines, check for engine compression immediately after engine stops; if no or little compression is noted immediately after engine stops; trouble may be caused by sticking intake or exhaust valve or cam followers. Refer to ENGINE SERVICE section and to engine repair data in the appropriate engine repair section of this manual for repair procedures.

Engine Overheats

When air cooled engines overheat, check for:

1. "Winterized" engine operated in warm temperatures.
2. Remove blower housing and shields and check for dirt or debris accumulated on or between cooling fins on cylinder.
3. Missing or bent shields or blower housing. (Never attempt to operate an air cooled engine without all shields and blower housing place.)
4. A too lean main fuel:air adjustment of carburetor.
5. Improper ignition spark timing. Check breaker-point gap, and on engine with unit type magneto, check magneto to engine timing. On battery ignition units with timer, check for breaker points opening at proper time.
6. Engines being operated under loads in excess of rated engine horsepower or at extremely high ambient (surrounding) air temperatures may overheat.
7. Two-stroke cycle engines being operated with an improper fuel-lubricat-

ing oil mixture may overheat due to lack of lubrication; refer to appropriate engine service section in this manual for recommended fuel-lubricating oil mixture.

Engine Surges When Running

Trouble with an engine surging is usually caused by improper carburetor adjustment or improper governor adjustment.

1. Refer to CARBURETOR paragraphs in the appropriate engine repair section and adjust carburetor as outlined.
2. If adjusting carburetor did not correct the surging condition, refer to GOVERNOR paragraph and adjust governor linkage.
3. If any wear is noted in governor linkage and adjusting linkage did not correct problem, renew worn linkage parts.
4. If trouble is still not corrected, remove and clean or overhaul carburetor as necessary. Also check for any possible air leaks between the carburetor to engine gaskets or air inlet elbow gaskets.

Engine Misses Out When Running (Two-Stroke Cycle Engines)

1. If engine misses out only at no load high idle speed, first be sure that engine is not equipped with an ignition cut-out governor. (If so equipped, engine will miss out at high speed due to cut-out action.) If not so equipped, refer to appropriate engine repair section and adjust carburetor as outlined in CARBURETOR paragraph. Some two-stroke cycle engines will miss out (four-cycle) when not under load, even though carburetor is adjusted properly. If a two-cycle engine fires evenly under normal load, it can usually be considered OK.

Special Notes on Engine Troubleshooting

SPECIAL APPLICATION ENGINES. An engine may be manufactured or modified to function as a power plant for a unique type of equipment. Troubleshooting and servicing must be performed while noting any out-of-the-ordinary features of the engine.

TWO-STROKE CYCLE ENGINES WITH REED VALVE. On two-stroke cycle engines, the incoming fuel:air mixture must be compressed in engine crankcase in order for the mixture to properly reach the engine cylinder. On engines utilizing reed type carburetor-to-crankcase intake valve, a bent or broken reed will not allow compression build up in the crankcase. Thus, if such an engine seems otherwise OK, remove and inspect the reed valve unit. Refer to appropriate engine repair section in this manual for information on individual two-stroke engine models.

TWO-STROKE CYCLE ENGINE EXHAUST PORTS. Two-stroke cycle engines, and especially those being operated on an overly rich fuel:air mixture or with too much lubricating oil mixed with the fuel, will tend to build up carbon in the cylinder exhaust ports. It is recommended that the muffler be removed on two-stroke cycle engines periodically and the carbon removed from the exhaust ports. Recommended procedure varies somewhat with different makes of engines; therefore, refer to CARBON paragraph of maintenance instructions in appropriate engine repair section of this manual.

On two-stroke cycle engines that are hard to start, or where complaint is loss of power, it is wise to remove the muffler and inspect the exhaust ports for carbon build up.

FOUR-STROKE CYCLE ENGINES WITH COMPRESSION RELEASE. Several different makes of four-stroke cycle engines now have a compression release that reduces compression pressure at cranking speeds, thus making it easier to crank the engine. Most models having this feature will develop full compression when crankshaft is turned in a reverse direction. Refer to the appropriate engine repair section in this manual for detailed information concerning the compression release used on different makes and models.

IGNITION SYSTEM SERVICE

The fundamentals of servicing ignition systems are outlined in the following paragraphs. Refer to appropriate heading for type of ignition system being inspected or overhauled.

BATTERY IGNITION SERVICE FUNDAMENTALS

Service of battery ignition systems used on small engines is somewhat simplified due to the fact that no distribution system is required as on automotive type ignition systems. Usually all components are readily accessible and while use of test instruments is sometimes desirable, condition of the system can be determined by simple checks. Refer to following paragraphs.

GENERAL CONDITION CHECK. Remove spark plug wire and if terminal is rubber covered, insert small screw or bolt in terminal. Hold uncovered end of wire terminal, or bolt inserted in terminal, about ⅛ inch (3 mm) away from engine, or connect spark plug wire to test plug. Crank engine while observing gap between spark plug wire terminal and engine; if a bright blue spark snaps across the gap, condition of the system, can be considered satisfactory. However, ignition timing may have to be adjusted. Refer to timing procedure in appropriate engine repair section.

VOLTAGE, WIRING AND SWITCH CHECK. If no spark, or a weak yellow-orange spark occurred when checking system as outlined in preceding paragraph, proceed with following checks:

Test battery condition with hydrometer or voltmeter. If check indicates a dead cell, renew the battery; recharge battery if a discharged condition is indicated.

NOTE: On models with electric starter or starter-generator unit, battery can be assumed in satisfactory condition if the starter cranks the engine freely.

If battery checks OK, but starter unit will not crank engine, a faulty starter unit is indicated and ignition trouble may be caused by excessive current draw of such a unit. If battery and starting unit, if so equipped, are in satisfactory condition, proceed as follows:

Remove battery lead wire from ignition coil and connect a test light of same voltage as the battery between the disconnected lead wire and engine ground. Light should go on when ignition switch is in "on" position and go off when switch is in "off" position. If not, renew switch and/or wiring and recheck for satisfactory spark. If switch and wiring check OK, but no spark is obtained, proceed as follows:

BREAKER POINTS AND CONDENSER. Remove breaker box cover and, using small screwdriver, separate and inspect breaker points. If burned or deeply pitted, renew breaker points and condenser. If point contacts are clean to grayish in color and are only slightly pitted, proceed as follows: Disconnect condenser and ignition coil lead wires from breaker point terminal and connect a test light and battery between terminal and engine ground (Fig. 5-5). Light should go on when points are closed and should go out when points are open. If light fails to go out when points are open, breaker arm insulation is defective and breaker points must be renewed. If light does not go on when points are in closed position, clean or renew the breaker points. In some instances, new breaker point contact surfaces may have an oily or wax coating or have foreign material between the surfaces so that proper contact is prevented. Check ignition timing and breaker point gap as outlined in appropriate engine repair section of this manual.

Connect test light and battery between condenser lead and engine ground; if light goes on, condenser is shorted out and should be renewed. Capacity of condenser can be checked if test instrument is available. It is usually good practice to renew the condenser whenever new breaker points are being installed if tester is not available.

IGNITION COIL. If a coil tester is available, condition of coil can be checked. However, if tester is not available, a reasonably satisfactory performance test can be made as follows:

Disconnect high tension wire from spark plug. Crank engine so that cam has allowed breaker points to close. With ignition switch on, open and close points with small screwdriver while holding high tension lead about ⅛ to ¼ inch (3-6 mm) away from engine ground. A bright blue spark should snap across the gap between spark plug wire and ground each time the points are opened. If no spark occurs, or spark is weak and yellow-orange, renewal of ignition coil is indicated.

Sometimes, an ignition coil may perform satisfactorily when cold, but fail after engine has run for some time and coil is hot. Check coil when hot if this condition is indicated.

FLYWHEEL MAGNETO SERVICE FUNDAMENTALS

In servicing a flywheel magneto ignition system, the mechanic is concerned with troubleshooting, service adjustments and testing magneto components. The following paragraphs outline the basic steps in servicing a flywheel type magneto. Refer to the appropriate engine section for adjustment and test specifications for a particular engine.

Troubleshooting

If the engine will not start and malfunction of the ignition system is suspected, make the following checks to find cause of trouble.

Check to be sure that the ignition switch (if so equipped) is in the "On" or "Run" position and that the insulation on the wire leading to the ignition switch is in good condition. The switch can be checked with the timing and test light shown in Fig. 5-6. Disconnect the lead from the switch and attach one clip of the test light to the switch terminal and the other clip to the engine. The light should go on when the switch is in the "Off" or "Stop" position, and should go off when the switch is in the "On" or "Run" position.

Inspect the high tension (spark plug) wire for worn spots in the insulation or breaks in the wire. Frayed or worn in-

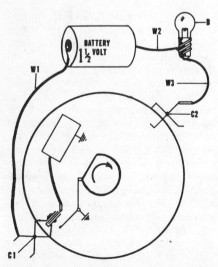

Fig. 5-6—Drawing showing a simple test lamp for checking ignition timing and/or breaker-point opening.

B. 1-1/2 volt bulb
C1. Spring clamp
C2. Spring clamp
W1. Wire
W2. Wire
W3. Wire

sulation can be repaired temporarily with plastic electrician's tape.

If no defects are noted in the ignition switch or ignition wires, remove and inspect the spark plug as outlined in the SPARK PLUG SERVICING section. If the spark plug is fouled or is in questionable condition, connect a spark plug of known quality to the high tension wire, ground the base of the spark plug to the engine and crank engine rapidly with the starter. If the spark across the electrode gap of the spark plug is a bright blue, the magneto can be considered in satisfactory condition.

NOTE: Some engine manufacturers specify a certain type spark plug and a specific test gap. Refer to appropriate engine service section; if no specific spark plug type or electrode gap is recommended for test purposes, use spark plug type and electrode gap recommended for engine make and model.

If spark across the gap of the test plug is weak or orange colored, or no spark occurs as engine is cranked, magneto should be serviced as outlined in following paragraphs.

Magneto Adjustments

BREAKER CONTACT POINTS. Adjustment of the breaker contact points affects both ignition timing and magneto edge gap. Therefore, the breaker contact point gap should be carefully adjusted according to engine manufacturer's specifications. Before adjusting the breaker contact gap, inspect contact points and renew if condition of contact surfaces is questionable. It is sometimes desirable to check the condition of points as follows: Disconnect the condenser and primary coil leads from the breaker-point terminal. Attach one clip of a test light (see Fig. 5-6) to the breaker-point terminal and the other clip of the test light to magneto ground. The light should be out when contact points are open and should go on when the engine is turned to close the breaker contact points. If the light stays on when points are open, insulation of breaker contact arm is defective. If light does not go on when points are closed, contact surfaces are dirty, oily or are burned.

Adjust breaker-point gap as follows unless manufacturer specifies adjusting breaker gap to obtain correct ignition timing. First, turn engine so that points are closed to be sure that the contact surfaces are in alignment and seat squarely. Then, crank engine so that breaker point opening is maximum

Fig. 5-7—On some engines, timing is adjustable by moving magneto stator plate in slotted mounting holes. Marks should be applied to stator plate and engine after engine is properly timed; marks usually appear on factory assembled stator plate and engine cylinder block.

and adjust breaker gap to manufacturer's specification. A wire type feeler gauge is recommended for checking and adjusting the breaker contact gap. Be sure to recheck gap after tightening breaker point base retaining screws.

IGNITION TIMING. On some engines, ignition timing is nonadjustable and a certain breaker-point gap is specified. On other engines, timing is adjustable by changing the position of the magneto stator plate (see Fig. 5-7) with a specified breaker-point gap or by simply varying the breaker-point gap to obtain correct timing. Ignition timing is usually specified either in degrees of engine (crankshaft) rotation or in piston travel before the piston reaches top dead center position. In some instances, a specification is given for ignition timing even though the timing may be nonadjustable; if a check reveals timing is incorrect on these engines, it is an indication of incorrect breaker-point adjustment or excessive wear of breaker cam. Also, on some engine, it may indicate that a wrong breaker cam has been installed or that the cam has been installed in a reversed position on engine crankshaft.

Some engines may have a timing mark or flywheel locating pin to locate the flywheel at proper position for the ignition spark to occur (breaker-points begin to open). If not, it will be necessary to measure piston travel as illustrated in Fig. 5-8 or install a degree indicating device on the engine crankshaft.

A timing light as shown in Fig. 5-6 is a valuable aid in checking or adjusting engine timing. After disconnecting the ignition coil lead from the breaker-point terminal, connect the leads of the timing light as shown. If timing is adjust-

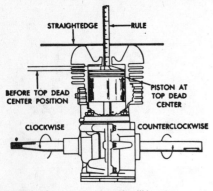

Fig. 5-8—On some engines, it will be necessary to measure piston travel with rule, dial indicator or special timing gauge when adjusting or checking ignition timing.

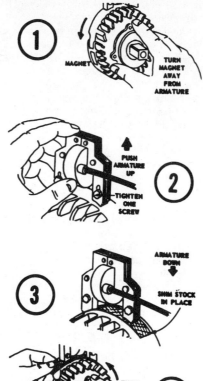

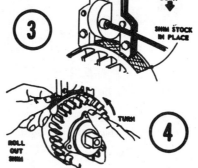

Fig. 5-9—Views showing adjustment of armature air gap when armature is located outside flywheel. Refer to Fig. 5-10 for engines having armature located inside flywheel.

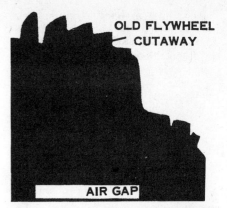

Fig. 5-10—Where armature core is located inside flywheel, check armature gap by using a cutaway flywheel unless other method is provided by manufacturer; refer to appropriate engine repair section. Where possible, an old discarded flywheel should be used to cutaway section for checking armature air gap.

able by moving the magneto stator plate, be sure that the breaker point gap is adjusted as specified. Then, to check timing, slowly crank engine in normal direction of rotation past the point at which ignition spark should occur. The timing light should be on, then go out (breaker points open) just as the correct timing location is passed. If not, turn engine to proper timing location and adjust timing by relocating the magneto stator plate or varying the breaker contact gap as specified by engine manufacturer. Loosen the screws retaining the stator plate or breaker points and adjust position of stator plate or points so that points are closed (timing light is on). Then, slowly move adjustment until timing light goes out (points open) and tighten the retaining screws. Recheck timing to be sure adjustment is correct.

ARMATURE AIR GAP. To fully concentrate the magnetic field of the flywheel magnets within the armature core, it is necessary that the flywheel magnets pass as closely to the armature core as possible without danger of metal to metal contact. The clearance between the flywheel magnets and the legs of the armature core is called the armature air gap.

On magnetos where the armature and high tension coil are located outside of the flywheel rim, adjustment of the armature air gap is made as follows: Crank the engine so that the flywheel magnets are located directly under the legs of the armature core and check the clearance between the armature core and flywheel magnets. If the measured clearance is not within manufacturers specifications, loosen the armature core mounting screws and place shims of thickness equal to minimum air gap specification between the magnets and armature core (Fig. 5-9). The magnets will pull the armature core against the shim stock. Tighten the armature core mounting screws and turn the flywheel

through several revolutions to be sure the flywheel does not contact the armature core.

Where the armature core is located under or behind the flywheel, the following methods may be used to check and adjust armature air gap: On some engines, slots or openings are provided in the flywheel through which the armature air gap can be checked. Some engine manufacturers provide a cut-away flywheel that can be installed temporarily for checking the armature air gap. A test flywheel can be made out of a discarded flywheel (see Fig. 5-10), or out of a new flywheel if service volume on a particular engine warrants such expenditure.

Another method of checking the armature air gap is to remove the flywheel and place a layer of plastic tape equal to the minimum specified air gap over the legs of the armature core. Reinstall flywheel and turn flywheel through several revolutions and remove flywheel; no evidence of contact between the flywheel magnets and plastic tape should be noticed. Then cover the legs of the armature core with a layer of tape of thickness equal to the maximum speci-

fied air gap; then, reinstall flywheel and turn flywheel through several revolutions. Indication of the flywheel magnets contacting the plastic tape should be noticed after the flywheel is again removed. If the magnets contact the first thin layer of tape applied to the armature core legs, or if they do not contact the second thicker layer of tape, armature air gap is not within specifications and should be adjusted.

NOTE: Before loosening armature core mounting screws, scribe a mark on mounting plate against edge of armature core so that adjustment of air gap can be gauged.

In some instances, it may be necessary to slightly enlarge the armature core mounting holes before proper air gap adjustment can be made.

MAGNETO EDGE GAP. The point of maximum acceleration of the movement of the flywheel magnetic field through the high tension coil (and therefore, the point of maximum current induced in the primary coil windings) occurs when the trailing edge of the flywheel magnet is slightly past the left hand leg of the armature core. The exact point of maximum primary current is determined by using electrical measuring devices, the distance between the trailing edge of the flywheel magnet and the leg of the armature core at this point is measured and becomes a service specification. This distance, which is stated either in thousandths of an inch or in degrees of flywheel rotation, is called the Edge Gap or "E" Gap.

For maximum strength of the ignition spark, the breaker points should just start to open when the flywheel magnets are at the specified edge gap position. Usually, edge gap is nonadjustable and will be maintained at the proper dimension if the contact breaker points are adjusted to the recommended gap and the correct breaker cam is installed. However, magnet edge gap can change (and peak intensity thereby reduced) due to the following:

a. Flywheel drive key sheared.
b. Flywheel drive key worn (loose).
c. Keyway in flywheel or crankshaft worn (oversized).
d. Loose flywheel retaining nut which can also cause any above listed difficulty.
e. Excessive wear on breaker cam.
f. Breaker cam loose on crankshaft.
g. Excessive wear on breaker point rubbing block or push rod so the points cannot be properly adjusted.

UNIT TYPE MAGNETO SERVICE FUNDAMENTALS

Improper functioning of the carburetor, spark plug or other components often causes difficulties that are thought to be an improperly functioning magneto. Since a brief inspection will often locate other causes of engine malfunction, it is recommended that one be certain that the magneto is at fault before opening the magneto housing. Magneto malfunction can easily be determined by simple tests outlined in following paragraph.

Troubleshooting

With a properly adjusted spark plug in good condition, the ignition spark should be strong enough to bridge a short gap in addition to the actual spark plug gap. With engine running, hold end of spark plug wire not more than $\frac{1}{16}$ inch (1.6 mm) away from spark plug terminal. Engine should not misfire.

To test the magneto spark if engine will not start, remove ignition wire from magneto end cap socket. Bend a short piece of wire so that when it is inserted in the end cap socket, other end is about $\frac{1}{8}$ inch (3.2 mm) from engine casting. Crank engine slowly and observe gap between wire and engine; a strong blue spark should jump the gap the instant that the impulse coupling trips. If a strong spark is observed, it is recommended that the magneto be eliminated as the source of engine difficulty and that the spark plug, ignition wire and terminals be thoroughly inspected.

If, when cranking the engine, the impulse coupling does not trip, the magneto must be removed from the engine and the coupling overhauled or renewed. It should be noted that if the impulse coupling will not trip, a weak spark will occur.

Magneto Adjustments and Service

BREAKER POINTS. Breaker points are accessible for service after removing the magneto housing end cap. Examine point contact surfaces for pitting or pyramiding (transfer of metal from one surface to the other); a small tungsten file or fine stone may be used to resurface the points. Badly worn or badly pitted points should be renewed. After points are resurfaced or renewed, check breaker point gap with rotor turned so that points are opened maximum distance. Refer to MAGNETO paragraph in appropriate engine repair section for point gap specifications.

When replacing the magneto end cap, both the end cap and housing mating surfaces should be thoroughly cleaned and a new gasket be installed.

CONDENSER. Condenser used in unit type magneto is similar to that used in other ignition systems. Refer to MAGNETO paragraph in appropriate engine repair section for condenser test specifications. Usually, a new condenser should be installed whenever the breaker points are renewed.

COIL. The ignition coil can be tested without removing the coil from the housing. The instructions provided with the coil tester should have coil test specifications listed.

ROTOR. Usually, service on the magneto rotor is limited to renewal of bushings or bearings, if damaged. Check to be sure that rotor turns freely and does not drag or have excessive end play.

MAGNETO INSTALLATION. When installing a unit type magneto on an engine, refer to MAGNETO paragraph in appropriate engine repair section for magneto to engine timing information.

SOLID-STATE IGNITION SERVICE FUNDAMENTALS

Because of differences in solid-state ignition construction, it is impractical to outline a general procedure for solid-state ignition service. Refer to the specific engine section for testing, overhaul notes and timing of solid-state ignition systems.

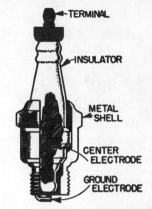

Fig. 5-11—Cross-sectional drawing of spark plug showing construction and nomenclature.

SPARK PLUG SERVICING

ELECTRODE GAP. The spark plug electrode gap should be adjusted by bending the ground electrode. Refer to Fig. 5-11. The recommended gap is listed in the SPARK PLUG paragraph in appropriate engine repair section of this manual. A wire type feeler gauge is recommended for setting electrode gap.

CLEANING AND ELECTRODE CONDITIONING. Spark plugs are usually cleaned by an abrasive action commonly referred to as "sand blasting." Actually, ordinary sand is not used, but a special abrasive which is nonconductive to electricity even when melted, thus the abrasive cannot short out the plug current. Extreme care should be used in cleaning the plugs after sand blasting, however, as any particles of abrasive left on the plug may cause damage to piston rings, piston or cylinder wall.

NOTE: Some engine manufacturers recommend that the spark plug be renewed rather than cleaned by an abrasive type cleaner because of possible engine damage from the abrasive.

After plug is cleaned by abrasive, and before gap is set, the electrode surfaces between the grounded and insulated electrodes should be cleaned and returned as nearly as possible to original shape by filing with a point file. Failure to properly dress the electrodes can result in high secondary voltage requirements and misfire of the plug.

PLUG APPEARANCE DIAGNOSIS. The appearance of a spark plug will be altered by use, and an examination of the plug tip can contribute useful

Fig. 5-12—Normal plug appearance in four-stroke cycle engine. Insulator is light tan to gray in color and electrodes are not burned. Renew plug at regular intervals as recommended by engine manufacturer.

Fig. 5-13—Appearance of four-stroke cycle spark plug indicating cold fouling. Cause of cold fouling may be use of a too-cold plug, excessive idling or light loads, carburetor choke out of adjustment, defective spark plug wire or boot, carburetor adjusted too "rich" or low engine compression.

Fig. 5-14—Appearance of four-stroke cycle spark plug indicating wet fouling; a wet, black oily film is over entire firing end of plug. Cause may be oil getting by worn valve guides, worn oil rings or plugged breather or breather valve in tappet chamber.

Fig. 5-15—Appearance of four-stroke cycle spark plug indicating overheating. Check for plugged cooling fins, bent or damaged blower housing, engine being operated without all shields in place or other causes of engine overheating. Also can be caused by too lean a fuel:air mixture or spark plug not tightened properly.

Fig. 5-16—Normal appearance of plug removed from a two-stroke cycle engine. Insulator is light tan to gray in color, few deposits are present and electrodes not burned.

information which may assist in obtaining better spark plug life. It must be remembered that the contributing factors differ in two-stroke cycle and four-stroke cycle engine operations, and although the appearance of two spark plugs may be similar, the corrective measures may depend on whether the engine is of two-stroke cycle or four-stroke cycle design. Figs. 5-12 through

Fig. 5-17—Appearance of plug from two-stroke cycle engine indicating wet fouling. A damp or wet black carbon coating is formed over entire firing end. Could be caused by a too-cold plug, excessive idling, improper fuel-lubricating oil mixture or carburetor adjustment too rich.

Fig. 5-18—Appearance of plug from two-stroke cycle engine indicating overheating. Insulator has gray or white blistered appearance and electrodes may be burned. Could be caused by use of a too-hot plug, carburetor adjustment too lean, "sticky" piston rings, engine overloaded, or cooling fins plugged causing engine to run too hot.

5-18 are provided by Champion Spark Plug Company to illustrate typical observed conditions. Refer to Figs. 5-12 through 5-15 for four-stroke cycle engines and to Figs. 5-16 through 5-18 for two-stroke cycle engines. Listed in captions are the probable causes and suggested corrective measures.

CARBURETOR SERVICE

GENERAL CARBURETOR SERVICE

The bulk of carburetor service consists of cleaning, inspection and adjustment. After considerable service it may become necessary to overhaul the carburetor and renew worn parts to restore original operating efficiency. Although carburetor condition affects engine operating economy and power, ignition and engine compression must also be considered to determine and correct causes of poor performance.

Before dismantling carburetor for cleaning or overhaul, clean all external surfaces and remove accumulated dirt and grease. Refer to appropriate engine repair section for carburetor exploded or cross-sectional views. Dismantle carburetor and note any discrepancies to assure correction during overhaul. Thoroughly clean all parts and inspect for damage or wear. Wash jets and passages and blow clear with clean, dry compressed air.

NOTE: Do not use a drill or wire to clean jets as the possible enlargement of calibrated holes will disturb operating balance of carburetor.

The measurement of jets to determine the extent of wear is difficult and new parts are usually installed to assure satisfactory results.

Carburetor manufacturers provide for many of their models an assortment of gaskets and other parts usually needed to do a correct job of cleaning and overhaul. These assortments are usually catalogued as Gasket Kits and Overhaul Kits respectively.

On float type carburetors, inspect float pin and needle valve for wear and renew if necessary. Check metal floats for leaks and where a dual type float is installed, check alignment of float sections. Check cork floats for loss of protective coating and absorption of fuel.

NOTE: Do not attempt to recoat cork floats with shellac or varnish or to resolder leaky metal floats. Renew part if defective.

Check the fit of throttle and choke valve shafts. Excessive clearance will cause improper valve plate seating and will permit dust or grit to be drawn into the engine. Air leaks at throttle shaft bores due to wear will upset carburetor calibration and contribute to uneven engine operation. Rebush valve shaft holes where necessary and renew dust seals. If rebushing is not possible, renew the body part supporting the shaft. Inspect throttle and choke valve plates for proper installation and condition.

Power or idle adjustment needles must not be worn or grooved. Check condition of needle seal packing or "O" ring and renew packing or "O" ring if necessary.

Reinstall or renew jets, using correct size listed for specific model. Adjust power and idle settings as described for specific carburetors in engine service section of this manual.

It is important that the carburetor bore at the idle discharge ports and in the vicinity of the throttle valve be free of deposits. A partially restricted idle port will produce a "flat spot" between idle and mid-range rpm. This is because the restriction makes it necessary to open the throttle wider than the designed opening to obtain proper idle speed. Opening the throttle wider than the design specified amount will uncover more of the port than was intended in the calibration of the carburetor. As a result an insufficient amount of the port will be available as a reserve to cover the transition period (idle to the mid-range rpm) when the high speed system begins to function.

When reassembling float-type carburetors, be sure float position is properly adjusted. Refer to CARBURETOR paragraph in appropriate engine repair section for float level adjustment specifications.

Setting or adjusting the metering control lever (metering diaphragm lever height) on diaphragm-type carburetors necessitates disassembly of the carburetor. Refer to the CARBURETOR section in appropriate engine repair section for adjusting the lever height.

TROUBLESHOOTING

Float-type Carburetor

Refer to Fig. 5-19 for schematic view of typical float-type carburetor showing location of parts. Normally encountered difficulties resulting from carburetor malfunction, along with possible causes of difficulty, for float-type carburetors are listed below.

ENGINE WILL NOT START OR HARD TO START. Could be caused by: (1) incorrect idle mixture screw adjustment, (2) restricted or plugged fuel filter or fuel line, (3) throttle shaft worn, (4) choke shaft worn or not functioning properly, (5) fuel inlet needle valve stuck closed, (6) damaged or worn float or hinge pin, (7) improper float height setting, (8) float chamber atmospheric vent hole restricted or plugged, (9) low speed fuel passages restricted or plugged.

CARBURETOR FLOODS. Could be caused by: (1) Choke not opening fully, (2) dirt or foreign particles preventing inlet fuel needle from seating, (3) damaged or worn fuel inlet needle and/or seat preventing proper seating of needle, (4) damaged or worn float or hinge pin, (5) improper float height setting, (6) restricted or plugged float chamber atmospheric vent hole, (7) Welch plug in fuel chamber is loose.

ENGINE RUNS LEAN. Could be caused by: (1) fuel tank vent plugged, (2) restricted fuel filter or fuel line, (3) high speed fuel passages restricted, (4) improper float height setting, (5) high speed mixture screw incorrectly adjusted, (7) leaky gaskets between carburetor and cylinder intake port.

ENGINE WILL NOT ACCELERATE SMOOTHLY. Could be caused by: (1) idle or main fuel mixture screws set too lean, (2), restricted low speed air bleed or fuel passages, (3) restricted tank vent, fuel filter or fuel line, (4) plugged air filter.

ENGINE STOPS WHEN DECELERATING. Could be caused by: (1) idle speed, idle mixture or high speed mixture screws incorrectly adjusted, (2) air leaks between carburetor and crankcase, (3) throttle shaft worn, (4) fuel inlet needle binding.

ENGINE WILL NOT IDLE. Could be caused by: (1) damaged or incorrect adjustment of idle fuel and/or idle speed screws, (2) idle discharge or air mixture ports plugged, (3) throttle and/or choke shaft worn, (4) fuel tank vent, filter or fuel line restricted, (5) damaged or worn float or hinge pin, (6) improper float height setting.

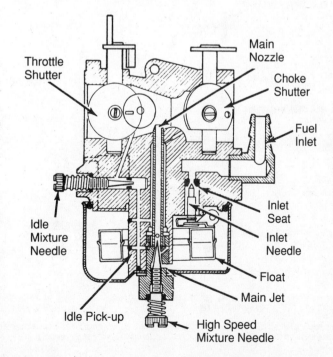

Throttle Shutter

Main Nozzle

Choke Shutter

Fuel Inlet

Idle Mixture Needle

Inlet Seat

Inlet Needle

Float

Main Jet

Idle Pick-up

High Speed Mixture Needle

Fig. 5-19—Cross-sectional view of a float-type carburetor showing location of major components.

ENGINE IDLES WITH LOW SPEED NEEDLE CLOSED. Could be caused by: (1) improper float height setting, (2) fuel inlet needle not seating due to wear or damage.

ENGINE RUNS RICH. Could be caused by: (1) plugged air filter, (2) low speed or high speed mixture screws incorrectly adjusted or damaged, (3) improper float height setting, (4) damaged or worn float or hinge pin, (5) choke not open fully, (6) fuel inlet needle valve leaking, (7) Welch plug leaking, (8) restricted or plugged air bleed passage.

ENGINE HAS LOW POWER UNDER LOAD. Could be caused by: (1) main mixture screw incorrectly adjusted, (2) plugged fuel tank vent, filter or fuel line, (3) leaky carburetor mounting gasket (4) plugged air filter, (5) throttle shaft worn, (6) air leaks between carburetor and intake manifold, (7) improper float height setting, (8) damaged or worn float or hinge pin, (9) high speed fuel passages restricted.

Diaphragm-type Carburetor

Refer to Fig. 5-20 for schematic view of typical diaphragm carburetor showing location of parts. Normally encoun-tered difficulties resulting from carbu-retor malfunction, along with possible causes of difficulty, for diaphragm-type carburetors are listed below.

ENGINE WILL NOT START OR HARD TO START. Could be caused by: (1) incorrect idle mixture screw adjust-ment, (2) restricted or plugged fuel filter or fuel line, (3) throttle shaft worn, (4) choke shaft worn or not functioning properly, (5) fuel inlet needle valve stuck closed, (6) metering lever worn, bent, binding, or set too low, (7) meter-ing diaphragm cover vent hole re-stricted or plugged, (8) metering diaphragm, gasket or cover leaking, (9) low speed fuel passages restricted or plugged.

CARBURETOR FLOODS. Could be caused by: (1) dirt or foreign particles preventing inlet fuel needle from seat-ing, (2) damaged or worn fuel inlet nee-dle and/or seat preventing proper seating of needle, (3) diaphragm lever spring not seated correctly on dia-phragm lever, (4) metering lever bind-ing or set too high, (5) hole in pump diaphragm, (6) Welch plug in fuel cham-ber is loose. Also, when fuel tank is located above carburetor, flooding can be caused by leaking fuel pump dia-phragm.

ENGINE RUNS LEAN. Could be caused by: (1) fuel tank vent plugged, (2) restricted fuel filter or fuel line, (3) high speed fuel passages restricted, (4) hole in fuel metering diaphragm, (5) meter-ing lever worn, binding, distorted or set too low, (6) high speed mixture screw incorrectly adjusted, (7) leak in pulse passage, (8) leaky gaskets between car-buretor and cylinder intake port. Also, check for leaking crankshaft seals, po-rous or cracked crankcase or other cause for air leak into crankcase.

ENGINE WILL NOT ACCELER-ATE SMOOTHLY. Could be caused by: (1) idle or main fuel mixture screws set too lean on models without accelerating pump, (2) inoperative accelerating pump, on carburetors so equipped, due to plugged channel, leaking diaphragm, stuck piston, etc., (3) restricted low speed fuel passage, (4) restricted tank vent, fuel filter or fuel line, (5) plugged air filter, (6) restricted vent hole in me-tering cover, (7) restricted pulse chan-nel, (8) defective pump diaphragm, (9) metering lever set too low, (10) defective

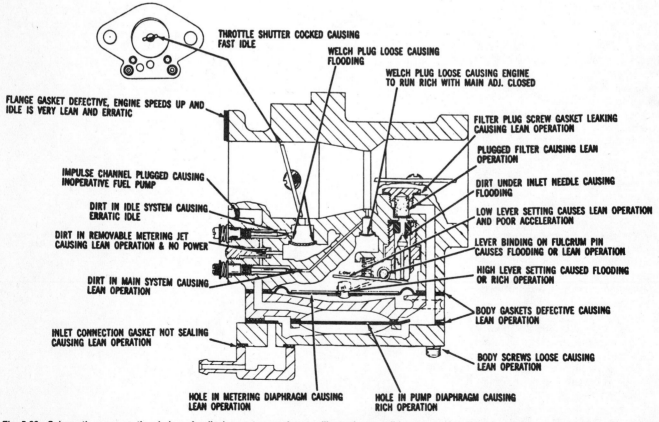

Fig. 5-20—Schematic cross-sectional view of a diaphragm-type carburetor illustrating possible causes of malfunction. Refer to appropriate engine repair section for adjustment information and for exploded and/or cross-sectional view of actual carburetors used.

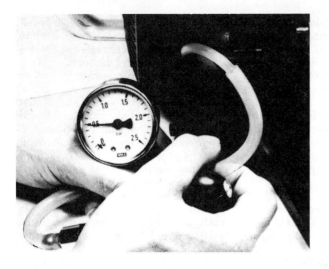

Fig. 5-21—View showing connection of pressure tester to fuel tank fuel line. Refer to text.

carburetors. With engine stopped and cooled, first adjust carburetor low speed and high speed mixture screws to equipment manufacturer's recommended initial settings. Remove fuel tank cap and withdraw fuel line out fuel tank opening. Remove strainer on end of fuel line and connect a suitable pressure tester as shown in Fig. 5-21. Pressurize system until 7 psi (48 kPa) is read on pressure gauge. Pressure reading must remain constant. If not, remove components as needed and connect pressure tester directly to carburetor inlet fitting as shown in Fig. 5-22. Pressurize carburetor until 7 psi (48 kPa) is read on pressure gauge. If pressure reading now remains constant, the fuel line is defective. If pressure reading decreases, then carburetor must be removed for further testing.

Connect pressure tester directly to carburetor inlet fitting and submerge carburetor assembly into a suitable container filled with a nonflammable solution or water as shown in Fig. 5-23. Pressurize carburetor until 7 psi (48 kPa) is read on pressure gauge. Observe carburetor and note location of leaking air bubbles. If air bubbles escape from around jet needles or venturi, then inlet needle or metering mechanism is defective. If air bubbles escape at impulse opening, then pump diaphragm is defective. If air bubbles escape from around fuel pump cover, then cover gasket or pump diaphragm is defective.

To check inlet needle and metering mechanism, first rotate low and high speed mixture screws inward until lightly seated. Pressurize system until 7 psi (48 kPa) is read on pressure gauge. If pressure reading does not remain constant, inlet needle is leaking. If pressure remains constant, depress metering diaphragm with a suitable length and thickness of wire through the vent hole in metering diaphragm cover. This will lift inlet needle off its seat and pressurize the metering chamber. A slight drop in pressure reading should be noted as metering chamber becomes pressur-

manifold or carburetor mounting gaskets, (11) metering lever set too low.

ENGINE STOPS WHEN DECELERATING. Could be caused by: (1) idle speed, idle mixture or high speed mixture screws incorrectly adjusted, (2) defective pump diaphragm, (3) pulse passage leaking or restricted, (4) air leaks between carburetor and crankcase, (5) throttle shaft worn, (6) metering lever set too high, (7) fuel inlet needle binding.

ENGINE WILL NOT IDLE. Could be caused by: (1) incorrect adjustment of idle fuel and/or idle speed screws, (2) idle discharge or air mixture ports plugged, (3) fuel channel plugged, (4) fuel tank vent, filter or fuel line restricted, (5) leaky gaskets between carburetor and cylinder intake ports.

ENGINE IDLES WITH LOW SPEED NEEDLE CLOSED. Could be caused by: (1) metering lever set too high or stuck, (2) fuel inlet needle not

seating due to wear or damage, (3) Welch plug covering idle ports not sealing properly.

ENGINE RUNS RICH. Could be caused by: (1) plugged air filter, (2) low speed or high speed mixture screws incorrectly adjusted or damaged, (3) metering lever worn, binding, distorted or set too high, (4) fuel pump diaphragm defective, (5) fuel inlet needle valve leaking, (6) Welch plug leaking, (7) faulty governor valve (if so equipped).

ENGINE HAS LOW POWER UNDER LOAD. Could be caused by: (1) main mixture screw incorrectly adjusted, (2) plugged fuel tank vent, filter or fuel line, (3) pulse channel leaking or restricted, (4) defective pump diaphragm, (5) plugged air filter, (6) air leaks between carburetor and crankcase, (7) metering lever distorted or set too low, (8) hole in metering diaphragm or gasket leaking, (9) faulty nozzle check valve.

PRESSURE TESTING. A hand pump and pressure gauge may be used to test fuel system for leakage when diagnosing problems with diaphragm

Fig. 5-22—View showing connection of pressure tester directly to carburetor inlet fitting. Refer to text.

Fig. 5-23—Submerge carburetor in a suitable container filled with solvent or water and pressure test as outlined in text.

ized. If no drop in pressure reading is noted, the inlet needle is sticking. If pressure does not hold after a slight drop, a defective metering mechanism or leaking high or low speed Welch plugs is indicated. To determine which component is leaking, submerge carburetor as previously outlined. Pressurize carburetor until 7 psi (48 kPa) is read on pressure gauge, then depress metering diaphragm as previously outlined. If bubbles escape from hole in metering diaphragm cover, metering diaphragm is defective. If bubbles escape from within venturi, determine which discharge port the air bubbles are escaping from to determine which Welch plug is leaking.

If low or high speed running problems are noted, the passage beneath the respective Welch plug may be restricted. To test idle circuit, adjust low speed mixture screw to recommended initial setting and rotate high speed mixture screw inward until lightly seated. Pressurize carburetor until 7 psi (48 kPa) is read on pressure gauge. Depress metering diaphragm as previously outlined. If pressure reading does not drop off or drops off very slowly, a restriction is indicated. To test high speed circuit, adjust high speed mixture screw to recommended initial setting and turn low

speed mixture screw inward until lightly seated. Pressurize carburetor and depress metering diaphragm as previously outlined and note pressure gauge. If pressure reading does not drop off or drops off very slowly, a restriction is indicated.

Refer to specific carburetor service section and repair defect or renew defective component as needed.

ADJUSTMENT

Initial setting for the mixture adjusting needles is listed in the specific engine section of this manual. Make final carburetor adjustment with engine warm and running. Make certain that engine air filter is clean before performing final adjustment, as a restricted air intake will affect the carburetor settings.

Adjust idle speed screw so that engine is idling at speed specified by equipment manufacturer. Adjust idle fuel mixture needle for best engine idle performance, keeping the mixture as rich as possible (turn needle out to enrich mixture). If necessary, readjust idle speed screw. To adjust main fuel needle, operate engine at wide-open throttle and find the rich and lean drop-off points, and set the mixture between

them. Main fuel needle may also be adjusted while engine is under load to obtain optimum performance. Do not operate engine with high speed mixture set too lean as engine damage may occur due to overheating.

If idle mixture is too lean and cannot be properly adjusted, consider the possibility of plugged idle fuel passages, expansion plug for main fuel check valve loose or missing, main fuel check valve not seating, improperly adjusted inlet control lever, leaking metering diaphragm or malfunctioning fuel pump.

If idle mixture is too rich, check idle mixture screw and its seat in carburetor body for damage. Check causes for carburetor flooding.

If high speed mixture is too lean and cannot be properly adjusted, check for dirt or plugging in main fuel passages, improperly adjusted metering lever, malfunctioning metering diaphragm or main fuel check valve. Also check for damaged or missing packing for high speed mixture screw and for malfunctioning fuel pump.

If high speed mixture is too rich, check high speed mixture screw and its seat for damage. Check for improperly adjusted metering lever or faulty fuel inlet needle valve. Check for faulty governor valve is carburetor is so equipped.

ENGINE SERVICE

DISASSEMBLY AND ASSEMBLY

Special techniques must be developed in repair of engines of aluminum alloy or magnesium alloy construction. Soft threads in aluminum or magnesium castings are often damaged by carelessness in over tightening fasteners or in attempting to loosen or remove seized fasteners. Manufacturer's recommended torque values for tightening screw fasteners should be followed closely.

NOTE: If damaged threads are encountered, refer to following paragraph, "REPAIRING DAMAGED THREADS."

A given amount of heat applied to aluminum or magnesium will cause it to expand a greater amount than will steel under similar conditions. Because of the different expansion characteristics, heat is usually recommended for easy installation of bearings, pins, etc., in aluminum or magnesium castings.

Sometimes, heat can be used to free parts that are seized or where an interference fit is used. Heat, therefore, becomes a service tool and the application of heat is one of the required service techniques. An open flame is not usually advised because it destroys the paint and other protective coatings and because a uniform and controlled temperature with open flame is difficult to obtain. Methods commonly used are heating in oil or water, with a heat lamp, electric hot plate, or in an oven or kiln.

The use of water or oil gives a fairly accurate temperature control but is somewhat limited as to the size and type of part that can be handled.

Thermal crayons are available which can be used to determine the temperature of a heated part. These crayons melt when the part reaches a specified temperature, and a number of crayons for different temperatures are available. Temperature indicating crayons are usually available at welding equipment supply houses.

The crankcase and combustion chambers of a two-stroke cycle engine must

be sealed against pressure and vacuum. To assure a perfect seal, nicks, scratches and warpage are to be avoided. Slight imperfections can be removed by using a fine-grit sandpaper. Flat surfaces can be lapped by using a surface plate or a smooth piece of plate glass, and a sheet of 120-grit sandpaper or lapping compound. Use a figure-eight motion with minimum pressure, and remove only enough metal to eliminate the imperfection. Bearing clearances, if any, must not be lessened by removing metal from the joint.

Use only the specified gaskets when reassembling, and use an approved gasket cement or sealing compound unless the contrary is stated. Seal all exposed threads and repaint or retouch with an approved paint.

REPAIRING DAMAGED THREADS

Fastener threads can be damaged for a variety of reasons, and in some cases, restoring the threads is required.

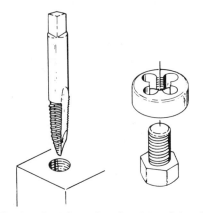

Fig. 6-1—A tap is used to clean internal threads (left view), and a die is used to clean external threads.

NOTE: Be sure to identify the thread size and type (US or metric) before attempting to restore the threads. The only positive method of identification is to use a screw pitch gauge or known fastener.

Often the threads can be cleaned by running a tap (for internal threads) or die (for external threads) through the threads. See Fig. 6-1. To clean or repair spark plug threads, a spark plug tap or "thread chaser" can be used.

If an internal thread is damaged, it may be necessary to install a thread repair insert. Thread repair inserts are available in a wide variety of US and metric thread sizes at auto supply stores and some hardware stores. Note that a specific size drill bit must be used, which must be purchased separately. To install a typical thread repair insert, proceed as follows:

Drill out the old threads using the drill bit recommended for the thread size being repaired (Fig. 6-2). Be sure

Fig. 6-2—First step in repairing damaged threads is to drill out old threads using exact size drill recommended in instructions provided with thread repair kit. Drill all the way through an open hole or all the way to bottom of blind hole, making sure hole is straight and that centerline of hole is not moved in drilling process.

Fig. 6-3—Special drill taps are provided in thread repair kit for threading drilled hole to correct size for outside of thread insert. A standard tap cannot be used.

Fig. 6-4—Turn new thread insert into hole using special tool.

the hole is straight and the centerline of the hole is not moved while drilling. Cut new threads in the hole using the tap provided in the kit (Fig. 6-3). This is a special tap that cuts threads to fit the outer threads on the thread repair insert. Turn the thread repair insert into the hole (Fig. 6-4) using the special tool until the top of the insert is a quarter to one-half turn below the surface (Fig. 6-5). Snap off the insert tang by pushing down on the tang; don't attempt to twist off the tang.

VALVE SERVICE FUNDAMENTALS

When overhauling engines, obtaining proper valve sealing is of primary importance. The following paragraphs cover fundamentals of servicing intake and exhaust valves, valve seats and valve guides.

VALVE TAPPET GAP. Specific settings and procedures for adjusting clearances between valve stem end and tappet are listed in the appropriate individual sections. Valve clearance should not be changed from the clear-

Fig. 6-5—A completed thread repair is shown above. Special tools are provided in thread repair kit for installation of thread insert.

Fig. 6-6—View showing one type of valve spring compressor used to remove and install valve spring and keeper. Position slotted-type valve keeper as shown when installing in engine.

ances listed. When the valves are closed, heat is transferred from the valve to the cylinder head, and valves may burn if valve clearance is set too tight. If the valve clearance is too loose, the engine will have decreased power. The "rattle" usually accompanying too much valve clearance is caused by metal parts of the valve system hitting. Although the noise is sometimes not objectionable, the constant pounding will result in increased wear and/or damage.

REMOVING AND INSTALLING VALVES. A valve spring compressor, one type of which is shown in Fig. 6-6, is a valuable aid in removing and installing the engine valves. This tool is used to hold spring compressed while removing or installing pin, collars or retainer from valve stem. Refer to Fig. 6-7 for views showing some of the different methods of retaining valve spring to valve stem.

VALVE REFACING. If valve face (Fig. 6-8) is slightly worn, burned or pitted, valve can usually be refaced providing proper equipment is available. Many shops will usually renew valves, however, rather than invest in somewhat costly valve refacing tools.

Before attempting to reface a valve, refer to specifications in appropriate engine repair section for valve face angle.

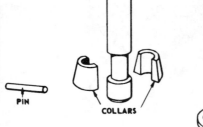

Fig. 6-7—Drawing showing three types of valve spring keepers used.

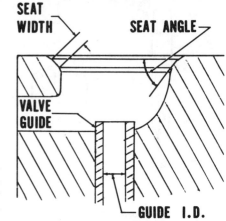

Fig. 6-10—Cross-sectional drawing of typical valve seat and valve guide as used on small engines. Valve guide may be integral part of cylinder block; on some models so constructed, valve guide ID may be reamed out and an oversize valve stem installed. On other models, a service guide may be installed after counter-boring cylinder block.

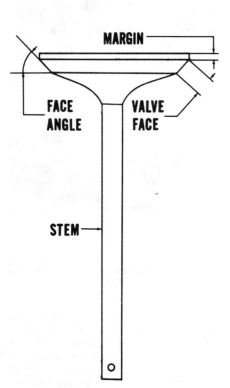

Fig. 6-8—Drawing showing typical four-stroke cycle engine valve. Face angle is usually 30 or 45°. On some engines, valve face is ground to an angle of 1/2 or 1 degree less than seat angle.

gin is less than manufacturer's minimum specification, or is less than one-half the margin of a new valve, renew the valve. Valves having excessive material removed in refacing operation will not give satisfactory service.

When refacing or renewing a valve, the seat should also be reconditioned. Note that valve seat is renewable on some engines. Then, the valve should be "lapped in" to its seat using a fine valve grinding compound. Refer to following paragraph "REFACING OR RENEWING VALVE SEATS."

REFACING OR RENEWING VALVE SEATS. Some engines have the valve seat machined directly in the cylinder block casting. The seat can be reconditioned by using a correct angle seat grinding stone or valve seat cutter. When reconditioning valve seat, care should be taken that only enough material is removed to provide a good seating area on valve contact surface. The width of seat should then be measured (Fig. 6-10) and if width exceeds manufacturer's maximum specifications, seat should be narrowed by using a stone or cutter with an angle 15° greater than seat angle and a second stone or cutter with an angle 15° less than seat angle. When narrowing seat, coat seat lightly with Prussian blue and check where seat contacts valve face by inserting valve in guide and rotating valve lightly against seat. Seat should contact approximately the center of valve face. By using only the narrow angle stone or cutter, seat contact will be moved toward outer edge of valve face.

Some engines have renewable valve seats. The seats are retained in cylinder block counterbore by an interference fit; that is, outside diameter of seat is slightly larger than counterbore in block. Refer to appropriate engine repair section in this manual for recommended method of removing old seat and install new seat. Refer to Fig. 6-11 for one method of installing new valve seats.

It sometimes occurs that a valve seat will become loose in counterbore, espe-

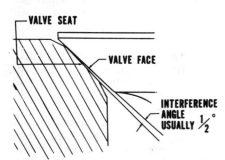

Fig. 6-9—Drawing showing line contact of valve face with valve seat when valve face is ground at smaller angle than valve seat; this is specified on some engines.

On some engines, manufacturer recommends grinding the valve face to an angle of 1/2 to 1 degree less than that of the valve seat. See Fig. 6-9. Also, nominal valve face angle may be either 30° or 45°.

After valve is refaced, measure thickness of valve "margin" (Fig. 6-8). If mar-

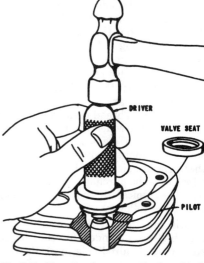

Fig. 6-11—View showing one method used to install valve seat insert. Refer to appropriate engine repair section for manufacturer's recommended method.

cially on engines with aluminum cylinder block. Some manufacturers provide oversize valve seat inserts (insert OD larger than standard part) so that if standard size insert fits loosely, counterbore can be cut oversize and a new insert be tightly installed. After installing valve seat insert in engines of aluminum construction, metal around seat should be peened as shown in Fig. 6-12. Where a loose insert is encountered and an oversize insert is not available, loose insert can usually be tightened by center-punching cylinder block material at three equally spaced points around insert, then peening completely around insert as shown in Fig. 6-12.

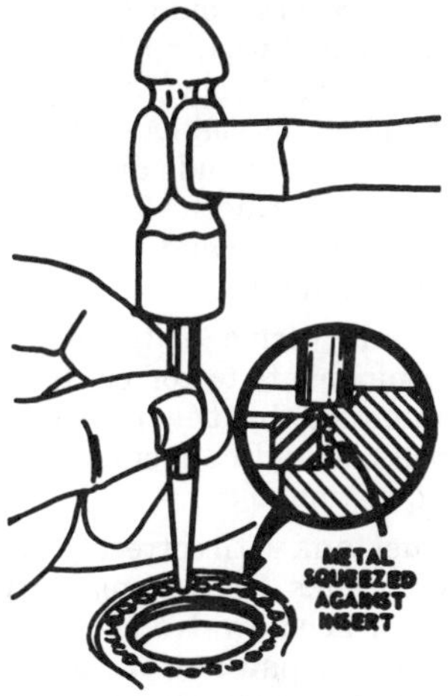

Fig. 6-12—It is usually recommended that on aluminum block engines, metal be peened around valve seat insert after insert is installed.

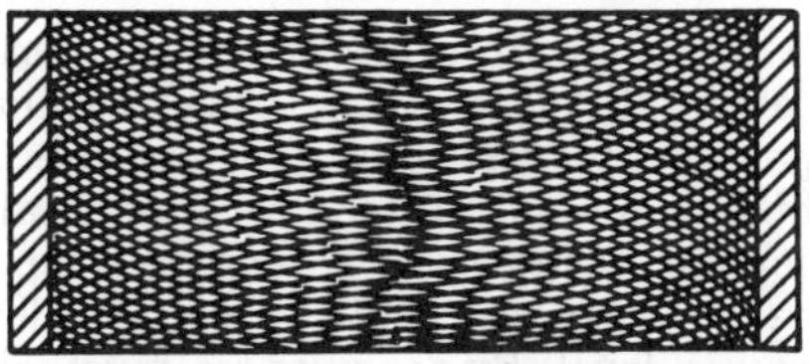

Fig. 6-14—Engine crankcase can be pressure tested for leaks by blocking off exhaust port and applying air pressure to intake port.

Fig. 6-13—A cross-hatch pattern as shown should be obtained when honing cylinder. Pattern is obtained by moving hone up and down cylinder bore as it is being turned by slow speed electric drill.

INSTALLING OVERSIZE PISTON AND RINGS

Some engine manufacturers have oversize piston and ring sets available for use in repairing engines in which cylinder bore is excessively worn and standard size piston and rings cannot be used. If care and approved procedure are used in oversizing cylinder bore, installation of an oversize piston and ring set should result in a highly satisfactory overhaul.

Cylinder bore may be oversized by using either a boring bar or a hone; however, if a boring bar is used, it is usually recommended cylinder bore be finished with a hone. Refer to Fig. 6-13. After honing is completed, clean cylinder bore thoroughly with warm soapy water, dry and lubricate with engine oil.

Where oversize piston and rings are available, it will be noted in appropriate engine repair section of this manual. Also, the standard bore diameter will be given. Before attempting to rebore or hone the cylinder to oversize, carefully measure the cylinder bore and examine for damage. It may be possible cylinder is excessively worn or damaged and boring or honing to larger oversize will not clean up the cylinder surface.

TWO-STROKE ENGINE CRANK-CASE PRESSURE TEST. Pressure testing the crankcase is an important part of troubleshooting and repair of two-stroke cycle engines that is often overlooked. An improperly sealed crankcase allows supplementary air to enter the engine and upsets the fuel:air mixture. This can cause the engine to be hard to start, run rough, have low power and overheat.

To test crankcase for leakage, a suitable hand pump and pressure gauge must be connected to the crankcase. Refer to Fig. 6-14. Remove muffler and carburetor. Fabricate a seal using gasket paper to cover the exhaust port, then reinstall muffler. Use either of the following methods to connect a hand pump and pressure gauge to crankcase. Fabricate an adapter plate and install in place of the carburetor to cover and

seal the intake manifold port. The adapter plate must have a nipple to connect pressure hose of tester. An alternate method of testing is as follows: Fabricate a seal using gasket paper to cover the intake port and reinstall carburetor. Connect a suitable hand pump and pressure gauge to carburetor pulse line, or plug carburetor end of pulse line and install pressure tester in spark plug hole. If testing leakage through spark plug hole, position piston at bottom dead center. If testing leakage through intake manifold or pulse line, position piston at top dead center. Apply pressure until gauge reads 7 psi (48 kPa) and close pump vent screw. Pressure should remain constant. If pressure drops more than 2 psi (14 kPa) in 30 seconds, leak must be located.

Crankcase seals may also be tested for a vacuum leak using a vacuum pump in place of pressure pump as outlined above. Actuate vacuum pump until gauge indicates vacuum of 7 psi (48 kPa). If vacuum reading remains constant or decreases no more than 2 psi (14 kPa), crankcase seals are in good condition. If seals fail to hold a vacuum, even if no pressure leak was indicated, they should be renewed.

If crankcase leakage is indicated, use a soap and water solution to check gaskets, crankcase seals, pulse line and castings for leakage.

LONG TERM STORAGE

A proper storage procedure can extend the life of an engine by preventing damage when the engine is not used. Exact procedures for storing depend on the type of equipment, length of storage, time of year stored and storage location. To obtain satisfactory results, storage must be coordinated with a regular maintenance program. The following outline lists procedures applicable for extended storage of most small engines.

ENTERING INTO STORAGE

Drain old oil from engine crankcase, gearboxes, chain cases, etc., while oil is warm. Refill with new approved oil specified by engine manufacturer.

Clean and dry all exterior surfaces. Remove all accumulated dirt and repair any damaged surface. Paint exposed surfaces to prevent rust.

Clean all cooling air passages and straighten, repair or renew any part

37

which would interfere with normal air flow. Remove shrouds and deflectors, then inspect and clean all cooling air passages.

Lubricate all moving parts requiring lubrication with approved oil or grease.

Inspect for worn or broken parts. Make necessary adjustments and repair all damage. Tighten all loose hardware.

Fuel should be either drained or treated with an approved stabilizer. All fuel should be drained from tank, filters, lines, pumps and carburetor unless specifically discouraged by manufacturer. Do not add fuel stabilizer to any fuel containing alcohol. Fuel containing alcohol will separate if permitted to sit for long period of time and internal parts may be extensively damaged by corrosion. Some manufacturers recommend coating inside of tank with a small amount of oil to deter rusting in tank. Filters should be serviced initially and water traps should be serviced regularly while in storage.

Loosen all drive belts and remove pressure from friction drive components. Inspect and note condition of drive belts. If condition is questionable, a new belt should be installed when removing equipment from storage.

Install new filter elements. Some filters can be cleaned and serviced, but most should be installed new at this time.

Pour a small amount (usually 1 tablespoon) of oil into cylinder of engine through spark plug hole. Crank engine with starter about 12 revolutions to distribute oil, then install spark plug and reconnect spark plug wire.

Install protective caps at ends of all disconnected lines. Seal openings of exhaust, air intake, engine dipstick and crankcase breather tube.

Remove battery and store in a cool, dry place. Do not permit battery to freeze and maintain fully charged, checking approximately every 30 days.

Store the engine in a dry, protected place. If necessary to store outside, cover engine to prevent entrance of water, but don't seal tightly. Sealing may cause condensation and accelerate rusting.

REMOVING FROM STORAGE

Check for obvious damage to covering and equipment. Remove any blocks used during storage.

Charge battery, then install in equipment making sure that battery is properly retained. Clean battery cables and battery posts, then attach cables to battery terminals.

Remove covers from exhaust, air intake, engine dipstick and crankcase breather tube. Remove any protective caps from lines disconnected during disassembly. Be sure ends are clean, then reconnect lines. Check all filters. Install new filters, or clean and service existing filters as required.

Adjust all drive belts and friction drive components to correct tension as recommended by the manufacturer. New belts should be installed if condition is questionable.

Fill fuel tank with correct type of fuel. Check for leaks. Gaskets may dry up or carburetor needle valve may stick during storage. Repair any problems before attempting to start. Drain water traps and check condition of fuel filters.

Check for worn or broken parts and repair before returning to service.

Lubricate all surfaces normally lubricated with oil or grease. Check cooling passages for restrictions such as insect, bird or animal nests. Check oil in all compartments such as engine crankcase, gearboxes, chain cases, etc., for proper level. Evidence of too much oil may indicate water settled below oil. Drain oil if contamination is suspected or if time of storage exceeds recommended time change interval. Fill to proper level with correct type of oil.

SERVICE SHOP TOOL BUYER'S GUIDE

This listing of service shop tools is solely for the convenience of users of this manual and does not imply endorsement or approval by Intertec Publishing Corporation of the tools and equipment listed. The listing is in response to many requests for information on sources for purchasing special tools and equipment. Every attempt has been made to make the listing as complete as possible at time of publication and each entry is made from the latest material available.

Special engine service tools such as seal drivers, bearing drivers, etc., which are available from the engine manufacturer are not listed in this section of the manual. Where a special service tool is listed in the engine service section of this manual, the tool is available from the central parts or service distributors listed at the end of most engine service sections, or from the manufacturer.

NOTE TO MANUFACTURERS AND NATIONAL SALES DISTRIBUTORS OF ENGINE SERVICE TOOLS AND RELATED SERVICE EQUIPMENT. To obtain either a new listing for your products, or to change or add to an existing listing, write to Intertec Publishing Corporation, Book Division, P.O. Box 12901, Overland Park, Kansas 66212.

ENGINE SERVICE TOOLS

Ammco Tools, Inc.
Wacker Park
North Chicago, Illinois 60064
Valve spring compressor, torque wrenches, cylinder hones, ridge reamers, piston ring compressors, piston ring expanders.

Black & Decker Mfg. Co.
701 East Joppa Road
Towson, Maryland 21204
Valve grinding equipment.

Bloom, Inc.
Route Four, Hiway 20 West
Independence, Iowa 50644
Engine repair stand with crankshaft straightening attachment.

Brush Research Mfg. Inc.
4642 East Floral Drive
Los Angeles, California 90022
Cylinder hones.

E-Z Lok
P.O. Box 2069
Gardena, California 90247
Thread repair insert kits for metal, wood and plastic.

Foley-Belsaw Company
Outdoor Power Equipment
Parts Division
6301 Equitable Road
P.O. Box 419593
Kansas City, Missouri 64141
Crankshaft straightener and repair stand, valve refacers, valve seat grinders, parts washers, cylinder hones, gages and ridge reamers, piston ring expanders and compressor, flywheel pullers, torque wrenches, rotary mower blade balancers.

Frederick Mfg. Co., Inc.
1400 C., Agnes Avenue
Kansas City, Missouri 64127
Crankshaft straightener.

Heli-Coil Products Division
Heli-Coil Corporation
Shelter Rock Lane
Danbury Connecticut 06810
Thread repair kits, thread inserts, installation tools.

K-D Tools
3575 Hempland Road
Lancaster, Pennsylvania 17604
Thread repair kits, valve spring compressors, reamers, micrometers, dial indicators, calipers.

Keystone Reamer & Tool Co.
Post Office Box 310
Millersburg, Pennsylvania 17061
Valve seat cutter and pilots, adjustable reamers.

Ki-Sol Corporation
100 Larkin Williams Ind. Court
Fenton, Missouri 63026
Cylinder hone, ridge reamer, ring compressor, ring expander, ring groove cleaner, torque wrenches, valve spring compressor, valve refacing equipment.

K-Line Industries, Inc.
315 Garden Avenue
Holland, Michigan 49423
Cylinder hone, ridge reamer, ring compressor, valve guide tools, valve spring compressor, reamers.

K.O. Lee Company
101 South Congress
Aberdeen, South Dakota 57401
Valve refacing, valve seat grinding, valve reseating and valve guide reaming tools.

Kwik-Way Mfg. Co.
500 57th Street
Marion, Iowa 52302
Cylinder boring equipment, valve facing equipment, valve seat grinding equipment.

Lisle Corporation
9807 East Main
Clarinda, Iowa 51632
Cylinder hones, ridge reamers, ring compressors, valve spring compressors.

Microdot, Inc.
P.O. Box 3001
800 So. State College Blvd.
Fullerton, California 92631
Thread repair insert kits.

Mighty Midget Mfg. Co., Div. of Kansas City Screw Thread Co.
2908 E. Truman Road
Kansas City, Missouri 64127
Crankshaft straightener.

Neway Manufacturing, Inc.
1013 No. Shiawassee
Corunna, Michigan 48817
Valve seat cutters.

Owatonna Tool Company
436 Eisenhower Drive
Owatonna, Minnesota 55060
Valve tools, spark plug tools, piston ring tools, cylinder hones.

Power Lawnmower Parts Inc.
1920 Lyell Avenue
P.O. Box 60860
Rochester, NY 14606-0860
Gasket cutter tool, gasket scraper tool, crankshaft cleaning tool, ridge reamer, valve spring compressor, valve seat cutters, thread repair kits, valve lifter, piston ring expander.

Precision Manufacturing & Sales Co., Inc.
2140 Range Rd., P.O. Box 149
Clearwater, Florida 33517
Cylinder boring equipment, measuring instruments, valve equipment, hones, porting, hand tools, test equipment, threading, presses, parts washers, milling machines, lathes, drill presses, glass

beading machines, dynos, safety equipment.

Sunnen Product Company
7910 Manchester Avenue
Saint Louis, Missouri 63143
Cylinder hones, rod reconditioning, valve guide reconditioning.

Rexnord Specialty
Fastener Division
3000 W. Lomita Blvd.
Torrance, California 90505
Thread repair insert kits (Keenserts) and installation tools.

Vulcan Tools
Div. of TRW, Inc.
2300 Kenmore Avenue
Buffalo, New York 14207
Cylinder hones, reamers, ridge removers, valve spring compressors, valve spring testers, ring compressor, ring groove cleaner.

Waters-Dove Manufacturing
Post Office Box 40
Skiatook, Oklahoma 74070
Crankshaft straightener, oil seal remover.

TEST EQUIPMENT AND GAGES

Allen Test Products
2101 North Pitcher Street
Kalamazoo, Michigan 49007
Coil and condenser testers, compression gages.

Applied Power, Inc., Auto Division
P.O. Box 27207
Milwaukee, Wisconsin 53227
Compression gage, condenser tester, tachometer, timing light, ignition analyzer.

AW Dynamometer, Inc.
131-1/2 East Main Street
Colfax, Illinois 61728
Engine test dynamometer.

B.C. Ames Company
131 Lexington
Waltham, Massachusetts 02254
Micrometer dial gages and indicators.

The Bendix Corporation
Engine Products Division
Delaware Avenue
Sidney, New York 13838
Condenser tester, magneto test equipment, timing light.

Burco
208 Delaware Avenue
Delmar, New York 12054
Coil and condenser tester, compression gage, carburetor tester, grinders, Pow-R-Arms, chain breakers, rivet spinners.

Dixson, Inc.
Post Office Box 1449
Grand Junction, Colorado 81501
Tachometer, compression gage, timing light.

Foley-Belsaw Company
Outdoor Power Equipment
Parts Division
6301 Equitable Road
P.O. Box 419593
Kansas City, Missouri 64141
Cylinder gage, amp/volt testers, condenser and coil tester, magneto tester, ignition testers for conventional and solid-state systems, tachometers, spark testers, compression gages, timing lights and gages, micrometers and calipers, torque wrenches, carburetor testers, vacuum gages.

Fox Valley Instrument Co.
Route 5, Box 390
Cheboygan, Michigan 49721
Coil and condenser tester, ignition actuated tachometer.

Graham-Lee Electronics, Inc.
4200 Center Avenue NE
Minneapolis, Minnesota 55421
Coil and condenser tester.

K-D Tools
3575 Hempland Road
Lancaster, Pennsylvania 17604
Diode tester and installation tools, compression gage, timing light, timing gages.

Ki-Sol Corporation
100 Larkin Williams Ind. Court
Fenton, Missouri 63026
Micrometers, telescoping gages, compression gages, cylinder gages.

K-Line Industries, Inc.
315 Garden Avenue
Holland, Michigan 49423
Compression gage, tachometers.

Merc-O-Tronic Instruments Corporation
215 Branch Street
Almont, Michigan 48003
Ignition analyzers for conventional solid-state and magneto systems, electric tachometers, electronic tachometer and dwell meter, power timing lights, ohmmeters, compression gages, mechanical timing devices.

Owatonna Tool Company
436 Eisenhower Drive
Owatonna, Minnesota 55060
Feeler gages, hydraulic test gages.

Power Lawnmower Parts Inc.
1920 Lyell Avenue
P.O. Box 60860
Rochester, NY 14606-0860
Condenser and coil tester, compression gage, flywheel magneto tester, ignition tester.

Prestolite Electronics Div.
An Allied Company
Post Office Box 931
Toledo, Ohio 43694
Magneto test plug.

Simpson Electric Company
853 Dundee Avenue
Elgin, Illinois 60120
Electrical and electronic test equipment.

L.S. Starrett Company
121 Crescent Street
Athol, Massachusetts 01331
Micrometers, dial gages, bore gages, feeler gages.

Stevens Instrument Company, Inc.
Post Office Box 193
Waukegan, Illinois 60085
Ignition analyzers, timing lights, tachometers, volt-ohmmeters, spark checkers, CD ignition testers.

Stewart-Warner Corporation
1826 Diversey Parkway
Chicago, Illinois 60614
Compression gage, ignition tachometer, timing light ignition analyzer.

P.A. Sturtevant Co., Division Dresser Ind.
3201 North Wolf Road
Franklin Park, Illinois 60131
Torque wrenches, torque multipliers, torque analyzers.

Sun Electric Corporation
Instrument Products Div.
1560 Trimble Road
San Jose, California 95131
Compression gage, hydraulic test gages, coil and condenser tester, ignition tachometer, ignition analyzer.

Westberg Mfg. Inc.
3400 Westach Way
Sonoma, California 95476
Ignition tachometer, magneto test equipment, ignition system analyzer.

SHOP TOOLS AND EQUIPMENT

A-C Delco Division, General Motors Corp.
400 Renaissance Center
Detroit, Michigan 48243
Spark plug cleaning and test equipment.

Applied Power, Inc., Auto. Division
Box 27207
Milwaukee, Wisconsin 53227
Arc and gas welding equipment, welding rods, accessories, battery service equipment.

Black & Decker Mfg. Co.
701 East Joppa Road
Towson, Maryland 21204
Air and electric powered tools.

Bloom, Inc.
Route 4, Hiway 20 West
Independence, Iowa 50644
Lawn mower repair bench with built-in engine cranking mechanism.

Campbell Chain Division
McGraw-Edison Co.
Post Office Box 3056
York, Pennsylvania 17402
Chain type engine and utility slings.

Champion Pneumatic Machinery Co.
1301 No. Euclid Avenue
Princeton, Illinois 61356
Air compressors.

Champion Spark Plug Co.
Post Office Box 910
Toledo, Ohio 43661
Spark plug cleaning and testing equipment, gap tools and wrenches.

Chicago Pneumatic Tool Co.
2200 Bleecker Street
Utica, New York 13503
Air impact wrenches, air hammers, air drills and grinders, nut runners, speed ratchets.

Clayton Manufacturing Company
415 North Temple City Boulevard
El Monte, California 91731
Steam cleaning equipment.

E-Z Lok
P.O. Box 2069
240 E. Rosecrans Avenue
Gardena, California 90247
Thread repair insert kits for metal, wood and plastic.

Foley-Belsaw Company
Outdoor Power Equipment
Parts Division
96301 Equitable Road
P.O. Box 419593
Kansas City, Missouri 64141
Torque wrenches, parts washers, micrometers and calipers.

G & H Products, Inc.
Post Office Box 770
St. Paris, Ohio 43027
Motorized lawn mower stand.

General Scientific Equipment Company
Limekiln Pike & Williams Avenue
Box 27309
Philadelphia, Pennsylvania 19118-0255
Safety equipment.

Graymills Corporation
3705 North Lincoln Avenue
Chicago, Illinois 60613
Parts washing stand.

Heli-Coil Products Div., Heli-Coil Corp.
Shelter Rock Lane
Danbury, Connecticut 06810
Thread repair kits, thread inserts and installation tools.

Ingersoll-Rand
253 E. Washington Avenue
Washington, New Jersey 07882
Air and electric impact wrenches, electric drills and screwdrivers.

Ingersoll-Rand Co.
Deerfield Industrial Park
South Deerfield, Massachusetts 01373
Impact wrenches, portable electric tools, battery powered ratchet wrench.

Jaw Manufacturing Co.
39 Mulberry Street, P.O. Box 213
Reading, Pennsylvania 19603
Files for renewal of damaged threads, hex nut size rethreader dies, flexible shaft drivers and extensions, screw extractors, impact drivers.

Jenny Division of Homestead Ind., Inc.
Box 348
Carapolis, Pennsylvania 15108-0348
Steam cleaning equipment, pressure washing equipment.

Keystone Reamer & Tool Co.
Post Office Box 310
Millersburg, Pennsylvania 17061
Adjustable reamers, twist drills, tape, dies, etc.

K-Line Industries, Inc.
315 Garden Avenue
Holland, Michigan 49423
Air and electric impact wrenches.

Microdot, Inc.
P.O. Box 3001
800 So. State College Blvd.
Fullerton, California 92631
Thread repair insert kits.

Owatonna Tool Company
436 Eisenhower Drive
Owatonna, Minnesota 55060
Bearing and gear pullers, hydraulic shop presses.

Power Lawnmower Parts Inc.
1920 Lyell Avenue
P.O. Box 60860
Rochester, NY 14606-0860
Flywheel puller, starter wrench and flywheel holder, blade sharpener, chain breakers, rivet spinners, gear pullers.

Pronto Tool Division, Ingersoll-Rand
2600 East Nutwood Avenue
Fullerton, California 92631
Torque wrenches, gear and bearing pullers.

Shure Manufacturing Corp.
1601 South Hanley Road
Saint Louis, Missouri 63144
Steel shop benches, desks, engine overhaul stand.

Sioux Tools, Inc.
2801-2999 Floyd Blvd.
Sioux City, Iowa 51102
Portable air and electric tools.

Rexnord Specialty
Fastener Division
3000 W. Lomita Blvd.
Torrance, California 90505
Thread repair insert kits (Keenserts) and installation tools.

Vulcan Tools
2300 Kenmore Avenue
Buffalo, New York 14207
Air and electric impact wrenches.

SHARPENING AND MAINTENANCE EQUIPMENT FOR SMALL ENGINE POWERED IMPLEMENTS

Bell Industries, Saw & Machine Division
Post Office Box 2510
Eugene, Oregon 97402
Saw chain grinder.

Desa Industries, Inc.
25000 South Western Ave.
Park Forest, Illinois 60466
Saw chain breakers and rivet spinners, chain files, file holder and filing guide.

Foley-Belsaw Company
Outdoor Power Equipment
Parts Division
6301 Equitable Road
P.O. Box 419593
Kansas City, Missouri 64141
Circular, band and hand saw filers, heavy duty grinders, saw setters, retoothers, circular saw vises, lawn mower sharpener, saw chain sharpening and repair equipment.

Granberg Industries
200 S. Garrard Blvd.
Richmond, California 94804
Saw chain grinder, file guides, chain breakers and rivet spinners, chain saw lumber cutting attachments.

Ki-Sol Corporation
100 Larkin Williams Ind. Court
Fenton, Missouri 63026
Mower blade balancer.

Magna-Matic Div., A.J. Karrels Co.
Box 348
Port Washington, Wisconsin 53074
Rotary mower blade balancer, "track" checking tool.

Omark Industries, Inc.
4909 International Way
Portland, Oregon 97222
Saw chain and saw bars, maintenance equipment, to include file holders, filing vises, rivet spinners, chain breakers, filing depth gages, bar groove gages, electric chain saw sharpeners.

Power Lawnmower Parts Inc.
1920 Lyell Avenue
P.O. Box 60860
Rochester, NY 14606-0860
Grinding wheels, parts cleaning brush, tube repair kits, blade balancer.

S.I.P. Grinding Machines
American Marsh Pumps
P.O. Box 23038
722 Porter Street
Lansing, Michigan 48909
Lawn mower sharpeners, saw chain grinders, lapping machine.

Specialty Motors Mfg.
641 California Way,
P.O. Box 157
Longview, Washington 98632
Chain saw bar rebuilding equipment.

MECHANIC'S HAND TOOLS

Channellock, Inc.
1306 South Main Street
Meadville, Pennsylvania 16335

John H. Graham & Company, Inc.
617 Oradell Avenue
Oradell, New Jersey 07649

Jaw Manufacturing Company
39 Mulberry Street
Reading, Pennsylvania 19603

K-D Tools
3575 Hempland Road
Lancaster, Pennsylvania 17604

K-Line Industries, Inc.
315 Garden Avenue
Holland, Michigan 49423

Millers Falls Division
Ingersoll-Rand
Deerfield Industrial Park
South Deerfield, Massachusetts 01373

New Britain Tool Company
Division of Litton Industrial Products
P.O. Box 12198
Research Triangle Park, N.C. 27709

Owatonna Tool Company
436 Eisenhower Drive
Owatonna, Minnesota 55060

Power Lawnmower Parts Inc.
1920 Lyell Avenue
P.O. Box 60860
Rochester, NY 14606-0860
Crankshaft wrench, spark plug wrench, hose clamp pliers.

Proto Tool Division
Ingersoll-Rand
2600 East Nutwood Avenue
Fullerton, California 92631

Snap-On Tools
2801 80th Street
Kenosha, Wisconsin 53140

Triangle Corporation—Tool Division
Cameron Road
Orangeburg, South Carolina 29115

Vulcan Tools
Division of TRW, Inc.
2300 Kenmore Avenue
Buffalo, New York 14207

J.H. Williams
Division of TRW, Inc.
400 Vulcan Street
Buffalo, New York 14207

SHOP SUPPLIES (CHEMICALS, METALLURGY PRODUCTS, SEALS, SEALERS, COMMON PARTS ITEMS, ETC.)

ABEX Corp.
Amsco Welding Products
Fulton Industrial Park
3-9610-14
P.O. Box 258
Wauseon, Ohio 43567
Hardfacing alloys.

Atlas Tool & Manufacturing Co.
7100 S. Grand Avenue
Saint Louis, Missouri 63111
Rotary mower blades.

Bendix Automotive Aftermarket
1094 Bendix Drive, Box 1632
Jackson, Tennessee 38301
Cleaning chemicals.

CR Industries
900 North State Street
Elgin, Illinois 60120
Shaft and bearing seals.

Clayton Manufacturing Company
415 North Temple City Blvd.
El Monte, California 91731
Steam cleaning compounds and solvents.

E-Z Lok
P.O. Box 2069,
240 E. Rosecrans Ave.
Gardena, California 90247
Thread repair insert kits for metal, wood and plastic.

Eutectic Welding Alloys Corp.
40-40 172nd Street
Flushing, New York 11358
Specialized repair and maintenance welding alloys.

Foley-Belsaw Company
Outdoor Power Equipment
Parts Division
6301 Equitable Road
P.O. Box 419593
Kansas City, Missouri 64141
Parts washers, cylinder head rethreaders, micrometers, calipers, gasket material, nylon rope, universal lawnmower blades, oil and grease products, bolt, nut, washer and spring assortments.

Frederick Manufacturing Co., Inc.
1400 C Agnes Street
Kansas City, Missouri 64127
Throttle controls and parts.

Heli-Coil Products Division
Heli-Coil Corporation
Shelter Rock Lane
Danbury, Connecticut 06810
Thread repair kits, thread inserts and installation tools.

King Cotton Cordage
617 Oradell Avenue
Oradell, New Jersey 07649
Starter rope, nylon rod.

Loctite Corporation
705 North Mountain Road
Newington, Connecticut 06111
Threading locking compounds, bearing mounting compounds, retaining compounds and sealants, instant and structural adhesives.

McCord Gasket Division
Ex-Cell-O Corporation
2850 West Grand Boulevard
Detroit, Michigan 48202
Gaskets, seals.

Microdot, Inc.
P.O. Box 3001
800 So. State College Blvd.
Fullerton, California 92631
Thread repair insert kits.

Permatex Industrial
705 N. Mountain Road
Newington, Connecticut 06111
Cleaning chemicals, gasket sealers, pipe sealants, adhesives, lubricants.

Power Lawnmower Parts Inc.
1920 Lyell Avenue
P.O. Box 60860
Rochester, NY 14606-0860
Grinding compound paste, gas tank sealer stick, shop aprons, rotary mower blades, oil seals, gaskets, solder, throttle controls and parts, nylon starter rope.

Radiator Specialty Co.
Box 34689
Charlotte, North Carolina 28234
Cleaning chemicals (Gunk), and solder seal.

Rexnord Specialty
Fastener Div.
3000 W. Lomita Blvd.
Torrance, California 90506
Thread repair insert kits (Keenserts) and installation tools.

Union Carbide Corporation
Home & Auto Products Division
Old Ridgebury Road
Danbury, CT 06817
Cleaning chemicals, engine starting fluid.

ACME

ACME NORTH AMERICA CORP.
5203 West 73rd Street
Minneapolis, MN 55439

Model	Bore	Stroke	Displacement
A 180 B,			
A 180 P	65 mm	54 mm	179 cc
	(2.56 in.)	(2.13 in.)	(10.92 cu. in.)
ALN 215 WB,			
ALN 215 WP	65 mm	65 mm	215 cc
	(2.56 in.)	(2.56 in.)	(13.12 cu. in.)
A 220 B,			
A 220 P	72 mm	54 mm	220 cc
	(2.83 in.)	(2.13 in.)	(13.43 cu. in.)

ENGINE IDENTIFICATION

All models are four-stroke, air-cooled, single-cylinder engines.

All models are horizontal crankshaft with intake and exhaust valves located in the cylinder block.

Models with the suffix "B" are gasoline fuel engines and models with the suffix "P" are kerosene fuel engines.

Engine model number is located on a plate mounted on the right side of the shroud as viewed from flywheel side. The engine serial number is stamped into the engine block approximately 3 inches (76.2 mm) below model number plate, just above oil fill plug (Fig. AC1).

Always furnish engine model and serial number when ordering parts.

MAINTENANCE

SPARK PLUG. Recommended spark plug for Models A 180 B, A 180 P, A 220 B and A 220 P is a Bosch RO10846

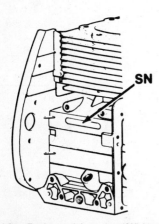

Fig. AC1—Engine serial number (SN) is stamped into engine block.

or equivalent. Recommended spark plug for Model ALN 215 WB is a Bosch W45T1 or equivalent.

On all models, spark plug should be removed, cleaned and inspected after every 100 hours of operation. Spark plug electrode gap should be 0.6-0.8 mm (0.024-0.032 in.) for all models.

CARBURETOR. Models ALN 215 WB and ALN 215 WP are equipped with the float type carburetor shown in Fig. AC2. Models A 180 B, A 180 P, A 220 B and A 220 P are equipped with the float type carburetor shown in Fig. AC4. Refer to the appropriate following paragraphs for model being serviced.

Models ALN 215 WB and ALN 215 WP. Initial adjustment of idle speed mixture screw (3—Fig. AC2) from a lightly seated position is 1¼ turns open. Main fuel mixture is controlled by fixed main jet (13). Fixed main jet size is #75 for Model ALN 215 WB and #80 for Model ALN 215 WP. Idle jet (7) size is #50 for both models.

Final adjustments are made with engine at operating temperature and running. Adjust engine idle speed to 1100 rpm at throttle stop screw (5). Adjust idle mixture screw (3) to obtain smoothest engine idle and smooth acceleration.

When installing new fuel inlet needle and seat, install seat and measure from point (A) to edge of fuel inlet needle seat. Measurement should be 33-37 mm (1.30-1.46 in.). Vary number of fiber washers between fuel inlet seat and carburetor body to obtain correct measurement.

To check carburetor float (12) level, carefully remove float bowl (15) and

measure distance from top of bowl (E—Fig. AC3) to fuel level (F) in bowl. Measurement should be 32-34 mm (1.26-1.34 in.). If float level is incorrect, fuel inlet needle seat must be shimmed as outlined in previous paragraph. Float weight should be 16.5 grams (0.6 oz.). If float is heavier than specified, it must be renewed.

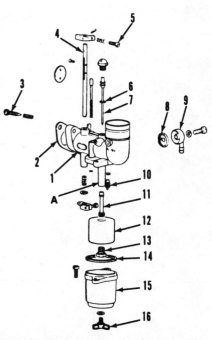

Fig. AC2—Exploded view of carburetor used on Models ALN 215 WB and ALN 215 WP.

1. Carburetor body	9. Filter housing
2. Gasket	10. Fuel inlet needle valve
3. Idle mixture screw	11. Nozzle
4. Throttle shaft	12. Float
5. Throttle stop screw	13. Main jet
6. Gasket	14. Gasket
7. Idle jet	15. Float bowl
8. Screen (filter)	16. Wing bolt

Illustrations courtesy Acme North America Corp.

Fuel filter screen (8–Fig. AC2) should be removed and cleaned after every 50 hours of operation.

Models A 180 B, A 180 P, A 220 B and A 220 P. Initial adjustment of idle speed mixture screw (3—Fig. AC4) from a lightly seated position is 1½ turns open. Main fuel mixture is controlled by fixed main jet (13). Fixed main jet size is #95 for Model A 180 B, #98 for Models A 180 P and A 220 B and #100 for Model A 220 P. Idle jet (7) size for all models is #35.

Final adjustments are made with engine at operating temperature and running. Adjust engine idle speed to 1000-1100 rpm at throttle stop screw (5). Adjust idle mixture screw (3) to obtain smoothest engine idle and smooth acceleration.

To check float level, remove carburetor float bowl (15). Invert carburetor body (1). Float height should be 15 mm (¹⁹⁄₃₂ in.) measured from carburetor body gasket surface to bottom of float (12). Float weight should be 8 grams (0.29 oz.). If float is heavier than specified, it must be renewed.

Fuel filter screen (8) should be removed and cleaned after every 50 hours of operation.

AIR FILTER. Models ALN 215 WB and ALN 215 WP are equipped with an oil bath type air filter (Fig. AC5). Air filter element should be removed and cleaned after every 8 hours of operation. Discard old oil and refill with new engine oil to level indicated on oil reservoir (5).

Models A 180 B, A 180 P, A 220 B and A 220 P are equipped with a paper element type air filter. Filter should be checked daily. After removing element, clean element by gently tapping element to dislodge dust and dirt or use very low air pressure and blow from the inside of the air filter element toward the outside. Element should be renewed after 100 hours of operation when engine is operated under normal conditions. Shorten element renewal intervals when engine is operated under adverse conditions.

GOVERNOR. Models ALN 215 WB and ALN 215 WP are equipped with a flyweight ball type centrifugal governor with governor assembly incorporated on camshaft gear. Models A 180 B, A 180 P, A 220 B, and A 220 P are equipped with a flyweight type governor. Governor gear and flyweight assembly is located on the crankcase cover and is driven by the crankshaft gear. On all models, the governor regulates engine speed via external linkage.

To adjust external linkage on Models ALN 215 WB and ALN 215 WP, first make certain all linkage moves freely with no binding or loose connections. Lock throttle control lever (5—Fig. AC6) in midposition with throttle lever locknut (6). Hook tension spring (3) in hole A for 3000 rpm engine speed, hole B for 3600 rpm engine speed and for special applications only, hole C for 4000 rpm engine speed. An alternate spring (part 551.107) is available. When tension spring 551.107 is hooked in hole A, engine speed is governed at 2400 rpm. Place throttle control lever (5) in full speed position. Loosen carburetor-to-governor lever rod adjustment lock (7) and push throttle fully open. With throttle held open, tighten rod adjustment lock (7). Governor and throttle linkage should work freely with no binding through entire operation range.

To adjust external linkage on Models A 180 B, A 180 P, A 220 B and A 220 P, first make certain all linkage moves freely with no binding or loose connections. Place throttle control lever (8—Fig. AC7) in full throttle position. Loosen clamp bolt (1). Insert screwdriver into slot of governor shaft (2) and rotate shaft clockwise as far as possible.

Fig. AC3—Measure distance from fuel level (F) to top of bowl (E). Distance should be 32-34 mm (1.26-1.34 in.).

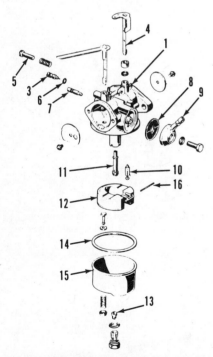

Fig. AC4—Exploded view of carburetor used on Models A 180 B, A 180 P, A 220 B and A 220 P.

1. Carburetor body	
3. Idle mixture screw	10. Fuel inlet needle valve
4. Throttle shaft	11. Nozzle
5. Throttle stop screw	12. Float
6. Gasket	13. Main jet
7. Idle jet	14. Gasket
8. Screen (filter)	15. Float bowl
9. Filter housing	16. Float pin

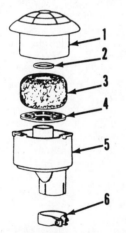

Fig. AC5—Oil bath air filter is used on Models ALN 215 WB and ALN 215 WP.

1, Cover	4. Plate
2. Gasket	5. Oil reservoir
3. Element	6. Clamp

Fig. AC6—View showing relative position and relationship of external governor linkage and throttle parts on Models ALN 215 WB and ALN 215 WP.

1. Carburetor-to-governor lever rod
2. Governor lever
3. Tension spring
4. Spring
5. Throttle control lever
6. Throttle lever locknut
7. Carburetor-to-governor lever rod adjustment lock

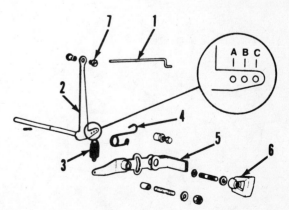

Tighten clamp bolt (1). Place throttle control lever (8) in idle position, then start and run engine. Engine idle should be adjusted to 1000-1100 rpm. Place throttle control (8) in full throttle position and adjust throttle stop screw (9) to obtain 2400, 3000 or 3600 rpm. Do not exceed maximum engine speed of 3600 rpm.

IGNITION SYSTEM. A breaker point ignition system is standard on Models ALN 215 WB and ALN 215 WP and a solid-state electronic ignition system is standard on Models A 180 B, A 180 P, A 220 B and A 220 P.

Models ALN 215 WB and ALN 215 WP. The breaker-point set and condenser are located on the left-hand side of crankcase as viewed from flywheel side. Ignition coil is located behind flywheel. Breaker points are actuated by plunger (1—Fig. AC8).

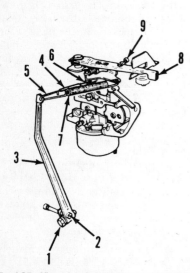

Fig. AC7—View showing relative position and relationship of external governor linkage and throttle parts on Models A 180 B, A 180 P, A 220 B and A 220 P.

1. Clamp bolt
2. Governor shaft
3. Governor lever
4. Spring
5. Governor lever-to-carburetor rod
6. Throttle lever
7. Spring
8. Throttle control lever
9. Throttle stop screw

Point gap should be checked and adjusted after every 400 hours of operation. To check point gap, remove cover (5) and use a suitable feeler gauge. Point gap should be 0.4-0.5 mm (0.016-0.020 in.).

Ignition timing should be set at 21° BTDC by aligning "AA" mark on flywheel with "PMS" mark on crankcase. Points should just begin to open at this position. Shift breaker point plate (3) if necessary. When "TDC" mark on flywheel is aligned with "PMS" mark on crankcase, piston is at top dead center.

Ignition coil is located behind the flywheel. Clearance between coil and flywheel magnets should be 0.6-0.8 mm (0.024-0.031 in.).

Models A 180 B, A 180 P, A 220 B and A 220 P. The solid-state electronic ignition system requires no regular maintenance. Use the correct feeler gauge to check clearance between ignition coil and flywheel. Clearance should be 0.40-0.45 mm (0.016-0.018 in.).

LUBRICATION. Check engine oil level after every 8 hours of operation. Maintain oil level at lower edge of fill plug opening.

Change oil after every 50 hours of operation. Manufacturer recommends oil with an API service classification SC or CC. Use SAE 40 oil for ambient temperatures above 10° C (50° F); SAE 30 oil for a temperature range of 0° to 10° C (32° to 50° F); SAE 20W-20 oil for a temperature range of −10° to 0° C (14° to 32° F) and SAE 10W oil for temperatures below −10° C (14° F).

Crankcase oil capacity for Models ALN 215 WB and ALN 215 WP is 0.75 L (0.793 qt.). Crankcase oil capacity for Models A 180 B, A 180 P, A 220 B and A 220 P is 0.6 L (0.634 qt.).

VALVE ADJUSTMENT. Valves should be adjusted after every 200 hours of operation. Valve stem-to-tappet clearance (cold) should be 0.10-0.15 mm (0.004-0.006 in.) for intake and ex-

haust valves on all models. If clearance is not as specified, remove or install shims (7 and 8—Fig. AC9) in shim holder (9) as necessary. Shims are available in 0.1 mm (0.004 in.) and 0.2 mm (0.008 in.) thicknesses.

CRANKCASE BREATHER. Models ALN 215 WB and ALN 215 WP are equipped with a crankcase breather which must be removed and cleaned after every 50 hours of operation. Breather is located on tube attached to valve chamber cover. Rubber valve must be installed as shown in Fig. AC10 to ensure crankcase vacuum.

Models A 180 B, A 180 P, A 220 B and A 220 P are equipped with a crankcase breather which is an integral part of the valve chamber cover. No regular maintenance is required.

GENERAL MAINTENANCE. Check and tighten all loose bolts, nuts or clamps prior to each day of operation. Check for fuel or oil leakage and repair if necessary.

Clean dust, dirt, grease or any foreign material from cylinder head and cylinder block cooling fins after every 100

Fig. AC9—View of valve system components.

1. Valves
2. Guides
3. Seals
4. Seats
5. Springs
6. Retainers
7. Shims (valve adjustment)
8. Shims (valve adjustment)
9. Shim holder (cap)
10. Tappets

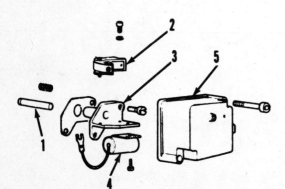

Fig. AC8—Breaker points and condenser on Models ALN 215 WB and ALN 215 WP are located on left-hand side of crankcase behind cover (5).

1. Plunger
2. Breaker points
3. Point plate
4. Condenser
5. Cover

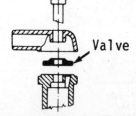

Fig. AC10—Breather valve must be installed as shown to ensure crankcase vacuum.

Illustrations courtesy Acme North America Corp.

hours of operation. Inspect fins for damage and repair if necessary.

REPAIRS

TIGHTENING TORQUES. Recommended tightening torque specifications are as follows:

Cylinder head:
A 180 B, A 180 P,
A 220 B, A 220 P . . . . 24.5 N•m
(18 ft.-lbs.)

ALN 215 WB,
ALN 215 WP . . . . 29 N•m
(22 ft.-lbs.)

Connecting rod:
A 180 B, A 180 P,
A 220 B, A 220 P . . . . 11.8 N•m
(9 ft.-lbs.)

ALN 215 WB,
ALN 215 WP . . . . 19 N•m
(14 ft.-lbs)

Crankcase cover:
A180 B, A 180 P,
A 220 B, A 220 P . . . . 11.8 N•m
(9 ft.-lbs.)

ALN 215 WB,
ALN 215 WP . . . . 15 N•m
(11 ft.-lbs.)

Flywheel (all models) . . . . 157 N•m
(116 ft.-lbs.)

CYLINDER HEAD. All models are equipped with an aluminum alloy cylinder head which should not be removed when engine is hot.

To remove the cylinder head from all models, first allow engine to cool, then remove fuel tank and cooling shrouds. Loosen head bolts evenly following sequence shown in Fig. AC11. Remove cylinder head.

Check cylinder head for warpage by placing head on a flat surface and using a feeler gauge to determine warpage. Warpage should not exceed 0.3-0.5 mm (0.012-0.020 in.).

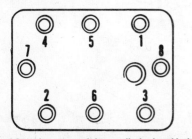

Fig. AC11—Loosen or tighten cylinder head bolts following sequence shown. Models A 180 B, A 180 P, A 220 B and A 220 P are equipped with two longer head bolts which must be installed in positions 1 and 3.

Always install a new head gasket. Tighten cylinder head bolts to specification listed under TIGHTENING TORQUES following sequence shown in Fig. AC11. Note that Models A 180 B, A 180 P, A 220 B and A 220 P have two longer head bolts which must be installed in positions 1 and 3 shown in Fig. AC11.

CONNECTING ROD. On all models, an aluminum alloy connecting rod rides directly on the crankpin journal.

To remove connecting rod, remove cylinder head and crankcase cover. Use care when Models ALN 215 WB and ALN 215 WP crankcase cover is removed as governor flyweight balls will fall out of ramps in camshaft gear. On all models, remove connecting rod cap and push connecting rod and piston assembly out of cylinder head end of block. Remove piston pin retaining rings and separate piston from connecting rod. Camshaft and lifters may be removed at this time if required.

On all models, clearance between piston pin and connecting rod pin bore should be 0.006-0.022 mm (0.0002-0.0009 in.).

Clearance between crankpin and connecting rod bearing bore on Models A 180 B, A 180 P, A 220 B and A 220 P should be 0.030-0.049 mm (0.0012-0.0019 in.). Clearance between crankpin and connecting rod bearing bore on Models ALN 215 WB and ALN 215 WP should be 0.040-0.064 mm (0.0016-0.0025 in.). On all models, if clearance exceeds specified dimension, renew connecting rod and/or crankshaft.

On all models, renew connecting rod if crankpin bearing bore is excessively worn or out-of-round more than 0.10 mm (0.004 in.).

On all models, connecting rod side play on crankpin should be 0.150-0.250 mm (0.0060-0.0100 in.).

On all models, piston should be installed on the connecting rod with arrow (1—Fig. AC12), on top of piston, facing toward side of connecting rod with mark (2). Connecting rod and connecting rod cap marks (2) must align when components are assembled.

When installing piston assembly in cylinder block, connecting rod and cap match marks must be toward crankcase cover side of engine and arrow on top of piston must be on side opposite the valves on engines with clockwise rotation. On engines with counterclockwise rotation, arrow on top of piston must be toward valve side of engine and connecting rod and cap match marks will face toward flywheel side of engine. On all models, tighten connecting rod bolts to specification listed under TIGHTEN-

ING TORQUES. On Models ALN 215 WB and ALN 215 WP, use heavy grease to retain governor flyweight balls in camshaft ramps during assembly. On Models A 180 B, A 180 P, A 220 B and A 220 P, crankcase cover has two longer bolts which should be installed in the upper right and lower left positions. On all models, tighten crankcase cover bolts to specification listed under TIGHTENING TORQUES.

CYLINDER AND CRANKCASE. Cylinder and crankcase are an integral casting of aluminum alloy on all models. A high density perlite cylinder sleeve is cast as an integral part of the cylinder block.

Standard cylinder bore diameter for Models A 180 B, A 180 P, ALN 215 WB and ALN 215 WP is 65.000-65.013 mm (2.5591-2.5596 in.). Standard cylinder bore diameter for Models A 220 B and A 220 P is 72.000-72.013 mm (2.8300-2.8305 in.). If cylinder bore is 0.06 mm (0.0024 in.) or more out-of-round or tapered, cylinder should be bored to nearest oversize for which piston and rings are available.

Crankshaft ball type main bearings should be a slight press fit in crankcase and crankcase cover. It may be necessary to slightly heat crankcase cover or crankcase assembly to remove or install main bearings. Renew bearings if they are loose, rough or damaged.

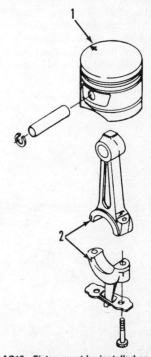

Fig. AC12—Piston must be installed on connecting rod so arrow (1) on top of piston faces toward side of connecting rod with mark (2). Connecting rod and connecting rod cap match marks (2) must align when components are assembled.

Models ALN 215 WB and ALN 215 WP are equipped with an oil slinger trough (2—Fig. AC13) installed in crankcase. If trough is removed, it must be securely repositioned prior to engine reassembly.

PISTON, PIN AND RINGS. Refer to previous CONNECTING ROD paragraphs for piston removal and installation procedure.

Standard piston diameter for Models A 180 B, A 180 P, ALN 215 WB and ALN 215 WP is 64.987-65.000 mm (2.5586-2.5591 in.). Standard piston diameter for Models A 220 B and A 220 P is 71.987-72.000 mm (2.8295-2.8300 in.). On all models, piston is a select fit at the factory.

Piston should be renewed and/or cylinder reconditioned if there is 0.013 mm (0.0005 in.) or more clearance between piston and cylinder bore.

Clearance between piston and piston rings in ring grooves should be 0.05 mm (0.002 in.) on all models.

Compression ring end gap on Models ALN 215 WB and ALN 215 WP should be 0.25-0.40 mm (0.010-0.016 in.). Compression ring end gap on Models A 180 B, A 180 P, A 220 B and A 220 P should be 0.25-0.45 mm (0.010-0.018 in.).

Oil control ring end gap on Models ALN 215 WB and ALN 215 WP should be 0.30-0.50 mm (0.012-0.020 in.). Oil control ring end gap on Models A 180 B, A 180 P, A 220 B and A 220 P should be 0.20-0.35 mm (0.008-0.014 in.).

Piston pin on all models should be a 0.004-0.012 mm (0.0002-0.0005 in.) interference fit in piston pin bore. It may be necessary to slightly heat piston to aid in pin removal and installation.

CRANKSHAFT. On all models, crankshaft is supported at each end in

ball bearing type main bearings (11 and 16—Fig. AC14) and crankshaft timing gear (15) is a press fit on crankshaft (13). Refer to previous CONNECTING ROD paragraphs for crankshaft removal procedure and crankpin-to-connecting rod bearing bore clearance.

Standard crankshaft crankpin journal diameter on Models A 180 B, A 180 P, A 220 B and A 220 P is 25.989-26.000 mm (1.0232-1.0236 in.). Standard crankshaft crankpin journal diameter on Models ALN 215 WB and ALN 215 WP is 29.985-30.000 mm (1.1805-1.1811 in.).

Standard crankshaft main bearing journal diameter on Models A 180 B, A 180 P, A 220 B and A 220 P is 25.002-25.015 mm (0.98433-0.98484 in.). Standard crankshaft main bearing journal diameter on models ALN 215 WB and ALN 215 WP is 30 mm (1.18 in.)

On all models, main bearings should be a slight press fit on crankshaft journal. Renew bearings if they are rough, loose or damaged. To prevent crankshaft damage, crankshaft should be supported on counterweights when pressing bearings or crankshaft timing gear onto crankshaft.

On all models, when installing crankshaft, make certain crankshaft and camshaft gear timing marks are aligned.

CAMSHAFT. Camshaft and camshaft gear on all models are an integral casting which rides in bearing bores in crankcase and crankcase cover.

On Models ALN 215 WB and ALN 215 WP, governor flyweight balls are located in ramps machined in face of camshaft gear.

On some models, a compression release mechanism is mounted on back

side of camshaft gear. The spring-loaded compression release mechanism should snap back against camshaft when weighted lever is pulled against spring tension and released. Spring is in the correct position when dimension (A—Fig. AC15) of pin projection is 0.5-0.6 mm (0.020-0.024 in.).

Inspect camshaft journals and lobes on all models. Renew camshaft if worn, scored or damaged.

Standard camshaft bearing journal diameter for Models A 180 B, A 180 P, A 220 B and A 220 P is 14.973-14.984 mm (0.589-0.590 in.) at each end. Standard intake lobe height is 23.275-23.325 mm (0.916-0.918 in.). Standard exhaust lobe height is 17.575-17.625 mm (0.692-0.694 in.). Standard camshaft bearing journal diameter for Models ALN 215 WB and ALN 215 WP is 16 mm (0.6299 in.) at each end. Standard intake and exhaust lobe height is 26 mm (1.0236 in.).

When installing camshaft on Models ALN 215 WB and ALN 215 WP, retain governor flyweight balls in ramps with heavy grease. On all models, make cer-

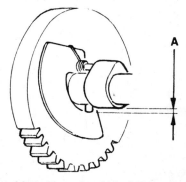

Fig. AC15—Dimension "A" of pin projection should be 0.5-0.6 mm (0.020-0.024 in.) for correct compression release mechanism operation.

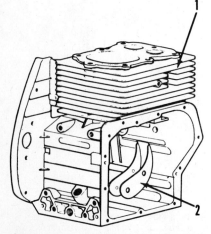

Fig. AC13—View of cylinder block (1) and oil slinger trough (2). If trough is removed, it must be securely repositioned prior to engine reassembly.

Fig. AC14—Exploded view of crankshaft, connecting rod and piston assembly.

1. Retaining rings
2. Piston pin
3. Compression rings
4. Oil control ring
5. Piston
6. Connecting rod
7. Connecting rod cap
8. Lockplate
9. Bolts
10. Seal
11. Main bearing
12. Key
13. Crankshaft
14. Crankshaft gear key
15. Crankshaft gear
16. Main bearing
17. Seal

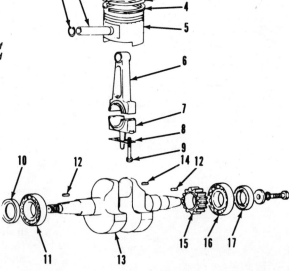

tain camshaft and crankshaft gear timing marks are aligned.

VALVE SYSTEM. Refer to VALVE ADJUSTMENT paragraphs in MAINTENANCE section for valve clearance adjustment procedure.

Valve face and seat angles on all models is 45°. Standard valve seat width is 1.2-1.3 mm (0.047-0.051 in.). If seat width is 2 mm (0.079 in.) or more, seat must be narrowed. If valve face margin is 0.5 mm (0.020 in.) or less, renew valve.

Standard exhaust valve stem diameter on all models is 6.955-6.970 mm (0.2738-0.2744 in.).

Standard intake valve stem diameter for Models A 180 B, A 180 P, A 220 B and A 220 P is 6.955-6.970 mm (0.2738-0.2744 in.). Standard intake valve stem diameter for Models ALN 215 WB and ALN 215 WP is 6.965-6.987 mm (0.2742-0.2751 in.).

Standard valve guide inside diameter on Models A 180 B, A 180 P, A 220 B and A 220 P is 7.015-7.025 mm (0.2762-0.2766 in.) for intake and exhaust valve guides. Standard valve guide inside diameter on Models ALN 215 WB and ALN 215 WP is 7.000-7.022 mm (0.2756-0.2764 in.). On all models, worn valve guides can be renewed using Acme puller 365109.

Standard valve spring free length for Models A 180 B, A 180 P, A 220 B and A 220 P is 34 mm (1.34 in.). If spring free length is 31 mm (1.22 in.) or less, renew spring. Standard valve spring free length for Models ALN 215 WB and ALN 215 WP is 35 mm (1.38 in.). If spring free length is 32 mm (1.26 in.) or less, renew spring.

ACME

ACME NORTH AMERICA CORP.
5203 West 73rd Street
Minneapolis, MN 55439

Model	Bore	Stroke	Displacement
AT 220B	72 mm	54 mm	220 cc
	(2.83 in.)	(2.13 in.)	(13.43 cu. in.)

Model AT 220B is a four-stroke, air-cooled, single-cylinder, overhead valve engine. Cylinder and crankcase are cast as a single unit.

Engine model number is located on a plate mounted on cooling shroud on the right side of engine as viewed from flywheel side. The engine serial number (SN—Fig. AC51) is stamped on the engine block.

MAINTENANCE

LUBRICATION. Check engine oil level prior to each operating interval. Maintain oil level at lower edge of fill plug opening. Change oil after every 50 hours of operation.

Engine oil should meet or exceed latest API service classification. Use SAE 40 oil for ambient temperatures above 86° F (30° C); SAE 30 oil for temperatures between 50° F (10° C) and 86° F (30° C); SAE 20W-20 oil for temperatures between 50° F (10° C) and 14° F (−10° C) and SAE 10W oil for temperatures below 14° F (−10° C).

Crankcase capacity is 0.6 L (0.63 qt.).

AIR CLEANER. Engine air filter should be cleaned and inspected after every 8 hours of operation, or more often if operating in extremely dusty conditions.

Filter elements may be cleaned by directing low pressure compressed air stream from inside filter toward the outside. Reinstall elements.

Filter elements should be renewed after every 50 hours of operation, or more often if operating in a severe environment.

FUEL FILTER. A fuel filter (8—Fig. AC52) is located inside the fuel inlet fitting on the carburetor. A fuel filter is also located inside the fuel tank on the shut-off valve. Both fuel filters should be removed and cleaned after every 100 hours of operation.

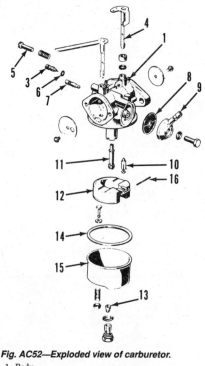

Fig. AC52—Exploded view of carburetor.

1. Body	
3. Idle mixture screw	10. Fuel inlet valve
4. Throttle shaft	11. Nozzle
5. Idle speed screw	12. Float
6. Gasket	13. Main jet
7. Idle jet	14. Gasket
8. Filter screen	15. Fuel bowl
9. Filter housing	16. Float pin

CRANKCASE BREATHER. The crankcase breather is located behind the tappet cover on the side of the engine. Regular maintenance is not required.

SPARK PLUG. Recommended spark plug is Champion L86, Bosch W95T1 or equivalent.

Spark plug should be removed, cleaned and inspected after 100 hours of operation. Spark plug electrode gap should be 0.6-0.8 mm (0.024-0.032 in.).

CARBURETOR. Refer to Fig. AC52 for an exploded view of float type carburetor used on engine.

Initial adjustment of idle speed mixture screw (3) from a lightly seated position is 1½ turns out. Adjust engine idle by rotating idle speed screw (5). Main fuel mixture is controlled by a fixed main jet (13). With engine at normal operating temperature, adjust idle mixture screw to obtain smoothest engine idle and acceleration without hesitation.

To check float level, remove carburetor fuel bowl (15) and invert carburetor. Float height should be 15 mm (¹⁹⁄₃₂ in.) measured from carburetor body gasket surface to bottom of float (12). Float weight should be 8 grams (0.29 oz.). If float is heavier than specified, it must be renewed.

GOVERNOR. The engine is equipped with a flyweight type governor located on the crankcase cover and driven by the camshaft gear. The governor regulates engine speed via external linkage.

To adjust governor, make sure all linkage moves freely without binding. Place throttle control lever (8—Fig. AC53) in full throttle position. Loosen clamp bolt (1). Insert screwdriver into slot of governor shaft (2) and rotate shaft clockwise as far as possible. Tighten clamp bolt (1). Adjust maximum governed speed by rotating throttle stop screw (9); do not exceed

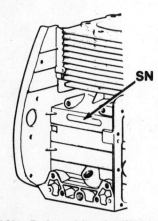

Fig. AC51—Engine serial number (SN) is stamped into engine block.

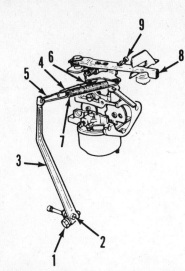

Fig. AC53—Drawing of governor and speed control linkage.

1. Clamp bolt
2. Governor shaft
3. Governor lever
4. Spring
5. Throttle rod
6. Throttle lever
7. Spring
8. Speed control lever
9. Maximum governed speed screw

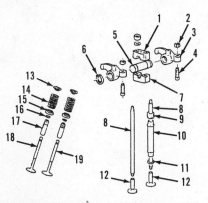

Fig. AC54—Exploded view of valve system.

1. Upper rocker shaft clamp
2. Jam nut
3. Rocker arm
4. Adjusting screw
5. Rocker arm shaft
6. Snap ring
7. Lower rocker shaft clamp
8. Push rod
9. Seal
10. Push rod tube
11. Seal
12. Tappet
13. Valve retainer
14. Valve spring
15. Valve spring seat
16. Clip
17. Valve guide
18. Exhaust valve
19. Intake valve

equipment manufacturer's recommended speed.

IGNITION SYSTEM. The engine is equipped with an electronic ignition system. Ignition timing is not adjustable. Air gap between ignition unit and flywheel should be 0.40-0.45 mm (0.016-0.018 in.).

VALVE ADJUSTMENT. Valve clearance should be checked, and adjusted if necessary, after every 300 hours of operation. To check clearance, remove rocker arm cover and position piston at top dead center of compression stroke. Use a feeler gauge to measure clearance between rocker arm (3—Fig.

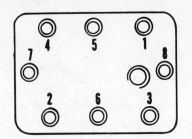

Fig. AC55—Loosen or tighten cylinder head screws in sequence shown. Note location of longer screws during disassembly. Tighten cylinder head screws to 24.5 N•m (18 ft.-lbs.).

AC54) and end of valve stem. Specified clearance is 0.10-0.15 mm (0.004-0.006 in.) for both valves.

To adjust clearance, loosen jam nut (2—Fig. AC54) and turn adjusting screw (4) in rocker arm (3) to obtain correct clearance. When correct clearance is obtained, hold adjustment screw in position while jam nut is tightened.

REPAIRS

TIGHTENING TORQUES. Recommended tightening torque specifications are as follows:

Connecting rod . . . . 11.8 N•m
(108 in.-lbs.)

Crankcase cover . . . . 11.8 N•m
(108 in.-lbs.)

Cylinder head . . . . 24.5 N•m
(18 ft.-lbs.)

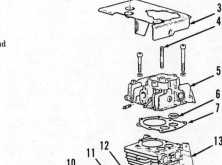

Fig. AC56—Exploded view of engine block.

1. Rocker arm cover
2. Gasket
3. Cooling shroud
4. Rocker shaft clamp stud
5. Cylinder head
6. Intake passage seal
7. Head gasket
8. Oil fill plug
9. Gasket
10. Crankcase cover
11. Gasket
12. Dowel pin
13. Crankcase/cyl. block
14. Mounting plate
15. Gasket
16. Drain plug
17. Spring
18. Breather valve
19. Gasket
20. Tappet cover
21. Breather tube
22. Gasket
23. Cover

Flywheel . . . . 157 N•m
(116 ft.-lbs.)

CYLINDER HEAD. The cylinder head is an aluminum alloy casting with hardened steel valve seat inserts. To remove the cylinder head, first remove cooling shrouds and rocker arm cover. Loosen and remove cylinder head screws in reverse order of sequence shown in Fig. AC55. Note location of longer screws so they can be reinstalled in correct location. Do not loosen cylinder head screws while engine is hot as cylinder head warpage may result.

Check cylinder head for distortion by placing head on a flat surface and using a feeler gauge to determine warpage. If cylinder head is warped more than 0.5 mm (0.020 in.), resurface or renew head.

Install a new head gasket when installing cylinder head. Be sure intake seal (6—Fig. AC56) and push rod tube and seals are installed correctly. Tighten cylinder head screws evenly to 24.5 N•m (18 ft.-lbs.) following sequence shown in Fig. AC55. Adjust valve clearance as outlined in previous VALVE ADJUSTMENT section.

VALVE SYSTEM AND PUSH RODS. Intake gases are routed through the side of the cylinder block, past push rod tube (10—Fig. AC54) into the cylinder head, then through a passage in the cylinder head to the intake valve. Valves are actuated by rocker arms mounted on a rocker shaft on the cylinder head.

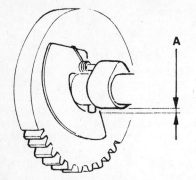

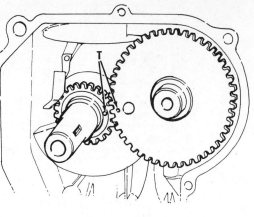

Fig. AC57—Compression release pin projection (A) should be 0.5-0.6 mm (0.020-0.024 in.) for proper operation.

Fig. AC58—Engine valve timing is correct when timing marks (T) on crankshaft gear and camshaft gear are aligned as shown above.

Valve face and seat angles should be 45° for intake and exhaust. Valve seat contact width should be 1.2-1.3 mm (0.047-0.051 in.).

Standard valve stem diameter is 6.995-6.970 mm (0.2738-0.2744 in.) for the intake and exhaust valves.

The intake valve guide is cast iron and the exhaust valve guide is bronze. Standard valve guide inside diameter is 7.015 mm (0.2762 in.). Renew guide if inside diameter is 7.097 mm (0.2794 in.) or more. Valve guides may be renewed using suitable removal and installation tools. Locating clips (16—Fig. AC54) are present on valve guides. Install guide so clip bottoms against cylinder head.

Valve spring free length should be 37.5 mm (1.48 in.). If spring free length is 35 mm (1.38 in.) or less, renew spring.

Maximum allowable clearance between rocker arm and rocker arm shaft is 0.15 mm (0.006 in.). If clearance exceeds specified dimension, renew rocker arm and/or rocker arm shaft.

Be sure seals (9 and 11) on intake push rod tube (10) are in good condition.

CAMSHAFT. The camshaft and camshaft gear are an integral casting which rides in bearing bores in crankcase and crankcase cover. The camshaft is accessible after removing the crankcase cover. If cylinder head has not been removed, rotate crankshaft so piston is at top dead center on compression stroke before removing camshaft.

A compression release mechanism is mounted on the back side of camshaft gear. The spring-loaded compression release mechanism should snap back against the camshaft when the weighted lever is pulled against spring tension and released. Spring is in the correct position when actuating pin protrusion (A—Fig. AC57) is 0.5-0.6 mm (0.020-0.024 in.).

Inspect camshaft journals and lobes and renew camshaft if worn, scored or damaged. Camshaft bearing journal diameters should be 14.973-14.984 mm

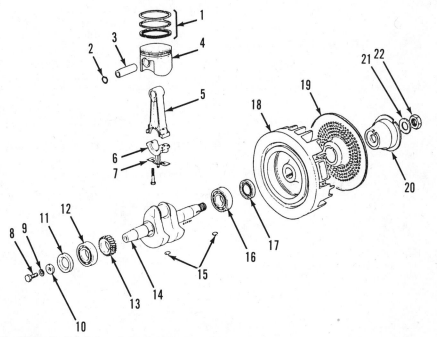

Fig. AC59—Exploded view of piston, rod and crankshaft assembly.

1. Piston rings	12. Ball bearing	17. Oil seal
2. Snap ring	13. Crankshaft gear	18. Flywheel
3. Piston pin	14. Crankshaft	19. Debris screen
4. Piston	15. Keys	20. Starter cup
5. Connecting rod	16. Ball bearing	21. Washer
6. Rod cap		22. Nut
7. Lockplate		
8. Screw		
9. Lockwasher		
10. Washer		
11. Oil seal		

(0.589-0.590 in.). Exhaust cam lobe height should be 17.575-17.625 mm (0.6919-0.6939 in.) and intake cam lobe height should be 23.275-23.325 mm (0.9163-0.9183 in.). Camshaft and tappets should be renewed as a set.

Be sure to align timing marks (T—Fig. AC58) on camshaft and crankshaft gears when installing camshaft. Tighten crankcase cover screws in a crossing pattern to 11.8 N·m (108 in.-lbs.).

PISTON, PIN AND RINGS. To remove the connecting rod and piston assembly, first remove cylinder head, crankcase cover and camshaft. Remove connecting rod cap and push connecting rod and piston assembly out of cylinder

head end of block. Remove piston pin retaining rings (2—Fig. AC59) and separate piston from connecting rod.

Compression ring end gap should be 0.25-0.45 mm (0.010-0.018 in.). Oil control ring end gap should be 0.20-0.35 mm (0.008-0.014 in.).

Piston pin should be 0.004-0.012 mm (0.0002-0.0005 in.) interference fit in piston pin bore. It may be necessary to heat piston slightly to aid in removal and installation of piston pin.

Standard piston diameter is 71.987-72.000 mm (2.8295-2.8300 in.). Piston is a select fit at the factory. Piston should be renewed and/or cylinder reconditioned if there is 0.08 mm (0.0032 in.) or more clearance between piston and cylinder bore.

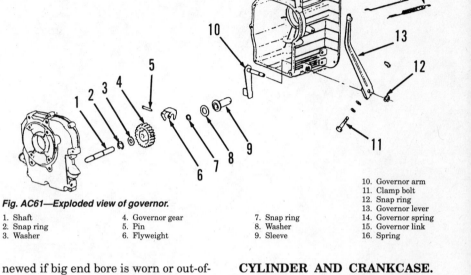

Fig. AC60—Piston must be installed on connecting rod with arrow (1) on piston crown facing toward side of connecting rod with mark (2). When installing piston assembly in cylinder, arrow on piston crown must be toward camshaft side of engine.

Fig. AC61—Exploded view of governor.

1. Shaft	4. Governor gear
2. Snap ring	5. Pin
3. Washer	6. Flyweight

7. Snap ring	10. Governor arm
8. Washer	11. Clamp bolt
9. Sleeve	12. Snap ring
	13. Governor lever
	14. Governor spring
	15. Governor link
	16. Spring

Piston must be installed on connecting rod with arrow (1—Fig. AC60) on piston crown facing toward side of connecting rod with mark (2). When installing piston assembly in cylinder, arrow on piston crown must be toward camshaft side of engine. Connecting rod and connecting rod cap marks (2) must align when components are assembled. Tighten connecting rod screws to 11.8 N•m (108 in.-lbs.). Be sure to align timing marks on camshaft and crankshaft gears when installing camshaft (Fig. AC58). Tighten crankcase cover screws in a crossing pattern to 11.8 N•m (108 in.-lbs.).

CONNECTING ROD. The aluminum alloy connecting rod rides directly on the crankpin journal. Connecting rod and piston are removed as a unit as outlined in PISTON, PIN AND RINGS section.

Clearance between the piston pin and connecting rod pin bore should be 0.006-0.022 mm (0.0002-0.0009 in.).

Connecting rod side play on crankpin journal should be 0.15-0.25 mm (0.006-0.010 in.). Connecting rod should be re-

newed if big end bore is worn or out-of-round more than 0.10 mm (0.004 in.). Specified clearance between connecting rod big end bore and crankshaft crankpin is 0.030-0.049 mm (0.0012-0.0019 in.).

GOVERNOR. The governor gear (4—Fig. AC61) and flyweight assembly rides on a stud (1) located in the crankcase cover. Components must move freely without binding.

CRANKSHAFT. The crankshaft is supported at each end by ball bearing type main bearings (12 and 16—Fig. AC59).

Standard clearance between crankpin journal and connecting rod bore is 0.030-0.049 mm (0.0012-0.0019 in.). Standard crankpin journal diameter is 25.989-26.000 mm (1.0232-1.0236 in.). Ball type main bearings should be a slight press fit on crankshaft journal. Renew bearings if they are rough, loose or damaged. Note that crankshaft should be supported on counterweights when pressing bearings or crankshaft timing gear onto or off of crankshaft to prevent bending crankshaft.

When installing crankshaft, make certain crankshaft and camshaft gear timing marks (T—Fig. AC58) are aligned.

CYLINDER AND CRANKCASE. Cylinder and crankcase are an integral casting of aluminum alloy with a high density perlite cylinder sleeve cast as an integral part of the cylinder block.

Standard cylinder bore diameter is 72.000-72.013 mm (2.8300-2.8305 in.). If cylinder bore is 0.06 mm (0.0024 in.) or more out-of-round or tapered, cylinder should be bored to nearest oversize for which piston and rings are available.

On all models, crankshaft ball type main bearings should be a slight press fit in crankcase and crankcase cover bores. Renew bearings if they are loose, rough or damaged.

REWIND STARTER. To disassemble starter, remove rope handle and allow rope to wind into starter. Disassembly is evident after inspection of unit and referral to Fig. AC62. Wear appropriate safety eyewear and gloves before disengaging pulley from starter as spring may uncoil uncontrolled. Place shop towel around pulley and lift pulley out of housing; spring should remain with pulley.

Inspect components for damage and excessive wear. Reverse disassembly procedure to install components. Rewind spring is installed in pulley so coils wind in counterclockwise direction from outer end. Wind rope around pulley in counterclockwise direction as viewed from flywheel side of pulley.

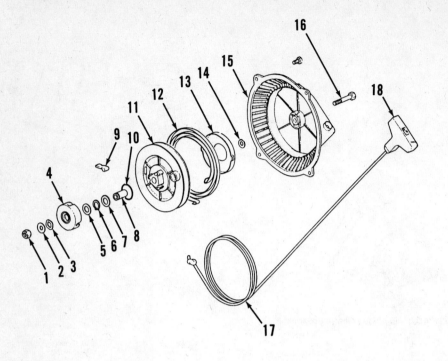

Fig. AC62—Exploded view of rewind starter.

1. Nut	10. Spring washer
2. Washer	11. Pulley
3. Spring washer	12. Rewind spring
4. Retainer	13. Spring cover
5. Washer	14. Washer
6. Snap ring	15. Starter housing
7. Washer	16. Bolt
8. Sleeve	17. Rope
9. Pawl	18. Rope handle

Illustrations courtesy Acme North America Corp.

SERVICING ACME ACCESSORIES

REWIND STARTER

Refer to Fig. AC25 for exploded view of rewind starter used on all models so equipped. Rewind spring (5) and spring housing (4) are serviced as an assembly only.

When installing rewind starter assembly on engine, install but do not tighten the six bolts retaining assembly to cooling shroud. Pull cable handle (7) until 150 mm (6 in.) of cable has been pulled from housing and starter dogs (9) have centered assembly. Hold tension on cable while the six bolts are tightened.

ALTERNATOR

Some models are equipped with a fixed armature type alternator mounted on engine with rotor as an integral part of the flywheel.

To test alternator output, disconnect rectifier leads and connect to an AC voltmeter with at least a 30 volt capacity. Start engine and refer to the following specifications for voltage output according to engine operation.

2400 rpm . . . . 20-22 volts
2800 rpm . . . . 23-25 volts
3200 rpm . . . . 26-28 volts
3600 rpm . . . . 29-30 volts

Rectifier may be checked by connecting ammeter between positive battery lead and the positive rectifier terminal. Connect 20 volt voltmeter to battery posts and use lights or other battery

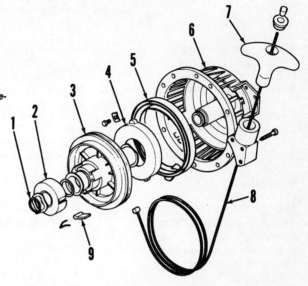

Fig. AC25—Exploded view of rewind starter assembly.

1. Snap ring
2. Starter dog housing
3. Cable pulley
4. Spring housing
5. Rewind spring
6. Housing
7. Handle
8. Cable
9. Starter dogs

drain method to lower battery voltage below 13 volts. Start engine and refer to the following specifications for aperage output according to engine speed.

1500 rpm . . . . 0.5 amp
2400 rpm . . . . 1.5 amp
3000 rpm . . . . 2.2 amp
3600 rpm . . . . 2.7 amp

If battery charge current is 0 amp with 12.5 volt or less battery voltage, renew rectifier.

CAUTION: Never operate engine with rectifier disconnected as rectifier will be damaged.

ACME SPECIAL TOOLS

The following special tools are available from Acme Central Parts Distributors or Acme Corporation.

Tool Description	Tool Number
Valve spring extractor	365110
Ignition coil positioning tool	365168
Valve guide check tool	365048
Valve guide puller	365109
Electrical tester	365180
Oil seal installation cone	365152
Engine flywheel and timing cover puller	365113

ACME CENTRAL PARTS DISTRIBUTORS

(Arranged Alphabetically by States)
**These franchised firms carry extensive stocks of repair parts.
Contact them for name of the nearest service distributor.**

Alaska Pump & Supply
Phone: (907) 563-3424
261 East 56th Avenue
Anchorage, Alaska 99518

Fessler Equipment
Phone: (907) 276-5335
2400 Commercial Drive
Anchorage, Alaska 99501

Waukesha-Alaska Corp.
Phone: (907) 345-6800
P.O. Box 111098
Anchorage, Alaska 99511

Engine Powered Products
Phone: (602) 258-9396
3040 North 27th Ave.
Phoenix, Arizona 85017

Scotsco Pro Power Products
Phone: (916) 383-3511
8806 Fruitridge Road
Sacramento, California 95826

Webb's Farm Supplies
Phone: (408) 475-1020
5381 Old San Jose Road
Soquel, California 95073

Central Power Engineering Corp.
Phone: (714) 676-0555
42169 Sarah Way
Temecula, California 92390

Scotsco Pro Power Products
Phone: (800) 525-1098
9162 S.E. 74th Avenue
Commerce City, Colorado 80022

Rep Co Sales, Inc.
Phone: (203) 322-9676
984 High Ridge Road
Stamford, Connecticut 06905

Kelly Tractor Co.
Phone: (305) 592-5360
P.O. Box 520775
Miami, Florida 33152

Roberts Supply, Inc.
Phone: (407) 657-5555
4203 Metric Drive
Winter Park, Florida 32792

Industrial Diesel Systems
Phone: (404) 428-4591
P.O. Box 669455
Marietta, Georgia 30066

Fauver Co.
Phone: (708) 682-5010
275 Commonwealth Drive
Carol Stream, Illinois 60188

Burlington Wholesale
Phone: (219) 546-9010
1533 East 3rd Road
Bremen, Indiana 46506

Fauver Co.
Phone: (317) 872-3060
7786 Moller Road
Indianapolis, Indiana 46268

Fauver Co.
Phone: (319) 366-6437
955 33rd Ave. S.W.
Cedar Rapids, Iowa 52404

Fauver Co.
Phone: (502) 267-7461
11400 Decimal Drive #1003
Louisville, Kentucky 40299

B&M Distributors
Phone: (800) 825-4565
11447 Cloverland Ave.
Baton Rouge, Louisiana 70809

The Engine Room
Phone: (508) 759-3921
57 Onset Ave.
Onset, Massachusetts 02532

Fauver Co.
Phone: (616) 957-5177
4550 40th Street
P.O. Box 320
Kentwood, Michigan 49508

Fauver Co.
Phone: (313) 585-5252
1500 East Avis Drive
Madison Heights, Michigan 48071

Fauver Co.
Phone: (517) 753-0474
361 Morlet Drive
Saginaw, Michigan 48601

Fauver Co.
Phone: (612) 943-1644
10286 West 70th Street
Eden Prairie, Minnesota 55344

Fauver Co.
Phone: (816) 452-4444
3939 N.E. 33rd Terrace, Suite H
Kansas City, Missouri 64117

Fauver Co.
Phone: (402) 392-0193
8031 W. Center Rd., Suite 202
Omaha, Nebraska 68124

Phillips Diesel Corporation
Phone: (505) 865-7332
P.O. Box 999
Los Lunas, New Mexico 87301

Proctor Equipment of NM
Phone: (505) 396-2450
P.O. Box 2005
Lovington, New Mexico 88260

Sea World Traders
Phone: (718) 421-4463
1223A Foster Avenue
Brooklyn, New York 11230

Amna Pump Corporation
Phone: (718) 784-2004
33-11 Green Point Ave.
Long Island Dity, New York 11101

Moriches Agricultural Supply
Phone: (516) 878-0264
640 Montauk
E. Moriches, New York 11940

Watermill Rental Center
Phone: (516) 726-6664
Montauk Highway
Water Mill, New York 11976

Asheville Opeco./Riverside
Phone: (704) 298-1988
501 Swannanoa River Road
Asheville, North Carolina 28805

Fauver Co.
Phone: (701) 280-1185
1131 Westrac Dr., Suite 202C
Fargo, North Dakota 58107

Fauver Co.
Phone: (513) 247-9900
11253 Williamson Road
Cincinnati, Ohio 45241

Fauver Co.
Phone: (513) 236-3554
6272 Executive Blvd.
Dayton, Ohio 45424

Fauver Co.
Phone: (614) 876-1261
4149 Weaver Court S.
Hillard, Ohio 43026

Fauver Co.
Phone: (419) 666-3404
6979 Wales Road
Northwood, Ohio 43619

Fauver Co.
Phone: (216) 923-8855
895 Hampshire Road, Suite C
Stow, Ohio 44224

Brown Engine & Equipment
Phone: (405) 632-2301
4315 S. Robinson
Oklahoma City, Oklahoma 73109

Cessco, Inc.
Phone: (800) 882-4959
P.O. Box 14579
Portland, Oregon 97314

GD Equipment
Phone: (717) 859-3533
Rd #2, Box 100
Ephrata, Pennsylvania 17522

Knox Auto Supply
Phone: (814) 797-1207
Box W, Miller St.
Knox, Pennsylvania 16232

Fauver Co.
Phone: (412) 733-3788
845 William Pitt Way
Pittsburg, Pennsylvania 15238

Jerry B. Leach Co.
Phone: (803) 537-2141
447 State Road
Cheraw, South Carolina 29520

M.T.A. Distributors
Phone: (615) 726-2225
2940 Foster Creighton
Nashville, Tennessee 37204

Industrial Engines & Accessories
Phone: (713) 784-1551
3660 Westchase Drive
Houston, Texas 77042

Ray Wright Pumps
Phone: (713) 487-0665
5514 Sycamor
Pasadena, Texas 77503

Memo Industrial Planning, Inc.
Phone: (817) 488-4444
1950 E. Continental Blvd.
South Lake, Texas 76092

Anchor Farms Equipment
Phone: (206) 376-5051
P.O. Box 1271
Eastsound, Washington 98245

Fauver Co.
Phone: (304) 346-3501
312 Peoples Building
Charleston, West Virginia 25301

Fauver Co.
Phone: (414) 781-1525
4475-C N. 124th St.
Brookfield, Wisconsin 53005

CANADIAN DISTRIBUTORS

Coast Dieselec, Ltd.
Phone: (604) 533-2601
#111 20120 64th Avenue
Langley, British Columbia V3A 4P7

Coast Dieselec, Ltd.
Phone: (604) 533-2601
#111 20120 64th Avenue
Langley, British Columbia V3A 4P7

Coast Dieselec, Ltd.
Phone: (416) 738-4710
Unit 12, 701 Millway Ave.
Concord, Ontario L4K 3S7

Terio Equipment
Phone: (418) 683-2952
575 Rue Marais
Bille Vanier, Quebec G1M 2Y2

Normand-Michel
Phone: (514) 453-4705
389, 24 'Eme Avenue
Ile Perrot, Quebec J7V 4N1

Povincial Diesel
Phone: (514) 937-9371
730 Rose De Lima
Montreal, Quebec H4C 2L8

BRIGGS & STRATTON

**BRIGGS & STRATTON CORPORATION
P.O. Box 702
Milwaukee, Wisconsin 53201**

BRIGGS & STRATTON
ENGINE IDENTIFICATION INFORMATION

Before servicing the engine and ordering parts, the Briggs & Stratton engine model and type numbers must be determined. Although engines may be similar in appearance, specific differences that affect service specifications and part configuration are only defined by the model and type numbers. Although rarely required, provide the code number when ordering parts.

Engine identification numbers, including the model number, type number and code number, are located on the blower housing around the flywheel. The numbers are stamped in an identification plate or directly in the metal. See Fig. BS1.

The engine model number identifies the basic engine family. Refer to the table in Fig. BS2 for a breakdown of Briggs & Stratton engine model numbers. As an example, an engine model number of 130202 would indicate that the engine has an approximate dis-

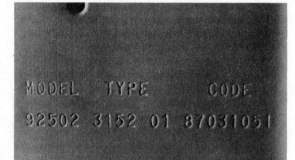

Fig. BS1—View of typical engine model information stamped in the blower housing.

placement of 13 cubic inches and the design series is "0". The engine is equipped with a horizontal crankshaft, Pulsa-Jet carburetor, plain main bearings and a rewind starter.

The type number specifies the parts configuration of the engine, as well as cosmetic details such as paint color and decals. The type number also determines governor speed settings depend-

ing on the engine's application, i.e., lawn mower, tractor, pump, generator, etc.

The code number provides information concerning the manufacturing of the engine. For instance, a code number of 90012201 indicates the engine was built in 1990, the first month of the year, on the 22nd day of the month, in manufacturing plant "01."

BRIGGS & STRATTON NUMERICAL MODEL NUMBER SYSTEM

	FIRST DIGIT AFTER DISPLACEMENT	SECOND DIGIT AFTER DISPLACEMENT	THIRD DIGIT AFTER DISPLACEMENT	FOURTH DIGIT AFTER DISPLACEMENT
CUBIC INCH DISPLACEMENT	BASIC DESIGN SERIES	CRANKSHAFT, CARBURETOR GOVERNOR	BEARINGS, REDUCTION GEARS & AUXILIARY DRIVES	TYPE OF STARTER
6	0	0 - Horizontal Diaphragm	0 - Plain Bearing	0 - Without Starter
8	1	1 - Horizontal Vacu-Jet	1 - Flange Mounting Plain Bearing	1 - Rope Starter
9	2			
10	3	2 - Horizontal Pulsa-Jet Pneumatic	2 - Ball Bearing	2 - Rewind Starter
11	4			
13	5	3 - Horizontal Flo-Jet Mechanical	3 - Flange Mounting Ball Bearing	3 - Elecric - 110 Volt, Gear Drive
14	6			
17	7	4 - Horizontal Flo-Jet	4 - Pressure Lube	4 - Electric Starter-Generator - 12 Volt, Belt Drive
19	8			
20	9			
23		5 - Vertical Vacu-Jet	5 - Gear Reduction (6 to 1) Gear Drive	5 - Electric Starter Only - 12 Volt, Gear Drive
24				
25				
30		6 -	6 - Gear Reduction (6 to 1) Reverse Rotation	6 - Alternator Only*
32				
		7 - Vertical Flo-Jet	7 - Pressure Lube	7 - Electric Starter, 12 Volt Gear Drive, with Alternator
		8 -	8 - Auxiliary Drive Perpendicular to Crankshaft	8 - Vertical-pull Starter
		9 - Vertical Pulsa-Jet	9 - Auxiliary Drive Parallel to Crankshaft	*Digit 6 formerly used for "Wind-Up" Starter on 60000, 80000 and 92000 Series

To identify Model 104772:

10	**4**	**7**	**7**	**2**
10 Cubic Inch	Design Series 4	Vertical Shaft– Flo-Jet Carburetor Mechanical Governor	Pressure Lube	Rewind Starter

Similarly, a Model 130207 is described as follows:

13	**0**	**2**	**0**	**7**
13 Cubic Inch	Design Series 0	Horizontal Shaft– Pulsa-Jet Carburetor	Plain Bearing DU	Electric Starter– 12-Volt Gear Drive With Alternator

Fig. BS2—Explanation of engine model numerical code used by Briggs & Stratton to identify engine and components.

BRIGGS & STRATTON
2-STROKE ENGINES

Model	Bore	Stroke	Displacement	Power Rating
62032	2.13 in.	1.75 in.	6.2 cu. in.	3.0 hp
	(54.0 mm)	(44.5 mm)	(102 cc)	(2.2 kW)

ENGINE INFORMATION

All models are two-stroke, single-cylinder engines utilizing a third-port scavenging system. The engine is equipped with a chrome plated cylinder bore. Refer to page 59 for Briggs & Stratton engine identification information.

MAINTENANCE

LUBRICATION. Manufacturer recommends mixing a good quality BIA or NMMA 2-cycle oil certified for TC-WII service with regular or low-lead gasoline at a 50:1 ratio. Do not use automotive (4-cycle) engine oil. The use of gasoline that contains alcohol is not recommended. However, if gasoline with alcohol is used, it must not contain more than 10 percent ethanol. Do not use gasoline containing methanol. If gasoline containing ethanol is used, it must be drained from the fuel system before

Champion J19LM or RJ19LM. Spark plug electrode gap should be 0.030 inch (0.76 mm).

CAUTION: Briggs & Stratton does not recommend using abrasive blasting to clean spark plugs as this may introduce some abrasive material into the engine which could cause extensive damage.

CARBURETOR. All models are equipped with the diaphragm type carburetor shown in Fig. BS10. The carburetor is equipped with an integral fuel pump.

Initial adjustment of carburetor fuel mixture screw (3—Fig. BS11) is 1½ turns counterclockwise from a lightly seated position. Make final adjustment with engine at operating temperature and running. Turn mixture screw (3) clockwise (leaner) until engine runs smoothly, then turn mixture screw

counterclockwise (richer) until engine just starts to run rough. This position will provide the best engine performance under load. Mixture screw should never be less than 1¼ turns open from a lightly seated position or improper engine lubrication will result.

To disassemble carburetor, refer to Fig. BS12, unscrew diaphragm cover screws and remove the diaphragm cover (1). Don't lose diaphragm springs (4). Remove the fuel pump diaphragm (2). Refer to Fig. BS13 and remove the fuel inlet valve assembly. Extract the retaining ring that holds the fuel inlet seat in the carburetor body. Pry out the Welch plug over the check valve as shown in Fig. BS14 with a 1/16-inch (1.5 mm) pin

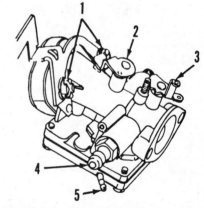

Fig. BS10—View of diaphragm type carburetor used on all models.

1. Mounting screws
2. Throttle lever
3. Choke lever
4. Fuel inlet
5. Primer inlet

50:1					
U.S.		**Imperial**		**Metric**	
Gasoline Gallons	2 Cycle Oil Ounces	Gasoline Gallons	2 Cycle Oil Ounces	Gasoline Liters	2 Cycle Oil Liters
1	2.5	1	3.2	4	.08
2	5	2	6.4	8	.16
5	13	5	16	20	.4

Fig. BS9—Gasoline/oil mixture chart.

storing the engine.

Always mix fuel in a separate container and add only mixed fuel to engine fuel tank. To assure thorough mixing of the oil and gasoline, fill the container partially with gasoline, then add the correct amount of oil per the chart shown in Fig. BS9. Shake the container to mix the oil and gasoline, then add remainder of gasoline and shake the container again.

SPARK PLUG. Recommended spark plug is Autolite 235 or 245 or

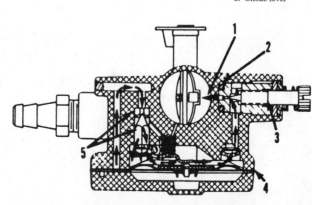

Fig. BS11—Cutaway view of carburetor.

1. Metering hole
2. "O" ring
3. Mixture screw
4. Diaphragm
5. Inlet needle & seat

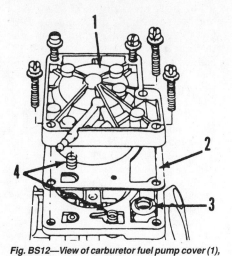

Fig. BS12—View of carburetor fuel pump cover (1), diaphragm (2), cup and spring (3) and valve springs (4).

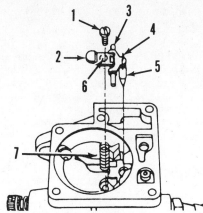

Fig. BS13—View of metering lever and fuel inlet valve assembly.

1. Screw		
2. Lever	5.	Fuel inlet valve
3. Pin	6.	Dimple
4. Clip	7.	Spring

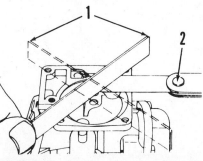

Fig. BS14—A 1/16-inch (1.5 mm) pin punch or Allen wrench can be used to pry out Welch plug over check valve.

punch or other suitable tool, then remove the check valve. Unscrew the fuel mixture screw approximately five turns then pull the mixture screw and seat assembly out of the body. Note the "O" ring which may remain on the screw seat or stay in the body.

Inspect for worn or damaged parts and renew as needed. Check the carburetor body for warpage with a straightedge as shown in Fig. BS15. Renew the body if warpage exceeds 0.003 inch (0.76 mm).

To reassemble the carburetor, install the fuel inlet valve seat so the grooved side faces toward the carburetor body. Install the retaining ring so the ring is 5/16 inch (7.9 mm) from top of hole as

shown in Fig. BS16. Install the fuel inlet valve assembly while being sure the retaining clip (4—Figure BS13) holds the

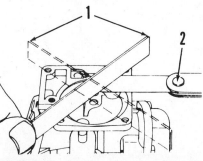

Fig. BS15—Check carburetor body warpage with straightedge (1) and feeler gauge (2). If warpage exceeds 0.003 inch (0.076 mm), renew carburetor.

Fig. BS16—Refer to text for proper installation of retainer ring (1) and fuel inlet valve seat (2).

5/16 in. (7.92 mm)

inlet valve against lever (2). Spring (7) sits in the spring well in the carburetor body and engages dimple (6) on lever (2). After installing the fuel inlet valve assembly, measure lever height as shown in Fig. BS17. The end of the lever should be $1/16$ inch (1.6 mm) below the face of the carburetor body. Carefully bend the lever to adjust the lever height. Assemble the remaining components while noting the following: Install the Welch plug above the check valve so the concave side of the plug is towards the check valve as shown in Fig. BS18; a $1/4$-inch (6 mm) round punch can be used to drive in the plug. Tighten diaphragm cover retaining screws in the sequence shown in Fig. BS19. Install the throttle plate so the notch is toward the metering holes in the carburetor bore as shown in Fig. BS20. Install the choke plate as shown in Fig. BS21 so the dimples on the plate are inward. Assemble fuel mixture screw components as shown in Fig. BS22. Note that the flats (7) on the seat (2) and on the carburetor body must align. Push the mixture screw assembly into the carburetor body until bottomed.

GOVERNOR. The engine is equipped with an air vane type gover-

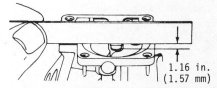

1.16 in. (1.57 mm)

Fig. BS17—Lever height should be 1.16 inch (1.57 mm). Measure as shown.

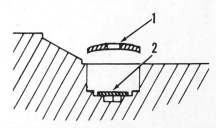

Fig. BS18—Install Welch plug (1) so concave side is towards check valve (2).

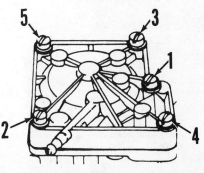

Fig. BS19—Tighten diaphragm cover retaining screws in sequence shown.

nor. Refer to Fig. BS23 for a view of the governor system. If adjustment of maximum no-load engine speed is necessary, tighten spring tension to increase speed or loosen spring tension to decrease speed. Adjust spring tension by bending the tang to which the governor spring is

Fig. BS20—Install the throttle plate so the notch is toward the metering holes in the carburetor bore as shown.

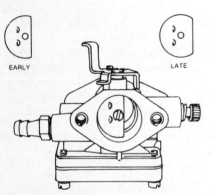

Fig. BS21—Install the choke plate so the dimples on the plate are inward.

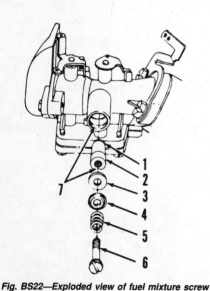

Fig. BS22—Exploded view of fuel mixture screw assembly.

1. "O" ring
2. Seat
3. Sealing washer
4. Metal washer
5. Spring
6. Mixture screw
7. Flat area

attached using tool 19229 as shown in Fig. BS24.

IGNITION SYSTEM. The engine is equipped with a Magnetron ignition system. The Magnetron ignition is a self-contained breakerless ignition system. Flywheel removal is not necessary except to check or service keyways or crankshaft key.

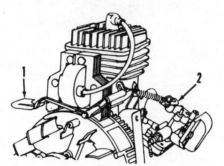

Fig. BS23—View of governor air vane (1) and throttle lever (2).

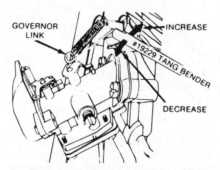

Fig. BS24—Use tool 19229 to adjust maximum engine speed. Bend tang as needed to adjust spring tension.

Fig. BS25—On models equipped with a removable Magnetron ignition module, the wire leads must be unsoldered to remove module. Refer to text.

To check spark, remove spark plug and connect spark plug cable to B&S tester 19051, then ground remaining tester lead to engine. Spin engine at 350 rpm or more. If spark jumps the 0.165 inch (4.2 mm) tester gap, system is functioning properly.

The ignition module cannot be tested. If the ignition module is suspected of faulty operation, then a new or serviceable module should be installed and the ignition system rechecked.

To remove armature and Magnetron module, remove flywheel shroud and armature retaining screws. Disconnect stop switch wire from module. The ignition module cannot be separated from the ignition coil on models with a one-piece coil and module. The module and ignition coil are available only as a unit assembly.

Some models are equipped with a removable ignition module (Figure BS25) that is mounted on the ignition coil. To

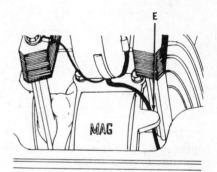

Fig. BS26—The ignition primary wire must be routed behind ear (E) on transfer port cover so wire will not contact flywheel.

separate the ignition module from the ignition coil, unscrew the module wire lead from the ignition coil armature. Using a 5/16-inch (8 mm) diameter rod, push in against the spring to release the primary and stop switch wires from the retainer hook. Unsolder the wires while being careful not to overheat the ignition module. Pull back the module retainer and dislodge the module from the ignition coil. Install the ignition module by reversing the disassembly procedure. Connect wires using 60/40 rosin core solder while being careful not to overheat the module. To prevent wire movement due to vibration, cement the wires to the ignition coil armature with RTV sealant.

When installing the ignition coil on the engine, note that the primary wire must be routed behind the ear (E—Fig. BS26) on the transfer port cover so the wire cannot contact the flywheel.

Ignition armature air gap should be 0.006-0.010 inch (0.15-0.25 mm). Ignition timing is not adjustable on these models.

COMPRESSION PRESSURE.

Compression reading for engine should be 90-110 psi (620-758 kPa) if engine is not equipped with a compression release or 80-90 psi (550-620 kPa) if en-

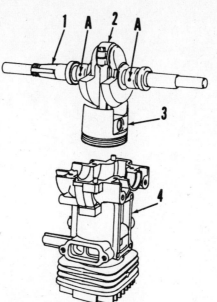

Fig. BS29—View showing crankshaft (1), connecting rod (2), piston (3), crankcase and cylinder (4).

gine is equipped with a compression release. A low compression reading could indicate worn cylinder bore, piston and/or rings. A higher than normal compression reading could indicate an excessive accumulation of carbon on the piston and combustion chamber.

REPAIRS

TIGHTENING TORQUES. Recommended tightening torque specifications are as follows:

Carburetor	100 in.-lbs.
	(11.3 N·m)
Connecting rod	55 in.-lbs.
	(6.2 N·m)
Crankcase	90 in.-lbs.
	(10.2 N·m)
Muffler	115 in.-lbs.
	(13 N·m)
Spark plug	170 in.-lbs.
	(19.2 N·m)
Starter clutch	30 ft.-lbs.
	(40.7 N·m)

CRANKCASE PRESSURE TEST. An improperly sealed crankcase can cause the engine to be hard to start, run rough, have low power and overheat. Refer to engine SERVICE SECTION TROUBLESHOOTING section of this manual for crankcase pressure test procedure. If crankcase leakage is indicated, pressurize crankcase and use a soap and water solution to check gaskets, seals, carburetor pulse line and casting for leakage.

CYLINDER AND CRANKCASE.

The lower crankcase half is secured to

the one-piece cylinder and upper crankcase half with screws. To separate the crankcase halves, remove the back plate, flywheel and crankcase cover screws. Using a soft-faced mallet, tap the pads on the lower crankcase half shown in Figs. BS27 and BS28. DO NOT attempt to separate the crankcase halves by prying. With the lower crankcase half removed, the crankshaft, rod and piston assembly (Fig. BS29) can be withdrawn from the crankcase/cylinder.

Remove the transfer port covers (13 and 16—Fig. BS30). Clean carbon from the cylinder head and ports. Inspect the cylinder for excessive wear and damage. Standard cylinder bore diameter is 2.124-2.125 inch (53.95-53.98 mm). If the cylinder bore diameter is 2.128 inch (54.05 mm) or more, renew cylinder. If cylinder bore is out-of-round 0.0025 inch (0.064 mm) or more, or if cylinder bore taper is 0.003 inch (0.076 mm) or more, renew cylinder.

Refer to Fig. BS31 for a view of the compression release valve used on some models. Clean all carbon and foreign material from cavity and valve. When assembling, apply a small amount of gasket-forming sealant on the cover (see Fig. BS31). Install the valve so the lettered side is down. Tighten retaining screws to 30 in.-lbs. (3.4 N·m) torque.

During assembly note that the transfer port covers (13 and 16—Fig. BS30) are different. The cover marked "MAG" must be installed on the flywheel side of the cylinder while the cover marked "PTO" must be installed on the output side of the cylinder.

No gasket is used between the crankcase halves. Before assembling the crankcase halves, clean the mating surfaces of the crankcase halves then apply a thin coat of gasket-forming sealant to the mating surfaces (Fig. BS32). When installing the crankshaft assembly into the crankcase/cylinder, note the locating pins (A—Fig. BS29) in the outer race of each main bearing. The pins must rest in notches in the crankcase as shown in Fig. BS33. Tighten crankcase screws to 90 in.-lbs. (10.2 N·m) torque. It is recommended that the crankcase be pressure tested for leakage before proceeding with remainder of engine reassembly.

PISTON, PIN AND RINGS. The piston is accessible after removing the crankshaft assembly as outlined in the previous section.

The piston pin is full floating and can be extracted after detaching the retaining clips at either end. The piston pin should be renewed if the outside diameter is 0.499 inch (12.67 mm) or less.

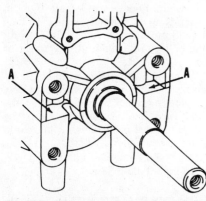

Fig. BS27—When separating crankcase halves, tap against pads shown with a soft-faced mallet.

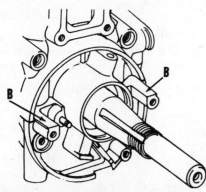

Fig. BS28—When separating crankcase halves, tap against pads shown with a soft-faced mallet.

Illustrations courtesy Briggs & Stratton Corp.

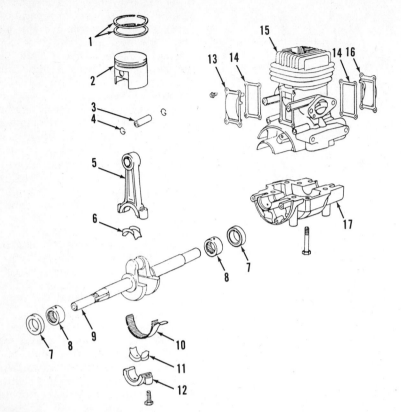

Fig. BS30—Exploded view of engine.

1. Piston rings		10. Roller bearings	14. Gasket
2. Piston	6. Upper bearing liner	11. Lower bearing liner	15. Cylinder/crankcase
3. Piston pin	7. Oil seal	12. Rod cap	16. Transfer port cover (PTO)
4. Retaining ring	8. Main bearing	13. Transfer port cover (MAG)	17. Crankcase half
5. Connecting rod	9. Crankshaft		

Piston and rings are available in standard size only.

Piston rings may be installed on piston with either side up. Piston may be installed on connecting rod in either direction.

CONNECTING ROD. The connecting rod rides on uncaged needle roller bearings on the crankpin. When detaching the connecting rod from the crankshaft be careful not to lose the bearing rollers.

A nonrenewable cartridge type roller bearing is located in the small end of the connecting rod. The connecting rod

must be renewed if the bearing is damaged or excessively worn.

When installing an old bearing on the crankpin, hold the rollers in place with petroleum jelly or grease. New bearing rollers are coated with wax that will hold the rollers in place on the crankpin.

The big end of the connecting rod is equipped with liners (C and B—Fig. BS34). Install the liner (B) in the rod cap so the hole in the liner fits the index pin (A) on the rod cap. Install the remaining liner (C) so the "V" ends of the liners will fit together properly when the rod and cap are mated.

The connecting rod and cap are equipped with match marks (D—Fig. BS34) which must be on same side when rod and cap are mated. Tighten connecting rod screws to 55 in.-lbs. (6.2 N•m) torque. Lubricate the crankpin bearing after assembly.

CRANKSHAFT AND MAIN BEARINGS. The crankshaft assembly is removed as outlined in the CYLINDER AND CRANKCASE section. When detaching the connecting rod from the crankshaft be careful not to lose the bearing rollers.

The crankshaft is supported by cartridge type needle roller bearings at both ends. Inspect crankshaft main bearing surfaces for damage and excessive wear. Renew crankshaft if main bearing journal diameter is 0.7515 inch (19.088 mm) or less, or if crankpin diameter is 0.7420 inch (18.847 mm) or less.

When installing the crankshaft assembly into the crankcase/cylinder, note the locating pins (A—Fig. BS29) in the outer race of each main bearing. The pins must rest in notches in the crankcase as shown in Fig. BS33.

Crankshaft end play should be 0.002-0.013 inch (0.05-0.33 mm).

CRANKSHAFT SEALS. Crankshaft seals must be installed so the spring side is toward the engine. Use a suitable driver to install seal.

REWIND STARTER. The engine is equipped with a rewind starter that is mounted on the blower housing. The

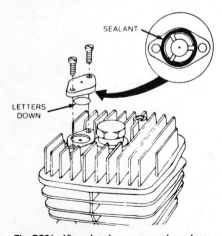

Fig. BS31—View showing compression release.

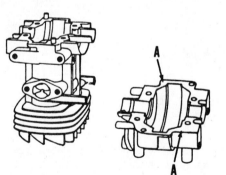

Fig. BS32—Apply sealer on crankcase mating surfaces indicated prior to assembling crankcase.

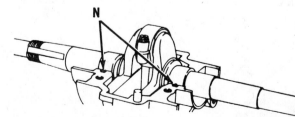

Fig. BS33—Locating pins in main bearings (A—Fig. BS29) must index in notches (N) of crankcase.

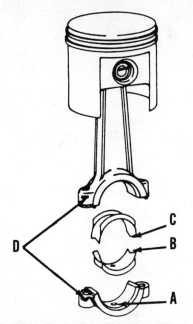

Fig. BS34—Bearing liner (B) hole must fit around index pin (A) on the rod cap. The "V" ends of bearing liners (B & C) must fit together properly when the rod and cap are mated. The match marks (D) on the connecting rod and cap must be on same side when rod and cap are mated.

Fig. BS35—View of starter clutch.

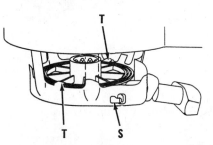

Fig. BS36—View of outer rewind spring end (S) and pulley retaining tangs (T).

Fig. BS37—Align inner rope hole (H) with rope outlet in housing before inserting rope.

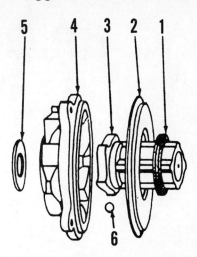

Fig. BS38—Exploded view of starter clutch. Refer to Fig. BS39 for cutaway view of ratchet (3).

1. Rubber seal
2. Ratchet cover
3. Ratchet
4. Clutch housing (flywheel nut)
5. Spring washer
6. Steel balls

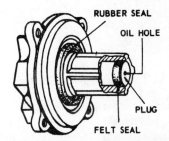

Fig. BS39—Cutaway view showing felt seal and plug in end of starter ratchet (3—Fig. BS38).

rope pulley drives a ball-type sprag clutch (Fig. BS35) mounted on the flywheel. When the rewind starter is operated, the flywheel rotates and one of the balls will engage a sprag. When the engine runs, the balls are thrown out by centrifugal force to a disengaged position.

To renew a broken rewind spring, proceed as follows: Grasp free outer end of spring (S—Fig. BS36) and pull broken end from starter housing. With blower housing removed, bend up tangs (T) and remove starter pulley from housing. Untie knot in rope and remove rope and inner end of broken spring from pulley. Install rewind spring by threading inner end of spring through notch in starter housing, then engage inner end of spring in pulley hub. Apply a small amount of grease on inner face of pulley and place pulley in housing. Bend the pulley retainer tangs (Fig. BS36) towards the pulley so the gap between the tang and the pulley is 1/16 inch (1.6 mm). Insert a 3/4 inch (19 mm) square bar in pulley hub and turn pulley in a counter-clockwise direction until the spring is tight. Rotate pulley clockwise until rope hole (H—Fig. BS37) in pulley is aligned with rope outlet hole in housing, then secure pulley so it cannot rotate. Hook a wire in inner end of rope and thread rope through guide and hole in pulley; then, tie a knot in rope and release the pulley allowing spring to wind rope into pulley groove.

To renew starter rope only, it is not generally necessary to remove starter pulley and spring. Wind up the spring and install new rope as outlined in preceding paragraph.

To disassemble starter clutch unit, refer to Fig. BS38 and proceed as follows: Remove starter ratchet cover (2). Lift ratchet (3) from housing and crankshaft and extract the steel balls (6). If necessary to remove housing (4), hold flywheel and unscrew housing in counterclockwise direction using B&S tool 19244. When installing housing, be sure spring washer (8) is in place on crankshaft with cup (concave) side towards flywheel; then, tighten housing securely. Inspect felt seal and plug in outer end of ratchet (Fig. BS39). Renew ratchet if seal or plug is damaged as these parts are not serviced separately. Lubricate the felt with oil and place ratchet on crankshaft. Insert the steel balls and install ratchet cover, rubber seal and rotating screen.

BRIGGS & STRATTON
2-STROKE ENGINES

Model	Bore	Stroke	Displacement	Power Rating
95700, 96700	60 mm	50 mm	141 cc	3.0 kW
	(2.36 in.)	(1.97 in.)	(8.6 cu. in.)	(4.0 hp)

NOTE: Metric fasteners are used throughout engine except threaded hole in pto end of crankshaft which is US threads.

ENGINE INFORMATION

All models are two-stroke, single-cylinder engines utilizing a third-port scavenging system. The engine may be equipped with a chrome plated or cast iron cylinder bore. Refer to table found on page 59 for an interpretation of Briggs & Stratton model numbers.

MAINTENANCE

LUBRICATION. Manufacturer recommends mixing a good quality BIA or NMMA 2-cycle oil certified for TC-WII service with regular or low-lead gasoline at a 50:1 ratio. Do not use an automotive (4-cycle) engine oil. The use of gasoline that contains alcohol is not recommended. However, if gasoline with alcohol is used, it must not contain more than 10 percent ethanol. Do not use gasoline containing methanol. If gasoline containing ethanol is used, it must be drained from the fuel system before storing the engine.

Always mix fuel in a separate container and add only mixed fuel to engine fuel tank. To assure thorough mixing of the oil and gasoline, fill the container partially with gasoline, then add the correct amount of oil per the chart shown in Fig. BS49. Shake the container to mix the oil and gasoline, then add remainder of gasoline and shake the container again.

AIR CLEANER. The air cleaner consists of a cover and a pleated paper cartridge, or a combination of a foam precleaner and a paper cartridge. Under normal operating conditions, filter elements should be cleaned and inspected after every 25 hours of engine operation, or after three months, whichever occurs first. Under extremely dusty conditions, service the filter more often. Tap filter gently to dislodge accumulated dirt. Filter may be washed using warm water and nonsudsing detergent directed from inside of filter to outside. DO NOT use petroleum-based cleaners or solvents to clean filter. DO NOT direct pressurized air towards filter. Let filter air dry thoroughly, then inspect filter and discard it if damaged or uncleanable. DO NOT apply oil to the paper cartridge or the foam precleaner. Clean the filter cover. Inspect and, if necessary, replace any defective gaskets.

FUEL FILTER. The fuel tank is equipped with a filter at the outlet. Check filter annually and periodically during operating season.

CARBURETOR. All models are equipped with the float type carburetor shown in Fig. BS50.

Adjustment. Idle speed at normal operating temperature should be 1200 rpm. Adjust idle speed by turning idle speed screw (IS). Idle mixture is controlled by idle jet (IJ) which is not adjustable. High speed mixture is controlled by main jet (MJ) which is not adjustable. Optional main jets are available for high altitude operation.

Overhaul. To disassemble carburetor, remove float bowl retaining screw (17—Fig. BS50), gasket (16) and float bowl (15). Remove float pin (13) by pushing against round end of pin towards the square end of pin. Remove float (12), fuel inlet needle (11) and clip.

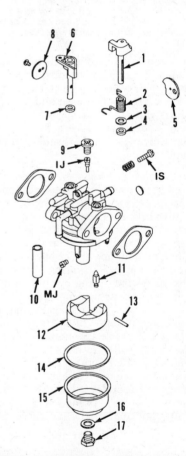

Fig. BS50—Exploded view of carburetor used on Models 95700 and 96700.

IJ. Idle mixture jet	8. Throttle plate
IS. Idle speed screw	9. Plug
MJ. Main jet	10. Vent tube
1. Choke shaft	11. Fuel inlet valve
2. Spring	12. Float
3. Plastic washer	13. Float hinge pin
4. Felt washer	14. Gasket
5. Choke plate	15. Fuel bowl
6. Throttle shaft	16. Gasket
7. Seal	17. Screw

50:1					
U.S.		Imperial		Metric	
Gasoline Gallons	2 Cycle Oil Ounces	Gasoline Gallons	2 Cycle Oil Ounces	Gasoline Liters	2 Cycle Oil Liters
1	2.5	1	3.2	4	.08
2	5	2	6.4	8	.16
5	13	5	16	20	.4

Fig. BS49—Gasoline/oil mixture chart.

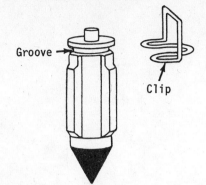

Fig. BS51—When installing fuel inlet valve, be sure round loop of clip fits around groove of valve and square portion of clip fits around float tab.

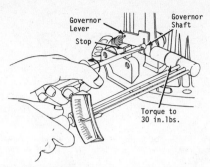

Fig. BS53—Follow procedure outlined in text for governor adjustment.

Remove throttle and choke shaft assemblies after unscrewing throttle and choke plate retaining screws.

Clean parts in a suitable carburetor cleaner. Do not use wire or drill bits to clean fuel passages as carburetor calibration may be affected if passages are enlarged. Inspect parts for wear or damage and renew as necessary.

When assembling the carburetor, note the following. Place a small drop of nonhardening sealant such as Permatex #2 or equivalent on throttle and choke plate retaining screws. The "U" bend end of choke return spring is on top and fits around choke shaft arm. Install throttle shaft seal with flat side towards carburetor body. Numbers on throttle plate must face out with throttle closed. Be sure inlet needle clip properly engages groove of inlet needle as shown in Fig. BS51 and square portion of clip fits around float tab. Float height is not adjustable.

GOVERNOR. Models 95700 and 96700 are equipped with the centrifugal type governor shown in Fig. BS52.

Maximum engine speed is limited by changing spring (4) and varying loca-

tion of the spring in holes in speed lever (6). Different tension governor springs are available which allows setting of governed speed from 2600 to 3600 rpm. Governor speed should be adjusted to desired speed specified by the equipment manufacturer.

If governor lever (7—Fig. BS52) is detached or moved on governor shaft, correct lever position on shaft must be established. Detach governor rod (1) from lever (7) and loosen clamp bolt (8). Use a screwdriver and turn slotted end of governor shaft counterclockwise as shown in Fig. BS53 as far as possible. Rotate the governor lever counterclockwise until it strikes the stop. Tighten clamp bolt to 3.4 N•m (30 in.-lbs.) torque.

SPARK PLUG. Recommended spark plug is Autolite 235 or 245 or Champion J19LM or RJ19LM. Spark plug electrode gap should be 0.76 mm (0.030 in.). Tighten spark plug to 18 N•m (160 in.-lbs.) torque.

CAUTION: Briggs & Stratton does not recommend using abrasive blasting to clean spark plugs as this may introduce some abrasive material into the engine which could cause extensive damage.

IGNITION SYSTEM. Models 95700 and 96700 are equipped with a Magnetron ignition system.

To check spark, remove spark plug and connect spark plug cable to B&S tester 19051, then ground remaining tester lead to engine. Spin engine at 350 rpm or more. If spark jumps the 4.2 mm (0.165 in.) tester gap, system is functioning properly.

To remove armature and Magnetron module, remove flywheel shroud and armature retaining screws. Disconnect stop switch wire from module. When installing armature, be sure stop switch wire and grommet are properly installed in cylinder fins. Position arma-

ture so air gap between armature legs and flywheel surface is 0.20-0.40 mm (0.008-0.016 in.).

If armature coils are suspected faulty, note the following specifications. Primary coil resistance should be 0.2-0.4 ohms and secondary coil resistance should be 2500-3500 ohms.

COMPRESSION PRESSURE. Compression reading for engine should be 620-820 kPa (90-120 psi). A low compression reading could indicate worn cylinder bore, piston and/or rings. A higher than normal compression reading could indicate an excessive accumulation of carbon on the piston and combustion chamber.

FLYWHEEL BRAKE. The engine is equipped with a band type flywheel brake. The brake should stop the engine within three seconds when the operator releases mower safety control and the speed control is in high speed position. Stopping time can be checked using B&S tool 19255.

To check brake band adjustment, remove starter and properly ground spark plug lead to prevent accidental starting. Turn flywheel nut using a torque wrench with brake engaged. Rotating flywheel nut at a steady rate in a clockwise direction should require at least 5.1 N•m (45 in.-lbs.) torque. An insufficient torque reading may indicate misadjustment or damaged components. Renew brake band if friction material of band is damaged or less than 0.76 mm (0.030 in.) thick.

To adjust brake, unscrew brake cable clamp retaining screw so hole (H—Fig. BS54) is vacant. (If a pop rivet secures the cable clamp, do not remove rivet; gauge tool can be inserted through hole in pop rivet.) Loosen screws (S) securing brake control bracket. Place bayonet end of B&S gauge 19256 (G) in hole (L)

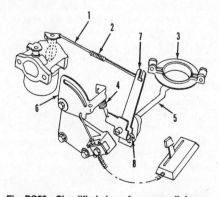

Fig. BS52—Simplified view of governor linkage.

1. Governor lever-to-carburetor rod
2. Spring
3. Flyweight assy.
4. Spring
5. Governor shaft
6. Speed lever
7. Governor lever
8. Clamp bolt

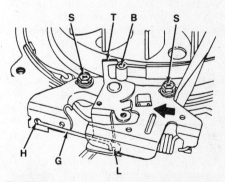

Fig. BS54—View of flywheel band brake and lever assembly.

B. Brake band
G. Gauge link
H. Hole
L. Hole
S. Retaining screws
T. Brake lever tang

of control lever (C). Push against control lever and insert opposite end of gauge (G) in hole (H), or in hole of pop rivet if used to secure cable. Apply pressure against control bracket in direction shown by arrow in Fig. BS54 until tension on gauge is just removed; then while holding pressure, tighten bracket screws (S) to 2.8-3.4 N·m (25-30 in.-lbs.) torque. As gauge is removed a slight friction should be felt and control lever should not move.

Brake band must be renewed if damaged or if friction material thickness is less than 0.76 mm (0.030 in.). Use tool 19229 to bend brake lever tang (T) away from band end (B—Fig. BS54). Release brake lever spring and remove brake band. Install new band with brake material side toward the flywheel. Reconnect brake spring, bend lever retaining tang over band end and adjust brake as outlined above.

REPAIRS

TIGHTENING TORQUES. Recommended tightening torque specifications are as follows:

Carburetor . . . . 5.6 N·m
  (50 in.-lbs.)
Crankcase . . . . 6.8 N·m
  (60 in.-lbs.)
Cylinder . . . . 12.4 N·m
  (110 in.-lbs.)
Flywheel nut . . . . 40.7 N·m
  (30 ft.-lbs.)
Muffler . . . . 9.6 N·m
  (85 in.-lbs.)
Spark plug . . . . 19.2 N·m
  (170 in.-lbs.)

CRANKCASE PRESSURE TEST. An improperly sealed crankcase can cause the engine to be hard to start, run rough, have low power and overheat. Refer to SERVICE SECTION TROUBLESHOOTING section of this manual for crankcase pressure test procedure. If crankcase leakage is indicated, pressurize crankcase and use a soap and water solution to check gaskets, seals, carburetor pulse line and casting for leakage.

PISTON, PIN, RINGS AND CYLINDER. The cylinder and head are one piece and may be removed after removing fuel tank, muffler guard, blower housing, carburetor and muffler. Unscrew four 5 mm Allen screws in cylinder base and carefully separate cylinder from crankcase. Remove and discard piston pin retaining clips. If possible, hand-push piston pin out of rod, otherwise, use a suitable puller to extract piston pin and separate piston from connecting rod. Do not use a hammer to

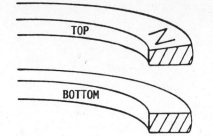

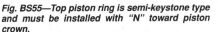

Fig. BS55—Top piston ring is semi-keystone type and must be installed with "N" toward piston crown.

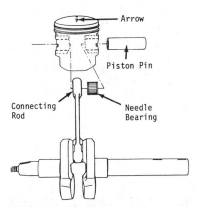

Fig. BS56—Install piston on rod so arrow on piston crown points toward exhaust port.

drive out piston pin as damage to connecting rod could result.

Piston ring end gap should not exceed 1.01 mm (0.040 in.). With ring placed 32 mm (1¼ in.) down in cylinder bore, use a feeler gauge to measure ring end gap.

Renew piston pin if diameter is 13.97 mm (0.550 in.) or less. Wear limit for piston pin hole in piston is 14.04 mm (0.553 in.).

Renew piston if it is scored or if diameter at skirt is 59.84 mm (2.356 in.) or less. Piston and rings are available in standard size only.

Wear limit for cylinder bore diameter is 60.14 mm (2.368 in.) for chrome plated bores and 60.17 mm (2.369 in.) for cast iron bores. Renew cylinder if cylinder bore has deep score marks or if chrome plating (if used) is damaged.

Piston rings are pinned and rings must be installed on piston so ring gap indexes with locating pin. Upper piston ring is semi-keystone type and must be installed with "N" toward top (crown) of piston as shown in Fig. BS55. Lower compression ring is rectangular and may be installed either side up.

Piston must be mated to connecting rod so arrow on piston crown will point toward the cylinder exhaust port (Fig. BS56). Use new piston pin retaining clips and install clips so gap of clip is toward either piston crown or crankshaft.

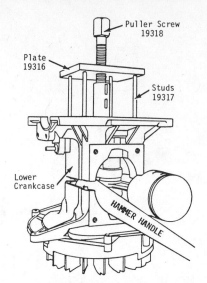

Fig. BS57—View showing B&S tools used to separate pto-side crankcase half from magneto-side crankcase half.

Lubricate piston and rings with clean 2 cycle engine oil, then slide cylinder with a new gasket onto piston and rings. Tighten cylinder base screws to 12.4 N·m (110 in.-lbs.).

CRANKSHAFT AND CRANKCASE. To disassemble crankcase assembly, remove cylinder and remaining components attached to crankcase and crankshaft. Remove four cap screws retaining crankcase halves. Place flywheel-side crankcase half on bottom with flywheel end of crankshaft inserted in flywheel as shown in Fig. BS57. Insert a small wooden hammer handle into the crankcase to prevent crankshaft rotation. Install B&S puller plate 19136 (with side marked "X" toward piston), puller screw 19138 and puller studs 19317 on lower crankcase as shown in Fig. BS57. Separate crankcase lower half from crankshaft by turning puller screw.

To separate upper crankcase half from crankshaft, thread the flywheel nut or other suitable nut on the end of the crankshaft to protect the threads. Install the B&S puller assembly as shown in Fig. BS58. Turn the puller screw until the crankcase disengages from the crankshaft.

Remove crankcase oil seals and if bearing on pto side is to be removed, remove governor crank. Use B&S tool assembly shown in Fig. BS59 to remove bearings from crankcase halves. Turn screw to force bearing from crankcase. Bearings must be discarded if removed from crankcase.

The crankshaft and connecting rod are a unit assembly and disassembly is not recommended. Renew crankshaft if diameter of main bearing journals is

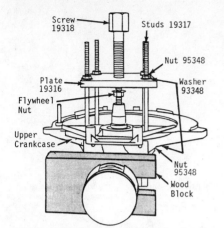

Fig. BS58—View showing B&S tools used to pull magneto-side crankcase half away from crankshaft assembly.

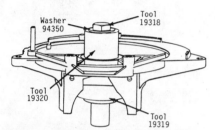

Fig. BS59—The main bearing in each crankcase half can be removed using the B&S tools shown above. Removal of magneto-side bearing is shown, removal of pto-side bearing is similar.

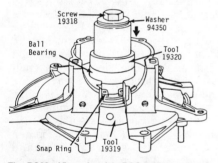

Fig. BS60—View showing B&S tools used to install main bearing in magneto-side crankcase half. Bearing must seat against snap ring.

24.95 mm (0.982 in.) or less, or if out-of-round 0.013 mm (0.0005 in.) or more.

To install magneto-side bearing, assemble B&S tools as shown in Fig. BS60 and turn screw until bearing seats against snap ring. To install pto-side bearing, assemble B&S tools as shown in Fig. BS61 and turn screw until bearing is seated against shoulder of crankcase half. Install oil seals using B&S tool assemblies shown in Fig. BS62. Flat side of seal must be towards outside of engine.

Pull crankshaft into magneto-side crankcase half using B&S tools 19314 and 19315 as shown in Fig. BS63.

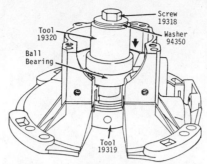

Fig. BS61—Assemble B&S tools as shown to install main bearing in pto-side crankcase half. Bearing must seat against shoulder of crankcase half.

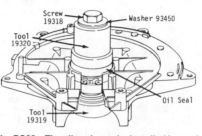

Fig. BS62—The oil seals can be installed in crankcase halves using the B&S tools shown above.

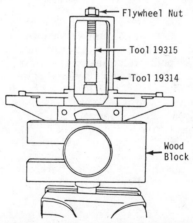

Fig. BS63—Install B&S tools 19314 and 19315 as shown and use flywheel nut or equivalent to pull magneto-side crankcase half onto crankshaft assembly. A wooden block is shown placed between piston and crankcase to prevent crankshaft rotation.

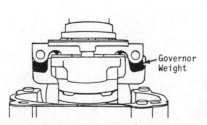

Fig. BS64—Governor weights are incorrectly positioned if they are not in the position shown above.

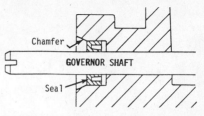

Fig. BS65—View of correctly installed governor shaft seal (seal lip facing inward).

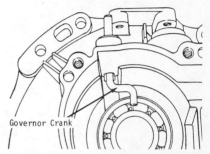

Fig. BS66—The governor crank must be positioned as shown for correct operation.

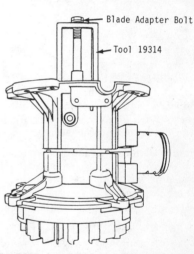

Fig. BS67—Install B&S tool 19314 and use the blade adapter bolt to pull the crankcase halves together.

NOTE: Crankshaft must not turn while pulling crankshaft into bearings.

Install crankcase gasket then install governor weight assembly as shown in Fig. BS64 (be sure weights appear as shown). Install governor shaft seal so lip is towards inside of crankcase and face is flush with chamfer (Fig. BS65). Install washer and "E" ring on governor shaft then position governor crank as shown in Fig. BS66.

Position pto-side crankcase half on crankshaft and magneto-side crankcase assembly as shown in Fig. BS67 and install B&S tool 19314. Pull halves together by turning blade adapter bolt; using one of the puller studs threaded into the crankcase will help align the halves. If excessive resistance is felt,

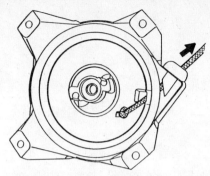

Fig. BS68—Exploded view of rewind starter.

1. Starter housing	6. Brake spring
2. Rewind spring	7. Retainer
3. Pulley	8. Screw (L.H. thread)
4. Dog spring	9. Starter rope insert
5. Dog	10. Starter rope grip

disassemble and check for complete seating of the main bearings. Tighten crankcase bolts to 6.8 N·m (60 in.-lbs.) torque. Trim crankcase gasket flush with cylinder mating surface.

It is recommended that the crankcase be pressure tested for leakage before proceeding with remainder of engine reassembly.

REWIND STARTER. Refer to Fig. BS68 for exploded view of rewind starter. The static guard cover above the starter must be removed for access to starter.

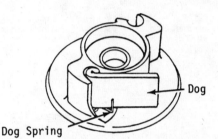

Fig. BS69—Route the starter rope as shown through pulley and rope outlet.

To install a new rope, proceed as follows. Rope length for models equipped with a band brake should be 22.6 cm (87 in.), 15.2 cm (60 in.) for engines not equipped with a band brake. Remove starter and pull old rope out as far as it will go. Extract old rope from pulley, then allow pulley to unwind. To install new rope, turn pulley counterclockwise until spring is tightly wound. Then rotate pulley clockwise until rope hole in pulley is aligned with rope outlet in housing. Pass rope through pulley hole and outlet as shown in Fig. BS69 and tie a temporary knot near handle end of rope. Release pulley and allow rope to wind onto pulley. Install rope handle, release temporary knot and allow rope to enter starter.

To disassemble starter, remove rope and allow pulley to totally unwind. Unscrew pulley retaining screw (screw has left-hand threads). Remove retainer (7—Fig. BS68) and brake spring (6). Wear appropriate safety eyewear and gloves before disengaging pulley (3) from starter as spring may uncoil uncontrolled. Place shop towel around pulley and lift pulley out of housing; spring should remain with pulley. Carefully extract spring from pulley.

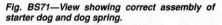

Fig. BS70—Apply a light coating of grease to pulley face. Outer end of rewind spring must engage spring anchor slot.

Fig. BS71—View showing correct assembly of starter dog and dog spring.

Inspect components for wear and damage which may prevent smooth operation. Apply a light coat of grease on pulley face where spring will rub. Install rewind spring (2) on pulley so spring is wound in a counterclockwise direction from outer end. Be sure outer end properly engages spring anchor slot of pulley (Fig. BS70). Install pulley and spring in starter housing and rotate pulley counterclockwise so inner spring end engages anchor of starter housing. Assemble remainder of components while noting correct assembly of dog and dog spring in Fig. BS71. Tighten pulley retaining screw (left-hand threads) to 3.4 N·m (30 in.-lbs.) torque.

Illustrations courtesy Briggs & Stratton Corp.

BRIGGS & STRATTON

4-STROKE ENGINES
(Except Europa, Quantum & Vanguard)

Model Series	Bore	Stroke	Displacement	Power Rating
60000	2.38 in. (60.3 mm)	1.50 in. (38.1 mm)	6.7 cu. in. (109 cc)	2.0 hp (1.5 kW)
80000, 82000, 83000	2.38 in. (60.3 mm)	1.75 in. (44.5 mm)	7.75 cu. in. (127 cc)	3 hp (2.2 kW)
90000, 91000, 92000, 93000, 94000, 95000	2.56 in. (65.0 mm)	1.75 in. (44.5 mm)	9.0 cu. in. (148 cc)	3.5 hp (2.6 kW)
96000	2.56 in. (65.0 mm)	1.75 in. (44.5 mm)	9.0 cu. in. (148 cc)	3.75 hp (2.8 kW)
110000, 111000, 112000, 113000, 114000	2.78 in. (70.6 mm)	1.88 in. (47.6 mm)	11.4 cu. in. (186 cc)	4.0 hp (3 kW)
130000, 131000, 132000, 133000, 135000	2.56 in. (65.0 mm)	2.44 in. (62.0 mm)	12.6 cu. in. (206 cc)	5.0 hp (3.8 kW)

ENGINE IDENTIFICATION

Engines covered in this section have aluminum cylinder blocks with either a plain aluminum cylinder bore or with a cast iron sleeve integrally cast into the block.

Refer to page 59 for Briggs & Stratton engine identification information.

MAINTENANCE

LUBRICATION. All engines are lubricated using splash lubrication. Oil in the oil pan is thrown onto internal engine parts by an oil dipper on the connecting rod for horizontal crankshaft engines or by a slinger driven by the camshaft gear for vertical crankshaft engines.

Some engines are equipped with a low-oil level system (Oil Gard) that uses a float which is located inside the crankcase and connected by a wire lead to the ignition. If there is insufficient oil in the crankcase, the ignition is grounded so the engine stops or will not start.

Engine oil should meet or exceed latest API service classification. Use SAE 30 oil for temperatures above 40° F (4° C); use SAE 10W-30 oil for temperatures between 0° F (−18° C) and 40° F (4° C); below 0° F (−18° C) use petroleum based SAE 5W-20 or a suitable synthetic oil. Briggs & Stratton states that multiviscosity oil should not be used above 40° F (4° C) as engine damage may result due to high running temperature of an air-cooled engine.

Fill engine with oil by pouring oil through oil fill plug opening or opening for the oil dipstick. If engine is equipped with an oil fill plug, unscrew plug and add oil until oil level is even with top threads in plug hole.

Some engines are equipped with an extended oil fill tube and a dipstick attached to oil fill cap. When checking the oil level, screw the dipstick into place until the cap bottoms on the filler tube, then unscrew the dipstick and observe oil level on dipstick. The oil level should be between the ADD and FULL marks on dipstick.

Briggs & Stratton specifies that the engine oil should be changed after the first five hours of operation, and then, if used normally, after every 50 hours of operation or seasonally, whichever is less. If the engine is run in severe conditions, such as under heavy load or in high ambient temperatures, the oil should be changed weekly or after every 25 hours of operation, whichever occurs first.

Approximate oil capacities are listed in the following table:

130000, 131000, 132000, 133000, 135000:

Horizontal crankshaft..... 1¼ pints (0.6 L)

Vertical crankshaft .... 1¾ pints (0.8 L)

All other models .... 1¼ pints (0.6 L)

AIR CLEANER. The air cleaner consists of a canister and the filter element it contains. The canister is secured by one or two wing nuts, screws or knobs. The filter element may be made of foam or paper, or a combination of both foam and paper. The recommended maintenance interval depends on the type of filter element.

Foam Type. Foam type filter elements should be cleaned, inspected and re-oiled after every 25 hours of engine operation, or after three months, whichever occurs first. Clean filter in kerosene or soapy water and squeeze until dry. Inspect filter for tears and holes or any other opening. Discard filter if it cannot be cleaned satisfactorily or if filter is torn or otherwise damaged. Pour clean engine oil into the filter, then squeeze filter to remove excess oil and distribute oil throughout filter. Clean filter canister. Inspect and, if necessary, replace any defective gaskets. Be sure filter element fits properly and any spacers are properly positioned during assembly. If air cleaner has a screen, the screen should be placed on top of the filter.

Paper Type. Paper type filter elements should be cleaned and inspected after every 25 hours of engine operation, or after three months, whichever occurs first. Tap filter gently to dislodge accumulated dirt. Filter may be washed using warm water and nonsudsing detergent, then rinse with clean water directed from inside of filter to outside. DO NOT use petroleum-based cleaners or solvents to clean filter. DO NOT direct pressurized air towards filter. Let filter air dry thoroughly, then inspect filter and discard it if damaged or uncleanable. Clean filter canister. Inspect and, if necessary, replace any defective gaskets.

Combination Foam and Paper Type. The combination type air cleaner consists of a foam type filter, known as the precleaner, wrapped around or in front of a paper type filter. The foam precleaner should be cleaned weekly or after every 25 hours of operation, whichever occurs first. The paper filter should be cleaned yearly or after every 100 hours of operation, whichever occurs first. Clean foam precleaner or paper filter using cleaning methods previously outlined for foam and paper filters.

CRANKCASE BREATHER. The crankcase breather is built into the tappet chamber cover. A fiber disc acts as a one-way valve. The breather allows vapor from the crankcase to be evacuated to the intake manifold, but blocks the return flow of air, thus maintaining a vacuum in the crankcase. The vacuum prevents oil from being forced out of the engine past the piston rings, oil seals and gaskets.

Clearance between fiber disc check valve and breather body (Fig. B1) should not exceed 0.045 inch (1.14 mm).

If it is possible to insert a 0.045 inch (1.14 mm) wire between disc and breather body, renew breather assembly. Do not use excessive force when measuring gap. Disc should not stick or bind during operation. Renew if distorted or damaged. Inspect breather tube for leakage.

SPARK PLUG. The original spark plug may be either 1½ inches or 2 inches long. Briggs & Stratton recommends Champion or Autolite spark plugs.

If a Champion spark plug is used and spark plug is 1½ inches long, recommended spark plug is J19LM or CJ8. Install a Champion RJ19LM or RCJ8 if a resistor type spark plug is required. If spark plug is 2 inches long, recommended spark plug is J19LM or J8C. Install a Champion RJ19LM or RJ8C if a resistor type spark plug is required.

If an Autolite spark plug is used and spark plug is 1½ inches long, recommended spark plug is 235. Install a Autolite 245 if a resistor type spark plug is required. If spark plug is 2 inches long, recommended spark plug is 295. Install a Autolite 306 if a resistor type spark plug is required.

NOTE: Briggs & Stratton does not recommend using abrasive blasting to clean spark plugs as abrasive material may enter engine.

Spark plug electrode gap should be 0.030 inch (0.76 mm). Tighten spark plug to 140-200 in.-lbs. (15.8-22.6 N•m) torque.

CARBURETOR. The engine may be equipped with either a suction type (Pulsa-Jet, Pulsa-Prime, Vacu-Jet) or a float type (Flo-Jet, Walbro) carburetor. The suction type carburetor is identified by its location on top of the fuel tank. Refer to the appropriate following section.

Suction Type Carburetors Except Pulsa-Prime. Pulsa-Jet and Vacu-Jet carburetors are suction type carburetors that are mounted on the fuel tank. The carburetors are differentiated by the presence of one fuel tube (Fig. B2) on Vacu-Jet carburetors and two fuel tubes on Pulsa-Jet carburetors. Some Pulsa-Jet carburetors are identified by the fuel pump (P) mounted on the side of the carburetor. If the carburetor has a primer bulb, it is a Pulsa-Prime carburetor, which is covered in the following section.

OPERATION. The Vacu-Jet carburetor has a fuel tube that extends into the fuel tank as shown in Fig. B3. Atmospheric pressure against the fuel forces fuel up the fuel tube due to the vacuum created in the carburetor bore when the engine runs. A check valve allows fuel to flow up the tube but prevents fuel from draining back into the fuel tank. The mixture screw controls the amount of fuel entering the carburetor bore at high speed. The fuel passes through two metering holes into the bore. The metering holes are different sizes and at idle only the smaller diameter hole passes fuel due to the position of the throttle plate. Fuel mixture in the Vacu-Jet is affected by the amount of

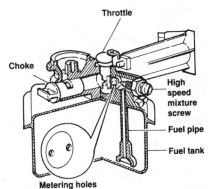

Fig. B3—Cutaway view of a typical Vacu-Jet carburetor. Inset shows fuel metering holes.

Fig. B1—Clearance between fiber disc valve and crankcase breather housing must be less than 0.045 inch (1.15 mm). A spark plug wire gauge may be used to check clearance as shown, but do not apply pressure against disc valve.

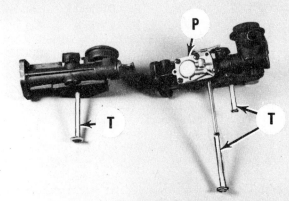

Fig. B2—Vacu-Jet carburetors are equipped with one fuel tube while Pulsa-Jet carburetors have two fuel tubes. Note fuel pump (P) found on the side of some Pulsa-Jet carburetors.

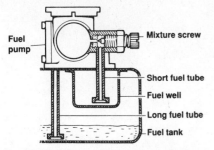

Fig. B4—Fuel flow in Pulsa-Jet carburetor. Fuel pump fills constant level fuel well below carburetor and excess fuel flows back into tank. Fuel is drawn from fuel well through tube to mixture screw.

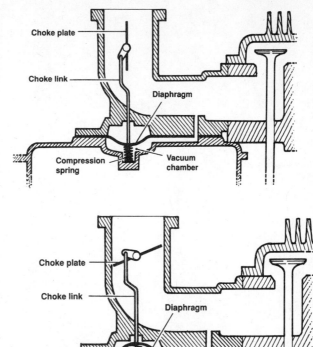

Fig. B6—Drawing of vacuum-operated automatic choke when engine starts. Lack of vacuum allows choke spring to close the choke plate.

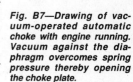

Fig. B7—Drawing of vacuum-operated automatic choke with engine running. Vacuum against the diaphragm overcomes spring pressure thereby opening the choke plate.

fuel in the fuel tank (fuel weight versus atmospheric pressure). Carburetor adjustments should be performed with a half-full fuel tank thereby resulting in a satisfactory mixture adjustment whether the tank is near empty or full.

The Pulsa-Jet carburetor operates similarly to the Vacu-Jet except in delivery of the fuel to the carburetor. The fuel tank for a Pulsa-Jet carburetor has a fuel well (Fig. B4) that always contains a constant amount of fuel. Fuel passes through the long fuel tube to a diaphragm type pump on the side of the carburetor which transfers fuel from the fuel tank to the fuel well. The short fuel tube passes fuel from the fuel well to the mixture screw. With a constant amount of fuel present in the fuel well, the fuel mixture in the carburetor remains the same regardless of the amount of fuel in the tank, unlike the Vacu-Jet. On some Pulsa-Jet carburetors, the fuel pump is a part of the fuel tank mating surface with the carburetor.

Most Pulsa-Jet and Vacu-Jet carburetors are equipped with a plate type choke valve. Some models are equipped with an automatic choke plate that is actuated by a link attached to a diaphragm between the carburetor and fuel tank. See Fig. B5. A compression spring works against the diaphragm holding the choke plate in closed position when the engine is not running. See Fig. B6. A vacuum passage leads from

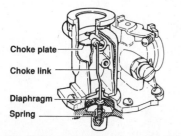

Fig. B5—Cutaway view of vacuum-operated automatic choke. See text and Figs. B6 and B7 for operation.

Fig. B8—View of bimetallic spring (S) used on some Pulsa-Jet and Vacu-Jet carburetors with an automatic choke. Refer to text.

the carburetor to a chamber under the diaphragm. When the engine starts, increased vacuum pulls down the diaphragm with the choke link, thereby opening the choke plate. See Fig. B7. This type automatic choke also operates if the engine loses engine speed under heavy load. The resulting loss of intake vacuum results in the choke plate partially closing, which provides a rich mixture so the engine will not stall. When engine speed increases, the increased vacuum returns the choke plate to the open position.

The automatic choke on some Pulsa-Jet and Vacu-Jet carburetors may be controlled by a bimetallic spring (S—Fig. B8) that is connected to the choke plate shaft. Air in the breather tube is

directed to the bimetallic spring cavity on the carburetor. Air temperature activates the spring which turns the choke shaft and plate. Cold air causes the spring to contract thereby closing the choke plate, while warm air opens the choke.

Some Pulsa-Jet and Vacu-Jet carburetors may be equipped with a slide type choke. When the choke slide is in (running position), a slot in the tube allows incoming air to flow into the carburetor. When the choke slide is out, incoming air is blocked and the mixture is enriched to enhance starting.

ADJUSTMENT. On all Pulsa-Jet and Vacu-Jet carburetors (except Pulsa-Jet with a fixed jet), the mixture screw controls high speed mixture setting. Turn mixture screw clockwise until it is lightly seated, then turn screw counterclockwise 1½ turns. Run engine until engine is at normal operating temperature. Be sure choke is open. Run engine at full throttle. Turn mixture screw clockwise until engine begins to stumble and note screw position. Turn mixture screw counterclockwise until engine begins to stumble and note screw position. Turn mixture screw clockwise to a position that is midway from the clockwise (lean) and counterclockwise (rich) positions. Adjust idle speed screw so engine speed is 1750 rpm. With engine running at idle, rapidly move speed

Fig. B9—Fixed jet Pulsa-Jet carburetors may be identified by the presence of threaded holes (H) in the inlet flange.

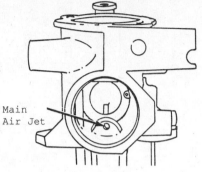

Fig. B10—On Pulsa-Jet carburetors with fixed main jet, the main jet air bleed can be removed to improve performance at high altitudes.

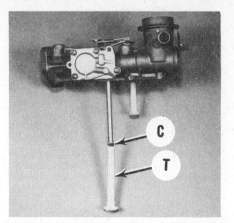

Fig. B12—To remove an extended plastic fuel tube (T), detach the retainer clip (C) then pull the plastic tube off the brass tube.

control to full throttle position. If engine stumbles or hesitates, slightly turn mixture screw counterclockwise and repeat test. Recheck idle speed and, if necessary, readjust idle speed screw.

The mixture screw on Pulsa-Jet carburetors with a fixed jet (carburetor is identified by threaded holes shown in Fig. B9) controls the idle mixture setting. Turn screw clockwise until it is lightly seated, then turn screw counterclockwise 1½ turns. Run engine until engine is at normal operating temperature. Be sure choke is open. Run engine with speed control in slow position and adjust idle speed screw so engine speed is 1750 rpm. Turn mixture screw clockwise until engine begins to stumble and note screw position. Turn idle mixture screw counterclockwise until engine begins to stumble and note screw position. Turn the idle mixture screw clockwise to a position that is midway from the clockwise (lean) and counterclockwise (rich) positions. With engine running at idle, rapidly move speed control to full throttle position. If engine stumbles or hesitates, slightly turn mixture screw counterclockwise and repeat test. Recheck idle speed and, if necessary, readjust idle speed screw.

NOTE: If engine performance is poor when operating at high altitudes, remove the Welch plug from the choke end of the carburetor and remove main air jet (Fig. B10).

R&R AND OVERHAUL. On suction type carburetors, it may be necessary to remove the carburetor and fuel tank as a unit from the engine before separating the carburetor from the fuel tank. Note the position of governor link and springs before removing the carburetor to ensure correct reassembly. Do not bend governor links or stretch springs.

Some engines are equipped with an automatic choke that uses a diaphragm (diaphragm is part of fuel tank gasket) and spring between the carburetor and fuel tank (Fig. B5). Before separating

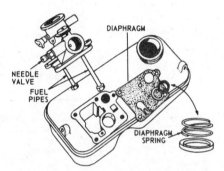

Fig. B11—View of fuel pump diaphragm and diaphragm spring found on the underside of some Pulsa-Jet carburetors.

carburetor from fuel tank, remove cover from carburetor body and detach the choke link (Fig. B5). Do not lose the choke spring after separating carburetor and fuel tank. The fuel pump on some Pulsa-Jet carburetors is located on the underside of the carburetor (fuel pump diaphragm is part of fuel tank gasket). The diaphragm spring and cup (Fig. B11) will be loose when the carburetor is separated from the fuel tank.

Note the following when disassembling the carburetor: DO NOT remove metal fuel feed tubes. Fuel feed tubes with a hex end can be unscrewed from the carburetor body. Fuel feed tubes with a round end can be removed by

grasping the tube with pliers and pulling the tube from the carburetor body.

NOTE: A new fuel feed tube must be the same length as the original tube. When ordering parts, measure and compare the old tube with the new tube.

If equipped with a brass fuel feed tube, the plastic pickup should be replaced if it cannot be cleaned in place with aerosol carburetor cleaner. If equipped with a long tube (Fig. B12), detach the retainer clip (C) then pull the plastic tube (T) from the brass tube. If equipped with a plastic pickup on the brass tube, drive the pickup off of the brass tube.

On Pulsa-Jet carburetors with a side-mounted fuel pump, unscrew and remove the pump cover (Fig. B13), then remove the diaphragm, spring and spring cup.

If equipped with an all-temperature choke, remove rubber elbow (E—Fig. B14), then force out the spring-shaft end of the choke with the spring by pushing against the inner end of the shaft. Note that the spring post on some

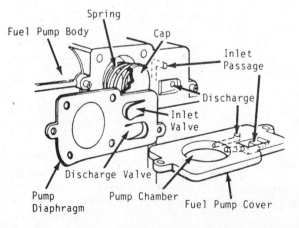

Fig. B13—Exploded view of Pulsa-Jet fuel pump mounted on side of carburetor.

Fig. B14—A rubber elbow (E) fits around the choke housing on carburetors with an all-temperature choke.

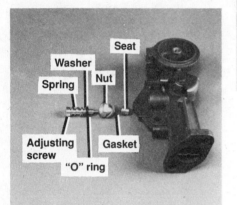

Fig. B15—Exploded view of fuel mixture screw assembly used on some carburetors.

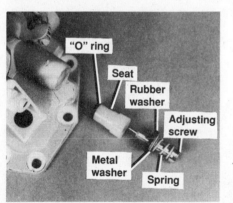

Fig. B16—Exploded view of fuel mixture screw assembly that uses a plastic seat.

carburetors must be dressed down to release the spring end.

Some carburetors are equipped with a plastic choke plate that is bonded to the plastic choke shaft. Use a tool with a sharp edge to cut through the bond along the edge of the shaft to separate shaft from plate.

Some carburetors are equipped with a removable idle mixture screw valve seat (Fig. B15).

Fig. B17—Some carburetors are equipped with a removable spiral insert (I).

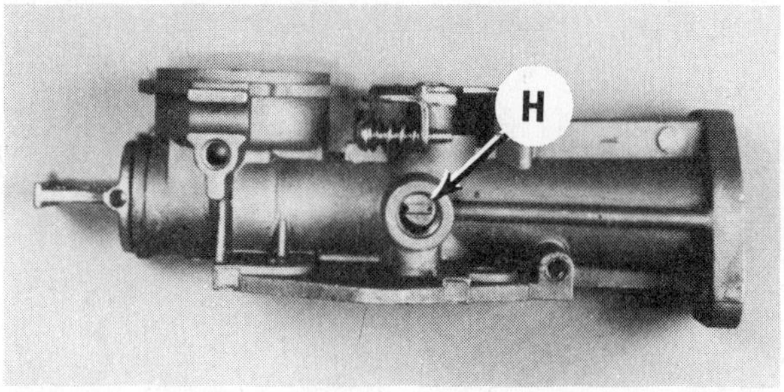

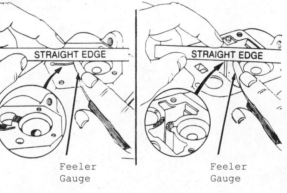

Fig. B18—Fuel metering holes (H) are accessible for cleaning after removing mixture screw. Be careful not to enlarge the holes when cleaning them.

Vacu-Jet Fuel Tank / Pulsa-Jet Fuel Tank — STRAIGHT EDGE — Feeler Gauge

Fig. B19—Use a straightedge and 0.002 inch (0.05 mm) feeler gauge to check machined surface of fuel tank for flatness. Measure at shaded areas shown in insets.

Some carburetors are equipped with a fuel mixture screw assembly that has a plastic seat (Fig. B16). Unscrew adjustment screw four or five turns then pull assembly out of carburetor.

If carburetor is equipped with Welch plugs, pierce the plug with a sharp punch, then pry out the plug. Be careful not to damage underlying metal.

If so equipped, spiral insert (I—Fig. B17) may be pulled from the carburetor bore.

Inspect the carburetor and components. Some carburetor bodies are made of plastic. They must not be soaked in carburetor cleaner for longer than 15 minutes. Metering holes (H—Fig. B18) in the mixture screw cavity are calibrated and should be cleaned with compressed air only. Do not enlarge or damage holes.

Discard any diaphragms that are torn, creased or otherwise damaged. On Pulsa-Jet carburetors, renew fuel pump diaphragm if flap valves (Fig. B13) are damaged. The body must be replaced if there is excessive throttle or choke shaft play as bushings are not available. All gaskets and seals should be replaced.

Check flatness of carburetor mounting surface on fuel tank. If surface is not flat, it is possible for gasoline to pass between the fuel tank surface and dia-

phragm and enter the carburetor vacuum passage. This will cause a rich air:fuel mixture. To check flatness, place a straightedge on the fuel tank and attempt to slide a 0.002 inch (0.05 mm) feeler gauge between the straightedge and the fuel tank surface as shown in Fig. B19. If the feeler gauge slides under the straightedge, discard the fuel tank. Do not attempt to flatten machined surface of fuel tank by filing. If the carburetor is not equipped with an automatic choke, Briggs & Stratton repair kit 391413 can be used to repair warped Pulsa-Jet fuel tank.

Note color of choke actuating diaphragm spring (S—Fig. B20) and measure spring length. Replace spring and diaphragm if colored red and length is not 1 1/8 to 1 7/32 inches (28.6-30.9 mm), if colored blue and length is not 1 5/16 to 1 3/8 inches (33.3-34.9 mm), or if colored

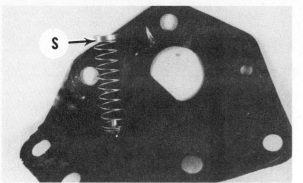

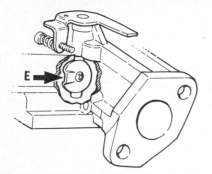

Fig. B20—Diaphragm spring (S) is color-coded to indicate length of spring.

Fig. B21—Install a throttle plate with a raised edge (E) so raised edge is toward mixture screw side of carburetor.

Fig. B23—If equipped with a choke plate that has dimples (D), install choke plate on choke shaft so dimples are up and hole in plate is toward throttle when choke is closed.

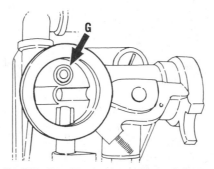

Fig. B22—Install a flat throttle plate so mark (M) is on fuel pump side of bore.

Fig. B24—If equipped with a spring-loaded valve on the choke plate, install choke plate so spring (G) is visible when choke plate is closed.

green and length is not 1⁷⁄₆₄ to 1³⁄₈ inches (28.2-34.9 mm).

Assemble the carburetor by reversing the removal procedure while noting the following: If so equipped, install throttle shaft seal so lip is out. If throttle plate has a raised edge (E—Fig. B21), install throttle plate with raised edge toward mixture screw side of carburetor. If throttle plate is flat, install throttle plate so mark (M—Fig. B22) is on fuel pump side of bore.

If so equipped, install spiral insert (I—Fig. B17) so end is parallel with fuel tank mounting surface and insert end is flush or just below mounting face of carburetor.

If carburetor is equipped with Welch plugs, apply a nonhardening sealer to

outer edge of plug and drive plug into hole with concave side toward carburetor. Be careful not to collapse the plug.

If equipped with a choke plate that has dimples (Fig. B23), install choke plate on choke shaft so dimples are up and hole in plate is toward throttle when choke is closed. Be sure to install the felt washer on the choke shaft before insertion into the carburetor.

If equipped with a spring-loaded valve (G—Fig. B24) on the choke plate, install choke plate so spring is visible when choke plate is closed.

If equipped with a manual choke using sector gears to rotate the choke, be sure choke operates properly after mating the gears.

If equipped with an all-temperature choke, position choke plate in closed position, then install choke shaft with bimetallic spring so outer spring end is above anchor pin (A—Fig. B25). Rotate spring end counterclockwise and attach to pin. If carburetor is plastic, use a warm soldering iron to flare end of spring anchor pin so spring end cannot slide off pin.

If equipped with a side-mounted fuel pump, install spring cup (C—Fig. B26) so smooth side is towards diaphragm. Tighten fuel pump cover retaining screws evenly in a crossing pattern.

If equipped with a brass fuel feed tube that has a plastic extension tube, heat the plastic tube in hot water before installation.

When installing the carburetor, note the following: If equipped with an all-temperature choke, use the following procedure when installing the carburetor on the fuel tank. Invert carburetor with choke diaphragm and spring in place. If equipped with a fuel pump, the fuel pump spring and spring cup (smooth side towards diaphragm) must be installed as well. Choke link should not be attached to choke lever. Place fuel tank on carburetor while guiding choke

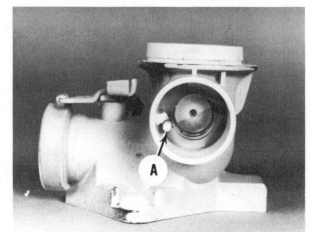

Fig. B25—If equipped with an all-temperature choke, position choke plate in closed position, then install choke shaft with bimetallic spring so outer spring end is above anchor pin (A).

Illustrations courtesy Briggs & Stratton Corp.

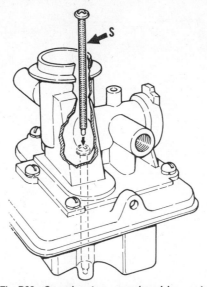

Fig. B26—Install fuel pump spring cup (C) so smooth side is towards diaphragm.

Fig. B27—While inverted, place fuel tank on carburetor while guiding choke diaphragm spring into well of fuel tank. See text.

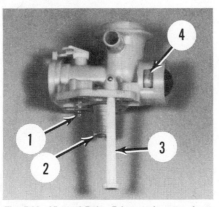

Fig. B28—On carburetors so equipped, be sure to install screw (S) located in carburetor bore.

Fig. B29—View of Pulsa-Prime carburetor showing location of fuel pump spring (1), screen (2), long fuel tube (3) and primer retainer tab (4).

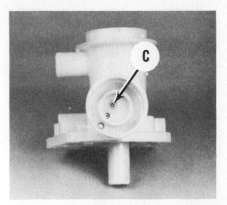

Fig. B30—View of check valve (C) on Pulsa-Prime carburetor.

Fig. B31—View of fuel pump diaphragm and valves (V).

diaphragm spring into well of fuel tank (Fig. B27). Install carburetor mounting screws, but do not tighten. Attach choke link to choke lever, hold choke plate in closed position and tighten carburetor mounting screws. Check choke operation.

On carburetors so equipped, be sure to install screw (S—Fig. B28) located in carburetor bore that is used to secure carburetor to fuel tank.

Pulsa-Prime Suction Type Carburetor. The Pulsa-Prime carburetor is mounted on the fuel tank and is equipped with a primer bulb.

OPERATION. The Pulsa-Prime carburetor has two fuel tubes. A screen (2—Fig. B29) covers the short fuel tube. The longer fuel tube (3) extends into the fuel tank. A diaphragm type fuel pump between the carburetor and fuel tank transfers fuel from the fuel tank into a fuel well that is a part of the fuel tank under the carburetor. The short fuel tube has a jet at the bottom. The tube extends into the fuel well and passes

fuel from the fuel well to the carburetor bore. There are no provisions for mixture adjustment on the Pulsa-Prime carburetor. Actuating the primer bulb forces fuel past check valves in the long fuel tube and primer cavity to prime the fuel pump as well as providing additional fuel for engine starting. Several pushes against the primer bulb are required to fill the fuel system if it is dry.

ADJUSTMENT. There are no adjustments on Pulsa-Prime carburetors.

R&R AND OVERHAUL. The fuel tank and carburetor should be removed and installed as a unit. Note the position of governor link and springs before removing the carburetor to ensure correct reassembly. Do not bend governor links or stretch springs.

After the fuel tank and carburetor are removed, the carburetor can be separated from the fuel tank. Carefully separate the carburetor from the fuel

tank so the diaphragm can be reused if undamaged.

When disassembling the carburetor, note the following: Use pliers to pull the long fuel tube out of the body. Press in tabs (4—Fig. B29) on primer bulb retainer ring and remove retainer ring and primer bulb. Using a suitable tool, extract the check valve seat (C—Fig. B30) in the primer cavity, then remove the check ball and spring. Do not deform the spring. The main jet in the bottom of the short fuel tube is permanently installed and should not be removed. Do not enlarge or damage the main jet hole. Replacement jets are not available.

The carburetor body is made of plastic. It must not be soaked in carburetor cleaner for longer than 15 minutes. Inspect carburetor and components. Discard diaphragm if torn, creased or otherwise damaged, also note condition of flap valves (V—Fig. B31). The body must be replaced if there is excessive throttle shaft play as bushings are not available. Replace long fuel tube or screen around short fuel tube if clogged or damaged.

When reassembling carburetor, note the following: Insert throttle plate in throttle shaft so hole (H—Fig. B32) in plate is out. Convex side of dimples in throttle plate should be toward breather tube side of carburetor.

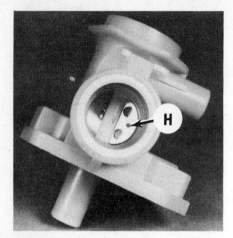

Fig. B32—Insert throttle plate in throttle shaft so hole (H) in plate is out.

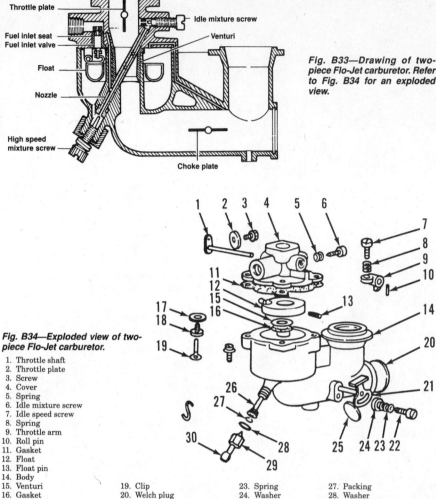

Fig. B33—Drawing of two-piece Flo-Jet carburetor. Refer to Fig. B34 for an exploded view.

Fig. B34—Exploded view of two-piece Flo-Jet carburetor.

1. Throttle shaft
2. Throttle plate
3. Screw
4. Cover
5. Spring
6. Idle mixture screw
7. Idle speed screw
8. Spring
9. Throttle arm
10. Roll pin
11. Gasket
12. Float
13. Float pin
14. Body
15. Venturi
16. Gasket
17. Fuel inlet seat
18. Fuel inlet valve
19. Clip
20. Welch plug
21. Choke shaft
22. Choke detent screw
23. Spring
24. Washer
25. Choke plate
26. Nozzle
27. Packing
28. Washer
29. Nut
30. High speed mixture screw

Install spring and check ball in primer cavity, then install ball seat so groove on seat is out (Fig. B30).

Moisten inside of primer cavity, then install primer bulb and retainer ring. Push in retainer ring until locking tabs engage slots on body.

Fuel pump spring (1—Fig. B29) fits around a peg on the underside of the carburetor.

To install the carburetor, place diaphragm on fuel tank, then place gasket on diaphragm. Install carburetor on fuel tank and evenly tighten screws in a crossing pattern. Apply a light coat of oil to "O" ring in carburetor bore, then install carburetor and fuel tank on engine.

Two-Piece Flo-Jet Carburetor. The two-piece Flo-Jet carburetor (Fig. B33) is a float type carburetor. The carburetor is identified by the fuel bowl that is integral with the carburetor body. Two castings make up the carburetor. The idle mixture screw and fuel inlet valve are located in the cover, which also serves as the attachment point for the float. The high speed mixture screw and nozzle are located in the body, although the nozzle protrudes into the cover to provide fuel for the idle circuit.

ADJUSTMENT. See Fig. B33 for location of idle mixture screw and high speed mixture screw. To adjust the mixture screws, proceed as follows: Turn idle mixture screw and high speed mixture screw clockwise until lightly seated. Turn idle mixture screw counterclockwise 1¼ turns. Turn high speed mixture screw counterclockwise 1½ turns. Run engine until at normal operating temperature. Be sure choke is open. Run engine with speed control in fast position. Turn high speed mixture screw clockwise until engine begins to

stumble and note screw position. Turn high speed mixture screw counterclockwise until engine begins to stumble and note screw position. Turn the high speed mixture screw clockwise to a position that is midway from the clockwise (lean) and counterclockwise (rich) positions. Run engine at idle and adjust idle speed screw so engine speed is 1750 rpm. Turn idle mixture screw clockwise until engine begins to stumble and note screw position. Turn idle mixture screw counterclockwise until engine begins to stumble and note screw position. Turn the idle mixture screw clockwise to a position that is midway from the clockwise (lean) and counterclockwise (rich) positions. With engine running at idle, rapidly move speed control to full throttle position. If engine stumbles or hesitates, slightly turn idle mixture screw counterclockwise and repeat test. Recheck idle speed and, if necessary, readjust idle speed screw.

OVERHAUL. Refer to Fig. B34 for an exploded view of carburetor. Before servicing the carburetor, check the

cover for distortion. Attempt to insert a 0.002 inch (0.05 mm) feeler gauge between the cover and body as shown in Fig. B35. If the feeler gauge can be inserted, then the body is warped or damaged and must be renewed.

When disassembling the carburetor, note that the nozzle (26—Fig. B34) is angled between the body and cover. The nozzle must be removed before the cover is removed, otherwise, the nozzle will be damaged.

Check the throttle shaft and bushing wear before removal. Play in bushings should not exceed 0.010 inch (0.25 mm), otherwise, the throttle shaft and/or bushings must be replaced.

The fuel inlet valve seat (17) may be a threaded or press-in type. If a threaded type, a screwdriver can be used to unscrew valve seat. If valve seat is a press-in type (identified by absence of a screwdriver slot) and replacement is required, thread a ¼-20 self-tapping screw or screw extractor into fuel inlet valve seat and remove valve seat.

Inspect carburetor and components. If throttle shaft bushings are worn and

Fig. B35—Attempt to insert a 0.002 inch (0.05 mm) feeler gauge between the cover and body as shown to check the cover for distortion.

Fig. B38—Removing the air jet from the right main air bleed hole (H) leans the high speed mixture, which may be desirable if the engine is operated at high altitudes. See text.

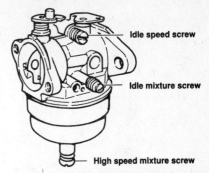

Fig. B39—View of Walbro LMS carburetor showing location of adjusting screws.

Fig. B36—Install throttle plate so dimples (D) are up.

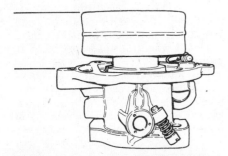

Fig. B37—Float should be parallel with fuel bowl mating surface on carburetor. Bend float tang to adjust float level.

must be replaced, use a ¼-inch tap or screw extractor to extract bushings. Push new bushings into carburetor, then insert throttle shaft and check for free shaft rotation. If binding occurs, use a ⁷⁄₃₂-inch drill bit as a line reamer by passing drill through both bushings.

When assembling carburetor, note the following: If equipped with a pressed-in fuel inlet valve seat, install valve seat so it is flush with carburetor body. Install throttle plate so dimples (D—Fig. B36) are up. If carburetor is equipped with a removable venturi, be sure to align holes in venturi and gasket during assembly.

After installing float on cover, check float level. Float should be parallel with fuel bowl mating surface on carburetor (Fig. B37). To adjust float level, bend float tang.

If Welch plug was removed, apply sealant to edge of Welch plug before installation, then swage carburetor body metal around edge of plug in several places to secure plug.

Walbro LMS Carburetor. The Walbro LMS carburetor is a float type carburetor. The carburetor can be identified by the letters "LMS" embossed on the carburetor mounting flange. Three versions of the LMS carburetor have been used. Some carburetors have both idle mixture and high speed mixture screws, some carburetors have a fixed jet instead of a high speed mixture screw, and some carburetors with a fixed jet use a primer system in place of the choke plate.

OPERATION. The Walbro LMS carburetor operates like other typical float type carburetors. If the engine is operated at high altitudes, improved engine performance may be obtained by removing the metal jet in the right main air bleed hole (H—Fig. B38), however, the air jet should be installed if engine is operated at lower altitudes. Removing the air jet leans the high speed mixture.

ADJUSTMENT. Refer to Fig. B39 for location of mixture adjusting screws. Some carburetors are not equipped with the adjustable high speed mixture screw on the bottom of the fuel bowl; a fixed main jet is located inside the carburetor.

To adjust mixture screws, proceed as follows: Turn idle mixture screw and high speed mixture screw, if so equipped, clockwise until they are lightly seated, then turn both screws

counterclockwise 1¼ turns. Run engine until normal operating temperature is attained. Be sure choke is open. Run engine with speed control in slow position and adjust idle speed screw so engine speed is 1750 rpm. Turn idle mixture screw clockwise until engine begins to stumble and note screw position. Turn idle mixture screw counterclockwise until engine begins to stumble and note screw position. Turn the idle mixture screw clockwise to a position that is midway from clockwise (lean) and counterclockwise (rich) positions.

If carburetor is equipped with a high speed mixture screw, run engine with speed control in fast position. Turn high speed mixture screw clockwise until engine begins to stumble and note screw position. Turn high speed mixture screw counterclockwise until engine begins to stumble and note screw position. Turn high speed mixture screw clockwise to a position that is midway from clockwise (lean) and counterclockwise (rich) positions. With engine running at idle, rapidly move speed control to full throttle position. If engine stumbles or hesitates, slightly turn idle mixture screw counterclockwise and repeat test. Recheck idle speed and, if necessary, readjust idle speed screw.

OVERHAUL. Disassembly of carburetor is evident after inspection and referral to Fig. B40. To remove the Welch plug, pierce the plug with a sharp punch, then pry out the plug, but do not damage underlying metal.

Inspect carburetor and renew any damaged or excessively worn components. The body must be replaced if there is excessive throttle or choke shaft play as bushings are not available. Apply fingernail polish or other suitable nonhardening sealant to outer edge of new Welch plug, then install plug using a pin punch that is slightly smaller in diameter than the plug. Be careful not to indent the plug; the plug should be flat after installation. Use a ³⁄₁₆-inch (5

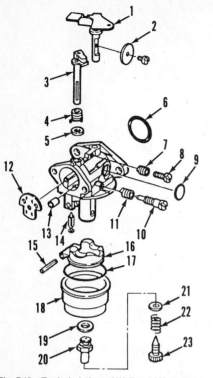

Fig. B40—Exploded view of Walbro LMS carburetor.

1. Throttle shaft
2. Throttle plate
3. Choke shaft
4. Washer
5. Seal
6. Gasket
7. Spring
8. Idle speed screw
9. Welch plug
10. Idle mixture screw
11. Spring
12. Choke plate
13. Main air jet
14. Fuel inlet valve
15. Float pin
16. Float
17. Gasket
18. Fuel bowl
19. Washer
20. Bowl retainer
21. "O" ring
22. Spring
23. High speed mixture screw

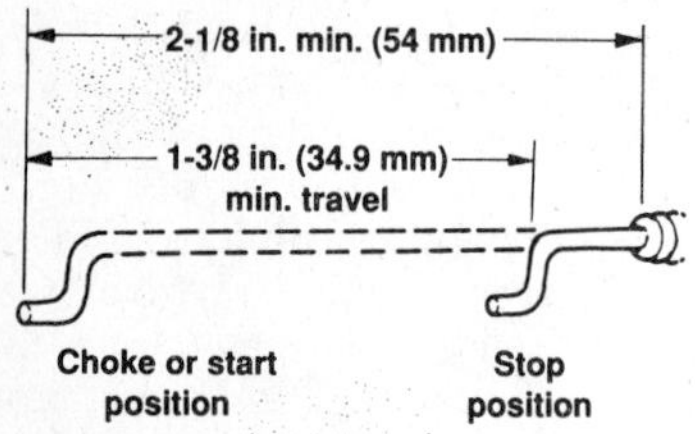

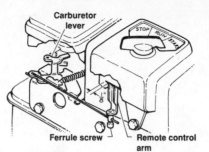

Fig. B42—Refer to text to adjust Choke-A-Matic control on models equipped with a dial speed control.

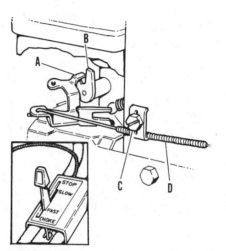

Fig. B43—If equipped with Choke-A-Matic control shown, refer to text for adjustment.

Tighten carburetor mounting screws to 75 in.-lbs. (8.5 N•m) torque.

CHOKE-A-MATIC CARBURETOR CONTROL. Some engines may be equipped with a control unit with which the carburetor choke, throttle and magneto grounding switch are operated from a single lever (Choke-A-Matic carburetors).

To check the operation of Choke-A-Matic engine control, proceed as follows: Move the speed control lever to the "CHOKE" or "START" position. Remove air cleaner and verify that choke slide or plate is completely closed. Move speed control lever to "RUN", "FAST" and "SLOW" positions; the choke should be open. Reinstall air cleaner. Start en-

gine. Move control lever to "STOP" position. The engine should stop running.

If equipped with a control cable between the remote control and engine, the remote control cable must move a certain distance for the remote control and carburetor to be synchronized. At full extension, control wire must extend 2⅛ inches (54 mm) from cable housing as shown in Fig. B41. The wire must travel at least 1⅜ inches (34.9 mm) from "CHOKE" or "START" to "STOP" positions.

If equipped with a dial type control (Fig. B42), adjust mechanism as follows: Move dial control to "START" position. Loosen ferrule screw shown in Fig. B42. Move carburetor lever fully clockwise. Position remote control arm so it is ⅛ inch (3.2 mm) from bracket. With carburetor lever and remote control arm in positions previously specified, tighten ferrule screw. Start engine. Move control lever to "STOP" position. The engine should stop running.

Several Choke-A-Matic configurations have been used on Briggs & Stratton engines. Refer to Figs. B43, B44 and B45 for some typical setups and the following list for adjustment points:

Fig. B43—On these units, lever (A) should just contact link or arm (B) when control is in "FAST" position. If not, loosen screw (C) and move control wire housing (D) as required, then tighten screw.

Fig. B44—On these units, adjustment is performed by bending choke link at point indicated by arrow so choke is closed when control is in "START" position.

Fig. B45—On these units, lever (A) should just contact choke lever (B) when lever (C) is in "FAST" detent. If not, loosen screws (D) and move control plate as required, then tighten screws.

GOVERNOR. Briggs & Stratton engines are equipped with either a pneumatic (air vane) or mechanical type governor system. Refer to Fig. B46 for a view of a typical air vane governor and speed control linkage used on vertical crankshaft engines. A view of a typical governor and speed control linkage used

Fig. B41—For proper operation of Choke-A-Matic controls, remote control wire must extend to dimension shown and have a minimum travel of 1-3/8 inches (34.9 mm).

mm) diameter rod to install the fuel inlet valve seat. The groove on the seat must be down (towards carburetor bore). Push in the seat until it bottoms. Install the choke plate so the numbers are visible when the choke is closed. Install the throttle plate so the numbers are visible and towards the idle mixture screw when the throttle is closed. The float level is not adjustable. If the float is not approximately parallel with the body when the carburetor is inverted, then the float, fuel valve and/or valve seat must be replaced. Tighten the fuel bowl retaining screw or high speed mixture nut to 50 in.-lbs. (5.6 N•m) torque.

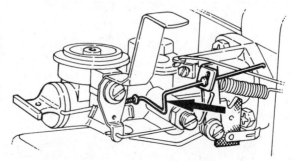

Fig. B44—Bend choke link at point indicated by arrow so choke is closed when control is in "START" position.

Illustrations courtesy Briggs & Stratton Corp.

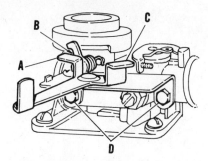

Fig. B45—If equipped with Choke-A-Matic control shown, refer to text for adjustment.

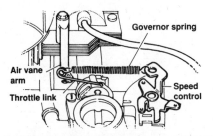

Fig. B46—View of a typical air vane type governor mechanism used on a vertical crankshaft engine.

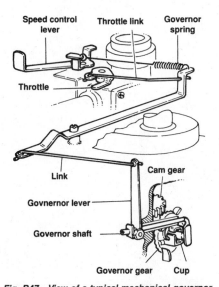

Fig. B47—View of a typical mechanical governor mechanism used on a vertical crankshaft engine.

on vertical crankshaft engines with a mechanical governor system is shown in Fig. B47. A view of a typical governor and speed control linkage used on horizontal crankshaft engines with a mechanical governor system is shown in Fig. B48.

NOTE: Observe the following when working on governor or speed control linkage. Before disconnecting linkage, mark linkage so it can be reassembled in its original configuration. Do not stretch governor spring during removal or installation. When attaching or detaching a governor spring end with a closed loop, use a twisting mo-

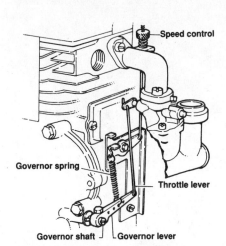

Fig. B48—View of a typical mechanical governor mechanism used on a horizontal crankshaft engine.

tion at spring end. Do not pull spring end with pliers or other tools as spring end may be deformed, which can affect governor operation.

ADJUSTMENT

AIR VANE GOVERNOR. Other than checking for disconnected or binding linkage, air vane governor system does not require adjustment.

MECHANICAL GOVERNOR. On some engines with a mechanical governor, idle speed is governed. Refer to previous sections in this chapter for adjustment of governed idle speed screw or tang.

On some engines, maximum governed speed is adjustable. The maximum governed speed is usually specified by the equipment manufacturer according to engine application.

On all engines with a mechanical governor, it may be necessary to adjust position of governor lever on governor shaft. Adjust position of governor lever assembly as follows: Loosen clamp bolt on governor lever (Fig. B49). Move governor lever (L) so carburetor throttle plate is in wide open position. Rotate governor shaft (S) as far as possible

clockwise on Models 80000 and 83000 or counterclockwise on all other models. Tighten clamp bolt.

IGNITION. All models are equipped with a Magnetron breakerless ignition system. All components are located outside the flywheel.

To check spark, remove spark plug and connect spark plug cable to B&S tester 19051, then ground remaining tester lead to engine. Spin engine at 350 rpm or more. If spark jumps the 0.165 inch (4.2 mm) tester gap, system is functioning properly.

Armature air gap should be 0.010-0.014 inch (0.25-0.36 mm) for Models 130000, 131000, 132000, 133000 and 135000 and 0.006-0.010 inch (0.15-0.25 mm). for all other models.

FLYWHEEL BRAKE. The engine may be equipped with a band type flywheel brake. The brake should stop the engine within three seconds when the operator releases the mower safety control and the speed control is in high speed position. Stopping time can be checked using tool 19255.

Refer to FLYWHEEL BRAKE in REPAIRS section for flywheel brake adjustment and service.

VALVE ADJUSTMENT. To correctly set tappet clearance, remove spark plug and breather/valve tappet chamber cover. Rotate crankshaft in normal direction (clockwise at flywheel) so piston is at top dead center on compression stroke. Continue to rotate crankshaft so piston is ¼ inch (6.4 mm) down from top dead center. This position places the tappets away from the compression release device on the cam lobes. Using a feeler gauge, measure clearance between intake and exhaust valve stem ends and the tappets.

The intake valve tappet gap should be 0.005-0.007 inch (0.13-0.18 mm). The exhaust valve tappet gap on Models 130000, 131000, 132000, 133000 and 135000 should be 0.009-0.011 inch (0.23-0.28 mm). The exhaust valve tap-

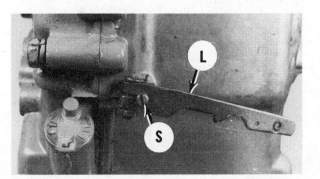

Fig. B49—The governor is adjusted by holding governor lever (L) while turning governor shaft (S) in direction specified in text.

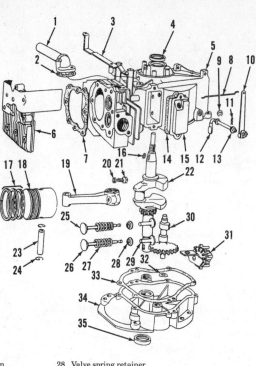

Fig. B50—Exploded view of typical vertical crankshaft engine. Mechanical governor components are not used on engines equipped with an air vane governor system.

1. Intake pipe
2. Gasket
3. Linkage lever
4. Oil seal
5. Cylinder block
6. Cylinder head
7. Cylinder head gasket
8. Governor link
9. "E" ring
10. Governor lever
11. Pin
12. Governor crank
13. Washer
14. Gasket
15. Breather & tappet chamber cover
16. Flywheel key
17. Piston rings
18. Piston
19. Connecting rod
20. Washer
21. Screw
22. Crankshaft

23. Piston pin
24. Clip
25. Intake valve
26. Exhaust valve
27. Valve spring

28. Valve spring retainer
29. Tappet
30. Camshaft
31. Governor & oil slinger assy.

32. Thrust washer
33. Gasket
34. Oil pan
35. Oil seal

pet gap for all other models should be 0.007-0.009 inch (0.18-0.23 mm).

NOTE: On some 90000 model series engines, the exhaust valve gap should be 0.005-0.007 inch (0.13-0.18 mm). These engines are identified by the specified gap stamped on the inside of the crankcase breather.

Valve tappet clearance is adjusted on all models by carefully grinding end of valve stem to increase clearance or by grinding valve seats deeper and/or renewing valve or tappet to decrease clearance.

CYLINDER HEAD. After 100 to 300 hours of engine operation, the cylinder head should be removed and any carbon or deposits should be removed.

REPAIRS

Refer to Fig. B50 for an exploded view of a typical vertical crankshaft engine or Fig. B51 for horizontal crankshaft engine.

TIGHTENING TORQUES. Recommended tightening torque specifications are as follows:
Connecting rod. . . . . 100 in.-lbs.
(11.3 N•m)

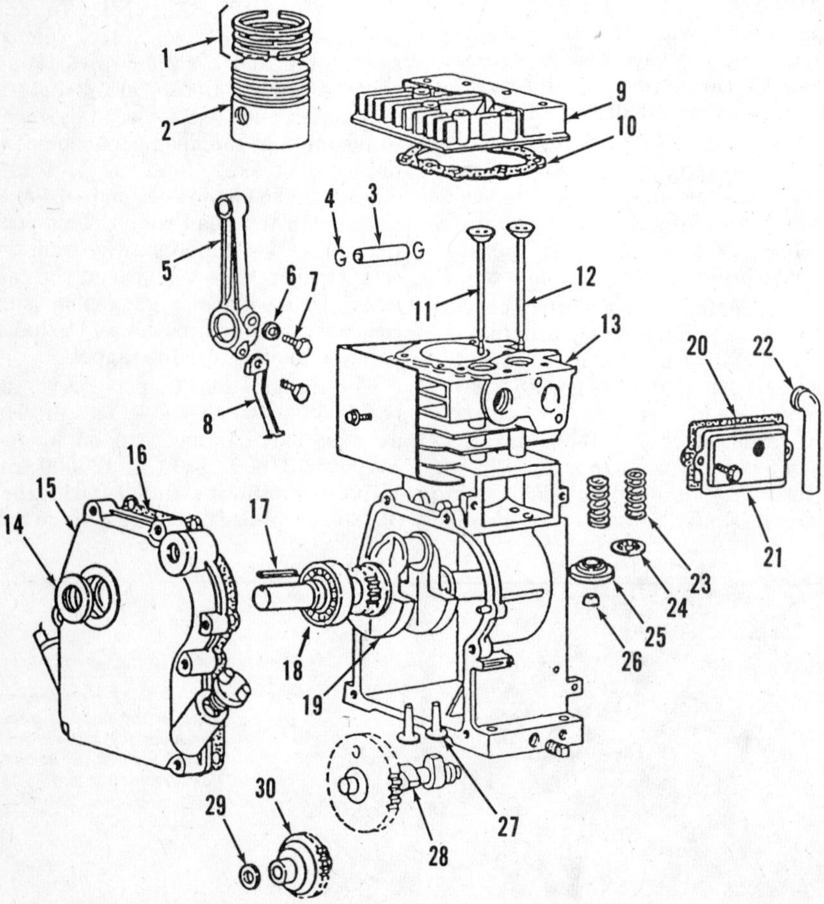

Fig. B51—Exploded view of typical horizontal crankshaft engine. Engine may be equipped with ball bearing (18).

1. Piston rings
2. Piston
3. Piston pin
4. Clip
5. Connecting rod
6. Washer
7. Screw
8. Oil dipper
9. Cylinder head
10. Cylinder head gasket
11. Exhaust valve
12. Intake valve
13. Cylinder block
14. Oil seal
15. Crankcase cover
16. Gasket

17. Key
18. Ball bearing
19. Crankshaft
20. Gasket
21. Breather & tappet chamber cover
22. Breather tube
23. Valve spring
24. Valve spring retainer
25. Valve rotater
26. Valve keys
27. Tappet
28. Camshaft
29. Thrust washer
30. Governor assy.

Fig. B52—Long cylinder head retaining screws (S) are used in positions shown.

Crankcase cover or oil pan:
  130000, 131000,
  132000, 133000,
  135000 . . . . 120 in.-lbs.
                         (13.6 N•m)
  All other models . . . . 85 in.-lbs.
                         (9.6 N•m)
Cylinder head . . . . 140 in.-lbs.
                         (15.8 N•m)
Flywheel nut or starter clutch:
  130000, 131000,
  132000, 133000,
  135000 . . . . 65 ft.-lbs.
                         (7.3 N•m)
  All other models . . . . 55 ft.-lbs.
                         (6.2 N•m)
Spark plug . . . . 140-200 in.-lbs.
                     (15.8-22.6 N•m)

CYLINDER HEAD. When removing cylinder head, note location and lengths of cylinder head retaining screws so they can be installed in their original positions.

Always install a new cylinder head gasket. Do not apply sealer to head gasket. Note position of three long screws (S—Fig. B52). Lubricate screws with graphite grease and tighten screws in sequence shown in Fig. B53 to 140 in.-lbs. (15.8 N•m) torque.

VALVE SYSTEM. Valve face and seat angles should be ground at 45°. Renew valve if margin is $\frac{1}{16}$ inch (1.6

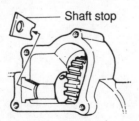

Fig. B53—Tighten cylinder head screws in sequence shown.

Cover plate

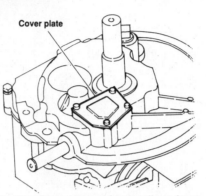

Fig. B54—Remove cover plate on oil pan for access to auxiliary drive shaft and gear.

mm) or less. Seat width should be $\frac{3}{64}$ to $\frac{1}{16}$ inch (1.2-1.6 mm). All models are equipped with seat inserts. Peen around insert to secure insert in engine.

The valves operate directly in valve guide bores in the aluminum crankcase. If original valve guide bores are excessively worn, service guides may be installed. Maximum allowable valve guide inside diameter is 0.266 inch (6.76 mm). If B&S valve guide gauge 19122 can be installed to a depth of $\frac{5}{16}$ inch (7.9 mm) in valve guide, then a new guide should be installed. Ream guide out first with B&S reamer 19064 to a depth approximately $\frac{1}{16}$ inch (1.6 mm) deeper than length of service guide. Press in service guide with B&S driver 19065, then ream using B&S finish reamer 19066. Reface valve seats after new guides are installed.

Some engines may be equipped with a heavy duty exhaust valve or exhaust valve seat made of Cobalite. The Cobalite exhaust valve is marked on the valve head with letters "TXS," "XS" or "PP-XS." The Cobalite exhaust valve

and valve seat can be installed on all engines. Some engines may be equipped with a valve rotator on the exhaust valve.

OIL SUMP REMOVAL, AUXILIARY PTO MODELS. On models equipped with an auxiliary pto, one of the oil sump (engine base) to cylinder retaining screws is installed in the recess in the sump for the pto auxiliary drive gear. To remove the oil sump, remove the cover plate (Fig. B54) and then remove the shaft stop (Fig. B55). The gear and shaft can then be moved as shown to allow removal of the oil pan retaining screws. Reverse procedure to reassemble.

CAMSHAFT. The camshaft rides directly in the aluminum bore of the crankcase, oil pan or crankcase cover. The camshaft gear and shaft are integral, except on models which have a camshaft constructed of metal and plastic.

On Models 60000 and 80000 fitted with an auxiliary drive gear on the camshaft, replace camshaft if journal diameter for end that fits into crankcase is 0.498 inch (12.65 mm) or less, or if journal diameter near worm drive gear is 0.751 inch (19.08 mm) or less. On Models 110000, 111000, 112000, 113000 and 114000, replace camshaft if journal diameter for end that fits into crankcase is 0.436 inch (11.07 mm) or less, or if journal diameter nearer cam gear is 0.498 inch (12.65 mm) or less. Replace camshaft on all other models if either camshaft journal is 0.498 inch (12.65 mm) or less.

Replace camshaft if cam lobe height of either cam lobe is equal or less than following dimension:
Cam Lobe Reject Height
  60000, 80000,
  82000, 83000, 90000,
  91000, 92000, 93000,
  94000, 95000, 96000 . . . . . 0.883 in.
                         (22.43 mm)

  110000, 111000,
  112000, 113000,
  114000 . . . . 0.870 in.
                         (22.10 mm)

  130000, 131000,
  132000, 133000,
  135000 . . . . 0.950 in.
                         (24.13 mm)

Shaft stop

Oil pan screw

Gear and shaft

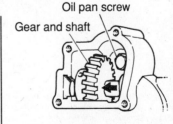

Fig. B55—For access to oil pan retaining screw, pull out shaft stop and move gear in direction of arrow.

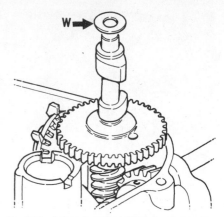

Fig. B56—View of yoke type compression release mechanism.

Fig. B59—A thrust washer (W) is located on the end of the auxiliary drive shaft. Shaft rotation determines washer location. See text.

Inspect camshaft bearing surfaces in crankcase and oil pan (on vertical crankshaft engines) or crankcase cover (on horizontal crankshaft engines). On Models 60000 and 80000 fitted with an auxiliary drive gear on camshaft, replace crankcase if camshaft bearing diameter in crankcase is greater than 0.504 inch (12.80 mm), or if camshaft bearing diameter in oil pan is 0.757 inch (19.23 mm) or more. On model series 110000, 111000, 112000, 113000 and 114000, replace crankcase, oil pan or crankcase cover if camshaft bearing diameter in crankcase is 0.443 inch (11.25 mm) or more, or if camshaft bearing diameter in oil pan or crankcase cover is greater than 0.504 inch (12.80 mm). Replace crankcase, oil pan or crankcase cover on all other models if camshaft bearing diameter is greater than 0.504 inch (12.80 mm).

Some camshafts may be equipped with a compression release mechanism: either the yoke type shown in Fig. B56 or the cam-weight type shown in Fig. B57. The mechanism should move freely without binding. No individual components are available.

Some camshafts are also equipped with "Easy-Spin" starting. The intake cam lobe is designed to hold the intake valve slightly open on part of the compression stroke. To check compression, the crankshaft must be turned backwards.

"Easy-Spin" camshafts (cam gears) can be identified by two holes drilled in web of gear. Where part number of an older cam gear and an "Easy-Spin" cam gear are same (except for an "E" following "Easy-Spin" part number), gears are interchangeable.

Some "Easy-Spin" camshafts are also equipped with a mechanically operated compression release on the exhaust lobe. With engine stopped or at cranking speed, the spring holds the actuator cam weight inward against the rocker cam. See Fig. B57. The rocker cam is held slightly above the exhaust cam surface, which in turn holds the exhaust valve open slightly during compression stroke. This compression release greatly reduces power needed for cranking.

When the engine starts and rpm is increased, the actuator cam weight moves outward overcoming spring pressure. See Fig. B58. The rocker cam is rotated below the cam surface to provide normal exhaust valve operation.

On some engines, a worm gear on the camshaft drives an auxiliary drive shaft that is used on self-propelled lawn mowers. The camshaft, worm gear and oil slinger are available only as a unit assembly. A thrust washer (W—Fig. B59) is used to accommodate additional axial thrust of the camshaft. The thrust washer is located at flywheel end of camshaft if auxiliary drive shaft rotation is clockwise. The thrust washer is located at worm gear end of camshaft if auxiliary shaft rotation is counterclockwise.

PISTON, RINGS AND PIN. Two different types of piston are used depending on the metal used for the cylinder bore. The piston used in an aluminum cylinder bore is chrome plated while the piston used in an iron sleeve is tin plated. The tin plated piston is marked with an "L" on the piston crown (Fig. B60). The pistons are not interchangeable.

Piston diameter sizes are not specified by Briggs & Stratton. The cylinder bore should be measured as outlined in the CYLINDER section.

To check piston ring grooves for wear, install new piston ring in top groove and measure side clearance between ring and ring groove land using a feeler gauge. Renew piston if top piston ring side clearance is 0.007 inch (0.18 mm) or more.

Piston pin is available in a standard size as well as 0.005 inch (0.13 mm) oversize. Renew piston pin if it is out-of-round by 0.0005 inch (0.013 mm) or

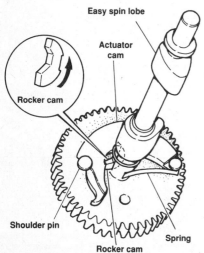

Fig. B57—View of cam-weight type of compression release mechanism. At cranking speed, spring holds actuator cam inward against the rocker cam and rocker cam is forced above exhaust cam surface.

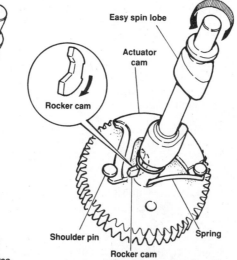

Fig. B58—When engine starts and rpm is increased, actuator cam weight moves outward allowing rocker cam to rotate below exhaust cam surface.

Illustrations courtesy Briggs & Stratton Corp.

Fig. B60—A tin plated piston is marked with an "L" on the piston crown. The piston is used in an iron cylinder bore.

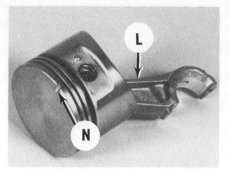

Fig. B62—If piston crown is notched (N), assemble connecting rod and piston as shown while noting relation of long side of rod (L) and notch (N) in piston crown.

Renew connecting rod on Model 60000 if big end diameter is equal to or greater than 0.876 inch (22.25 mm). Renew connecting rod on all other models if big end diameter is equal to or greater than 1.001 inch (25.43 mm).

The connecting rod should be replaced or the piston pin hole in the connecting rod should be reamed to accept an oversize piston pin if the hole diameter is equal to or greater than 0.492 inch (12.50 mm).

The piston pin hole in the connecting rod, as well as the piston, can be enlarged with a reamer, presuming it is standard size, to accept a piston pin that is 0.005 inch (0.13 mm) larger in diameter than the standard piston pin.

Tighten connecting rod screws to 100 in.-lbs. (11.3 N•m) torque.

GOVERNOR. Engines that use a mechanical governor are equipped with a flyweight assembly that is driven by the camshaft gear.

The flyweight assembly on most horizontal crankshaft engines is mounted on a stub shaft on the crankcase cover. See Fig. B63. The flyweight assembly is available only as a unit assembly. The flyweight assembly can be removed by pulling it off of the shaft. A thrust washer (W—Fig. B64) is located between the flyweight assembly and crankcase cover boss. Wire clip (C—Fig. B65) fits around the shaft and holds the

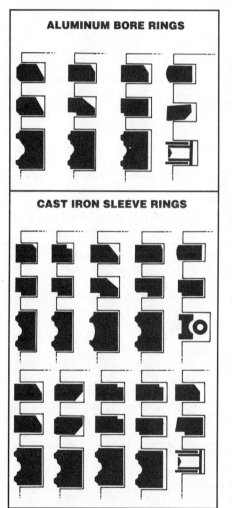

Fig. B61—View showing correct location of piston rings for typical ring sets used on engines in this section.

Discard piston ring on engines with an iron sleeve if ring end gap is 0.030 inch (0.76 mm) or more for a compression ring or 0.035 inch (0.89 mm) for the oil control ring. Discard piston ring on aluminum bore engines if ring end gap is 0.035 inch (0.89 mm) or more for a compression ring or 0.045 inch (1.14 mm) for the oil control ring.

When installing piston rings, note that top side (toward the piston crown) of some compression rings is marked with a dot to identify the top side of the ring. Several piston ring configurations have been used. Refer to the information supplied with the new piston rings for proper ring placement. The chart in Fig. B61 shows the correct location of some piston ring combinations.

The oil control ring may be a one-piece ring or an assembled unit. The assembled oil control ring may consist of two pieces—the control ring plus an expander ring, or three pieces—two rails and an expander ring.

New pistons and piston rings are available in standard size as well as various oversizes to fit a resized cylinder. Be sure the proper size is available before machining or purchasing components.

The piston and connecting rod can be assembled in either direction, except on pistons with a notch used in vertical crankshaft engines. If the piston on a vertical crankshaft engine has a notch in the piston crown, then the piston and connecting rod must be assembled so the notch (N—Fig. B62) and long side of the rod (L) are positioned as shown in Fig. B62.

CONNECTING ROD. The connecting rod rides directly on the crankshaft crankpin. The mating surfaces of the rod and cap are serrated and must be correctly assembled, otherwise the cap will be improperly aligned on the rod. Inspect the connecting rod for excessive wear and damage.

Fig. B63—View of governor flyweight assembly (G) on the crankcase cover of a horizontal crankshaft engine.

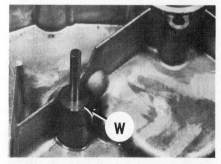

Fig. B64—A thrust washer (W) is located on the governor stub shaft adjacent to the cover boss.

more. The standard size piston pin should be renewed if it is worn to a diameter of 0.489 inch (12.42 mm) or less.

The piston should be renewed or the piston pin hole in the piston should be reamed to accept an oversize piston pin if the hole diameter is equal to or greater than 0.492 inch (12.50 mm).

Fig. B65—Wire clip (C) fits in a groove on the stub shaft to hold the flyweight assembly in place.

Fig. B66—The spool should fit between the flyweight fingers as shown.

Fig. B67—View of governor and oil slinger assembly used on vertical crankshaft engines so equipped.

hub in position. The stub shaft is not removable.

Inspect flyweight assembly for broken components, worn gear teeth and excessive clearance on stub shaft. When installing flyweight assembly, note that flange on thrust sleeve must engage flyweight fingers as shown in Fig. B66.

The flyweight assembly on vertical crankshaft engines is mounted on end of camshaft along with oil slinger (Fig. B67). The oil slinger and flyweight assembly are only available as a unit assembly. Inspect flyweight assembly for broken components.

CRANKSHAFT AND MAIN BEARINGS. Some crankshafts are equipped with a removable crankshaft gear that is driven by a pin or key in the crankshaft. It may be necessary to pry off gear. Note that gear must be in-

Fig. B68—Install crankshaft gear so timing mark (M) is out.

stalled on crankshaft so side with timing mark (Fig. B68) is visible (chamfered side is towards crankshaft counterweight). A crankshaft drive pin is not removable, while a crankshaft drive key can be removed.

Inspect all mating surfaces on crankshaft for indications of scoring, scuffing and other damage. Renew crankshaft on Model 60000 if crankpin journal diameter is equal to or less than 0.870 inch (22.10 mm). Renew crankshaft on all other models if crankpin journal diameter is equal to or less than 0.996 inch (25.30 mm).

On engines so equipped, inspect ball bearing(s) for roughness, pitting, galling and excessive wear. Bearings are a press fit on crankshaft. The crankshaft should be replaced if bearings are loose on journals. Ball bearing removal and installation can be performed by a shop equipped with a press. The new bearing must be installed so bearing side with metal shield is towards counterweight. An alternate method for bearing installation, is to warm bearing in oil that is heated to no more than 250° F (121° C). The bearing must not contact the container. While hot, the bearing can be slipped down the crankshaft into place.

Renew crankshaft on Models 130000, 131000, 132000, 133000 and 135000 if the pto crankshaft journal diameter is equal to or less than 0.998 inch (25.35 mm), or if the flywheel end crankshaft journal diameter is equal to or less than 0.873 inch (22.17 mm). Renew crankshaft on all other models if either crankshaft main journal diameter is equal to or less than 0.873 inch (22.17 mm).

NOTE: All models equipped with an auxiliary drive unit have a rejection size for the main bearing journal at the output end of 0.998 inch (25.35 mm).

If not equipped with a ball bearing main bearing, the crankshaft rides directly in the aluminum of the crankcase and oil pan (on vertical crankshaft engines) or crankcase cover (on horizontal crankshaft engines), or the crankshaft

rides in a renewable bushing. On some engines, a bushing can be installed if the aluminum is excessively worn or damaged. Most engines can be repaired using a bushing, but check parts availability on a specific engine.

Tools for reaming and installing service bushings are available from Briggs & Stratton. If the bearings are scored, out-of-round 0.0007 inch (0.018 mm) or more, or are worn larger than the reject size, ream the bearing and install a service bushing.

Main bearing reject size on Models 130000, 131000, 132000, 133000 and 135000 is 1.003 inches (25.48 mm) for the crankcase cover or oil pan bearing and 0.878 inch (22.30 mm) for the crankcase bearing. Main bearing reject size on all other models is 0.878 inch (22.30 mm) for either bearing.

NOTE: All models equipped with an auxiliary drive unit have a rejection size for the main bearing in the oil pan of 1.003 inch (25.48 mm).

Install steel-backed aluminum service bushing as follows. Prior to installing bushing, use a chisel to make an indentation in inside edge of bearing bore in crankcase. Install bushing so oil notches are properly aligned and bushing is flush with outer face of bore. Stake bushing into previously made indentation and finish ream bushing. Do not stake where bushing is split.

When installing "DU" type bushing, stake bushing at oil notches in crankcase, but locate bushing so bushing split is not aligned with an oil notch. Bushing should be 1/32 inch (0.8 mm) below face of crankcase bore.

Crankshaft end play is 0.002-0.030 inch (0.05-0.76 mm) on 92500 and 92900 models with a next-to-last digit of "5" in the code number. Crankshaft end play is 0.002-0.008 inch (0.05-0.20 mm) for all other models. At least one 0.015 inch crankcase gasket must be in place when measuring end play. Additional gaskets in several sizes are available to aid in end play adjustment. If end play is excessive, place shims between crankshaft gear and crankcase on plain bearing models, or on flywheel side of crankshaft if equipped with a ball bearing.

Refer to VALVE TIMING section for proper timing procedure when installing crankshaft.

CYLINDER. If cylinder bore wear is 0.003 inch (0.08 mm) or more or is 0.0025 inch (0.06 mm) or more out-of-round, cylinder must be bored to next larger oversize.

The standard cylinder bore sizes for each model are given in the following table.

STANDARD CYLINDER BORE SIZES

Model	Cylinder Diameter
60000, 80000, 82000, 83000	2.3740-2.3750 in. (60.30-60.33 mm)
90000, 91000, 92000, 93000, 94000, 95000, 96000	2.5615-2.5625 in. (65.06-65.09 mm)
110000, 111000, 112000, 113000, 114000	2.7802-2.7812 in. (70.62-70.64 mm)
130000, 131000, 132000, 133000, 135000	2.5615-2.5625 in. (65.06-65.09 mm)

A rigid hone is recommended for resizing cylinders. Operate hone at 300-700 rpm and with an up and down movement that will produce a 45° crosshatch pattern. Always check availability of oversize piston and ring sets before honing cylinder.

NOTE: A chrome piston ring set is available for slightly worn standard bore cylinders. No honing or cylinder deglazing is required for these rings. The cylinder bore can be a maximum of 0.005 inch (0.01 mm) oversize when using chrome rings.

VALVE TIMING. On engines equipped with ball bearing mains, align the timing mark on the cam gear with the timing mark on the crankshaft counterweight as shown in Fig. B69. On engines with plain bearings, align the timing mark on the camshaft gear with the timing mark on the crankshaft gear as shown in Fig. B70. On engines with a plastic camshaft gear, the dimple on the crankshaft gear tooth should be positioned between the gear teeth on the

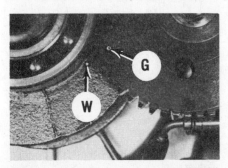

Fig. B69—Align timing mark on cam gear (G) with mark on crankshaft counterweight (W) on ball bearing equipped models.

Fig. B70—Align timing marks on cam gear and crankshaft gear on plain bearing models.

Fig. B71—On engines with a plastic camshaft gear, the dimple on the crankshaft gear tooth should be positioned between the gear teeth on the camshaft gear marked by the two lines.

camshaft gear marked by the two lines (Fig. B71). Note that other marks (such as "O") on the camshaft gear face are molding impressions.

AUXILIARY PTO. To remove auxiliary pto shaft and gear, refer to Fig. B72 and remove shaft stop, if not previously removed. Drive pin out of gear and shaft, then withdraw shaft.

Some model series 110000 engines are equipped with a clutch mechanism on the camshaft to engage the pto shaft assembly. When spring tang of clutch sleeve (4—Fig. B73) is pushed, worm should lock. With spring released, worm should rotate freely in both directions. When assembling clutch and worm, upper end of spring (3) must engage hole in camshaft while lower end of spring must engage hole in clutch sleeve (4). Install lower copper washer (8) so gray coated side is next to thick thrust

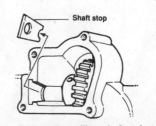

Fig. B72—To remove auxiliary shaft and gear, pull out shaft stop, if not previously removed, and drive pin out of gear hub.

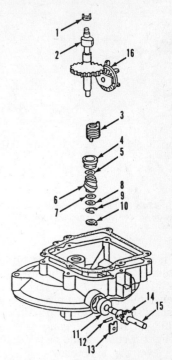

Fig. B73—Exploded view of auxiliary pto shaft assembly used on some model series 110000 engines.

1. Clip washer	9. "E" ring
2. Camshaft assy.	10. Washer
3. Spring	11. Seal
4. Clutch sleeve	12. Pin
5. Thrust washer	13. Shaft stop
6. Worm gear	14. Worm gear
7. Thrust washer (thick)	15. Pto shaft
8. Copper washer	16. Governor & oil slinger assy.

washer (7). Worm gear end play should be 0.004-0.017 inch (0.10-0.43 mm). Be sure clip of washer (10) properly engages camshaft bearing boss.

FLYWHEEL BRAKE. The engine may be equipped with a band type flywheel brake. To check brake band adjustment, remove starter and properly ground spark plug lead to prevent accidental starting. On electric start models, remove battery. Turn flywheel nut using a torque wrench with brake engaged. Rotating flywheel nut at a steady rate in a clockwise direction should require at least 45 in.-lbs. (5.08 N•m) torque. An insufficient torque reading may indicate misadjustment or damaged components. Renew brake band if friction material of band is damaged or less than 0.030 inch (0.76 mm) thick.

To adjust brake, unscrew brake cable retaining screw so hole (H—Fig. B74) is vacant. (If a pop rivet secures the cable, do not remove rivet; gauge tool can be inserted through hole in rivet.) Loosen screws (S) securing brake control bracket. Place bayonet end of gauge 19256 (G) in hole (L) of control lever (C). Push against control lever and insert opposite end of gauge (G) in hole (H), or

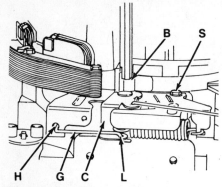

Fig. B74—To adjust flywheel brake, loosen bracket screws (S) and position gauge (G) as outlined in text.

in hole of pop rivet if used to secure cable. Apply pressure against bracket until tension on gauge is just removed, then while holding pressure, tighten bracket screws (S) to 25-30 in.-lbs. (2.82-3.39 N•m) torque. As gauge is removed a slight friction should be felt and control lever should not move.

Brake band must be renewed if damaged or if friction material thickness is less than 0.030 inch (0.76 mm). Use tool 19229 to bend retaining tang away from band end (B—Fig. B74).

FLYWHEEL. The flywheel is secured to the crankshaft by a retaining nut or by the starter clutch on engines equipped with a rewind starter on the blower housing. To remove flywheel, remove blower housing and any other components so flywheel is accessible. If equipped with flywheel brake, disconnect brake spring. Secure the flywheel from rotating using Briggs & Stratton or aftermarket flywheel holder, then unscrew starter clutch or retaining nut. Use a suitable puller to remove flywheel from crankshaft.

NOTE: Do not attempt to hold the flywheel by inserting a tool between the fins as damage to flywheel may result.

The tapered portion of flywheel and crankshaft mating surfaces must be clean and smooth with no damage. Renew flywheel if any cracks are evident or any fins are broken. Be sure the keyway in crankshaft and flywheel is not damaged or worn. If flywheel key is replaced, a key made of aluminum must be installed. DO NOT install a steel key.

Install flywheel and key and tighten retaining nut or starter clutch to torque specified in TIGHTENING TORQUES section.

Fig. B75—View of starter clutch used on engines with a rewind starter mounted on the blower housing.

REWIND STARTER. Rewind Starter On Blower Housing. Most "L" head Briggs & Stratton engines are equipped with a rewind starter that is mounted on the blower housing. In some cases, a detachable plastic ring surrounds the starter housing. The rope pulley drives a ball-type sprag clutch (Fig. B75) mounted on the flywheel. When the rewind starter is operated, the flywheel rotates and one of the balls will engage a sprag. When the engine runs, the balls are thrown out by centrifugal force to a disengaged position.

To renew a broken rewind spring, proceed as follows: Grasp free outer end of spring (S—Fig. B76) and pull broken end from starter housing. With blower housing removed, bend up tangs (T) and remove starter pulley from housing. Untie knot in rope and remove rope and inner end of broken spring from pulley. Install rewind spring by threading in-

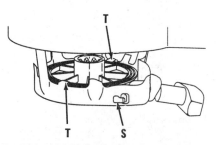

Fig. B76—View of outer rewind spring end (S) and pulley retaining tangs (T).

Fig. B77—Align inner rope hole (H) with rope outlet in housing before inserting rope.

ner end of spring through notch in starter housing, then engage inner end of spring in pulley hub. Apply a small amount of grease on inner face of pulley and place pulley in housing. Bend the pulley retainer tangs (T—Fig. B76) towards the pulley so the gap between the tang and the pulley is 1/16 inch (1.6 mm). Insert a 3/4 inch bar in pulley hub and turn pulley in a counterclockwise direction until spring is tight. Rotate pulley so rope hole (H—Fig. B77) in pulley is aligned with rope outlet hole in housing, then secure pulley so it cannot rotate. Hook a wire in inner end of rope and thread rope through guide and hole in pulley; then, tie a knot in rope and release the pulley allowing spring to wind rope into pulley groove.

To renew starter rope only, it is not generally necessary to remove starter pulley and spring. Wind up the spring and install new rope as outlined in preceding paragraph.

To disassemble starter clutch unit, refer to Fig. B78 and proceed as follows: Remove rotating screen (2) and starter ratchet cover (4). Lift ratchet (5) from housing and crankshaft and extract the steel balls (6). If necessary to remove housing (7), hold flywheel and unscrew housing in counterclockwise direction using B&S tool 19224. When installing housing, be sure spring washer (8) is in place on crankshaft with cup (concave) side towards flywheel; then, tighten housing securely. Inspect felt seal and plug in outer end of ratchet (Fig. B79). Renew ratchet if seal or plug is damaged as these parts are not serviced separately. Lubricate the felt with oil and place ratchet on crankshaft. Insert the steel balls and install ratchet cover, rubber seal and rotating screen.

Vertical-Pull Rewind Starter. Some "L" head Briggs & Stratton en-

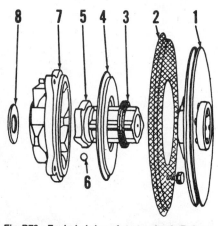

Fig. B78—Exploded view of starter clutch. Refer to Fig. 79 for cutaway view of ratchet (5).

1. Rope pulley	6. Steel balls
2. Debris screen	7. Clutch housing
3. Rubber seal	(flywheel nut)
4. Ratchet	8. Spring washer

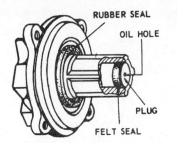

Fig. B79—Cutaway view showing felt seal and plug in end of starter ratchet (5—Fig. B78).

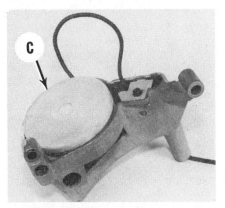

Fig. B80—Pull out rope as shown and rotate pulley counterclockwise to release spring tension. Cover (C) encloses rewind spring.

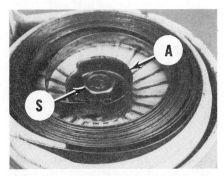

Fig. B81—Anchor screw (S) retains spring anchor (A).

gines are equipped with a vertical-pull rewind starter that is mounted on the side of the engine. When the starter rope is pulled, the starter gear inside the starter housing travels up a helix to engage the ring gear on the flywheel.

To renew rope or spring, first remove all spring tension from rope. Extract rope from starter as shown in Fig. B80, then wind rope and pulley counterclockwise three turns to remove spring tension. Carefully pry off the plastic spring cover (C). Refer to Fig. B81 and remove anchor screw (S) and spring anchor (A). Carefully remove spring from housing. Unscrew and remove rope guide (G—Fig. B82), then remove pulley and gear assembly (Fig. B83).

It is not necessary to remove the gear retainer unless pulley or gear is dam-

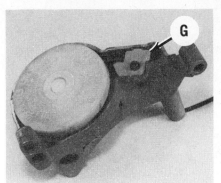

Fig. B82—Unscrew rope guide for access to friction link (L—Fig. B85).

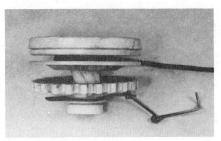

Fig. B83—View of pulley, gear and friction link assembly.

aged and renewal is necessary. Clean and inspect parts. The friction link should move the gear to both extremes of its travel. If not, renew the linkage assembly.

Install new spring by hooking end in retainer slot and winding until spring is coiled in housing (Fig. B84). Insert new rope through the housing and into the pulley. Tie a small knot, heat seal the knot and pull it tight into the recess in pulley. Install pulley assembly in the housing with friction link (L—Fig. B85) in groove in casting as shown. Install rope guide. Rotate pulley counterclockwise until rope is fully wound. Hook free end of spring to anchor, install anchor screw and tighten to 75-90 in.-lbs. (8-11 N•m) torque. Lubricate spring with a small amount of engine oil. Snap the plastic spring cover in place. Preload spring by pulling out about one foot of rope as shown in Fig. B86, then winding rope and pulley two or three turns clockwise. Check starter operation.

ELECTRIC STARTER MOTOR.
Two types of electric starter motors have been used.

The starter shown in Fig. B87 is used on System 3 and 4 engines using 6 or 12-volt batteries. At a no-load minimum speed of 800 rpm measured at end of helix (2), 6-volt starter should draw no more than 18 amps and 12-volt starter should draw no more than 8 amps. Do not run starter for more than five seconds during testing. Starter pinion and helix must move without binding. Do not apply oil to helix (2) or nylon gear (3). Renew brushes if length (L—Fig. B88) is $5/64$ inch (2.0 mm) or less.

When assembling starter, install fiber washer (11—Fig. B89), gray plastic washers (15) and steel washer (12) on armature shaft. Install sufficient gray plastic washers (9—Fig. B87) at drive

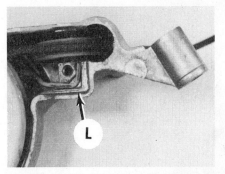

Fig. B85—Friction link (L) must fit in groove in housing.

Fig. B86—With rope pulled out as shown, rotate pulley clockwise to apply tension to rewind spring.

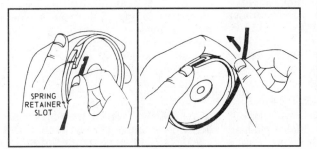

Fig. B84—Hook outer end of spring in retainer slot, then coil the spring counterclockwise in the housing.

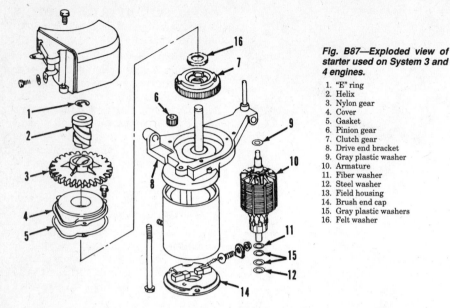

Fig. B87—Exploded view of starter used on System 3 and 4 engines.

1. "E" ring
2. Helix
3. Nylon gear
4. Cover
5. Gasket
6. Pinion gear
7. Clutch gear
8. Drive end bracket
9. Gray plastic washer
10. Armature
11. Fiber washer
12. Steel washer
13. Field housing
14. Brush end cap
15. Gray plastic washers
16. Felt washer

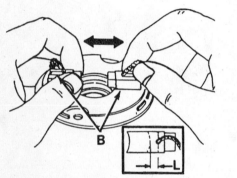

Fig. B88—Minimum brush length (L) is 5/64 inch (2.0 mm).

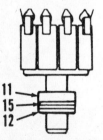

Fig. B89—Install gray plastic washers (15) between fiber washer (11) and steel washer (12).

end of armature so armature shaft end play is 0.005-0.025 inch (0.13-0.63 mm). Be sure notches in end cap, housing and end bracket are aligned. Apply approximately ¾ ounce of gear lubricant under clutch (7). Tap end cap (14) to seat bearings and tighten through-bolts to 25 in.-lbs. (2.8 N•m) torque. Tighten starter cover screws to 25 in.-lbs. (2.8 N•m) torque.

The starter shown in Fig. B90 is used on 130000 series engines with 12-volt and 110-volt versions available. No-load test of 12-volt starter should indicate a

Fig. B90—Exploded view of electric starter used on 130000 series engines.

1. Drive housing	7. Gasket
2. Washer	8. Drive end bracket
3. Retainer	9. Thrust washer
4. Gear	10. Armature
5. Drive assy.	11. Field housing
6. Spring washer	12. Brush end cap

minimum tachometer reading of 5600 rpm at pinion gear (4) with a maximum ammeter reading of 6 amps. No-load test of 110-volt starter should indicate a minimum tachometer reading of 8300 rpm at pinion gear (4) with a maximum ammeter reading of 1½ amps.

WARNING: Exercise particular caution when testing 110-volt starter and starter circuit as damaged compo-

nents may result in a dangerous short circuit.

Do not clamp starter housing (11) in a vise or strike housing of the 110-volt starter as ceramic magnets may crack. Make match marks on end bracket (8), housing (11) and end cap (12) before disassembly so they can be reassembled in their original positions. Renew brushes if length is ¼ inch (6.4 mm) or less.

ALTERNATOR. Some Briggs & Stratton engines use an alternator adjacent to the flywheel to provide electrical energy. Current from the alternator may be used to power lights or directed to a battery and other electrical components. A rectifier (diode) may be installed in the electrical system to provide direct current.

Identification. The alternator must be identified before testing the system. Model series 90000, 91000, 92000, 93000, 94000, 110000, 111000, 112000, 113000 and 114000 may be equipped with a System 3 or System 4 alternator as shown in Fig. B91. Model series 130000, 131000, 132000, 133000 and 135000 may be equipped with the 1.2 amp alternator shown in Fig. B92 or the 1½ amp alternator shown in Fig. B93. Refer to the following sections for testing procedures.

System 3 and System 4. Refer to Fig. B91. To test alternator output, disconnect black stator lead at white connector. Connect red lead of a DC ammeter to stator lead and black ammeter lead to engine. With engine running, alternator output should be at least 0.5 amps DC at 2800 rpm. If alternator output is zero or low and stator air gap is correct, replace stator.

The air gap between the stator and the flywheel magnets should be 0.010 inch (0.25 mm).

1.2 Amp Alternator. Refer to Fig. B92. The 1.2 amp alternator provides battery charging current. The stator is

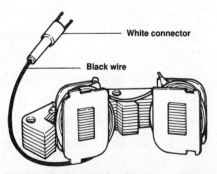

Fig. B91—Drawing of alternator used on System 3 and System 4 engine models.

Illustrations courtesy Briggs & Stratton Corp.

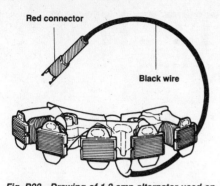

Fig. B92—Drawing of 1.2 amp alternator used on some 130000, 131000, 132000, 133000 and 135000 engine models.

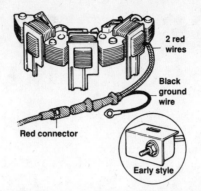

Fig. B93—Drawing of 1-1/2 amp alternator used on some 130000, 131000, 132000, 133000 and 135000 engine models.

located under the flywheel. A 12 ampere hour battery should be used for warm temperature operation and a 24 ampere hour battery should be used for cold temperatures.

To test alternator output, disconnect black stator lead at red connector. Connect red lead of a DC ammeter to stator lead and black ammeter lead to engine. With engine running, alternator output should be at least 1.0 amps DC at 3600 rpm. If alternator output is zero or low, replace stator.

1½ Amp Alternator. Refer to Fig. B93. The 1½ amp alternator provides battery charging current. The stator is located under the flywheel. A 12 ampere hour battery should be used for warm temperature operation and a 24 ampere hour battery should be used for cold temperatures.

The solid-state rectifier on early models is adjacent to starter drive housing, while on later models the rectifier diodes are contained in the stator wire near the connector. Only later inline type rectifier is available and may be substituted for early type.

To test alternator output, disconnect charging lead from charging terminal on early models, or disconnect stator lead at connector on later models. Connect red lead of a DC ammeter to charging terminal or stator lead and black ammeter lead to engine. With engine running, alternator output should be at least 1.2 amps DC at 3600 rpm. Test rectifier if output is low or zero.

Check for a faulty rectifier as follows:
1. With engine stopped, disconnect lead from output terminal on early mod-els, or disconnect stator connector on later models.
2. Insert a pin in one of the stator wires and connect one ohmmeter lead to pin and other ohmmeter lead to output terminal on early models or to end of connector on later models.
3. Check for continuity.
4. Reverse ohmmeter leads and again check for continuity. Ohmmeter should show a continuity reading (low ohms) for one direction only.
5. Repeat test on other stator wire. If tests show no continuity in either direction or continuity in both directions, rectifier is faulty and must be replaced.
6. With a pin inserted in one of the stator wires, connect one ohmmeter lead to pin and other ohmmeter lead to engine ground.
7. Check for continuity.
8. Reverse ohmmeter leads and again check for continuity. Ohmmeter should show a continuity reading (low ohms) for one direction only.
9. Repeat test on other stator wire.
10. Replace stator lead if tests indicate rectifier diodes are faulty.

The old stator leads must be cut and the new stator leads must be attached using rosin core solder. Mechanical fasteners must not be used to splice the old and new stator leads together.

If rectifier tests satisfactorily, replace stator.

BRIGGS & STRATTON
4-STROKE QUANTUM ENGINES

Model	Bore	Stroke	Displacement	Power Rating
100700	2.56 in.	1.94 in.	10 cu.in.	3.5 hp
	(65.1 mm)	(49.3 mm)	(164 cc)	(2.6 kW)
121000	2.69 in.	2.04 in.	11.57 cu.in.	3.5 hp
	(68.0 mm)	(51.8 mm)	(190 cc)	(2.6 kW)
122000	2.69 in.	2.04 in.	11.57 cu.in.	4.0 hp
	(68.0 mm)	(51.8 mm)	(190 cc)	(3.0 kW)
123000, 124000				
125000, 126000	2.69 in.	2.04 in.	11.57 cu.in.	5 hp
	(68.0 mm)	(51.8 mm)	(190 cc)	(3.8 kW)

ENGINE INFORMATION

The 100700, 121000, 122000, 123000, 124000, 125000 and 126000 Quantum models are air-cooled, four-stroke, single-cylinder engines. The engine has a vertical crankshaft and utilizes a compression release. Refer to page 59 for a table that outlines the Briggs & Stratton model numbering system.

MAINTENANCE

LUBRICATION. All models are lubricated by an oil slinger that is gear-driven by the camshaft.

It is recommended that the oil be changed after first eight hours of operation and after every 50 hours of operation or at least once each operating season. Change oil weekly or after every 25 hours of operation if equipment undergoes severe usage.

Engine oil should meet or exceed latest API service classification. Use SAE 30 oil for temperatures above 40° F (4° C); use SAE 10W-30 oil for temperatures between 0° F (−18° C) and 100° F (38° C); below 20° F (−7° C) use petroleum based SAE 5W-20 or a suitable synthetic oil.

Crankcase capacity is 1¼ pints (0.6 liters).

AIR CLEANER. The paper type filter element should be cleaned and inspected after every 25 hours of engine operation, or after three months, whichever occurs first. Tap filter gently to dislodge accumulated dirt. Filter may be washed using warm water and non-sudsing detergent directed from inside of filter to outside. DO NOT use petroleum-based cleaners or solvents to clean filter. DO NOT direct pressurized air towards filter. Let filter air dry thoroughly, then inspect filter and discard it if damaged or uncleanable. Clean filter canister.

FUEL FILTER. The fuel tank is equipped with a filter at the outlet, and an inline filter may also be installed in the fuel line. Check filters periodically during operating season and clean or renew as necessary.

CRANKCASE BREATHER. The crankcase breather is built into the tappet chamber cover. A fiber disc acts as a one-way valve. The breather allows vapor from the crankcase to be evacuated to the intake manifold, but blocks the return flow of air, thus maintaining a vacuum in the crankcase. The vacuum prevents oil from being forced out of the engine past the piston rings, oil seals and gaskets.

Clearance between fiber disc check valve and breather body (Fig. B201) should not exceed 0.045 inch (1.14 mm). If it is possible to insert a 0.045 inch (1.14 mm) wire between disc and breather body, renew breather assembly. Do not use excessive force when measuring gap. Disc should not stick or

Fig. B201—Clearance between fiber disc valve and crankcase breather housing must be less than 0.045 inch (1.15 mm). A spark plug wire gauge may be used to check clearance as shown, but do not apply pressure against disc valve.

bind during operation. Renew if distorted or damaged. Inspect breather tube for leakage.

SPARK PLUG. Recommended spark plug is Champion RJ19LM or J19LM. Specified spark plug electrode gap is 0.030 inch (0.76 mm). Tighten spark plug to 165 in.-lbs. (19 N•m) torque.

CAUTION: Briggs & Stratton does not recommend using abrasive blasting to clean spark plugs as this may introduce some abrasive material into the engine which could cause extensive damage.

CARBURETOR. Exploded view of carburetor used on all models is shown in Fig. B202.

Adjustment. Initial setting of idle mixture screw (9—Fig. B203) is 1¼ turn out. With engine at normal operating temperature and equipment control lever in "SLOW" position, adjust idle speed screw (7) so engine idles at 1400 rpm. With engine running at idle speed, turn idle mixture screw clockwise until engine speed just starts to drop. Note screw position. Turn idle mixture screw counterclockwise until engine speed just starts to drop again. Note screw position, then turn screw to midpoint between the noted screw positions. If engine will not accelerate cleanly, slightly enrichen mixture by turning idle mixture screw counterclockwise. If necessary, readjust idle speed screw.

High speed mixture is controlled by a fixed main jet; on some models the jet is a part of the fuel bowl retaining screw (19—Fig. B202). No optional sizes are available. If engine does not run properly at high altitude, remove main air

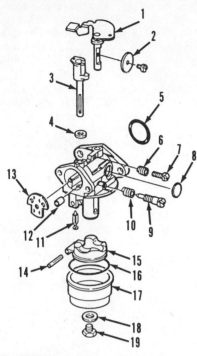

Fig. B202—Exploded view of carburetor.

1. Throttle shaft	11. Fuel inlet valve
2. Throttle plate	12. Main air jet
3. Choke shaft	13. Choke plate
4. Washer	14. Float pin
5. Gasket	15. Float
6. Spring	16. Gasket
7. Idle speed screw	17. Fuel bowl
8. Welch plug	18. Gasket
9. Idle mixture screw	19. Screw
10. Spring	

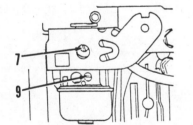

Fig. B203—View showing location of idle speed screw (7) and idle mixture screw (9).

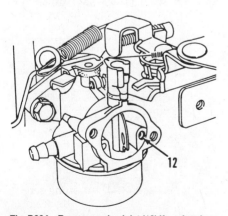

Fig. B204—Remove main air jet (12) if engine does not run properly at high altitudes.

jet (12—Fig. B204) and adjust mixture for smooth operation.

Overhaul. Disassembly of carburetor is evident after inspection of carburetor and referral to exploded view in Fig. B202. A 5/32 inch (4 mm) diameter punch ground flat at the end makes a suitable tool for removing Welch plug (8). Clean and inspect components and discard any parts which are damaged or excessively worn.

When reassembling carburetor, note the following: Do not deform Welch plug (8) during installation; it should be flat. Seal outer edges of plug with nonhardening sealer. Install choke and throttle plates so numbers are on outer face when choke or throttle plate is in closed position. Install fuel inlet seat using B&S driver 19057 or a suitable tool so grooved face of seat is down. Float height is not adjustable. Tighten fuel bowl retaining nut to 50 in.-lbs. (5.6 N•m) torque. Tighten carburetor mounting nuts to 90 in.-lbs. (10.2 N•m) torque.

CHOKE-A-MATIC CARBURETOR CONTROLS. The engine may be equipped with a control unit that controls the throttle, choke and grounding switch.

The remote control wire must travel at least 1 3/8 inches (35 mm) for proper Choke-A-Matic operation. See Fig. B205.

Model 100700. To check operation of Choke-A-Matic carburetor control, move control lever to "CHOKE" position. Carburetor choke plate must be completely closed. Move control lever to "STOP" position. Magneto grounding switch should be making contact. With the control lever in "RUN", "FAST" or "SLOW" position, carburetor choke should be completely open. On units with remote control, synchronize movement of remote lever to carburetor control lever by loosening screw (C—Fig. B206) and moving control wire housing (D) as required; then, tighten screw to clamp the housing securely. Check for proper operation.

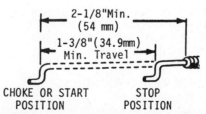

Fig. B205—For proper operation of Choke-A-Matic controls, remote control wire must extend to dimension shown and have a minimum travel of 1-3/8 inches (34.9 mm).

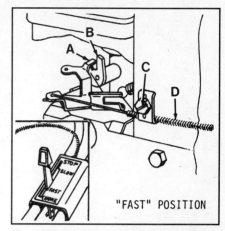

Fig. B206—On Choke-A-Matic controls shown, lever (A) should just contact choke shaft arm (B) when control is in "FAST" position. If not, loosen screw (C) and move control wire housing (D) as required, then tighten screw.

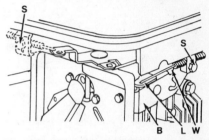

Fig. B207—View of Choke-A-Matic controls used on 121000, 122000, 123000, 124000, 125000 and 126000 models. Cable housing screw may be located at either location (S).

All Other Models. If equipment control lever has detents, proceed as follows: Move equipment control lever to "FAST" position and loosen cable clamp screw (S—Fig. B207). Move cable so 1/8 inch (3 mm) holes in governor control lever (L) and bracket (B) are aligned and tighten cable clamp screw (S). Check for proper operation. If equipped with stop switch, switch (W) should contact governor control lever when equipment control lever is in "STOP" position.

If equipment control lever does not have detents proceed as follows: Move equipment control lever to "CHOKE" position and loosen cable clamp screw (S—Fig. B207). Move cable so carburetor choke plate is closed and tighten clamp screw (S). Check for proper operation. If equipped with stop switch, switch (W) should contact governor control lever when equipment control lever is in "STOP" position.

GOVERNOR. Model 107000. These engines are equipped with an air vane type governor. See Fig. B208 for view of governor control linkage. No adjustment is required. Governor mechanism should not bind. Renew any damaged or worn components.

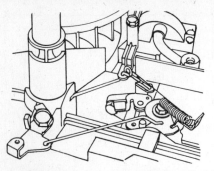

Fig. B208—View of governor linkage on Model 100700.

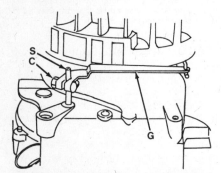

Fig. B209—To adjust governor on Models 121000, 122000, 123000, 124000, 125000 and 126000, move equipment control lever to "FAST" position and place an 1/8-inch rod through holes in governor control lever (L) and bracket (B), then refer to Fig. B210.

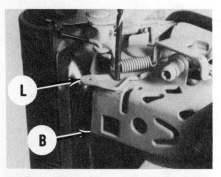

Fig. B210—The governor on Models 121000, 122000, 123000, 124000, 125000 and 126000 is adjusted by holding governor lever (G) while turning governor shaft (S) counterclockwise as outlined in text.

All Other Models. These engines are equipped with a mechanical governor. The governor gear and flyweight assembly is driven by the camshaft and located inside the crankcase. Refer to REPAIRS section for service information.

To adjust governor, move equipment control lever to "FAST" position and place an 1/8 inch (3 mm) diameter rod through holes in governor control lever (L—Fig. B209) and bracket (B). Loosen clamp bolt (C—Fig. B210) then rotate governor shaft (S) counterclockwise until it stops. Hold shaft and tighten clamp bolt to 35-45 in.-lbs. (4-5 N•m) torque.

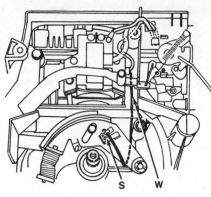

Fig. B211—View showing location of stop switch (S) and proper routing of stop switch wire (W) on Model 100700.

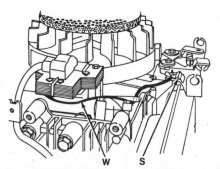

Fig. B212—View showing location of stop switch (S) and proper routing of stop switch wire (W) on Models 121000, 122000, 123000, 124000, 125000 and 126000.

IGNITION. All models are equipped with a Magnetron breakerless ignition system. All components are located outside the flywheel.

To check spark, remove spark plug and connect spark plug cable to B&S tester 19051, then ground remaining tester lead to engine. Spin engine at 350 rpm or more. If spark jumps the 0.165 inch (4.2 mm) tester gap, system is functioning properly.

To remove armature and Magnetron module, remove flywheel shroud and armature retaining screws. Disconnect stop switch wire from module. When reinstalling armature, position armature so air gap between armature legs and flywheel surface is 0.006-0.010 inch (0.15-0.25 mm).

On models equipped with a flywheel brake, stop switch (S—Figs. B211 and B212) should ground ignition system when flywheel brake is activated. Note in Figs. B209 and B210 correct routing for stop switch wire (W) of models equipped with flywheel brake system.

VALVE ADJUSTMENT. To correctly set tappet clearance, remove spark plug and breather/valve tappet chamber cover. Rotate crankshaft in normal direction (clockwise at flywheel) so piston is at top dead center on compression stroke. Continue to rotate crankshaft so piston is 1/4 inch (6.4 mm) down from top dead center. Using a feeler gauge, measure clearance between intake and exhaust valve stem ends and the tappets.

Valve tappet clearance (engine cold) should be 0.005-0.007 inch (0.13-0.18 mm) for intake valve and 0.007-0.009 inch (0.18-0.23 mm) for exhaust valve. Adjust tappet clearance by grinding end of valve stem to increase clearance or grind valve seat deeper to decrease clearance.

CYLINDER HEAD. Manufacturer recommends that after every 100-300 hours of operation the cylinder head is removed and cleaned of deposits.

REPAIRS

TIGHTENING TORQUES. Recommended tightening torque specifications are as follows:

Alternator. . . . . 25 in.-lbs.
(2.8 N•m)
Carburetor mounting nuts . 90 in.-lbs.
(10.2 N•m)
Connecting rod. . . . . 100 in.-lbs.
(11.3 N•m)
Crankcase cover. . . . . 85 in.-lbs.
(9.6 N•m)
Cylinder head . . . . 140 in.-lbs.
(16 N•m)
Spark plug . . . . 165 in.-lbs.
(19 N•m)

CYLINDER HEAD. Lubricate cylinder head screws with graphite grease before installation. Do not apply sealant to the head gasket. Tighten screws hand-tight, then tighten in sequence shown in Fig. B213 or Fig. B214 to 140 in.-lbs. (16 N•m) torque.

VALVE SYSTEM. Valve face and seat angles should be ground at 45°. Renew valve if margin is 1/16 inch (1.6

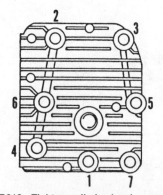

Fig. B213—Tighten cylinder head screws on Model 100700 in sequence shown.

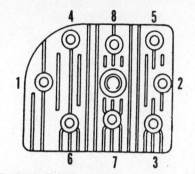

Fig. B214—Tighten cylinder head screws on Models 121000, 122000, 123000, 124000, 125000 and 126000 in sequence shown.

mm) or less. Seat width should be $\frac{3}{64}$ to $\frac{1}{16}$ inch (1.2-1.6 mm).

The valves operate directly in valve guide bores in the aluminum crankcase. If original valve guide bores are excessively worn, service guides may be installed. Maximum allowable valve guide inside diameter is 0.266 inch (6.76 mm). If B&S valve guide gauge 19122 can be installed to a depth of $\frac{5}{16}$ inch (7.9 mm) in valve guide, then a new guide should be installed. Ream guide out first with B&S reamer 19064 to a depth approximately $\frac{1}{16}$ inch (1.6 mm) deeper than length of service guide.

Fig. B216—Exploded view of Model 121000, 122000, 123000, 124000, 125000 and 126000 engine. Components in inset are used on models with an auxiliary pto.

1. Nut
2. Starter cup
3. Screen
4. Flywheel
5. Baffle
6. Gasket
7. Intake manifold tube
8. Gasket
9. Seal
10. Governor lever
11. Push nut
12. Washer
13. Cylinder block
14. Gasket
15. Breather & tappet cover
16. Head gasket
17. Cylinder head
18. Shroud
19. Governor shaft
20. Key
21. Crankshaft
22. Camshaft
22A. Camshaft, auxiliary pto gear
23. Tappets
24. Key
25. Connecting rod
26. Piston
27. Piston rings
28. Valve retainers
29. Valve springs
30. Intake valve
31. Exhaust valve
32. Piston pin
33. Clips
34. Governor & oil slinger
35. Gasket
36. Crankcase
37. Seal
38. Washer
39. Shaft stop
40. Roll pin
41. Auxiliary pto shaft
42. Seal

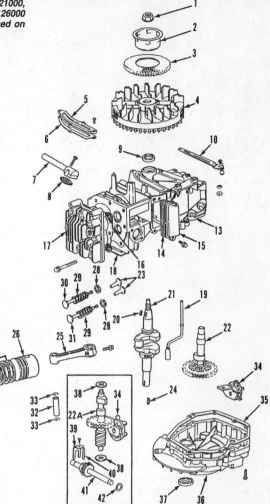

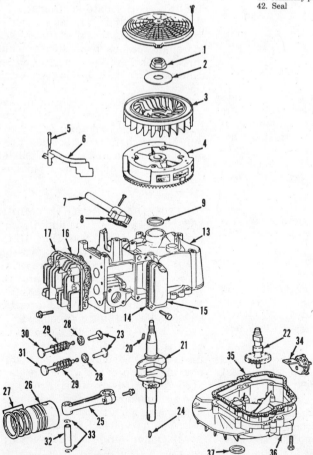

Fig. B215—Exploded view of Model 100700 engine.

1. Nut
2. Washer
3. Fan
4. Flywheel
5. Pivot pin
6. Governor air vane
7. Intake manifold tube
8. Gasket
9. Seal
13. Cylinder block
14. Gasket
15. Breather & tappet cover
16. Head gasket
17. Cylinder head
20. Key
21. Crankshaft
22. Camshaft
23. Tappets
24. Key
25. Connecting rod
26. Piston
27. Piston rings
28. Valve retainers
29. Valve springs
30. Intake valve
31. Exhaust valve
32. Piston pin
33. Clips
34. Governor & oil slinger
35. Gasket
36. Crankcase
37. Seal

Press in service guide with B&S driver 19065, then ream using B&S finish reamer 19066. Reface valve seats after new guides are installed.

CAMSHAFT. Some models are equipped with a camshaft (22—Figs. B215 and B216) constructed of metal and plastic. The camshaft and gear should be inspected for wear and damage on journals, cam lobes and gear teeth. Renew camshaft if bearing journals are 0.498 inch (12.65 mm) or less.

When installing camshaft, align timing mark on camshaft gear with mark on crankshaft gear as shown in Fig. B217.

PISTON, PIN, RINGS AND CONNECTING ROD. The connecting rod and piston may be removed as an assembly after removing the cylinder head and lower crankcase.

Renew piston if it shows visible signs of wear, scoring or scuffing. If, after cleaning carbon from top ring groove, a new top ring has a side clearance of

Fig. B217—Align timing marks (M) on camshaft gear and crankshaft gear.

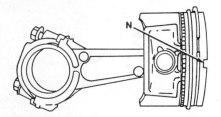

Fig. B218—Notch (N) in piston crown must be toward flywheel.

0.007 inch (0.18 mm) or more, renew the piston. Renew piston or hone piston pin hole to 0.005 inch (0.13 mm) oversize if pin hole is 0.0005 inch (0.013 mm) or more out-of-round, or is worn to a diameter of 0.491 inch (12.47 mm).

If piston pin is 0.0005 inch (0.013 mm) or more out-of-round, or is worn to a diameter of 0.489 inch (12.42 mm), renew piston pin. A piston pin that is 0.005 inch (0.13 mm) oversize is available.

Piston and piston rings are available in standard size and oversizes of 0.010, 0.020 and 0.030 inch.

Connecting rod reject size for crankpin hole is 1.001 inch (25.43 mm). Renew rod if piston pin hole is scored or out-of-round more than 0.0005 inch (0.013 mm) or pin hole is 0.492 inch (12.50 mm) or greater. The connecting rod piston pin hole can be reamed to accept a 0.005 inch (0.13 mm) oversize piston pin. A connecting rod with 0.020 inch (0.51 mm) undersize big end diameter is available to accommodate a

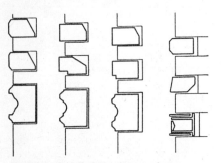

Fig. B219—Install piston rings as shown.

worn crankpin (machining instructions are included with new rod).

Install piston on connecting rod so notch in piston crown will be toward flywheel as shown in Fig. B218. Be sure that offset connecting rod is positioned as shown and that match marks on rod and cap are aligned. Refer to Fig. B219 and install piston rings according to type of ring set to be installed on piston.

Tighten connecting rod screws to 100 in.-lbs. (11.3 N•m) torque.

CRANKSHAFT AND MAIN BEARINGS. The crankshaft rides directly in the crankcase bores. Rejection sizes for crankshaft are: pto-end bearing journal 1.060 inch (26.92 mm); flywheel-end bearing journal 0.873 inch (22.17 mm); crankpin 0.996 inch (25.30 mm). A connecting rod with 0.020 inch (0.51 mm) undersize big end diameter is available to accommodate a worn crankpin (machining instructions are included with new rod).

The crankcase main bearing bore rejection size is 0.878 inch (22.30 mm). The crankcase cover or oil pan main bearing bore rejection size is 1.065 inch (27.05 mm). A service bushing is available for installation in the flywheel-side crankcase if the bearing bore requires service. No bushing is available for the pto-side crankcase so it must be renewed if bearing bore is damaged or worn excessively.

Install oil seals so lip is towards inside of crankcase.

Crankshaft end play should be 0.002-0.030 inch (0.05-0.76 mm) with 0.015 inch crankcase gasket installed. Additional gaskets of several thicknesses are available for end play adjustment. If end play is excessive, a thrust washer which can be installed on pto end of crankshaft is available. Tighten crankcase screws to 85 in.-lbs. (9.6 N•m) torque.

CYLINDER. If cylinder bore wear is 0.003 inch (0.76 mm) or more, or out-of-round is 0.0025 inch (0.04 mm) or greater, cylinder must be rebored to next oversize.

Standard cylinder bore diameter is 2.5615-2.5625 inch (65.06-65.09 mm) for Model 100700 and 2.6875-2.6885 inch (68.26-68.29 mm) for all other models.

AUXILIARY PTO. Models 121000, 122000, 123000, 124000, 125000 and 126000 may be equipped with an auxiliary pto. Auxiliary pto shaft (41—Fig. B220) is driven by a worm gear on the camshaft. On these models, the camshaft (22A), oil slinger (34) and worm

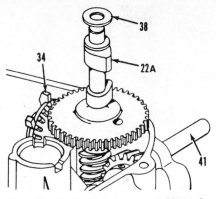

Fig. B220—View of auxiliary pto assembly used on some Models 121000, 122000, 123000, 124000, 125000 and 126000.

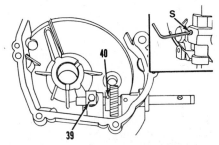

Fig. B221—Drive out roll pin (40) and unfasten shaft stop (39) to remove auxiliary pto shaft.

gear are available only as a unit assembly. Note that a thrust washer (38) is used to accommodate additional axial thrust of camshaft. Thrust washer is located at flywheel end of camshaft if pto shaft rotation is clockwise, while thrust washer is located at gear end of camshaft if pto shaft rotation is counterclockwise.

To remove auxiliary pto shaft and gear, unscrew Allen screw (S—Fig. B221) on bottom of crankcase. Insert a punch through the screw hole and drive out pin (40) retaining gear. Unfasten shaft stop (39) and withdraw auxiliary pto shaft.

FLYWHEEL BRAKE. Model 100700. The engine is equipped with a pad type flywheel brake (Fig. B211). The brake should stop the engine within three seconds when the operator releases mower safety control and the speed control is in high speed position. Stopping time can be checked using tool 19255. Adjustment is not required.

The brake mechanism is located under the flywheel. If brake operation is unsatisfactory, remove flywheel and renew any components which are damaged or excessively worn. Brake pad and arm must be renewed if the pad is damaged or friction material thickness is less than 0.090 inch (2.3 mm).

All Other Models. The engine is equipped with a pad type flywheel

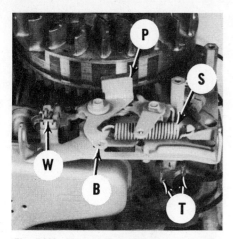

Fig. B222—View of flywheel brake mechanism used on Models 121000, 122000, 123000, 124000, 125000 and 126000 showing bracket (B), brake pad (P), brake spring (S), electric starter motor terminals (T) and stop switch (W).

brake (Fig. B222). The brake should stop the engine within three seconds when the operator releases mower safety control and the speed control is in high speed position. Stopping time can be checked using tool 19255. Adjustment is not required.

The brake pad is available only as part of the bracket assembly. Minimum allowable brake pad thickness is 0.090 inch (2.3 mm).

FLYWHEEL. The flywheel is secured to the crankshaft by a retaining nut. To remove flywheel, remove blower housing and any other components so flywheel is accessible. If equipped with flywheel brake, disconnect brake spring. Secure the flywheel from rotating using Briggs & Stratton or aftermarket flywheel holder, then unscrew retaining nut. Use a suitable puller to remove flywheel from crankshaft.

NOTE: Do not attempt to hold the flywheel by inserting a tool between

the fins as damage to flywheel may result.

The tapered portion of flywheel and crankshaft mating surfaces must be clean and smooth with no damage. Renew flywheel if any cracks are evident or any fins are broken. Be sure the keyway in crankshaft and flywheel is not damaged or worn. If flywheel key is replaced, a key made of aluminum must be installed. DO NOT install a steel key.

Install flywheel and key and tighten retaining nut to 55 ft.-lbs. (75 N•m) torque.

REWIND STARTER. Model 100700. Refer to Fig. B223 for an exploded view of starter. To disassemble starter, unfasten starter from engine. Pull out rope as far as possible, hold pulley, move rope off pulley, then slowly allow pulley to unwind so spring tension is released. Remove decal from starter cover (3) and unscrew cover screw (1); screw has left-hand threads. Bend out tang (T—Fig. B224) in cover, rotate cover counterclockwise to disengage spring end from cover and remove cover. Wear appropriate safety eyewear and gloves before disengaging spring from pulley as spring may uncoil uncon-

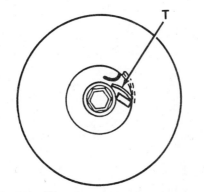

Fig. B224—Tang (T) engages inner end of rewind spring.

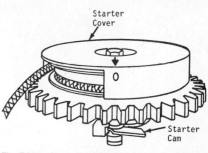

Fig. B225—Arrow or "O" on starter cover must be positioned in line with starter cam on engines with rope handle on top of engine.

trolled. Using a suitable pair of pliers, detach spring from pulley and carefully allow spring to uncoil. Remove remainder of components.

Inspect components and renew any which are damaged or excessively worn. Assemble as shown in Fig. B223. Be sure steel washer (10) is installed first on starter shaft (11), then the plastic washer (9). Install brake spring (8) so looped end of spring is toward and between fingers of starter shaft (11). Attach clip (7) to ends of brake spring. Install rewind spring so outer end is attached to notch in periphery of pulley and coil direction is clockwise from outer spring end. Before installing cover (3), apply a light coat of grease to inner face. Before tightening cover screw (left-hand threads), rotate cover clockwise so arrow or "O" on cover is aligned with starter cam (Fig. B225) if rope handle is situated on top of engine, or rotate cover clockwise so arrow or "O" on cover is 90° from starter cam (Fig. B226) if starter rope is situated or runs through a rope guide adjacent to cylinder head. Tighten cover screw to 55 in.-lbs. (6.2 N•m) torque. Turn gear against spring tension until spring is tightly wound. Align pulley rope knot pocket and rope outlet in cover, then install rope in pulley and through cover opening. Tie a temporary knot in rope end and allow rope to wind onto pulley. Install starter on engine and install rope handle.

All Other Models. Refer to Fig. B227 for exploded view of rewind

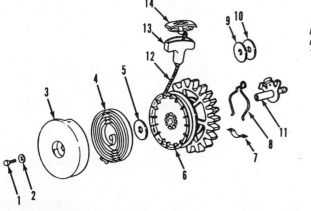

Fig. B223—Exploded view of rewind starter used on Model 100700.

1. Screw (L.H.)
2. Washer
3. Cover
4. Rewind spring
5. Washer
6. Pulley & gear
7. Clip
8. Brake spring
9. Plastic washer
10. Steel washer
11. Starter shaft
12. Rope
13. Handle
14. Insert

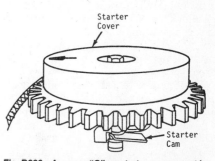

Fig. B226—Arrow or "O" on starter cover must be positioned 90° from starter cam on engines with rope adjacent to cylinder head.

Illustrations courtesy Briggs & Stratton Corp.

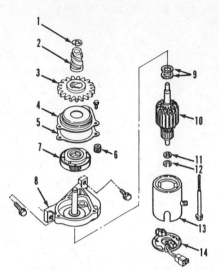

Fig. B227—Exploded view of rewind starter used on Models 121000, 122000, 123000, 124000, 125000 and 126000.

1. Starter housing	8. Plastic washer
2. Spring cover	9. Spring
3. Rewind spring	10. Steel washer
4. Pulley	11. Pin
5. Springs	12. Rope
6. Dogs	13. Handle
7. Retainer	14. Insert

starter. To remove starter proceed as follows: Remove starter cover, detach fuel line, remove fuel tank and oil fill tube. Detach starter from shroud.

To install a new rope, proceed as follows. Remove starter and extract old rope from pulley. Allow pulley to unwind, then turn pulley counterclockwise until spring is tightly wound. Rotate pulley clockwise until rope hole in pulley is aligned with rope outlet in housing. Pass rope through pulley hole and housing outlet and tie a temporary knot near handle end of rope. Release pulley and allow rope to wind onto pulley. Install rope handle, release temporary knot and allow rope to enter starter.

To disassemble starter, remove rope and allow pulley to totally unwind. Position a suitable hollow sleeve support under pulley. Using pin punch, drive out retainer pin (11). Remove retainer (7), dogs (6), springs (5) and brake spring (9). Wear appropriate safety eyewear and gloves before disengaging pulley from starter housing as spring may uncoil uncontrolled. Place shop towel around pulley and lift pulley out of housing; spring should remain with pulley. Do not attempt to separate spring from pulley as they are a unit assembly.

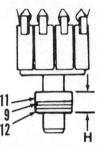

Fig. B228—Exploded view of electric starter motor used on some 121000, 122000, 123000, 124000, 125000 and 126000 models.

1. "E" ring	8. Drive end bracket
2. Helix	9. Plastic washers
3. Pinion	10. Armature
4. Cover	11. Fiber washer
5. Gasket	12. Steel washer
6. Gear	13. Field housing
7. Clutch	14. End cap

Inspect components for damage and excessive wear. Reverse disassembly procedure to install components. Be sure inner end of rewind spring engages spring retainer adjacent to housing center post. Pin (11) should be driven or pressed until flush with retainer (7). Install rope as previously outlined.

ELECTRIC STARTER. Some models may be equipped with the 12-volt DC electric starter motor shown in Fig. B228. The starter pinion engages gear teeth on the flywheel.

Starter should draw no more than 8 amps at a no-load minimum speed of 800 rpm measured at end of helix (2). Starter pinion and helix must move without binding. Do not apply oil to helix or pinion. Do not engage starter for more than five seconds during testing.

Remove gear assembly at drive end of starter, then unscrew through-bolts to disassemble remainder of starter. Be sure positive terminal remains with end cap assembly (14) when detaching housing (13). Renew brushes if length (L—Fig. B229) is ⁵⁄₆₄ inch (2.0 mm) or less.

When assembling starter, install fiber washer (11) and steel washer (12) on armature shaft with needed number of gray plastic washers (9) between fiber washer and steel washer to obtain a stack height (H—Fig. B230) equal to 0.200-0.225 inch (5.08-5.71 mm). Install sufficient gray plastic washers (9—Fig.

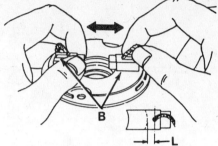

Fig. B229—Minimum brush length (L) is 5/64 inch (2.0 mm).

Fig. B230—Install gray plastic washers (9) between fiber washer (11) and steel washer (12) so stack height (H) is 0.200-0.225 inch (5.08-5.71 mm).

B228) at drive end of armature so armature shaft end play is 0.005-0.025 inch (0.13-0.63 mm). Be sure notches in end cap (14), housing (13) and end bracket (8) are aligned. Apply approximately ¾ ounce of gear lubricant under clutch (7). Tap end cap (14) to seat bearings and tighten through-bolts to 25 in.-lbs. (2.8 N•m) torque. Tighten starter cover screws to 25 in.-lbs. (2.8 N•m) torque. Tighten ⁵⁄₁₆-inch starter mounting screw to 140 in.-lbs. (15.8 N•m) torque. Tighten ¼-inch starter mounting screw to 90 in.-lbs. (10.2 N•m) torque.

ALTERNATOR. Models equipped with an electric starter are also equipped with an alternator which provides charging current for the starter battery.

Air gap between alternator stator legs and flywheel should be 0.007 inch (0.18 mm). Loosen alternator mounting screws and move alternator to obtain desired air gap. Tighten screws to 25 in.-lbs. (2.8 N•m) torque.

Alternator output can be checked by connecting a suitable ammeter to output lead of alternator and to engine ground. Alternator output with engine running at 3200 rpm should be no less than 0.4 amps DC. If low or no output is found, check to be sure stator air gap is set correctly. If air gap is within specification, renew stator assembly.

BRIGGS & STRATTON
4-STROKE VANGUARD ENGINES

Model	Bore	Stroke	Displacement	Power Rating
85400	60.9 mm	42.0 mm	126 cc	3.0 kW
	(2.44 in.)	(1.653 in.)	(7.7 cu.in.)	(4.0 hp)
115400	68.1 mm	50.0 mm	182 cc	4.1 kW
	(2.68 in.)	(1.968 in.)	(11.1 cu.in.)	(5.5 hp)

NOTE: Metric fasteners are used throughout engine except threaded hole in pto end of crankshaft, flange mounting holes and flywheel puller holes which are US threads.

ENGINE INFORMATION

The 85400 and 115400 Vanguard models are air-cooled, four-stroke, single-cylinder engines. The engine has a horizontal crankshaft and utilizes an overhead valve system. Refer to page 59 for a table that outlines the Briggs & Stratton model numbering system.

MAINTENANCE

LUBRICATION. All engines are lubricated using splash lubrication. Oil in the oil pan is thrown onto internal engine parts by an oil dipper on the connecting rod.

The engine is equipped with a low-oil level system (Oil Gard) that stops the engine or prevents starting when the oil level in the crankcase is low. The system consists of a module attached to the engine behind the control panel cover and a float attached to the inside of the crankcase. When the pin on the float contacts the crankcase due to a low oil level, the ignition primary circuit is grounded through the module to the float pin.

Change oil after first eight hours of operation and after every 50 hours of operation or at least once each operating season. Change oil weekly or after every 25 hours of operation if equipment undergoes severe usage.

Engine oil level should be maintained at full mark on dipstick attached to oil fill plug. Engine oil should meet or exceed latest API service classification. Use SAE 30 oil for temperatures above 40° F (4° C); use SAE 10W-30 oil for temperatures between 0° F (–18° C) and 40° F (4° C); below 20° F (–7° C) use petroleum based SAE 5W-20 or a suitable synthetic oil.

Crankcase capacity is 0.6 L (20 fl. oz.) on 85400 series engines and 0.7 L (24 fl. oz.) on 115400 series engines. Fill engine with oil so oil level reaches, but does not exceed, full mark on dipstick.

AIR CLEANER. The air cleaner consists of a canister and the filter element it contains. The filter element is made of paper and a foam precleaner is located in front.

The foam precleaner should be cleaned weekly or after every 25 hours of operation, whichever occurs first. The paper filter should be cleaned yearly or after every 100 hours of operation, whichever occurs first.

Tap paper filter gently to dislodge accumulated dirt. Filter may be washed using warm water and nonsudsing detergent directed from inside of filter to outside. DO NOT use petroleum-based cleaners or solvents to clean filter. DO NOT direct pressurized air towards filter. Let filter air dry thoroughly, then inspect filter and discard it if damaged or uncleanable. Clean filter canister. Inspect and, if necessary, replace any defective gaskets.

Clean foam precleaner in soapy water and squeeze until dry. Inspect filter for tears and holes or any other opening. Discard precleaner if it cannot be cleaned satisfactorily or if precleaner is torn or otherwise damaged. Pour clean engine oil into the precleaner, then squeeze to remove excess oil and distribute oil throughout.

FUEL FILTER. The fuel tank is equipped with a filter at the outlet, and an inline filter may also be installed in the fuel line. Check filters periodically during operating season and clean or renew as necessary.

CRANKCASE BREATHER. The crankcase breather, which provides a vacuum for the crankcase, is built into the valve cover. Vapor from the crankcase is evacuated through a tube to the intake manifold. The breather system must operate properly or excessive oil consumption may result.

There should be no leakage when air is blown through the breather tube and air should pass easily when vacuum is applied to the breather tube. Breather valve and valve cover are available only as a unit assembly. Inspect breather tube for leakage.

SPARK PLUG. Recommended spark plug is Champion RC12YC or Autolite 3924. Specified spark plug electrode gap is 0.76 mm (0.030 in.). Note that spark plug wire is routed under breather tube.

CAUTION: Briggs & Stratton does not recommend using abrasive blasting to clean spark plugs as this may introduce some abrasive material into the engine which could cause extensive damage.

The spark plug boot contains a threaded terminal (T—Fig. B301) that secures the spark plug wire in the boot. Unscrew terminal so plug wire can be withdrawn from boot. When assembling

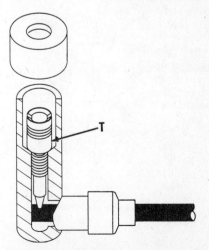

Fig. B301—Spark plug terminal (T) is threaded and penetrates spark plug wire when screwed into spark plug boot.

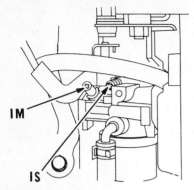

Fig. B302—Drawing showing location of idle mixture screw (IM) and idle speed screw (IS).

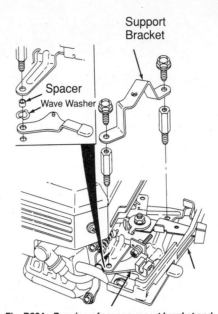

Fig. B304—Drawing of cover support bracket and speed control lever assembly used on Model 115400.

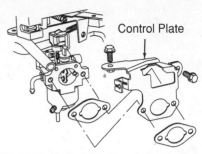

Fig. B305—Control plate fits on carburetor mounting studs.

boot and wire, turn terminal so it penetrates wire and bottoms in boot.

CARBURETOR. Adjustment. During normal operation, idle speed is controlled by the governor. After carburetor adjustments are completed, adjust governed idle speed as outlined in GOVERNOR section.

Set idle speed for carburetor adjustment procedure by turning idle speed screw (IS—Fig. B302) so engine idles at 1200 rpm at normal operating temperature. Idle mixture is controlled by idle mixture screw (IM) and idle jet (11—

Fig. B303). Note that idle mixture screw is equipped with a limiting cap which must be removed for initial setting. Initial setting of idle mixture screw is 1½ turns out from a lightly seated position. Reinstall limiting cap so shouldered portion of cap is down.

To adjust idle mixture, turn idle mixture screw clockwise to lean mixture until engine speed just starts to slow, then turn screw counterclockwise to enrich mixture just until engine speed begins to slow. Turn idle screw to halfway point between lean and rich positions. Turn idle screw counterclockwise in small increments to enrich mixture if engine will not accelerate without stumbling. Idle mixture jet is not adjustable.

High speed operation is controlled by main jet (16—Fig. B303) and is not adjustable.

R&R And Overhaul. When removing carburetor on 115400 series engines, note that the cover support bracket (Fig. B304) must be removed so the choke lever can be disengaged from the carburetor. A speed control plate bracket (Fig. B305) fits on the carburetor mounting studs of all models.

To disassemble carburetor, remove fuel bowl retaining screw (25—Fig. B303) and fuel bowl (23). Remove float pin (21) by forcing against small end of pin toward large end of pin. Remove float (20) and fuel inlet valve (18). Unscrew main jet (16), then unscrew nozzle (17).

NOTE: Main jet protrudes into nozzle cavity and nozzle will be damaged if nozzle is removed first.

Remove throttle and choke shaft assemblies after unscrewing throttle and choke plate retaining screws. Remove idle mixture screw (12) and idle mixture jet (11).

When assembling the carburetor, note the following. Install throttle shaft seal (4) with flat side toward carburetor. Numbers on throttle plate (1) must face out and be on fuel inlet side of carburetor. Install choke stop plate (7) on choke shaft so indented segment of plate is opposite "CHOKE" on shaft and stop tab on plate is down. See view of installed choke shaft in Fig. B306. Numbers on choke plate (5—Fig. B303) must face out and be on fuel inlet side of carburetor. Insert small end of float pin first as shown in Fig. B307 and tap into place.

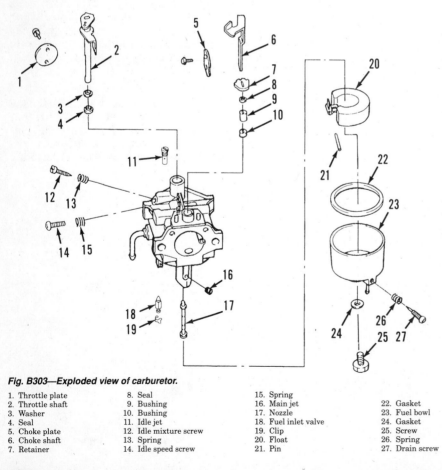

Fig. B303—Exploded view of carburetor.

1. Throttle plate
2. Throttle shaft
3. Washer
4. Seal
5. Choke plate
6. Choke shaft
7. Retainer
8. Seal
9. Bushing
10. Bushing
11. Idle jet
12. Idle mixture screw
13. Spring
14. Idle speed screw
15. Spring
16. Main jet
17. Nozzle
18. Fuel inlet valve
19. Clip
20. Float
21. Pin
22. Gasket
23. Fuel bowl
24. Gasket
25. Screw
26. Spring
27. Drain screw

Float should be parallel with body when carburetor is inverted as shown in Fig. B308. Float height is not adjustable; replace components necessary so float is parallel.

REMOTE CARBURETOR CONTROLS. The engine may be equipped

with a control unit that controls the throttle, choke and grounding switch.

The remote control wire must travel at least 34.9 mm (1⅜in.) for proper operation. See Fig. B309.

GOVERNOR. The engine is equipped with a mechanical, flyweight type governor. Refer to Figs. B310 and B311 for drawings of governor linkage.

NOTE: Nuts (T—Fig. B310 or B311) and governor shaft (S) have left-hand threads.

To adjust governor linkage, proceed as follows: Loosen governor lever clamp nut (T—Fig. B310 or B311), rotate governor lever (L) so throttle plate is fully open and hold lever in place. Turn governor shaft (S) counterclockwise as far as possible, then retighten nut (T).

If internal governor assembly must be serviced, refer to REPAIRS section.

To adjust governed idle speed, the carburetor should first be adjusted as outlined in CARBURETOR section. Then adjust governed idle speed by rotating governed idle speed screw (I—Fig. B312 or B313) so engine idles at 1400 rpm with speed control in idle position.

Maximum governed speed is adjusted by turning screw (M—Fig. B312

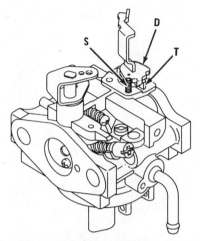

Fig. B306—Install choke shaft and retainer on Model 85400 so detents (D) are toward detent spring (S) and tab (T) is down.

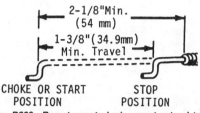

Fig. B309—Remote control wire must extend to dimension shown and have a minimum travel of 34.9 mm (1-3/8 in.).

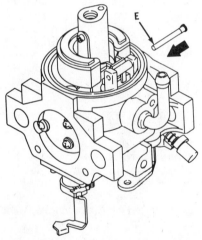

Fig. B307—Insert float pin so small end (E) is positioned as shown.

Fig. B310—Drawing of governor linkage on Model 85400. Attach governor spring (G) to inner hole (N) on 50 Hz generators. Attach governor spring to outer hole (O) for all other applications.

    G. Governor spring
    I. Governed idle speed screw
    K. Link
    L. Governor lever
    M. Maximum governed speed screw
    N. Inner spring hole
    O. Outer spring hole
    R. Link spring
    S. Governor shaft
    T. Nut

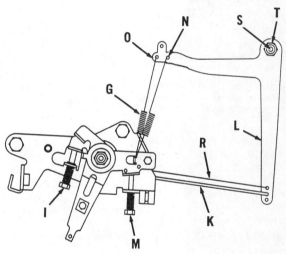

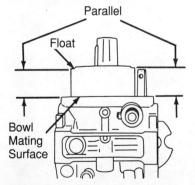

Fig. B308—With carburetor inverted, float should be parallel to fuel bowl mating surface.

Fig. B311—Drawing of governor linkage on Model 115400. Attach governor spring (G) to inner hole (N) on 50 Hz generators. Attach governor spring to outer hole (O) for all other applications.

    G. Governor spring
    I. Governed idle speed screw
    K. Link
    L. Governor lever
    M. Maximum governed speed screw
    N. Inner spring hole
    O. Outer spring hole
    R. Link spring
    S. Governor shaft
    T. Nut

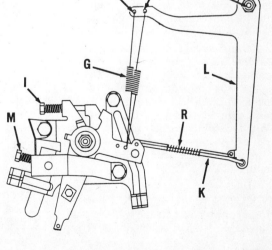

Illustrations courtesy Briggs & Stratton Corp.

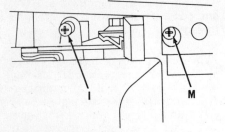

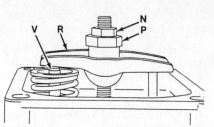

Fig. B312—Drawing showing location of governed idle speed screw (I) and maximum governed speed screw (M) on Model 85400.

Fig. B314—Loosen locknut (N) and turn pivot nut (P) to adjust clearance between rocker arm (R) and valve stem end (V). Valve clearance should be 0.05-0.10 mm (0.002-0.004 in.) for both valves.

Cylinder head	24.8 N•m
	(220 in.-lbs.)
Flywheel nut	61.2 N•m
	(45 ft.-lbs.)
Rocker arm cover	5.1 N•m
	(45 in.-lbs.)
Rocker arm locknut	4.0 N•m
	(35 in.-lbs.)
Rocker arm stud	19.8 N•m
	(175 in.-lbs.)

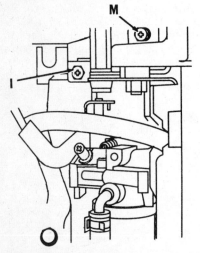

Fig. B313—Drawing showing location of governed idle speed screw (I) and maximum governed speed screw (M) on Model 115400.

or B313) and should be set to engine speed recommended by equipment manufacturer. For generator applications, no-load maximum governed speed should be 3150 rpm on 50 Hz generators and 3750 rpm on 60 Hz generators.

IGNITION SYSTEM. The engine is equipped with a Magnetron ignition system.

To check spark, remove spark plug and connect spark plug cable to B&S tester 19051 or 19368, then ground remaining tester lead to engine. Spin engine at 350 rpm or more. If spark jumps the 4.2 mm (0.165 in.) tester gap, system is functioning properly.

To remove armature and Magnetron module, remove air cleaner cover, control panel and blower housing. Detach spark plug boot from spark plug wire as noted in SPARK PLUG section. Disconnect stop switch wire from module. When installing armature, position armature so air gap between armature legs and flywheel surface is 0.30-0.51 mm (0.012-0.020 in.).

VALVE ADJUSTMENT. Remove rocker arm cover. Remove spark plug. Rotate crankshaft so piston is at top

dead center on compression stroke. Then, using a suitable measuring device inserted through spark plug hole, rotate crankshaft clockwise as viewed at flywheel end so piston is 6.4 mm (¼ in.) below TDC to prevent interference by the compression release mechanism with the exhaust valve. Clearance between rocker arm pad and valve stem end (Fig. B314) should be 0.05-0.10 mm (0.002-0.004 in.) for intake and exhaust. Loosen locknut (N) and turn rocker arm pivot nut (P) to obtain desired clearance. Hold pivot nut and tighten locknut to 4.0 N•m (35 in.-lbs.) torque. Tighten valve cover screws to 5.1 N•m (45 in.-lbs.) torque.

CYLINDER HEAD. Manufacturer recommends that after every 100-300 hours of operation the cylinder head is removed and cleaned of deposits.

REPAIRS

TIGHTENING TORQUES. Recommended tightening torque specifications are as follows:

Air cleaner base	5.1 N•m
	(45 in.-lbs.)
Connecting rod	10.2 N•m
	(90 in.-lbs.)
Crankcase cover	19.8 N•m
	(175 in.-lbs.)

FLYWHEEL. A suitable puller should be used to remove flywheel. Briggs & Stratton states that only the steel flywheel key supplied by Briggs & Stratton is recommended. Tighten flywheel nut to 61.2 N•m (45 ft.-lbs.) torque.

CYLINDER HEAD. To remove the cylinder head, first remove blower housing, carburetor, muffler, exhaust manifold and rocker arm cover. Loosen locknuts (3—Fig. B315) and remove rocker arm pivot nuts (4), rocker arms (5) and push rods (9); mark all parts so they can be returned to original location. Unscrew cylinder head screws and remove cylinder head (8).

Note the following when reinstalling cylinder head: Do not apply sealer to cylinder head gasket. Tighten cylinder head bolts progressively in three steps using sequence shown in Fig. B316 until final torque reading of 24.8 N•m (220 in.-lbs.) is obtained.

VALVE SYSTEM. Valves are actuated by rocker arms mounted on a stud threaded into the cylinder head. Depress valve springs so valve keys (11—Fig. B315) can be removed. Release spring pressure and remove retainer, spring and valve from cylinder head. Clean cylinder head thoroughly, then

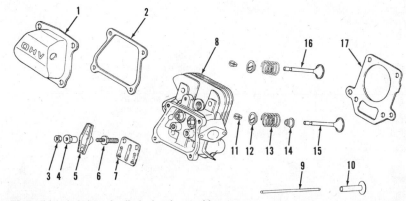

Fig. B315—Exploded view of cylinder head assembly.

1. Rocker cover	
2. Gasket	6. Stud
3. Locknut	7. Push rod guide
4. Pivot nut	8. Cylinder head
5. Rocker arm	9. Push rod

10. Valve lifter (tappet)	14. Valve seal
11. Valve keys	15. Intake valve
12. Valve retainer	16. Exhaust valve
13. Valve spring	17. Head gasket

Illustrations courtesy Briggs & Stratton Corp.

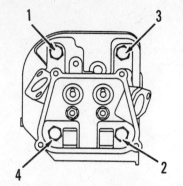

Fig. B316—Tighten cylinder head bolts in steps following sequence shown above.

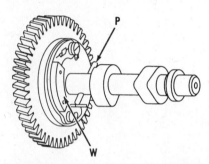

Fig. B317—Drawing of camshaft and compression release mechanism. Pin (P) and weight (W) must move freely.

check for cracks, distortion or other damage.

Valve face and seat angles are 45° for intake and exhaust. Specified seat width is 0.79-1.19 mm (0.031-0.047 in.). Minimum allowable valve margin is 0.38 mm (0.015 in.).

The cylinder head is equipped with nonrenewable valve guides for both valves. Maximum allowable inside diameter of guide is 6.10 mm (0.240 in.).

Rocker arm studs (6—Fig. B315) are threaded into cylinder head. When installing studs, tighten to 19.8 N·m (175 in.-lbs.) torque.

Install a new valve seal (14—Fig. B315) on intake valve.

CAMSHAFT. Camshaft and camshaft gear are an integral casting which is equipped with a compression release mechanism. The compression release pin (P—Fig. B317) extends at cranking speed to hold the exhaust valve open slightly thereby reducing compression pressure. The camshaft rides directly in crankcase and crankcase cover, except on Model 115400 with an extended camshaft. The extended camshaft provides

a 2:1 reduction drive and is supported in the crankcase cover by a ball bearing.

To remove camshaft, remove rocker arm cover and disengage push rods from rocker arms. Drain engine oil, then remove crankcase cover (2—Fig. B318). Rotate crankshaft so timing marks on crankshaft and camshaft gears are aligned (this will position valve tappets out of way). Withdraw camshaft.

Renew camshaft if either camshaft bearing journal diameter is 14.922 mm (0.5875 in.) or less. Renew camshaft if lobes are excessively worn or scored. Inspect compression release mechanism and check for proper operation. Pin and weight should move freely.

Inspect camshaft bearing surfaces in crankcase and crankcase cover. Renew crankcase or crankcase cover if camshaft bearing diameter is 15.06 mm (0.593 in.) or greater. On engines equipped with a ball bearing in the crankcase cover, remove the bearing by pressing to inside of cover. Install bearing so it seats against shoulder in cover.

Install camshaft while aligning timing marks (Fig. B319) on crankshaft and camshaft gears. Install crankcase cover and tighten cover screws to 19.8 N·m (175 in.-lbs.) torque in sequence shown in Fig. B320 or B321. Reassemble remainder of components.

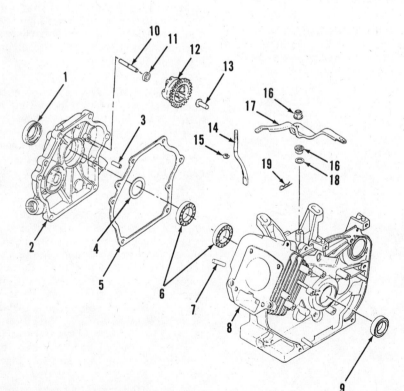

Fig. B318—Exploded view of crankcase and governor.

1. Oil seal	6. Ball bearings	11. Thrust washer	15. Snap ring
2. Crankcase cover	7. Dowel pin	12. Governor gear/flyweight assy.	16. Locknuts
3. Dowel pin	8. Crankcase	13. Thrust sleeve	17. Governor lever
4. Thrust washer	9. Oil seal	14. Governor shaft	18. Washer
5. Gasket	10. Governor shaft		19. Clip

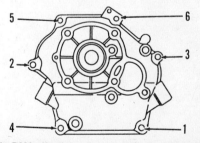

Fig. B319—Install camshaft so timing marks on camshaft and crankshaft gears are aligned.

Fig. B320—Use tightening sequence shown above when tightening crankcase cover screws on Model 85400.

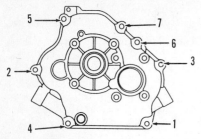

Fig. B321—Use tightening sequence shown above when tightening crankcase cover screws on Model 115400.

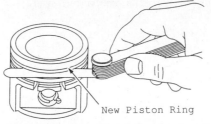

Fig. B323—To check piston ring groove wear, install a new ring in piston groove and measure side clearance between ring and ring land using a feeler gauge. Refer to text for wear limit specification.

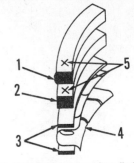

Fig. B324—Drawing of correct installation of piston rings. On Model 115400, top ring (1) does not have an identification mark (5) and can be installed either side up.

1. Top ring, chrome plated
2. Second ring, phosphate coated (black)
3. Oil control rings
4. Expander
5. Identification mark

PISTON, PIN AND RINGS. To remove piston (2—Fig. B322) and rod (5) assembly, drain engine oil and remove cylinder head as previously outlined. Clean pto end of crankshaft and remove any burrs or rust. Unscrew fasteners and remove crankcase cover from cylinder block. Remove camshaft. Remove carbon or any ring wear ridge from top of cylinder before removing piston. Unscrew connecting rod cap screws and push piston and rod out top of engine.

Thoroughly clean the piston and connecting rod, then check for wear or damage. Make sure that the oil drain holes in the oil control ring groove are open.

Maximum allowable piston ring end gap is 0.76 mm (0.030 in.) for compression rings and oil ring rails. Renew piston if ring side clearance exceeds 0.18 mm (0.007 in.) with a new piston ring installed in groove (Fig. B323). Oversize as well as standard size piston and rings are available.

Piston pin is a slip fit in piston and rod. Renew piston if piston pin bore diameter is 13.54 mm (0.533 in.) or greater on Model 85400 or 16.05 mm (0.632 in.) or greater on Model 115400. Renew piston pin if diameter is 13.472 mm (0.5304 in.) or less on Model 85400 or 15.98 mm (0.629 in.) or less on Model 115400.

If top piston ring has an identification mark on one side of the ring, install marked side toward piston crown, otherwise, top ring may be installed with either side up. Install second piston ring with marked side toward piston crown. Refer to Fig. B324.

When assembling piston and rod, note relation of notch in piston crown and long side of rod as shown in Fig. B325 ("MAG" is marked on flywheel side of some rods). Install piston and rod assembly in engine with notch on piston crown toward flywheel. Install rod cap so it meshes properly with rod and dipper points down, then tighten rod screws to 10.2 N•m (90 in.-lbs.) torque.

Install camshaft and remainder of components as previously outlined.

CONNECTING ROD. The connecting rod rides directly on crankpin. Connecting rod and piston are removed as an assembly as outlined in previous section.

Connecting rod reject size for crankpin hole is 26.06 mm (1.026 in.) for Model 85400 and 30.05 mm (1.183 in.) for Model 115400. Renew connecting rod if piston pin bore diameter is 13.54 mm (0.533 in.) or greater on Model 85400 or 16.05 mm (0.632 in.) or greater

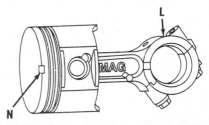

Fig. B325—Assemble rod and piston so long side of rod (L) and notch (N) in piston crown are positioned as shown. Piston must be installed so notch (N) in piston crown is toward flywheel side of engine.

on Model 115400. A connecting rod with 0.51 mm (0.020 in.) undersize big end diameter is available to accommodate a worn crankpin (machining instructions are included with new rod).

GOVERNOR. The governor gear and flyweight assembly is located on the inside of the crankcase cover. The flyweights on the governor gear (12—Fig. B318) force the sleeve (13) against governor arm and shaft (14) in the crankcase. The governor shaft and arm transfer governor action to the external governor linkage.

To remove governor gear, pry up on opposite sides of gear until gear unsnaps. The governor shaft can be removed by driving or pressing out the shaft from the outside of the crankcase cover as shown in Fig. B326. Do not attempt to twist governor shaft out of boss. Twisting will enlarge hole. Install governor shaft so top is 30.99-31.04 mm (1.220-1.222 in.) above boss as shown in Fig. B327. Install governor gear by pushing down until it snaps into place on shaft.

Governor gear and flyweight assembly is available only as a unit assembly.

CRANKSHAFT AND MAIN BEARINGS. The crankshaft is sup-

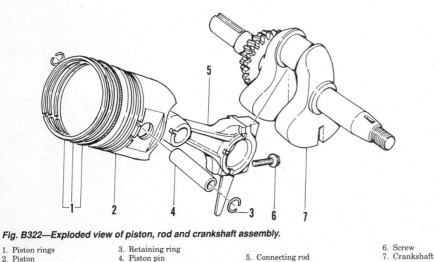

Fig. B322—Exploded view of piston, rod and crankshaft assembly.

1. Piston rings
2. Piston
3. Retaining ring
4. Piston pin
5. Connecting rod
6. Screw
7. Crankshaft

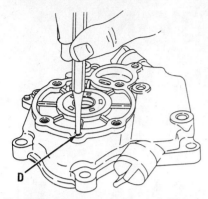

Fig. B326—Drive out governor shaft (D) toward inside of crankcase cover.

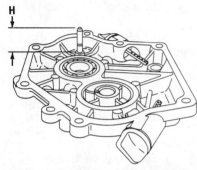

Fig. B327—Install governor shaft so top is 30.99-31.04 mm (1.220-1.222 in.) above boss (H).

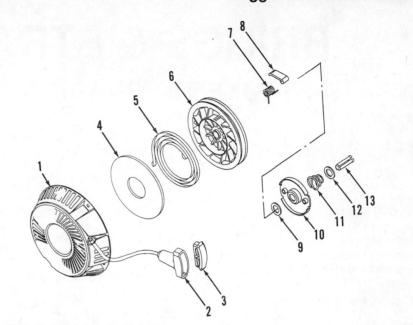

Fig. B328—Exploded view of rewind starter.

1. Starter housing			
2. Handle	5. Rewind spring	8. Dogs	11. Spring
3. Insert	6. Pulley	9. Plastic washer	12. Steel washer
4. Spring cover	7. Springs	10. Retainer	13. Pin

ported by ball bearings located in the crankcase and crankcase cover.

Crankshaft reject size for crankpin is 25.98 mm (1.023 in.) for Model 85400 and 29.98 mm (1.1803 in.) for Model 115400. A connecting rod with 0.51 mm (0.020 in.) undersize big end diameter is available to accommodate a worn crankpin (machining instructions are included with new rod).

Crankshaft end play should be 0.025-0.20 mm (0.001-0.008 in.). Adjust end play by installing required thrust washers (4—Fig. B318).

Remove main bearings by pressing ball bearing to inside of crankcase or crankcase cover. Install bearing so it bottoms against shoulder in crankcase or crankcase cover.

CYLINDER. If cylinder bore wear is 0.76 mm (0.003 in.) or more, or out-of-round is 0.038 mm (0.0015 in.) or greater, cylinder must be rebored to next oversize.

Standard cylinder bore diameter is 62.00-62.03 mm (2.441-2.442 in.) for Model 85400 and 68.00-68.02 mm (2.677-2.678 in.) for Model 115400. Standard and oversize pistons and rings are available.

REWIND STARTER. Refer to Fig. B328 for exploded view of rewind starter.

To install a new rope, proceed as follows. Remove starter and extract old rope from pulley. Allow pulley to unwind, then turn pulley counterclockwise until spring is tightly wound. Rotate pulley clockwise until rope hole in pulley is aligned with rope outlet in housing. Pass rope through pulley hole and housing outlet and tie a temporary knot near handle end of rope. Release pulley and allow rope to wind onto pul-

ley. Install rope handle, release temporary knot and allow rope to enter starter.

To disassemble starter, remove rope and allow pulley to totally unwind. Position a suitable hollow sleeve support under pulley. Using a pin punch, drive out retainer pin (13). Remove retainer (10), brake spring (11), washers (9 and 12), dogs (8) and springs (7). Wear appropriate safety eyewear and gloves before disengaging pulley from starter housing as spring may uncoil uncontrolled. Place shop towel around pulley and lift pulley out of housing; spring should remain with pulley. Do not attempt to separate spring from pulley as they are a unit assembly.

Inspect components for damage and excessive wear. Reverse disassembly procedure to install components. Be sure inner end of rewind spring engages spring retainer adjacent to housing center post. Pin (13) should be driven or pressed until flush with retainer (10). Install rope as previously outlined.

BRIGGS & STRATTON

4-STROKE VANGUARD OHV ENGINES

Model	Bore	Stroke	Displacement	Rated Power
104700	65.1 mm	49.3 mm	164 cc	3.7 kW
	(2.56 in.)	(1.94 in.)	(10 cu.in.)	(5.0 hp)

NOTE: Metric fasteners are used throughout engine except threaded hole in pto end of crankshaft, flange mounting holes and flywheel puller holes which are US threads.

ENGINE INFORMATION

The 104700 Vanguard models are air-cooled, four-stroke, single-cylinder engines. The engine has a vertical crankshaft and utilizes an overhead valve system. Refer to page 59 for a table that outlines the Briggs & Stratton model numbering system.

MAINTENANCE

LUBRICATION. The engine is lubricated by oil supplied by a rotor type oil pump located in the bottom of the crankcase.

Periodically check oil level; do not overfill. Oil dipstick should be screwed in until bottomed for correct oil level reading. Change engine oil after first eight hours of operation and every 50 hours thereafter under normal operating conditions. Recommended oil change interval is 25 hours if severe service is encountered. Note location of three drain plugs in Fig. B401.

The engine may be equipped with a spin-on type oil filter. If so equipped, manufacturer recommends changing oil filter after every 100 hours of operation. Filter should be changed more fre-

quently if engine is operated in a severe environment.

Engine oil should meet or exceed latest API service classification. Use SAE 30 oil for temperatures above 40° F (4° C); use SAE 10W-30 oil for temperatures between 0° F (−18° C) and 40° F (4° C); below 0° F (−18° C) use petroleum based SAE 5W-20 or a suitable synthetic oil.

Crankcase capacity is 0.9 liter (29 fl. oz.) if equipped with an oil filter, 0.7 liter (24 fl. oz.) if not equipped with a filter.

A low oil pressure switch may be located on the sump or on the oil filter housing, if so equipped. Switch should be closed at zero pressure and open at 28-41 kPa (4-6 psi). If switch does not open before engine reaches 2000 rpm, cause must be identified.

AIR CLEANER. The air cleaner consists of a canister and the filter element it contains. The filter element is made of paper. A foam precleaner surrounds the filter element.

The foam precleaner should be cleaned weekly or after every 25 hours of operation, whichever occurs first. The paper filter should be cleaned yearly or after every 100 hours of operation, whichever occurs first.

Tap paper filter gently to dislodge accumulated dirt. Filter may be washed using warm water and nonsudsing detergent directed from inside of filter to outside. DO NOT use petroleum-based cleaners or solvents to clean filter. DO NOT direct pressurized air towards filter. Let filter air dry thoroughly, then inspect filter and discard it if damaged or uncleanable. Clean filter canister. Inspect and, if necessary, replace any defective gaskets.

Clean foam precleaner in soapy water and squeeze until dry. Inspect filter for tears and holes or any other opening. Discard precleaner if it cannot be cleaned satisfactorily or if precleaner is torn or otherwise damaged. Pour clean engine oil into the precleaner, then squeeze to remove excess oil and distribute oil throughout.

FUEL FILTER. The fuel tank is equipped with a filter at the outlet and an inline filter may also be installed in fuel line. Check filters annually and periodically during operating season.

CRANKCASE BREATHER. The engine is equipped with a crankcase breather that provides a vacuum for the crankcase. Vapor from the crankcase is evacuated to the intake manifold. A fiber disk acts as a one-way valve to maintain crankcase vacuum. The breather system must operate properly or excessive oil consumption may result.

The breather valve is located in the top of the crankcase. Remove the flywheel for access to breather cover. Remove cover and inspect the fiber disk valve (V—Fig. B402) should be renewed if warped, damaged or excessively worn. Inspect breather tube (T) for cracks and damage which can cause leakage. Tighten breather cover screws to 6 N•m (55 in.-lbs.) torque.

SPARK PLUG. Recommended spark plug is either an Autolite 3924 or Champion RC12YC. Specified spark plug electrode gap is 0.76 mm (0.030 in.). Tighten spark plug to 19 N•m (165 in.-lbs.) torque.

CAUTION: Briggs & Stratton does not recommend using abrasive blasting to clean spark plugs as this may

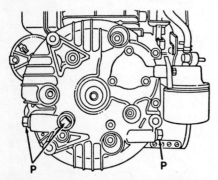

Fig. B401—View of three crankcase drain plugs (P).

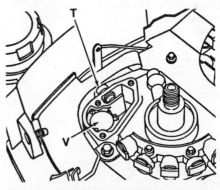

Fig. B402—Crankcase breather is located beneath flywheel. Be sure breather tube (T) and disc valve (V) are not damaged and seal properly.

Illustrations courtesy Briggs & Stratton Corp.

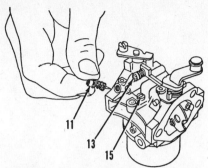

Fig. B403—Idle speed is adjusted by turning screw (13) and idle mixture is adjusted by turning idle mixture screw (11). Idle mixture jet (15) is not adjustable.

introduce some abrasive material into the engine which could cause extensive damage.

CARBURETOR. The engine is equipped with the float type carburetor shown in Fig. B404.

Adjustment. Idle speed at normal operating temperature should be 1500 rpm. Adjust idle speed by turning idle speed screw (13—Fig. B403). Idle mixture is controlled by idle jet (15) and idle mixture screw (11). To adjust idle mixture, turn idle mixture screw clockwise to lean mixture until engine speed just starts to slow, then turn screw counterclockwise to enrich mixture just until engine speed begins to slow. Turn idle screw to halfway point between lean and rich positions. Turn idle screw counterclockwise in small increments to enrich mixture if engine will not accelerate without stumbling. Idle mixture jet is not adjustable. High speed operation is controlled by main jet (22—Fig. B404) and is not adjustable. An optional main jet is available for high altitude operation.

Overhaul. To disassemble carburetor, remove fuel bowl retaining screw (28—Fig. B404) and fuel bowl (26). Remove float pin (23) by pushing against round end of pin toward the square end of pin. Remove float (24) and fuel inlet needle (20). Remove throttle and choke shaft assemblies after unscrewing throttle and choke plate retaining screws. Remove idle mixture screw (11), idle mixture jet (15), main jet (22), main fuel nozzle (21) and air bleeds (17 and 18).

When assembling the carburetor, note the following. Place a small drop of nonhardening sealant such as Permatex #2 or equivalent on throttle and choke plate retaining screws. Numbers on choke plate (16) must face out and be on fuel inlet side of carburetor. Install throttle shaft seal (10) with flat side toward carburetor. Numbers on throttle

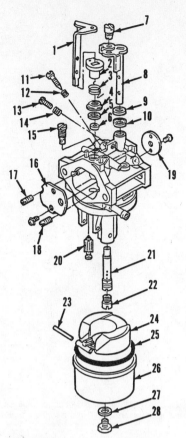

Fig. B404—Exploded view of carburetor.

1. Choke shaft & lever	15. Idle mixture jet
2. Bushing	16. Choke plate
3. Spring	17. Idle air bleed
4. Washer	18. Main air bleed
5. Seal	19. Throttle plate
6. Bushing	20. Fuel inlet valve
7. Link retainer	21. Main fuel nozzle
8. Throttle shaft & lever	22. Main jet
9. Washer	23. Float pin
10. Seal	24. Float
11. Idle mixture screw	25. Gasket
12. Spring	26. Fuel bowl
13. Idle speed screw	27. Washer
14. Spring	28. Screw

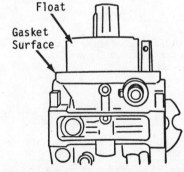

Fig. B405—Float must be parallel with gasket surface when carburetor is inverted. Float height is not adjustable so components must be replaced if not parallel.

plate (19) must face out and be on fuel inlet side of carburetor. Be sure groove of fuel inlet valve (20) engages slot in float tab. Float should be parallel with body when carburetor is inverted as shown in Fig. B405. Float height is not

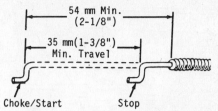

Fig. B406—Control wire must be capable of travel shown above for proper operation.

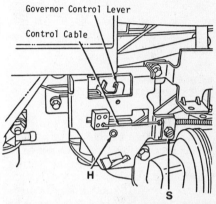

Fig. B407—Loosen cable clamp screw (S) and align holes (H) of governor lever and bracket to synchronize remote control and carburetor as outlined in text.

adjustable; replace components necessary so float is parallel.

If removed, install spacer between carburetor and manifold so circular opening is toward carburetor and "D" shaped opening is toward engine. Tighten carburetor mounting nuts to 5.1 N•m (45 in.-lbs.) torque.

CHOKE-A-MATIC CARBURETOR CONTROLS. The engine may be equipped with a control unit that controls the throttle, choke and grounding switch.

The remote control wire must travel at least 35 mm (1⅜ in.) for proper Choke-A-Matic operation. See Fig. B406.

To check operation of Choke-A-Matic carburetor control, move control lever to "CHOKE" position. Carburetor choke plate must be completely closed. Move control lever to "STOP" position. Magneto grounding switch should be making contact. With the control lever in "RUN", "FAST" or "SLOW" position, carburetor choke should be completely open. On units with remote control, synchronize movement of remote lever to carburetor control lever by moving remote control lever to "FAST" position. Loosen cable clamp screw (S—Fig. B407) and move cable so holes (H) in governor lever and bracket are aligned. Retighten clamp screw. Check for proper operation.

Illustrations courtesy Briggs & Stratton Corp.

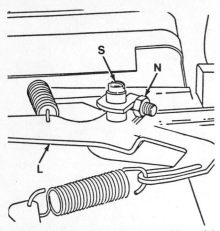

Fig. B408—View of governor shaft and lever. Adjust linkage as outlined in text.

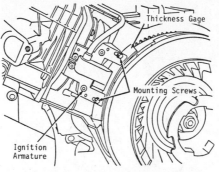

Fig. B409—Adjust air gap between ignition armature and flywheel to 0.20-0.30 mm (0.008-0.012 in.).

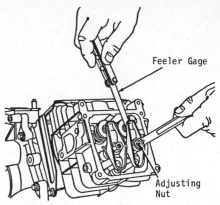

Fig. B412—Use a feeler gauge to check clearance between valve stem and rocker arm.

GOVERNOR. The engine is equipped with a mechanical, flyweight type governor. To adjust governor linkage, proceed as follows: Loosen governor lever clamp nut (N—Fig. B408), rotate governor lever (L) so throttle plate is fully open and hold lever in place. Turn governor shaft (S) clockwise as far as possible, then tighten nut (N) to 4 N•m (35 in.-lbs.) torque.

If internal governor assembly must be serviced, refer to REPAIRS section.

IGNITION SYSTEM. The engine is equipped with a Magnetron ignition system.

To check spark, remove spark plug and connect spark plug cable to B&S tester 19051, then ground remaining tester lead to engine. Spin engine at 350 rpm or more. If spark jumps the 4.2 mm (0.165 in.) tester gap, system is functioning properly.

To remove armature and Magnetron module, remove flywheel shroud and armature retaining screws. Disconnect stop switch wire from module. When installing armature, position armature so air gap between armature legs and flywheel surface (Fig. B409) is 0.20-0.30 mm (0.008-0.012 in.).

FLYWHEEL BRAKE. The engine is equipped with a flywheel brake that utilizes a pad which is forced against the flywheel's circumference. The brake should stop the engine within three seconds when the operator releases mower safety control and the speed control is in high speed position. Stopping time can be checked using tool 19255.

To check brake adjustment, remove starter and properly ground spark plug lead to prevent accidental starting. Turn flywheel nut using a torque wrench with brake engaged. Rotating flywheel nut at a steady rate in a clock-

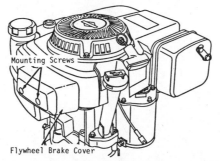

Fig. B410—The flywheel brake cover may be removed after unscrewing two mounting screws.

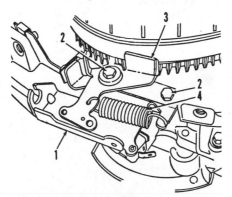

Fig. B411—Flywheel rotation is stopped when brake pad (3) contacts flywheel.

1. Brake lever bracket
2. Mounting screws
3. Brake lever & pad
4. Brake spring

wise direction should require at least 3.4 N•m (30 in.-lbs.) torque.

An insufficient torque reading may indicate misadjustment or damaged components. Remove flywheel brake cover shown in Fig. B410 for access to brake mechanism (Fig. B411). Inspect components for damage and excessive wear. Renew brake arm and pad assembly (3—Fig. B411) if friction material of pad is damaged or less than 2.28 mm (0.090 in.) thick. If condition of mechanism is satisfactory, reduce clearance between pad and flywheel by repositioning control cable housing (1). Recheck brake action.

VALVE ADJUSTMENT. Remove rocker arm cover. Remove spark plug. Rotate crankshaft so piston is at top dead center on compression stroke. Then, using a suitable measuring device inserted through spark plug hole, rotate crankshaft clockwise as viewed at flywheel end so piston is 6.35 mm (0.250 in.) below TDC to prevent interference by the compression release mechanism with the exhaust valve. Clearance between rocker arm pad and valve stem end (Fig. B412) should be 0.08-0.12 mm (0.003-0.005 in.) for intake and exhaust. Loosen lock screw and turn rocker arm adjusting nut to obtain desired clearance. Tighten lock screw to 5.6 N•m (50 in.-lbs.) torque.

CYLINDER HEAD. Manufacturer recommends that after every 100-300 hours of operation the cylinder head is removed and cleaned of deposits.

REPAIRS

TIGHTENING TORQUES. Recommended tightening torque specifications are as follows:

Carburetor mounting nuts . . . 5.1 N•m (45 in.-lbs.)
Connecting rod . . . . 10 N•m (90 in.-lbs.)
Cylinder head . . . . 8.5 N•m (75 in.-lbs.)
Flywheel nut . . . . 88 N•m (65 ft.-lbs.)
Oil pan . . . . 16 N•m (140 in.-lbs.)
Oil pump cover . . . . 6.2 N•m (55 in.-lbs.)
Rocker arm lock screw . . . . . . 5.6 N•m (50 in.-lbs.)
Rocker arm stud . . . . 9.6 N•m (85 in.-lbs.)

CYLINDER HEAD. To remove the cylinder head, first remove blower housing, carburetor, muffler, exhaust manifold and rocker arm cover. Loosen lock

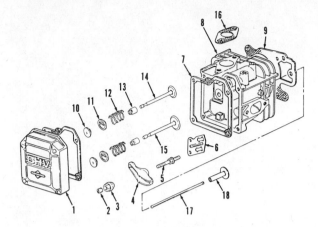

Fig. B413—Exploded view of cylinder head assembly.

1. Rocker cover
2. Lock screw
3. Adjusting nut
4. Rocker arm
5. Stud
6. Push rod guide
7. Gasket
8. Cylinder head
9. Head gasket
10. Valve cap
11. Valve retainer
12. Valve spring
13. Valve seal
14. Exhaust valve
15. Intake valve
16. Exhaust gasket
17. Push rod
18. Cam follower (tappet)

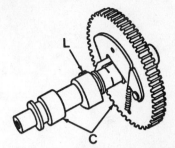

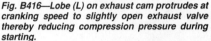

Fig. B416—Lobe (L) on exhaust cam protrudes at cranking speed to slightly open exhaust valve thereby reducing compression pressure during starting.

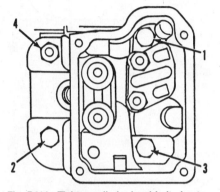

Fig. B414—Tighten cylinder head bolts in steps following sequence shown above.

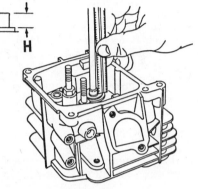

Fig. B415—Valve guide height (H) should be 3.05-3.81 mm (0.120-0.150 in.).

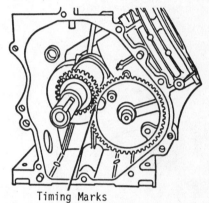

Timing Marks

Fig. B417—Install camshaft so timing marks on camshaft and crankshaft gears are aligned.

screws (2—Fig. B413) and remove rocker arm adjusting nuts (3), rocker arms (4) and push rods (17); mark all parts so they can be returned to original location. Unscrew cylinder head bolts and remove cylinder head (8). Clean cylinder head thoroughly, then check for cracks, distortion or other damage.

Note the following when reinstalling cylinder head: Do not apply sealer to cylinder head gasket. Tighten cylinder head bolts progressively in three steps using sequence shown in Fig. B414 until final torque reading of 8.5 N•m (75 in.-lbs.) is obtained.

VALVE SYSTEM. Valves are actuated by rocker arms mounted on a stud threaded into the cylinder head. Remove valve wear caps (10—Fig. B413). Depress valve springs until slot in valve spring retainer (11) can be aligned with end of valve stem. Release spring pressure and remove retainer, spring and valve from cylinder head. Clean cylinder head thoroughly, then check for cracks, distortion or other damage.

Valve face and seat angles are 45° for intake and exhaust. Specified seat width is 0.8-1.2 mm (0.031-0.047 in.). Minimum allowable valve margin is 0.38 mm (0.015 in.).

The cylinder head is equipped with renewable valve guides for both valves. Maximum allowable inside diameter of guide is 6.10 mm (0.240 in.). Use B&S tool 19367 to remove and install guides. Guides may be installed either way up. Top of guide should protrude 3.05-3.81 mm (0.120-0.150 in.) as shown in Fig. B415. Use B&S tools 19345 and 19346 to ream valve guide to correct size.

Rocker arm studs (5—Fig. B413) are screwed into cylinder head. Hardening sealant should be applied to threads contacting cylinder head. When installing studs, tighten to 9.6 N•m (85 in.-lbs.) torque.

CAMSHAFT. Camshaft and camshaft gear (C—Fig. B416) are an integral casting which is equipped with a compression release mechanism. The compression release lobe (L) extends at cranking speed to hold the exhaust valve open slightly thereby reducing compression pressure.

To remove camshaft proceed as follows: Drain crankcase oil and remove engine from equipment. Clean pto end of crankshaft and remove any burrs or rust. Remove rocker arm cover, rocker arms and push rods; mark all parts so they can be returned to original position. Unscrew fasteners and remove oil

pan. Rotate crankshaft so timing marks on crankshaft and camshaft gears are aligned (this will position valve tappets out of way). Withdraw camshaft and remove tappets.

Reject size for bearing journal at flywheel end of camshaft is 12.65 mm (0.498 in.); reject size for pto end bearing journal is 17.45 mm (0.687 in.). Reject size for camshaft lobes is 28.85 mm (1.136 in.). With compression release lobe (L—Fig. B416) fully extended, lobe protrusion should be 0.51-0.64 mm (0.020-0.025 in.). If not, renew camshaft. Compression release mechanism must operate freely without binding.

Reverse removal procedure to reassemble components. Install camshaft while aligning timing marks (Fig. B417) on crankshaft and camshaft gears. Be sure governor arm is in proper position to contact governor, and camshaft end will properly engage oil pump drive. Install oil pan and tighten cover screws to 16 N•m (140 in.-lbs.) torque in sequence shown in Fig. B418. Do not force mating of oil pan with crankcase. Reassemble remainder of components.

PISTON, PIN AND RINGS. To remove piston and rod assembly, drain engine oil and remove engine from equipment. Remove cylinder head and camshaft as previously outlined. Un-

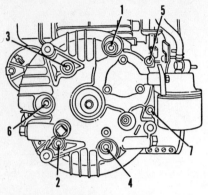

Fig. B418—Use tightening sequence shown above when tightening oil pan screws.

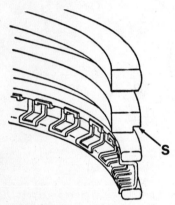

Fig. B419—Install second compression piston ring so step (S) is toward piston skirt.

screw connecting rod screws and remove piston and rod.

Insert each piston ring (one at a time) squarely in top of cylinder and use a feeler gauge to measure ring end gap. Maximum allowable piston ring end gap is 0.76 mm (0.030 in.) for compression rings and 1.65 mm (0.065 in.) for oil ring. Check piston skirt for score marks and wear and renew as necessary. To check piston ring grooves for wear, insert a new ring in piston ring groove and use a feeler gauge to measure side clearance between ring and piston land. Renew piston if ring side clearance exceeds 0.10 mm (0.004 in.) for compression rings and 0.20 mm (0.008 in.) for oil ring. Piston and rings are available in oversizes as well as standard size.

Piston pin is a slip fit in piston and rod. Renew piston if piston pin bore diameter is 14.02 mm (0.552 in.) or greater. Renew piston pin if diameter is 14.00 mm (0.551 in.) or less. A piston pin 0.012 mm (0.0005 in.) oversize is available.

Top piston ring may be installed with either side up. Second piston ring must be installed with stepped edge (S—Fig. B419) toward piston skirt.

When assembling piston and rod, note relation of notch in piston crown

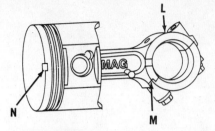

Fig. B420—Assemble rod and piston so long side of rod (L) and notch (N) in piston crown are positioned as shown. Piston must be installed so notch (N) in piston crown is toward flywheel side of engine.

and long side of rod as shown in Fig. B420 ("MAG" is marked on flywheel side of some rods). Install piston and rod assembly in engine with notch on piston crown toward flywheel. Install rod cap so match marks (M—Fig. B420) are aligned and tighten rod screws to 10 N•m (90 in.-lbs.) torque.

Install camshaft and cylinder head as previously outlined.

CONNECTING ROD. The connecting rod rides directly on crankpin. Connecting rod and piston are removed as an assembly as outlined in previous section.

Connecting rod reject size for crankpin hole is 31.33 mm (1.233 in.). Renew connecting rod if piston pin bore diameter is 14.02 mm (0.552 in.) or greater. A connecting rod with 14.02 (0.552 in.) undersize big end diameter is available to accommodate a worn crankpin (machining instructions are included with new rod).

GOVERNOR. The governor gear and flyweight assembly (30—Fig. B421) is located on the inside of the oil pan. The plunger in the gear assembly contacts the governor arm and shaft (16—Fig. B422) in the crankcase. The governor shaft and arm transfer gover-

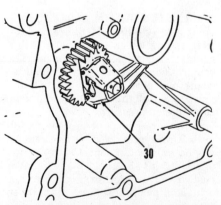

Fig. B421—Governor gear and flyweight assembly (30) is mounted inside the oil pan.

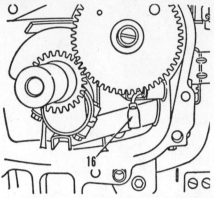

Fig. B422—Governor shaft (16) must be positioned as shown before installing oil pan.

nor action to the external governor linkage.

To gain access to the governor gear assembly, drain crankcase oil and remove engine from equipment. Clean pto end of crankshaft and remove any burrs or rust. Unscrew fasteners and remove oil pan.

Flyweight assembly must operate freely for proper governor action. The governor shaft and arm ride in a bushing (7—Fig. B423) in the crankcase. The bushing should be renewed if worn excessively. B&S reamer 19333 will size bushing to desired diameter.

To reassemble, position governor gear assembly in oil pan. Be sure that governor arm is in proper position to contact governor as shown in Fig. B422 and that camshaft end will properly engage oil pump drive. Install oil pan and tighten retaining screws to 16 N•m (140 in.-lbs.) torque in sequence shown in Fig. B418. Do not force mating of oil pan with crankcase. Reassemble remainder of components.

CRANKSHAFT AND MAIN BEARINGS. To remove crankshaft, first remove flywheel and camshaft. Rotate crankshaft so piston is at top dead center and remove connecting rod cap. Rotate crankshaft so it will clear connecting rod and withdraw crankshaft from crankcase.

Renew crankshaft if main bearing journal diameter is 29.94 mm (1.179 in.) or less. Wear limit for crankpin is 31.19 mm (1.228 in.). A connecting rod with 0.51 mm (0.020 in.) undersize big end diameter is available to accommodate a worn crankpin (machining instructions are included with new rod).

Main bearing located in crankcase is a renewable bushing. The pto end of crankshaft rides directly in the aluminum bore of oil pan. If crankcase bushing diameter is 30.07 mm (1.184 in.) or greater, renew bushing. When installing new main bearing, be sure that oil

Fig. B423—Exploded view of engine crankcase and cylinder assembly.

1. Oil seal
2. Breather cover
3. Breather valve
4. Gasket
5. Crankcase
7. Governor shaft bushing
8. Seal
9. Washer
10. Cotter pin
11. Washers
12. Governor lever
13. Dowel pin
14. Dowel pin
16. Governor shaft
17. Crankshaft
18. Key
19. Connecting rod
20. Piston pin
21. Clips
22. Piston
23. Piston rings
24. Key
25. Gear
26. Snap ring
27. Camshaft
28. Compression release spring
29. Oil pump drive pin
30. Governor gear & flyweight
31. Snap ring
32. Washer
33. Oil screen
34. Spring
35. Oil pressure relief valve
36. Gasket
37. Oil pan
38. Oil seal
39. Drain plug
40. Oil pump inner rotor
41. Outer rotor
42. "O" ring
43. Oil pump cover

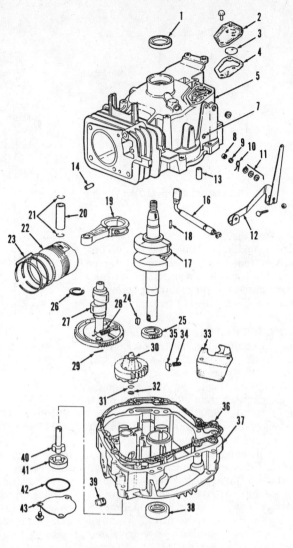

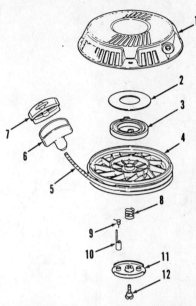

Fig. B425—Exploded view of rewind starter.

1. Starter housing	7. Insert
2. Spring cover	8. Brake spring
3. Rewind spring	9. Dog springs
4. Pulley	10. Dogs
5. Rope	11. Retainer
6. Handle	12. Screw

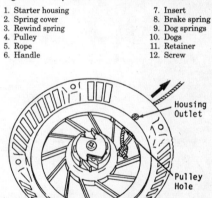

Fig. B426—Rope should extend through hole in pulley and housing outlet.

Fig. B424—Remove oil pump cover (C) for access to oil pump components.

hole in bearing is aligned with oil hole in crankcase. Renew oil pan if diameter of main bearing bore is 30.07 mm (1.184 in.) or greater.

CYLINDER. If cylinder bore wear exceeds 0.076 mm (0.003 in.) or if bore is out-of-round more than 0.038 mm (0.0015 in.), then cylinder should be bored to the next oversize. Standard

cylinder bore diameter is 65.06-65.09 mm (2.5615-2.5625 in.).

OIL PUMP. The rotor type oil pump is located in the oil pan and is driven by the camshaft.

Remove engine from equipment for access to oil pump cover (C—Fig. B424). Remove cover and extract pump rotors (40 and 41—Fig. B423). Mark rotors so they can be reinstalled in their original position. Renew any components which are damaged or excessively worn. Tighten oil pump cover screws to 6.2 N•m (55 in.-lbs.) torque.

REWIND STARTER. Refer to Fig. B425 for exploded view of rewind starter.

To install a new rope, proceed as follows. First remove starter from engine. Pull old rope out part way, untie knot and remove insert (7) and handle (6). Then pull rope out as far as it will go and while holding rewind pulley, extract old rope from pulley. Allow pulley to unwind. Rope length should be 234 cm (92

in.). Turn pulley counterclockwise until spring is tightly wound, then rotate pulley clockwise until rope hole in pulley is aligned with rope outlet in housing. Pass new rope through pulley hole and housing outlet as shown in Fig. B426 and tie a temporary knot near handle end of rope. Release pulley and allow rope to wind onto pulley. Install rope handle, release temporary knot and allow rope to enter starter.

To disassemble starter, remove rope handle and allow pulley to totally unwind. Unscrew pulley retaining screw (12—Fig. B425). Remove retainer (11), dogs (10), springs (9) and brake spring (8). Wear appropriate safety eyewear and gloves before disengaging pulley from starter as spring may uncoil uncontrolled. Place shop cloth around pulley and lift pulley out of housing; spring should remain with pulley. Do not attempt to separate spring from pulley as they are a unit assembly.

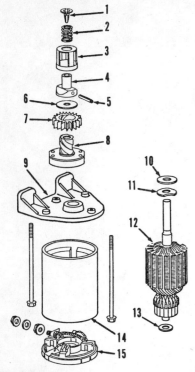

Fig. B427—Exploded view of optional electric starter motor.

1. Spring retainer	9. Drive end cap
2. Return spring	10. Thrust washer
3. Cover	11. Spring washer
4. Clutch retainer	12. Armature
5. Roll pin	13. Washer
6. Washer	14. Starter housing
7. Pinion gear	15. Brush end cap
8. Helix	

Inspect components for damage and excessive wear. Reverse disassembly procedure to install components. Be sure inner end of rewind spring engages spring retainer adjacent to housing center post. Tighten screw (12) to 7.9 N•m (70 in.-lbs.) torque. Install rope as previously outlined.

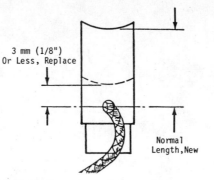

Fig. B428—Renew brush if worn to 3 mm (1/8 in.) or less.

When installing the starter, position starter on blower housing then pull out starter rope until dogs engage starter cup. Continue to place tension on rope and tighten starter mounting screws to 6.2 N•m (55 in.-lbs.) torque.

ELECTRIC STARTER. The engine may be equipped with a 12-volt DC electric starter motor. Refer to Fig. B427 for an exploded view of starter motor.

Under no load, the starter should draw no more than 20 amps at a minimum speed of 5000 rpm.

When disassembling starter, do not clamp starter housing (14) in a vise or strike housing as ceramic magnets may crack. Place alignment marks on drive end cap (9) and starter housing before disassembly. Drive roll pin (5) out of starter drive retainer (4) and remove starter drive assembly. Remove through-bolts and withdraw drive end cap (9), brush end cap (15) and armature (12) from field housing (14).

Install new brushes if existing brushes are worn to less than 3 mm (⅛ in.) as shown in Fig. B428. Brush spring pressure should be at least 113-

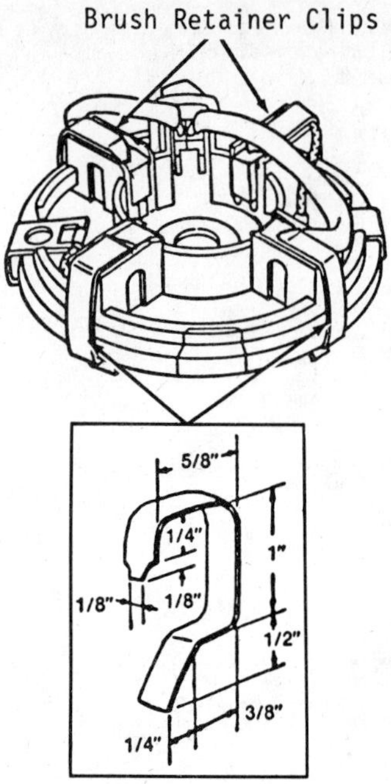

Fig. B429—Use suitable metal, such as a rewind starter spring, to construct brush retainer clips for use in assembling starter motor.

170 grams (4-6 ounces). Minimum allowable commutator diameter is 31.24 mm (1.230 in.).

When reassembling starter, brush retainer clips shown in Fig. B429 may be constructed from spring steel (such as scrap rewind starter springs) and used to retain brushes in end cap. Lubricate bearings in both end caps with a small amount of 20W oil. Notch in starter housing must index with protruding terminal on brush end cap (15—Fig. B427). Apply Lubriplate to inside diameter of gear (7). Install gear with chamfered edges of teeth toward end of shaft. A new roll pin (5) should be installed. Pin must be centered in shaft.

BRIGGS & STRATTON
4-STROKE EUROPA ENGINES

Model	Bore	Stroke	Displacement	Power Rating
99700	2.56 in.	1.74 in.	8.9 cu.in.	5 hp
	(65.0 mm)	(44.2 mm)	(147 cc)	(3.7 kW)

ENGINE INFORMATION

The 99700 Europa models are air-cooled, four-stroke, single-cylinder engines. The engine has a vertical crankshaft and utilizes an overhead valve system. Refer to page 59 for a table that outlines the Briggs & Stratton model numbering system.

MAINTENANCE

LUBRICATION. The engine is lubricated by oil supplied by a rotor type oil pump located in the bottom of the crankcase, as well as by a slinger driven by the camshaft gear.

Change oil after first eight hours of operation and after every 50 hours of operation or at least once each operating season. Change oil weekly or after every 25 hours of operation if equipment undergoes severe usage.

Engine oil level should be maintained at full mark on dipstick. Engine oil should meet or exceed latest API service classification. Use SAE 30 oil for temperatures above 40° F (4° C); use SAE 10W-30 oil for temperatures between 0° F (–18° C) and 40° F (4° C); below 20° F (–7° C) use petroleum based SAE 5W-20 or a suitable synthetic oil.

Crankcase capacity is 22 fl. oz. (0.65 L). Fill engine with oil so oil level reaches, but does not exceed, full mark on dipstick.

AIR CLEANER. The air cleaner consists of a canister and the filter element it contains. The filter element is made of paper. A foam precleaner surrounds the filter element.

The foam precleaner should be cleaned weekly or after every 25 hours of operation, whichever occurs first. The paper filter should be cleaned yearly or after every 100 hours of operation, whichever occurs first.

Tap paper filter gently to dislodge accumulated dirt. Filter may be washed using warm water and nonsudsing detergent directed from inside of filter to outside. DO NOT use petroleum-based cleaners or solvents to clean filter. DO

NOT direct pressurized air towards filter. Let filter air dry thoroughly, then inspect filter and discard it if damaged or uncleanable. Clean filter canister. Inspect and, if necessary, replace any defective gaskets.

Clean foam precleaner in soapy water and squeeze until dry. Inspect filter for tears and holes or any other opening. Discard precleaner if it cannot be cleaned satisfactorily or if precleaner is torn or otherwise damaged. Pour clean engine oil into the precleaner, then squeeze to remove excess oil and distribute oil throughout.

CRANKCASE BREATHER. The engine is equipped with a crankcase breather that provides a vacuum for the crankcase. Vapor from the crankcase is evacuated to the air cleaner housing. A fiber disk acts as a one-way valve to maintain crankcase vacuum. The breather system must operate properly or excessive oil consumption may result.

The crankcase breather is built into the tappet cover. A fiber disc acts as a one-way valve. Clearance between fiber disc valve and breather body should not exceed 0.045 inch (1.14 mm). If it is possible to insert a 0.045 inch (1.14 mm) diameter wire (W—Fig. B501) between disc and breather body, renew breather assembly. Do not use excessive force when measuring gap. Disc should not stick or bind during operation. Renew if

distorted or damaged. Inspect breather tube for leakage.

SPARK PLUG. Recommended spark plug is Champion RC12YC or Autolite 3924. Specified spark plug electrode gap is 0.030 inch (0.76 mm).

CAUTION: Briggs & Stratton does not recommend using abrasive blasting to clean spark plugs as this may introduce some abrasive material into the engine which could cause extensive damage.

CARBURETOR. The engine is equipped with a Walbro LMS float type carburetor.

The Walbro LMS carburetor operates like other typical float type carburetors. If the engine is operated at high altitudes, improved engine performance may be obtained by removing the metal jet in the right main air bleed hole (H—Fig. B502), however, the air jet should be installed if engine is operated at lower altitudes. Removing the air jet leans the high speed mixture.

Adjustment. Initial setting of idle mixture screw (9—Fig. B503) is 1¼ turns out from a lightly seated position. Run engine until normal operating temperature is attained. Be sure choke is open. Run engine with speed control in slow position and adjust idle speed

Fig. B501—Clearance between fiber disc valve and crankcase breather housing must be less than 0.045 inch (1.15 mm). A spark plug wire gauge (W) may be used to check clearance as shown, but do not apply pressure against disc valve.

Fig. B502—Removing the air jet from the right main air bleed hole (H) leans the high speed mixture, which may be desirable if the engine is operated at high altitudes. See text.

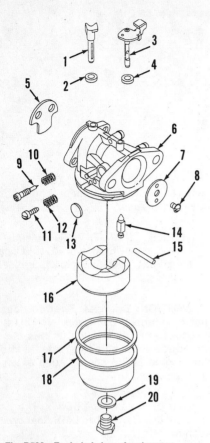

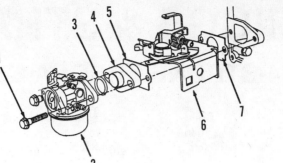

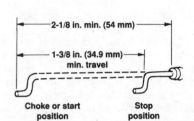

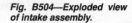

Fig. B504—Exploded view of intake assembly.

1. Screws
2. Carburetor
3. "O" ring
4. Adapter
5. Gasket
6. Speed control bracket
7. Gasket

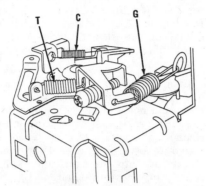

Fig. B505—Drawing of speed control assembly showing location of choke spring (C), throttle spring (T) and governor spring (G).

Fig. B506—For proper operation of Choke-A-Matic controls, remote control wire must extend to dimension shown and have a minimum travel of 1-3/8 inches (34.9 mm).

Fig. B503—Exploded view of carburetor.

1. Choke shaft	11. Idle speed screw
2. Washer	12. Spring
3. Throttle shaft	13. Welch plug
4. Washer	14. Fuel inlet valve
5. Choke plate	15. Pin
6. Body	16. Float
7. Throttle plate	17. Gasket
8. Screw	18. Fuel bowl
9. Idle mixture screw	19. Gasket
10. Spring	20. Main jet nut

screw so engine speed is 1500 rpm. Turn idle mixture screw clockwise until engine begins to stumble and note screw position. Turn idle mixture screw counterclockwise until engine begins to stumble and note screw position. Turn the idle mixture screw clockwise to a position that is midway from clockwise (lean) and counterclockwise (rich) positions. With engine running at idle, rapidly move speed control to full throttle position. If engine stumbles or hesitates, slightly turn idle mixture screw counterclockwise and repeat test. Recheck idle speed and, if necessary, readjust idle speed screw.

High speed mixture is controlled by the jet in the fuel bowl retaining nut (20—Fig. B503) and is not adjustable. A nut with a jet for high altitude operation is available.

R&R And Overhaul. When removing or installing carburetor, note location of components shown in Fig. B504. Also note the location of governor linkage and springs to ensure correct reas-

sembly. Tighten carburetor mounting screws to 75 in.-lbs. (8.5 N•m) torque.

To disassemble carburetor refer to Fig. B503 and remove bowl nut (20) and float bowl (18). Remove float pin (15), float (16) and fuel inlet valve (14). If necessary to remove inlet valve seat, use a stiff wire with a hook on one end to pull seat from carburetor body. To remove the Welch plug (13), pierce the plug with a sharp pin punch, then pry out the plug, but do not damage underlying metal. Remove idle mixture needle (9). Remove retaining screws from throttle plate (7) and choke plate (5) and withdraw throttle and choke shafts.

Inspect carburetor and renew any damaged or excessively worn components. The body must be replaced if there is excessive throttle or choke shaft play as bushings are not available. Install Welch plug while being careful not to indent the plug; the plug should be flat after installation. Apply a nonhardening sealant around the plug. Use a 3/16 inch (5 mm) diameter rod to install the fuel inlet valve seat. The groove on the seat must be down (towards carburetor bore). Push in the seat until it bottoms. Install the choke plate so the numbers are visible when the choke is closed. Install the throttle plate so the numbers are visible and towards the idle mixture screw when the throttle is closed. The float level is not adjustable. If the float is not approximately parallel

with the body when the carburetor is inverted, then the float, fuel valve and/or valve seat must be replaced. Tighten the high speed mixture nut to 50 in.-lbs. (5.6 N•m) torque.

Refer to Fig. B505 for view of speed control panel.

CHOKE-A-MATIC CARBURETOR CONTROL. The engine may be equipped with a control unit with which the carburetor choke, throttle and magneto grounding switch are operated from a single lever (Choke-A-Matic).

If equipped with a control cable between the remote control and engine, the remote control cable must move a certain distance for the remote control and carburetor to be synchronized. At full extension, control wire must extend 2 1/8 inches (54 mm) from cable housing as shown in Fig. B506. The wire must travel at least 1 3/8 inches (34.9 mm) from "CHOKE" or "START" to "STOP" positions.

To synchronize the control cable and speed control, loosen cable clamp screw (W—Fig. B507). Move the equipment speed control to the fast position. Move the speed control lever (L) so a 1/8 inch (3 mm) diameter rod (R) can be inserted in holes in the lever and control bracket (B). Retighten clamp screw (W).

GOVERNOR. Remove blower housing and fuel tank for access to governor linkage. To adjust governor linkage, loosen governor lever clamp nut (N—Fig. B508). Move speed control lever (L—Fig. B507) to fast position and insert a 1/8 inch (3 mm) diameter rod (R)

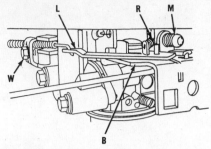

Fig. B507—Drawing of speed control assembly. Refer to text for adjustments.

Fig. B508—Drawing of governor lever (L), shaft (S) and clamp nut (N). Refer to text for adjustment.

into holes in the lever and control bracket (B). Rotate governor shaft (S—Fig. B508) counterclockwise as far as possible, hold shaft and tighten clamp nut.

Maximum governed speed is adjusted by turning screw (M—Fig. B507). Maximum governed speed is usually specified by the equipment manufacturer according to engine application.

IGNITION. The engine is equipped with a Magnetron breakerless ignition system. All components are located outside the flywheel.

To check spark, remove spark plug and connect spark plug cable to B&S tester 19051, then ground remaining tester lead to engine. Spin engine at 350 rpm or more. If spark jumps the 0.165 inch (4.2 mm) tester gap, system is functioning properly.

Armature air gap should be 0.006-0.012 inch (0.15-0.30 mm).

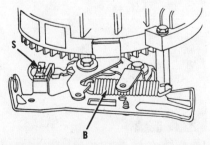

Fig. B509—Drawing of flywheel brake assembly used on some engines showing location of stop switch (S) and brake spring (B).

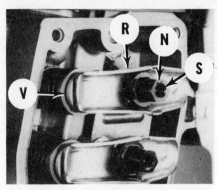

Fig. B510—Loosen lock screw (S) and rotate adjusting nut (N) to adjust valve clearance. Clearance between rocker arm (R) and valve stem cap (V) should be 0.005-0.007 inch (0.12-0.18 mm) for intake and exhaust.

On models equipped with a flywheel brake, stop switch (S—Fig. B509) should ground ignition system when flywheel brake is activated.

VALVE ADJUSTMENT. Remove rocker arm cover. Remove spark plug. Rotate crankshaft so piston is at top dead center on compression stroke. Insert a suitable measuring device through spark plug hole, then rotate crankshaft clockwise as viewed at flywheel end so piston is ¼ inch (6.4 mm) below TDC to prevent interference by the compression release mechanism with the exhaust valve. Measure clearance between rocker arm (R—Fig. B510) and valve stem cap (V) using a feeler gauge. Clearance should be 0.005-0.007 inch (0.12-0.18 mm) for intake and exhaust. Loosen lock screw (S) and turn rocker arm pivot nut (N) to obtain desired clearance.

NOTE: Pivot nut is 10 mm and lock screw is either Torx design or metric Allen design.

After clearance is adjusted, hold pivot nut and retighten lock screw to 45 in.-lbs. (5.1 N•m) torque. Tighten valve cover screws to 85 in.-lbs. (9.6 N•m) torque.

CYLINDER HEAD. Manufacturer recommends that after every 100-300 hours of operation the cylinder head is removed and cleaned of deposits.

REPAIRS

TIGHTENING TORQUES. Recommended tightening torque specifications are as follows:

Connecting rod . . . . 100 in.-lbs. (11.3 N•m)
Cylinder head . . . . 160 in.-lbs. (18.1 N•m)
Flywheel nut . . . . 60 ft.-lbs. (81.6 N•m)
Oil pan . . . . 85 in.-lbs. (9.6 N•m)
Oil pump cover . . . . 80 in.-lbs. (9 N•m)
Rocker arm cover . . . . 85 in.-lbs. (9.6 N•m)
Rocker arm lock screw . . . . 45 in.-lbs. (5.1 N•m)
Rocker arm stud . . . . 110 in.-lbs. (12.4 N•m)

CYLINDER HEAD. To remove the cylinder head, first remove fuel tank, air cleaner, blower housing, carburetor, speed control bracket, muffler, and rocker arm cover. Loosen lock screws (10—Fig. B511) and remove rocker arm adjusting nuts (11), rocker arms (12) and push rods (15); mark all parts so they can be returned to original location. Unscrew cylinder head screws and remove cylinder head (17). Clean cylinder head thoroughly, then check for cracks, distortion or other damage.

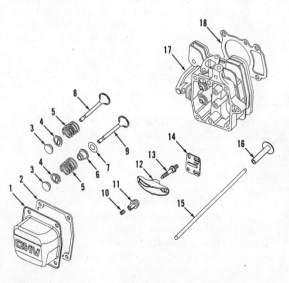

Fig. B511—Exploded view of cylinder head assembly.

1. Rocker cover
2. Gasket
3. Valve cap
4. Valve retainer
5. Valve spring
6. Valve seal
7. Gasket
8. Exhaust valve
9. Intake valve
10. Lock screw
11. Adjusting nut
12. Rocker arm
13. Stud
14. Push rod guide
15. Push rod
16. Valve lifter (tappet)
17. Cylinder head
18. Head gasket

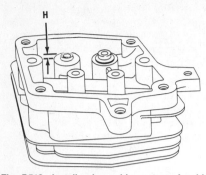

Fig. B512—Install valve guide so top of guide protrudes (H) above boss 0.120-0.150 inch (3.05-3.81 mm).

When reinstalling cylinder head, do not apply sealer to cylinder head gasket. Tighten cylinder head bolts progressively in three steps using a crossing pattern until final torque reading of 160 in.-lbs. (18.1 N•m) is obtained.

VALVE SYSTEM. Valves are actuated by rocker arms mounted on a stud threaded into the cylinder head. Remove valve wear caps (3—Fig. B511). Depress valve springs until slot in valve spring retainer (4) can be aligned with end of valve stem. Release spring pressure and remove retainer, spring and valve from cylinder head.

Valve face and seat angles are 45° for intake and exhaust. Specified seat width is 0.047-0.063 inch (1.2-1.6 mm). Minimum allowable valve margin is 0.015 inch (0.38 mm).

The cylinder head is equipped with renewable valve guides for both valves. Renew valve guide if inside diameter is 0.240 inch (6.10 mm) or more. Use B&S tool 19367 to remove and install guides. Guides may be installed either way up. Top of guide should protrude 0.120-0.150 inch (3.05-3.81 mm) as shown in Fig. B512. Use B&S tools 19345 and 19346 to ream valve guide to correct size.

Install push rod guide plate (P—Fig. B513) so "TOP" mark is toward flywheel side of cylinder head. Rocker arm studs

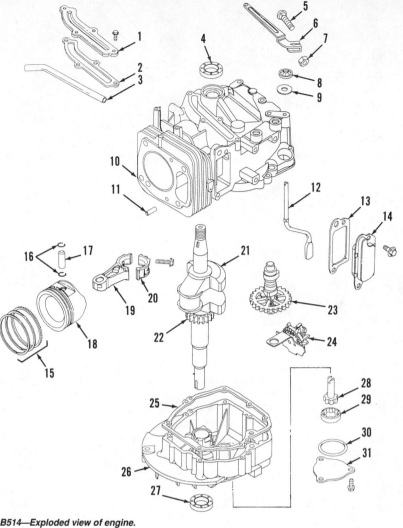

Fig. B514—Exploded view of engine.

1. Cover	9. Washer	17. Piston pin	25. Gasket
2. Gasket	10. Crankcase	18. Piston	26. Oil pan
3. Breather tube	11. Dowel pin	19. Connecting rod	27. Oil seal
4. Oil seal	12. Governor shaft	20. Rod cap	28. Oil pump inner rotor
5. Clamp bolt	13. Gasket	21. Crankshaft	29. Outer rotor
6. Governor lever	14. Tappet cover	22. Gear	30. "O" ring
7. Nut	15. Piston rings	23. Camshaft	31. Cover
8. Push nut	16. Retaining rings	24. Governor	

(D) are screwed into cylinder head. Hardening sealant should be applied to threads contacting cylinder head. When installing studs, tighten to 110 in.-lbs. (12.4 N•m) torque.

Install gasket (7—Fig. B511) and valve seal (6) on intake valve. Do not lubricate wear caps (3).

CAMSHAFT. To remove camshaft (23—Fig. B514) proceed as follows: Remove engine from equipment and drain crankcase oil. Clean pto end of crankshaft and remove any burrs or rust. Remove rocker arm cover, rocker arms and push rods; mark all parts so they can be returned to original position. Unscrew fasteners and remove oil pan. Remove governor assembly (24). Rotate crankshaft so timing marks on crankshaft and camshaft gears (Fig. B515)

are aligned (this will position valve tappets out of way). Withdraw camshaft and remove tappets.

Renew camshaft if either camshaft bearing journal diameter is 0.615 inch

Fig. B513—Install push rod guide plate (P) so "TOP" is toward flywheel side of head.

Fig. B515—Install camshaft so timing marks (M) on camshaft and crankshaft gears are aligned. Pin (P) in camshaft end must engage slot in oil pump drive shaft when installing crankcase cover.

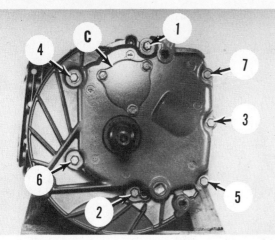

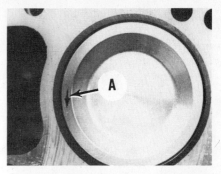

Fig. B516—Tighten oil pan retaining screws in sequence shown.

Fig. B519—Install piston so arrow (A) on piston crown points toward flywheel side of engine.

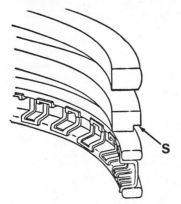

Fig. B517—Install second compression piston ring so step (S) is toward piston skirt.

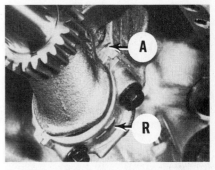

Fig. B520—Install rod cap so arrow on cap (R) points in same direction as arrow (A) on rod.

(15.62 mm) or less. Renew camshaft if lobes are excessively worn or scored.

Reverse removal procedure to reassemble components. Install camshaft while aligning timing marks (M—Fig. B515) on crankshaft and camshaft gears. Install governor assembly on camshaft. Note that roll pin in end of camshaft (Fig. B515) must engage oil pump drive shaft in oil pan during assembly. Install oil pan and apply nonhardening sealant such as Permatex 2 to screw (4—Fig. B516). Tighten cover screws to 85 in.-lbs. (9.6 N·m) torque in sequence shown in Fig. B516. Do not force mating of oil pan with crankcase. Reassemble remainder of components.

PISTON, PIN AND RINGS. To remove piston and rod assembly, drain engine oil and remove engine from equipment. Remove cylinder head and camshaft as previously outlined. Unscrew connecting rod screws and remove piston and rod.

Insert each piston ring (one at a time) squarely in top of cylinder and use a feeler gauge to measure ring end gap. Maximum allowable piston ring end gap is 0.030 inch (0.76 mm) for compression rings and 0.060 inch (1.52 mm) for oil ring rails. To check piston ring grooves for wear, insert a new ring in piston top groove and use a feeler gauge to measure side clearance between ring and piston land. Renew piston if ring side clearance exceeds 0.005 inch (0.12 mm) with a new piston ring installed in groove. Oversize as well as standard size piston and rings are available.

Piston pin is a slip fit in piston and rod. Renew piston if piston pin bore diameter is 0.552 inch (14.02 mm) or greater. Renew piston pin if diameter is 0.551 inch (14.00 mm) or less.

Top piston ring may be installed with either side up. Second piston ring must be installed with stepped edge (S—Fig. B517) toward piston skirt.

When assembling piston and rod, note that arrow on piston crown and "MAG" on connecting rod must be on same side. See Fig. B518. Install piston

and rod assembly in engine with arrow on piston crown toward flywheel as shown in Fig. B519. Install rod cap so arrow on cap (R—Fig. B520) points in same direction as arrow (A) on rod. Tighten rod screws to 100 in.-lbs. (11.3 N·m) torque.

Install camshaft and cylinder head as previously outlined.

CONNECTING ROD. The connecting rod rides directly on crankpin. Connecting rod and piston are removed as an assembly as outlined in previous section.

Connecting rod reject size for crankpin hole is 1.127 inch (28.63 mm). Renew connecting rod if piston pin bore diameter is 0.5525 inch (14.03 mm) or greater. A connecting rod with 0.020 inch (0.51 mm) undersize big end diameter is available to accommodate a worn crankpin (machining instructions are included with new rod).

GOVERNOR. The engine is equipped with a mechanical governor that is driven by the camshaft gear. Remove engine from the equipment, drain oil and remove oil pan for access to governor unit.

The flyweight assembly is mounted on end of camshaft along with oil slinger (Fig. B521). The oil slinger and flyweight assembly are only available as a

Fig. B518—Assemble piston and rod so arrow (A) on piston crown and "MAG" on side of rod are positioned as shown.

Fig. B521—View of governor and oil slinger assembly.

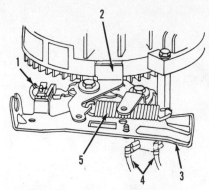

Fig. B523—View of flywheel brake assembly.

1. Stop switch wire
2. Brake pad & arm
3. Brake bracket
4. Starter interlock switch wires
5. Brake spring

unit assembly. Inspect flyweight assembly for broken components.

CRANKSHAFT AND MAIN BEARINGS. The crankshaft rides directly in the crankcase bores. Rejection sizes for crankshaft are: pto-end bearing journal 1.060 inch (26.92 mm); flywheel-end bearing journal 0.873 inch (22.17 mm); crankpin 1.122 inch (28.50 mm). A connecting rod with 0.020 inch (0.51 mm) undersize big end diameter is available to accommodate a worn crankpin (machining instructions are included with new rod).

The crankcase main bearing bore rejection size is 0.878 inch (22.30 mm). The oil pan main bearing bore rejection size is 1.065 inch (27.05 mm). A service bushing is available for installation in the crankcase if the bearing bore requires service. No bushing is available for the oil pan so it must be renewed if bearing bore is damaged or worn excessively.

Install oil seals so lip is toward inside of crankcase.

Crankshaft end play should be 0.002-0.034 inch (0.05-0.86 mm) with 0.015 inch crankcase gasket installed. Additional gaskets of several thicknesses are available for end play adjustment. If end play is excessive, renew the oil pan. Tighten oil pan screws to 85 in.-lbs. (9.6 N•m) torque.

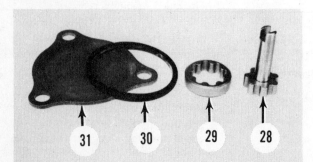

Fig. B522—View of oil pump components.

28. Oil pump inner rotor
29. Outer rotor
30. "O" ring
31. Cover

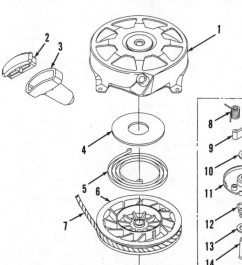

Fig. B524—Exploded view of rewind starter.

1. Starter housing
2. Insert
3. Handle
4. Spring cover
5. Rewind spring
6. Pulley
7. Rope
8. Springs
9. Pawls
10. Plastic washer
11. Retainer
12. Spring
13. Steel washer
14. Pin

CYLINDER. If cylinder bore wear is 0.003 inch (0.76 mm) or more, or out-of-round is 0.0015 inch (0.038 mm) or greater, cylinder must be rebored to next oversize.

Standard cylinder bore diameter is 2.5615-2.5625 inch (65.06-65.09 mm).

OIL PUMP. A rotor type oil pump driven by the camshaft is located in the bottom of the oil pan.

Remove engine from equipment for access to oil pump cover (C—Fig. B516). Remove cover and extract pump rotors (28 and 29—Fig. B522). Mark rotors so they can be reinstalled in their original

position. Renew any components which are damaged or excessively worn. Tighten oil pump cover screws to 80 in.-lbs. (9 N•m) torque.

FLYWHEEL BRAKE. Some engines are equipped with a pad type flywheel brake (2—Fig. B523). The brake should stop the engine within three seconds when the operator releases mower safety control and the speed control is in high speed position. Stopping time can be checked using tool 19255.

Adjustment may be possible on some applications by adjusting control cable position so brake pad is closer to flywheel.

To service flywheel brake, remove fuel tank, dipstick and oil fill tube, blower housing and rewind starter. Disconnect brake spring (5—Fig. B523). Disconnect stop switch wire (1) and safety interlock wires (4) if so equipped. Remove two retaining screws from brake pad arm and brake bracket.

The brake pad is available only as part of the bracket assembly. Minimum allowable brake pad thickness is 0.090 inch (2.3 mm). When installing brake assembly, tighten retaining screws to 40 in.-lbs. (4.5 N•m) torque.

REWIND STARTER. Refer to Fig. B524 for exploded view of rewind starter.

To install a new rope, proceed as follows. Remove fuel tank, dipstick and oil fill tube, and blower housing for access to starter assembly. Extract old rope from starter pulley. Allow pulley to unwind, then turn pulley counterclockwise until spring is tightly wound. Rotate pulley clockwise until rope hole in pulley is aligned with rope outlet in housing. Be sure that the new rope is the same diameter and length as the old rope. Pass new rope through pulley hole and housing outlet and tie a temporary knot near handle end of rope. Release pulley and allow rope to wind onto pulley. Install rope handle, release temporary knot and allow rope to enter starter.

To disassemble starter, remove rope and allow pulley to totally unwind. Position a suitable hollow sleeve support under pulley. Using a pin punch, drive out retainer pin (14). Remove retainer (11), brake spring (12), washers (10 and 13), pawls (9) and springs (8). Wear appropriate safety eyewear and gloves before disengaging pulley from starter housing as spring may uncoil uncontrolled. Place shop towel around pulley and lift pulley out of housing; spring should remain with pulley. Do not attempt to separate spring from pulley as they are a unit assembly.

Inspect components for damage and excessive wear. Reverse disassembly procedure to install components. Be sure inner end of rewind spring engages spring retainer adjacent to housing center post. Pin (14) should be driven or pressed until flush with retainer (11). Install rope as previously outlined.

BRIGGS AND STRATTON
CENTRAL PARTS DISTRIBUTORS

(Arranged Alphabetically by States)
**These franchised firms carry extensive stocks of repair parts.
Contact them for name of the nearest service distributor.**

BEBCO, Inc.
Phone (205) 251-4600
2221 Second Avenue, South
Birmingham, Alabama 35233

Power Equipment Company
Phone (602) 272-3936
#7 North 43rd Avenue
Phoenix, Arizona 85107

Pacific Western Power
Phone (415) 692-3254
1565 Adrain Road
Burlingame, California 94010

Power Equipment Company
Phone (805) 684-6637
1045 Cindy Lane
Carpinteria, California 93013

Pacific Power Equipment Company
Phone (303) 744-7891
1441 W. Bayaud Avenue #4
Denver, Colorado 80223

Spencer Engine, Inc.
Phone (813) 253-6035
1114 W. Cass St.
Tampa, Florida 33606

Sedco, Inc.
Phone (404) 925-4706
4305 Steve Reynolds Blvd.
Norcross, Georgia 30093

Small Engine Clinic, Inc.
Phone (808) 488-0711
98019 Kam Highway
Aiea, Hawaii 96701

Midwest Engine Warehouse
Phone (708) 833-1200
515 Roman Road
Elmhurst, Illinois 60126

Commonwealth Engine, Inc.
Phone (502) 267-7883
11421 Electron Drive
Louisville, Kentucky 40229

Delta Power Equipment
Phone (504) 465-9222
755 E. Airline Highway
Kenner, Louisiana 70062

Atlantic Power
Phone (508) 543-6911
77 Green Street
Foxboro, Massachusetts 02035

Wisconsin Magneto, Inc.
Phone (612) 780-5585
8010 Ranchers Road
Minneapolis, Minnesota 55432

Diamond Engine Sales
Phone (314) 652-2202
3134 Washington
St. Louis, Missouri 63103

Original Equipment, Inc.
Phone (406) 245-3081
905 Second Avenue, North
Billings, Montana 59101

Midwest Engine Warehouse of Omaha
Phone (402) 339-4700
7706-30 "I" Plaza
Omaha, Nebraska 68127

Atlantic Power
Phone (908) 356-8400
650 Howard Avenue
Somerset, New Jersey 08873

Power Equipment Company
Phone (505) 345-8851
7209 Washington Street, North East
Albuquerque, New Mexico 87109

AEA, Inc.
Phone (704) 377-6991
700 West 28th Street
Charlotte, North Carolina 28206

Central Power Systems
Phone (614) 876-3533
2555 International Street
Columbus, Ohio 43228

Engine Warehouse, Inc.
Phone (405) 946-7800
4200 Highline Blvd.
Oklahoma City, Oklahoma 73108

Brown & Wiser, Inc.
Phone (503) 692-0330
9991 South West Avery Street
Tualatin, Oregon 97062

Three Rivers Engine Distributors
Phone (412) 321-4111
1411 Beaver Avenue
Pittsburgh, Pennsylvania 15233

Automotive Electric Corporation
Phone (901) 345-0300
3250 Millbranch Road
Memphis, Tennessee 38116

Grayson Company, Inc.
Phone (214) 630-3272
1234 Motor Street
Dallas, Texas 75207

Engine Warehouse, Inc.
Phone (713) 937-4000
7415 Empire Central Drive
Houston, Texas 77040

Frank Edwards Company
Phone (801) 972-0128
1284 South 500 West
Salt Lake City, Utah 84101

RBI Corporation
Phone (804) 550-2210
101 Cedar Ridge Drive
Ashland, Virginia 23005

Wisconsin Magneto, Inc.
Phone (414) 445-2800
4727 North Teutonia Avenue
Milwaukee, Wisconsin 53209

CANADIAN DISTRIBUTORS

Briggs & Stratton Canada, Inc.
Phone (403) 435-9265
9519 49th Avenue
Edmonton, Alberta T6E 5Z5

Briggs & Stratton Canada, Inc.
Phone (604) 520-1294
1360 Cliveden Avenue
Delta, British Columbia V3M 6K2

Briggs & Stratton Canada, Inc.
Phone (204) 633-5400
89 Paramount Road
Winnipeg, Manitoba R2X 2W6

Briggs & Stratton Canada, Inc.
Phone (416) 625-6557
1815 Sismet Road
Mississauga, Ontario L4W 1P9

Briggs & Stratton Canada, Inc.
Phone (514) 366-6891
112-116 Lindsay Avenue
Dorval, Quebec H9P 2T8

BRIGGS & STRATTON SPECIAL TOOLS

The following special tools are available from Briggs & Stratton Central
Parts Distributors.

TOOL KITS

19158—Main bearing service kit for engine Series 60000, 80000, 82000, 83000, 90000, 91000, 92000, 93000, 94000, 95000, 96000, 100700, 110000, 111000, 112000, 113000, 114000, 121000, 122000, 123000, 124000, 125000, 126000, 130000, 131000 and 132000. Includes tool numbers 19094, 19095, 19096, 19097, 19099, 19100, 19101, 19123, 19124 and 16166.

291661—Dealer service tool kit. Includes tool numbers 19051, 19055, 19056, 19057, 19058, 19061, 19062, 19063, 19064, 19065, 19066, 19069, 19070, 19114, 19122, 19151, 19165, 19167, 19191, 19203.

PLUG GAUGES

19055—Check breaker plunger hole on Series 60000, 80000, 82000, 90000, 91000, 92000, 93000, 94000, 95000, 96000, 110000, 111000, 112000, 113000, 130000 and 131000.

19122—Check valve guide bore on Series 60000, 80000, 82000, 83000, 90000, 91000, 92000, 93000, 94000, 95000, 96000, 110000, 111000, 112000, 113000, 114000, 121000, 122000, 123000, 124000, 125000, 126000, 130000, 131000, 132000, 133000 and 135000.

19164—Check camshaft bearings on Series 60000, 80000, 82000, 83000, 90000, 91000, 92000, 93000, 94000, 95000, 96000, 100700, 110000, 111000, 112000, 113000, 114000, 121000, 122000, 123000, 124000, 125000, 126000, 130000, 131000, 132000, 133000 and 135000.

19166—Check main bearing bore on Series 60000, 80000, 82000, 83000, 90000, 91000, 92000, 93000, 94000, 95000, 96000, 100700, 110000, 111000, 112000, 113000, 114000, 121000, 122000, 123000, 124000, 125000, 126000, 130000, 131000, 132000, 133000 and 135000.

19375—Check pto main bearing bore on Series 99700 and 104700.

19381—Check valve guide bore on Series 85400, 99700, 104700 and 115400.

19387—Check camshaft bearings on Series 99700.

REAMERS

19064—Ream valve guide bore to install bushing on Series 60000, 80000, 82000, 83000, 90000, 91000, 92000, 93000, 94000, 95000, 96000, 100700, 110000, 111000, 112000, 113000, 114000, 121000, 122000, 123000, 124000, 125000, 126000, 130000, 131000, 132000, 133000 and 135000.

19066—Finish ream valve guide bushing on same models as 19064 reamer.

19095—Finish ream flywheel main bearing on Series 100700, 121000, 122000, 123000, 124000, 125000, 126000, 130000, 131000, 132000, 133000 and 135000 and both main bearings on Series 60000, 80000, 82000, 83000, 90000, 91000, 92000, 93000, 94000, 95000, 96000, 110000, 111000, 112000, 113000 and 114000.

19099—Ream counterbore for main bearings on same models as 19095 reamer.

19172—Ream counterbore for pto main bearing on Series 130000, 131000, 132000, 133000 and 135000.

19173—Finish ream main bearings on same models as 19172 reamer.

19346—Finish ream valve guide on Series 85400, 99700, 104700 and 115400.

GUIDE BUSHINGS FOR VALVE GUIDE REAMERS

19191—For Series 60000, 80000, 82000, 83000, 90000, 91000, 92000, 93000, 94000, 95000, 96000, 110000, 111000, 112000, 113000, 114000, 121000, 122000, 123000, 124000, 125000, 126000, 130000, 131000, 132000, 133000 and 135000.

19345—For Series 85400, 99700, 104700 and 115400.

PILOTS

19096—Pilot for main bearing reamer on Series 60000, 80000, 82000, 83000, 90000, 91000, 92000, 93000, 94000, 95000, 96000, 99700, 100700, 110000, 111000, 112000, 113000, 114000, 121000, 122000, 123000, 124000, 125000, 126000, 130000, 131000, 132000, 133000 and 135000.

19126—Expansion pilot for valve seat counterbore cutter on Series 60000, 80000, 82000, 83000, 90000, 91000,

92000, 93000, 94000, 95000, 96000, 100700, 110000, 111000, 112000, 113000, 114000, 121000, 122000, 123000, 124000, 125000, 126000, 130000, 131000, 132000, 133000 and 135000.

19395—Main bearing removal and installation on Series 85400 and 115400.

19396—Main bearing removal and installation on Series 115400.

DRIVERS

19057—To install breaker plunger bushing on Series 60000, 80000, 82000, 90000, 91000, 92000, 93000, 94000, 95000, 96000, 110000, 111000, 112000, 113000, 114000, 130000 and 131000.

19065—To install valve guide bushings on Series 60000, 80000, 82000, 83000, 90000, 91000, 92000, 93000, 94000, 95000, 96000, 100700, 110000, 111000, 112000, 113000, 114000, 121000, 122000, 123000, 124000, 125000, 126000, 130000, 131000, 132000, 133000 and 135000.

19124—To install main bearing bushings on Series 60000, 80000, 82000, 83000, 90000, 91000, 92000, 93000, 94000, 95000, 96000, 99700, 100700, 110000, 111000, 112000, 113000, 114000, 121000, 122000, 123000, 124000, 125000, 126000, 130000, 131000, 132000, 133300 and 135000.

19136—To install valve seat inserts.

19204—To install governor shaft bushing on Series 104700.

19274—To install valve guides on Series 104700.

19349—Remove and install main bearings on Series 104700.

19367—To install valve guide bushings on Series 99700 and 104700.

19367—Remove and install main bearings on Series 85400.

GUIDE BUSHINGS FOR MAIN BEARING REAMERS

19094—For pto main bearing pilot on Series 60000, 80000, 82000, 83000, 90000, 91000, 92000, 93000, 94000, 95000, 96000, 100700, 110000, 111000, 112000, 113000, 114000, 121000, 122000, 123000, 124000, 125000, 126000, 130000, 131000, 132000, 133000 and 135000. For flywheel main bearing pilot on Models 80590, 80790,

92590, 92990, 110900, 111900 and 112900.

19100—For pto main bearing reamer on Series 60000, 80000, 82000, 83000 and 90000.

19101—For flywheel main bearing reamer on Series 60000, 80000, 82000, 83000, 90000, 91000, 92000, 93000, 94000, 95000, 96000, 100700, 110000, 111000, 112000, 113000, 114000, 121000, 122000, 123000, 124000, 125000, 126000, 130000, 131000 and 132000.

19168—For pto main bearing pilot on Series 130000, 131000, 132000, 133000 and 135000.

19170—For pto main bearing reamer on horizontal Series 130000, 131000, 132000, 133000 and 135000.

19186—For pto main bearing reamer on vertical Series 130000, 131000, 132000, 133000 and 135000.

19373—For pto main bearing reamer on Series 99700.

FLYWHEEL HOLDER

19167—For Series 60000, 62000, 80000, 82000, 83000, 90000, 91000, 92000, 93000, 94000, 95000, 95700, 96000 and 96700.

19167—For Series 85400, 99700, 104700 and 115400.

VALVE SPRING COMPRESSOR

19063—For all engines except Model 104700.

19347—For Series 104700.

PISTON RING COMPRESSOR

19070—For Series 60000, 80000, 82000, 83000, 85400, 90000, 91000, 92000, 93000, 94000, 95000, 96000, 99700, 100700, 104700, 110000, 111000, 112000, 113000, 114000, 115400, 121000, 122000, 123000, 124000, 125000, 126000, 130000, 131000, 132000, 133000 and 135000.

FLYWHEEL PULLERS

19069—For Series 60000, 62000, 80000, 82000, 83000, 90000, 91000, 92000, 93000, 94000, 95000, 95700, 96000, 96700, 99700, 100700, 110000, 111000, 112000, 113000, 114000, 121000, 122000, 123000, 124000, 125000 and 126000.

19203—For Series 85400, 104700 and 115400.

CRANKCASE SUPPORT JACK

19123—For Series 60000, 80000, 82000, 83000, 90000, 91000, 92000, 93000, 94000, 95000, 96000, 99700, 100700, 110000, 111000, 112000, 113000, 114000, 121000, 122000, 123000, 124000, 125000, 126000, 130000, 131000, 132000, 133000 and 135000.

IGNITION SPARK TESTER

19051—For all engines.

VALVE SEAT REPAIR TOOLS

19137—T-handle for expansion pilots.

19138—Valve seat puller kit.

19182—Puller nut adapter.

PULLER KIT

19332—Remove and install seals, bearings and crankshaft on Models 95700 and 96700.

STARTER CLUTCH WRENCH

19244—All engines equipped with rewind starter clutch.

CRAFTSMAN

SEARS ROEBUCK & CO.
Sears Tower
Chicago, IL 60684

The following Craftsman engines were manufactured by Tecumseh for Sears Roebuck & Co. Locate the Craftsman model in the cross-reference table and note the respective Tecumseh model. Follow the service procedures outlined in this manual for the Tecumseh model when servicing the Craftsman engine.

TWO-STROKE MODELS

Craftsman Model Number	Tecumseh Model Number
200.701001	AH600
200.701011	TC200
200.701021	TC300
200.701031	TC300
200.701041	TCH300
200.701051	TC300
200.711001	HSK600
200.711011	HSK840
200.711021	TCH300
200.711031	TC300

FOUR-STROKE ENGINES
Vertical Crankshaft

Craftsman Model Number	Tecumseh Model Number
143.401012	TVS75
143.404022	TVS75
143.404032	TVS90
143.404042	TVS105
143.404082	TVS105
143.404092	TVS105
143.404122	TVS120
143.404132	TVS105
143.404142	TVS105
143.404152	TVS120
143.404162	TVS105
143.404172	TVS105
143.404182	TVS120
143.404202	TVS105
143.404222	TVS105
143.404232	TVS105
143.404242	TVS105
143.404252	TVS105
143.404282	TVS105
143.404292	TVS120
143.404312	TVS105
143.404322	TVS105

Craftsman Model Number	Tecumseh Model Number
143.404332	TVS105
143.404342	TVS90
143.404352	TVS90
143.404362	TVS105
143.404372	TVS105
143.404382	TVS105
143.404392	TVS105
143.404402	TVS120
143.404412	TVS105
143.404422	TVS105
143.404432	TVS105
143.404442	TVS105
143.404452	TVS105
143.404462	TVS105
143.404472	TVS120
143.404482	TVS120
143.404502	TVS90
143.404532	TVS90
143.406082	TVM125
143.414012	TVS90
143.414022	TVS105
143.414032	TVS90
143.414042	TVS90
143.414052	TVS90
143.414062	TVS105
143.414072	TVS105
143.414082	TVS90
143.414092	ECV100
143.414102	ECV100
143.414112	ECV100
143.414122	ECV100
143.414132	ECV100
143.414142	ECV100
143.414152	ECV100
143.414162	ECV100
143.414172	TVS90
143.414182	TVS90
143.414192	ECV100
143.414202	ECV100
143.414212	TVS90
143.414232	TVS90
143.414242	TVS90
143.414252	TVS90
143.414262	ECV100
143.414272	ECV100
143.414282	TVS90
143.414292	TVS105
143.414302	TVS120
143.414312	TVS105
143.414322	TVS105
143.414332	TVS90
143.414342	TVS105
143.414352	TVS120
143.414362	TVS105

Craftsman Model Number	Tecumseh Model Number
143.414372	TVS105
143.414382	TVS105
143.414392	TVS120
143.414402	TVS105
143.414412	TVS105
143.414422	TVS120
143.414432	OVRM50
143.414442	TVS105
143.414452	TVS105
143.414462	TVS105
143.414472	TVS120
143.414482	TVS105
143.414492	TVS120
143.414502	TVS90
143.414512	TVS90
143.414522	TVS90
143.414532	TVS90
143.414542	TVS105
143.414552	TVS120
143.414562	TVS105
143.414572	TVS120
143.414582	TVS105
143.414592	TVS105
143.414602	TVS105
143.414612	TVS90
143.414622	TVS120
143.414632	TVS105
143.414642	TVS120
143.414652	TVS105
143.414662	TVS105
143.414672	TVS105
143.414682	ECV100
143.414692	TVS100
143.416052	TVM125
143.416062	TVM125
143.424012	TVS90
143.424022	TVS105
143.424032	TVS90
143.424042	TVS105
143.424052	TVS90
143.424062	TVS120
143.424072	TVS120
143.424082	TVS105
143.424102	TVS120
143.424112	TVS100
143.424122	TVS100
143.424132	TVS100
143.424142	TVS105
143.424152	TVS120
143.424162	TVS105
143.424172	TVS120
143.424182	TVS100
143.424202	TVS90
143.424312	TVS105

Craftsman Model Number	Tecumseh Model Number
143.424322	TVS105
143.424332	TVS120
143.424342	TVS120
143.424352	TVS105
143.424362	TVS90
143.424372	TVS90
143.424382	TVS105
143.424392	TVS105
143.424402	TVS120

Horizontal Crankshaft

Craftsman Model Number	Tecumseh Model Number
143.804062	HS40
143.804072	HS50
143.804082	H30
143.804092	H30
143.804102	H35
143.804112	H30

Craftsman Model Number	Tecumseh Model Number
143.814012	HS40
143.814022	HS50
143.814032	H30
143.814042	H30
143.814052	H35
143.814062	H35
143.814072	H30
143.824012	H30
143.824022	H30

HOMELITE

HOMELITE DIVISION OF TEXTRON
P.O. Box 7047
14401 Carowinds Blvd.
Charlotte, NC 28217

Model	Bore	Stroke	Displacement
420, 420E	2.125	1.748 in.	6.2 cu. in.
	(54.0 mm)	(44.4 mm)	(102 cc)

ENGINE INFORMATION

This section covers service for the engine used on Homelite Model 420 and 420E snowthrowers.

MAINTENANCE

LUBRICATION. The engine is lubricated by mixing oil with unleaded gasoline (gasoline blended with alcohol is not recommended). Recommended oil is Homelite two-stroke oil mixed at ratio designated on oil container. If Homelite oil is not available, a good quality oil formulated for air-cooled two-stroke engines at 40:1 or 50:1 ratios may be used at Homelite specified 32:1 ratio. Two-stroke oils formulated at less than 32:1 ratio or automotive (4-stroke) oils should not be used.

The 32:1 ratio is obtained by mixing 4 ounces (120 mL) of oil with 1 gallon (3.8 L) of gasoline.

SPARK PLUG. Recommended spark plug is a Champion CJ8Y spark plug. Electrode gap should be 0.025 inch (0.6 mm). Tighten spark plug to 120-180 in.-lbs. (13.6-20.3 N•m).

CARBURETOR. Initial setting of idle speed mixture screw (29—Fig. HL1) and high speed mixture screw (27) is one turn out from a lightly seated position. Perform carburetor adjustments with engine at normal operating temperature. Run engine at idle speed, then turn idle mixture screw clockwise until engine begins to stumble and note screw position. Turn idle mixture screw counterclockwise until engine begins to stumble and note screw position. Turn the idle mixture screw clockwise to a position that is midway from the clockwise (lean) and counterclockwise (rich) positions. With engine running at idle, rapidly move speed control to full throttle position. If engine stumbles or hesitates, turn idle mixture screw counterclockwise slightly and retest. Turn high speed mixture screw clockwise until engine begins to stumble and note screw position. Turn high speed

mixture screw counterclockwise until engine begins to stumble and note screw position. Turn the high speed mixture screw clockwise to a position that is midway from the clockwise (lean) and counterclockwise (rich) positions. Check engine operation under load. If engine stalls when put under load, turn high speed mixture screw approximately ⅛ turn counterclockwise (rich). If engine smokes excessively or "4-cycles" under load, turn high speed mixture screw approximately ⅛ turn clockwise (lean). Do not set high speed mixture screw too lean as engine damage may result.

To disassemble carburetor, refer to Fig. HL1 and remove pump cover (1) and fuel pump diaphragm (3). Remove cover (46) and fuel metering diaphragm (45). Remove screw (41), pin (40), metering lever (39) and fuel inlet valve (37). Remove circuit plate (43) and gasket. Remove fuel mixture screws (27 and 29). Remove retaining screws from choke plate (4) and throttle plate (19). Remove "E" ring (10) and withdraw choke shaft (5), being careful not to lose the detent ball (11) and spring (12). Remove retainer (21) and withdraw throttle shaft (18). Walbro tool number 500-16 is available for removal of Welch

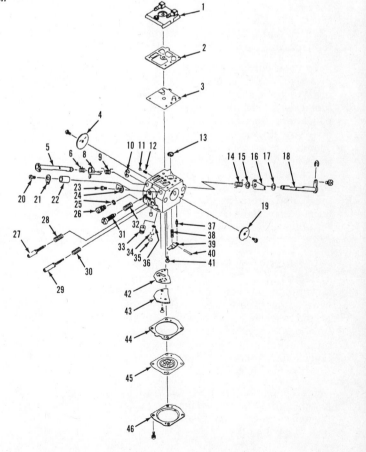

Fig. HL1—Exploded view of Walbro HDA carburetor.

1. Pump cover
2. Gasket
3. Fuel pump diaphragm & valves
4. Choke plate
5. Choke shaft
6. Spring
8. Arm
9. Spring
10. "E" ring
11. Choke detent ball
12. Spring
13. Fuel inlet screen
14. Spring
15. Dust seal
16. Arm
17. Washer
18. Throttle shaft
19. Throttle plate
20. Screw
21. Retainer
22. Spacer
23. Screw
24. Throttle stop
25. Gasket
26. Governor
27. High speed mixture screw
28. Spring
29. Idle mixture screw
30. Spring
31. Idle speed screw
32. Spring
33. Check valve
34. Retainer
35. Welch plug
36. Check valve screen
37. Fuel inlet valve
38. Spring
39. Metering lever
40. Pin
41. Screw
42. Gasket
43. Circuit plate
44. Gasket
45. Fuel metering diaphragm
46. Cover

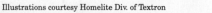

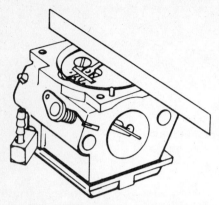

Fig. HL2—Metering lever should just touch a straightedge laid on carburetor body as shown.

plugs. A sharp pin punch can also be used to remove Welch plugs. Be careful not to contact the metal or damage the passageways behind the plugs.

Clean and inspect all components. Inspect diaphragms (3 and 45) for defects which may affect operation. Examine fuel inlet valve and seat. Inlet valve (37) is renewable, but carburetor body must be renewed if seat is damaged or excessively worn. Wires or drill bits should not be used to clean passages as fuel flow may be altered. Screens should be clean. Compressed air should not be used to clean main nozzle as check valve may be damaged. Fuel mixture screws must be renewed if grooved or broken. Discard carburetor body if mixture screw seats are damaged or excessively worn. Be sure choke and throttle plates fit shafts and carburetor bore properly.

To reassemble carburetor, reverse disassembly procedure. When installing Welch plugs, apply small amount of fuel resistant sealant, such as fingernail polish, around edge of plug. Use suitable size driver to flatten plug; plug should not be indented when properly installed. Wipe off any excess sealant. Check metering lever height as shown in Fig. HL2. Metering lever should just touch a straightedge positioned on carburetor body. Bend lever (39—Fig. HL1) to obtain correct lever height.

IGNITION. A solid-state ignition system is used. The ignition system is serviced by renewing the faulty component, however, be sure all wires are connected properly and the ignition switch functions correctly before renewing ignition module (8—Fig. HL3). The ignition module is accessible after removing flywheel fan housing (10).

Ignition timing is fixed and not adjustable. Air gap between ignition module and flywheel should be 0.008-0.012 inch (0.2-0.3 mm).

GOVERNOR. The engine is equipped with an air vane governor. Maximum governed engine speed is adjusted by moving governor spring attachment tab (16—Fig. HL3). Maximum governed speed should be 4600-4800 rpm.

COMPRESSION PRESSURE. Specified compression pressure is 95-120 psi (656-828 kPa). Minimum allowable compression pressure is 90 psi (621 kPa).

REPAIRS

CRANKCASE PRESSURE TEST. An improperly sealed crankcase can cause the engine to be hard to start, run rough, have low power and overheat. Refer to the SERVICE SECTION TROUBLE-SHOOTING section of this manual for crankcase pressure test procedure. If crankcase leakage is indicated, pressurize crankcase and use a soap and water solution to check gaskets, seals, carburetor pulse line and casting for leakage.

CYLINDER, PISTON, PIN AND RINGS. To remove cylinder, remove engine from equipment. Remove starter, carburetor, flywheel and shroud. The cylinder (1—Fig. HL3) may be removed after unscrewing four screws in bottom of crankcase (25). Tap crankcase with plastic mallet to break seal between mating surfaces, then separate cylinder from crankcase. Be careful when removing cylinder as crankshaft assembly will be loose in crankcase. Care should be taken not to damage mating surfaces of cylinder and crankcase. Extract piston pin retaining ring (5) and push piston pin (6) out of piston to separate piston from connecting rod.

Inspect piston and cylinder for wear, scoring or other damage and renew as

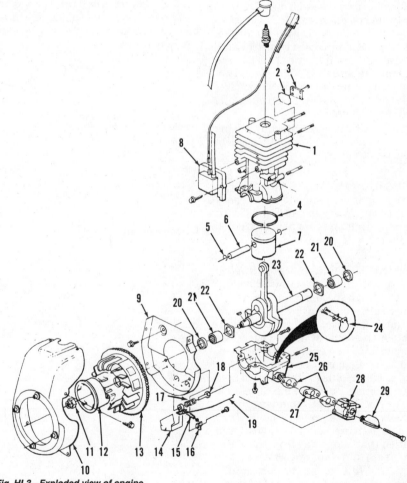

Fig. HL3—Exploded view of engine.

1. Cylinder	9. Fan shroud	16. Spring attachment	23. Crankshaft assy.
2. Gasket	10. Fan housing	17. Spring	24. Reed valve petal
3. Cover	11. Nut	18. Governor pivot	25. Crankcase
4. Piston ring	12. Starter hub	19. Throttle rod	26. Gasket
5. Retaining ring	13. Flywheel	20. Oil seal	27. Spacer
6. Piston pin	14. Air vane	21. Needle bearing	28. Carburetor
7. Piston	15. Governor spring	22. Thrust washer	29. Shield
8. Ignition module			

necessary. Use a wooden scraper to remove carbon from cylinder ports. Clean mating surfaces of cylinder and crankcase. Mating surfaces should be flat and free of nicks and scratches.

Piston must be installed so closed end of piston pin is toward exhaust port. Lubricate piston and cylinder with engine oil before installing piston in cylinder.

Apply a thin coat of Loctite Gasket Eliminator to cylinder and crankcase mating surfaces before assembly. Be sure oil seals, bearings and thrust washers are properly positioned when installing cylinder on crankcase. Thrust washers (22) must be installed in recess in cylinder and crankcase. Oil seals (20) should be recessed $\frac{1}{8}$ inch (3.2 mm). Tighten crankcase screws evenly to 100-125 in.-lbs. (11.3-14.1 N·m). Make certain that crankshaft does not bind when rotated. Crankshaft end play should be 0.002-0.037 inch (0.05-0.94 mm).

After cylinder and crankcase are assembled, it is recommended that the crankcase be pressure tested as outlined in the ENGINE SERVICE section on page 37 of this manual.

CRANKSHAFT AND CONNECTING ROD.
The crankshaft and connecting rod are serviced as an assembly. To remove crankshaft, remove cylinder as previously outlined, then lift crankshaft assembly out of crankcase.

Inspect crankshaft bearings (21—Fig. HL3) and renew if scored or worn. Crankshaft and connecting rod must be renewed as an assembly if worn or damaged. Refer to CYLINDER, PISTON,

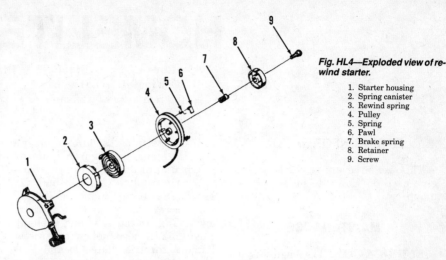

Fig. HL4—Exploded view of rewind starter.

1. Starter housing
2. Spring canister
3. Rewind spring
4. Pulley
5. Spring
6. Pawl
7. Brake spring
8. Retainer
9. Screw

PIN AND RINGS section for crankcase assembly instructions.

REED VALVE. The engine is equipped with a reed valve induction system. The reed petal (24—Fig. HL3) is located on the inside of the crankcase. Renew reed petal if cracked, bent or otherwise damaged. Do not attempt to straighten a bent reed petal. Seating surface for reed petal should be flat, clean and smooth.

REWIND STARTER. Refer to Fig. HL4 for exploded view of rewind starter.

To disassemble starter, remove rope handle and allow rope to wind into starter. Unscrew retainer screw (9). Remove retainer (8), brake spring (7), pawl (6) and spring (5). Remove pulley with spring. Wear appropriate safety eyewear and gloves before disengaging spring canister (2) and rewind spring (3) from pulley as spring may uncoil uncontrolled.

To reassemble, reverse the disassembly procedure. If replacing the starter rope, be sure that new rope is the same diameter and length, 75 inches (19 cm), as the original rope. Spring (3) should be lightly greased. Install pawl (6) and spring (5) so the spring end forces the pawl toward the center of the pulley. With starter assembled, except for rope, install rope as follows: Turn pulley counterclockwise until tight, then allow pulley to unwind so hole in pulley aligns with rope outlet in starter housing. Insert rope through starter housing and pulley hole, tie a knot in inner rope end, allow rope to wind onto pulley and install rope handle.

HOMELITE

Model	Bore	Stroke	Displacement
AP-125	$1^5/_{16}$ in.	$1^1/_8$ in.	1.53 cu. in.
	(33 mm)	(29 mm)	(25 cc)

ENGINE INFORMATION

This section covers service for the engine used on Homelite Model AP-125 pump.

MAINTENANCE

LUBRICATION. The engine is lubricated by mixing oil with unleaded gasoline (gasoline blended with alcohol is not recommended). Recommended oil is Homelite two-stroke oil mixed at ratio designated on oil container. If Homelite oil is not available, a good quality oil formulated for air-cooled two-stroke engines at 40:1 or 50:1 ratios may be used at Homelite specified 32:1 ratio. Two-stroke oils formulated at less than 32:1 ratio or automotive (4-stroke) oils should not be used.

The 32:1 ratio is obtained by mixing 4 ounces (120 mL) of oil with 1 gallon (3.8 L) of gasoline.

AIR FILTER. The air filter element should be cleaned periodically in warm, soapy water. Renew filter element if excessively dirty or damaged.

SPARK PLUG. Recommended spark plug is a Champion DJ7Y spark plug. Electrode gap should be 0.025 inch (0.6 mm).

CARBURETOR. Initial setting of idle mixture screw (I—Fig. HL40) and high speed mixture screw (H) is $1^1/_4$ turns out from a lightly seated position (later models are not equipped with a high speed mixture screw). Operate engine until it reaches normal operating temperature before making final carburetor adjustment. Adjust idle mixture screw so engine accelerates cleanly. Adjust high speed mixture screw (if so equipped) with engine at full throttle and full load so engine produces optimum power. If engine slows down or stalls under load, turn high speed mixture screw approximately $1/_8$ turn counterclockwise (rich). If engine smokes excessively or "4-cycles" under load, turn mixture screw approximately $1/_8$ turn clockwise (lean). Do not adjust high speed mixture screw too lean as engine damage may result.

To disassemble carburetor, refer to Fig. HL41 and remove cover (1) and fuel pump diaphragm (3). Remove cover (29), primer plate (27) and bulb (28), and fuel metering diaphragm (26). Remove screw (21), pin (22), metering lever (25), spring (24) and fuel inlet valve (23). Remove fuel mixture screws. Remove throttle plate (7) and choke plate (18) retaining screws. Remove shaft retainers and withdraw throttle shaft (10) and choke shaft (11) from carburetor body. Be careful not lose detent ball (20) when withdrawing choke shaft. Walbro tool number 500-16 is available for removal of Welch plugs (16 and 17). A sharp pin punch can also be used to remove Welch plugs. Be careful not to contact the metal or damage the passage behind the plugs.

Clean and inspect all components. Inspect diaphragms (3 and 26) for defects which may affect operation. Examine fuel inlet valve and seat. Inlet valve (23) is renewable, but carburetor body must be renewed if seat is damaged or excessively worn. Wires or drill bits should not be used to clean passages as fuel flow may be altered. Screens should be clean. Compressed air should not be used to clean main nozzle as check valve may be damaged. Fuel mixture screws must be renewed if grooved or broken. Discard carburetor body if mixture screw seats are damaged or excessively worn. Be sure choke and throttle plates fit shafts and carburetor bore properly.

To reassemble carburetor, reverse disassembly procedure. When installing Welch plugs, apply small amount of fuel resistant sealant, such as fingernail polish, around edge of plug. Use suitable size driver to flatten plug; plug should not be indented when properly installed. Wipe off any excess sealant. Check metering lever height using Walbro gauge 500-13 as shown in Fig. HL42. Bend metering lever (25—Fig. HL41) to obtain correct lever height.

IGNITION. A solid-state ignition system is used. The ignition system is serviced by renewing the faulty compo-

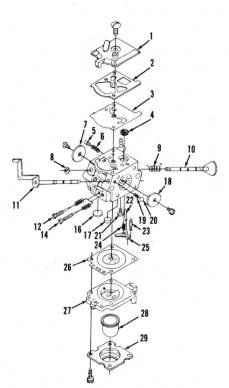

Fig. HL41—Exploded view of Walbro diaphragm carburetor. Later models are not equipped with a high speed mixture screw.

1. Cover	16. Welch plug
2. Gasket	17. Welch plug
3. Fuel pump diaphragm	18. Choke plate
4. Screen	19. Spring
5. Piston	20. Detent ball
6. Spring	21. Screw
7. Throttle plate	22. Pin
8. "E" ring	23. Fuel inlet valve
9. Spring	24. Spring
10. Throttle shaft	25. Metering lever
11. Choke shaft	26. Metering diaphragm
12. Idle mixture screw	27. Plate
14. High speed	28. Primer bulb
mixture screw	29. Cover

Fig. HL40—View showing location of idle mixture screw (I) and high speed mixture screw (H). Later models are not equipped with a high speed mixture screw.

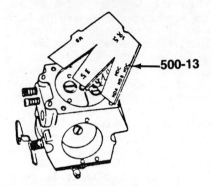

Fig. HL42—Using Walbro tool 500-13 as shown, metering lever on Walbro carburetor should just touch leg on tool.

nent, however, be sure all wires are connected properly and the ignition switch functions correctly before renewing ignition module (7—Fig. HL43). The ignition module is accessible after removing starter and fan housing (1).

Ignition timing is fixed and not adjustable. Air gap between ignition module and flywheel is adjustable. To adjust air gap, loosen module retaining screws and place a 0.0125 inch (0.32 mm) shim between flywheel and module. Position module against shim and tighten module retaining screws.

MUFFLER. Spark screen in muffler should be cleaned or replaced periodically. Muffler and cylinder exhaust ports should be cleaned periodically to prevent loss of power due to carbon buildup. Remove muffler and scrape free of carbon. With muffler removed, turn crankshaft so piston is at top dead center and carefully remove carbon from exhaust ports with a wooden scraper. Be careful not to damage chamfered edges of exhaust ports or scratch piston. DO NOT run engine with muffler removed.

REPAIRS

CRANKCASE PRESSURE TEST. An improperly sealed crankcase can cause the engine to be hard to start, run rough, have low power and overheat. Refer to SERVICE SECTION TROUBLE-SHOOTING section of this manual for crankcase pressure test procedure. If crankcase leakage is indicated, pressurize crankcase and use a soap and water solution to check gaskets, seals, carburetor pulse line and casting for leakage.

CYLINDER. The cylinder (17—Fig. HL43) is mounted on the crankcase with three retaining screws. Cylinder can be removed after first removing en-

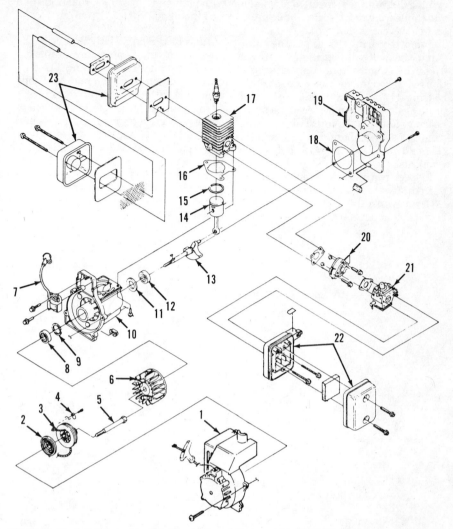

Fig. HL43—Exploded view of engine. A washer is used between drive shaft (5) and flywheel (6) on some models.

1. Starter housing	7. Ignition coil & module	13. Crankshaft
2. Rewind spring	8. Bearing	14. Piston & rod assy.
3. Pulley	9. Snap ring	15. Piston ring
4. Retainer	10. Crankcase	16. Gasket
5. Drive shaft	11. Seal	17. Cylinder
6. Flywheel	12. Bearing	18. Gasket

19. Cover	
20. Heat insulator	
21. Carburetor	
22. Air filter assy.	
23. Muffler assy.	

gine from pump assembly. Remove starter and fan housing (1), muffler assembly and carburetor. Remove retaining screws and withdraw cylinder, being careful not to damage piston or connecting rod.

The cylinder bore is chrome plated. Inspect bore for flaking, scoring and other damage. Renew cylinder if excessively worn or damaged.

PISTON, PIN, RING AND CONNECTING ROD. The piston and connecting rod (14—Fig. HL43) can be removed after first removing the cylinder (17) as previously outlined and the crankcase end cover (19). Slide connecting rod off crankshaft crankpin and withdraw piston and connecting rod.

The piston, pin and connecting rod are available only as an assembly. The piston is equipped with a single piston ring. Inspect piston and ring for scuffing, wear or other damage and renew as necessary. There is no piston ring locating pin in the piston ring groove.

When installing cylinder onto piston, locate piston ring so gap is towards the center of the exhaust port (away from the intake port).

CRANKSHAFT AND CRANKCASE. The crankshaft is supported by two ball bearings at the drive end and unsupported at the crankpin end. See Fig. HL43.

To remove crankshaft (13), first remove engine from pump assembly. Remove starter and fan housing, flywheel, cylinder, crankcase cover, and piston and connecting rod. Use a suitable press to remove crankshaft from crankcase. To remove bearings, force out the outer (small) sealed ball bearing (8) using a 9/16 inch (14 mm) diameter rod. Insert the rod through the inner (large) ball bearing (12) and main seal (11) and press against the inner race of the outer bearing. Remove the internal snap ring from the bearing bore. Press main seal (11) and inner bearing (12) out of the crankcase.

Inspect bearings and replace as required. Always install a new main seal (11) when installing bearings. Service replacement crankcases are supplied with both main bearings, seal and snap ring installed.

REWIND STARTER. To service the rewind starter, remove the pump assembly and starter housing (1—Fig. HL43). Pull starter rope and hold rope pulley (3) with notch in pulley adjacent to rope outlet. Pull rope back through outlet so it engages notch in pulley and allow pulley to completely unwind. Unscrew pulley retaining screw and clamp. Remove rope pulley being careful not to dislodge rewind spring in housing. Care must be taken if rewind spring is removed to prevent injury if spring is allowed to uncoil uncontrolled.

Rewind spring (2) is wound in clockwise direction in starter housing. Rope is wound on rope pulley in clockwise direction as viewed with pulley in housing. To place tension on rewind spring, pass rope through rope outlet in housing and install rope handle. Pull rope out and hold rope pulley so notch on pulley is adjacent to rope outlet. Pull rope back through outlet between notch on pulley and housing. Turn rope pulley clockwise to place tension on spring (approximately two turns). Release pulley and check starter action. Do not place more tension on rewind spring than is necessary to draw rope handle against housing.

HONDA

AMERICAN HONDA MOTOR CO., INC.
100 W. Alondra Blvd.
Gardena, California 90247

Model	Bore	Stroke	Displacement	Power Rating
G100	46 mm	46 mm	76 cc	1.6 kW
	(1.84 in.)	(1.84 in.)	(4.6 cu.in.)	(2.2 hp)
GV100	50 mm	46 mm	90 cc	1.6 kW
	(1.97 in.)	(1.84 in.)	(5.5 cu.in.)	(2.2 hp)
G150	64 mm	45 mm	144 cc	2.6 kW
	(2.5 in.)	(1.8 in.)	(8.8 cu.in.)	(3.5 hp)
G200, GV200	67 mm	56 mm	197 cc	3.7 kW
	(2.6 in.)	(2.2 in.)	(12.0 cu.in.)	(5.0 hp)

NOTE: Metric fasteners are used throughout engine.

ENGINE INFORMATION

All models are four-stroke, single-cylinder, air-cooled engines. Models GV100 and GV200 are vertical crankshaft engines. All other models are horizontal crankshaft engines.

Engine model number is cast into side of crankcase. Engine serial number is stamped into crankcase (Fig. HN1). Always furnish engine model and serial number when ordering parts.

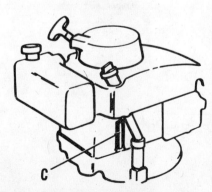

Fig.HN1—Engine serial number may be located at one of the locations shown by arrows (A, B or C).

MAINTENANCE

LUBRICATION. Engine oil level should be checked prior to operating engine. Maintain oil level at top of reference marks (Fig. HN2) when checked with cap not screwed in, but just touching first threads.

Oil should be changed after the first 20 hours of engine operation and after every 100 hours thereafter.

Engine oil should meet or exceed latest API service classification. Use SAE 30 oil for temperatures above 50° F (10° C); use SAE 10W-30 oil for temperatures between 0° F (−18° C) and 85° F (30° C); below 32° F (0° C) SAE 5W-30 may be used.

Crankcase capacity is 0.4 L (0.42 qt.) for Model GV100, 0.45 L (0.48 qt.) for Model G100, and 0.7 L (0.74 qt.) for all other models.

The reduction gear unit on horizontal crankshaft engines may be lubricated by oil from the engine or by a separate oil supply. If there is no oil level screw on the reduction case, then the unit is

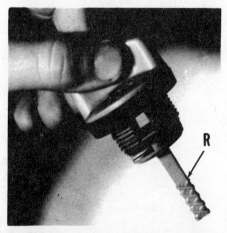

Fig. HN2—Do not screw in oil plug and gauge when checking oil level. Maintain oil level at top edge of reference marks (R) on dipstick.

lubricated by oil from the engine. The reduction gear unit with an automatic clutch is equipped with an oil dipstick to check oil level in the case. Oil level is checked similar to checking engine oil by unscrewing the oil fill plug. Maintain oil level at top of notch on gauge when checked with plug not screwed in, but just touching first threads. Fill with same oil recommended for engine. Capacity is 0.15 L (0.3 pt.) for Model G150 and 0.5 L (0.53 qt.) for Model G200.

To check oil level on reduction units equipped with an oil level screw plug but no dipstick, add oil so level reaches oil level screw hole.

AIR CLEANER. Engine air filter should be cleaned and inspected after every 50 hours of operation, or more often if operating in extremely dusty conditions.

Remove foam and paper air filter elements from air filter housing. Foam element should be washed in a mild detergent and water solution, rinsed in clean water and allowed to air dry. Soak foam element in clean engine oil. Squeeze out excess oil.

Paper element may be cleaned by directing low pressure compressed air stream from inside filter toward the outside. Reinstall elements.

Some engines may be equipped with an oil bath type filter system in addition to the foam type element. Fill the oil compartment to the level indicated on the side of the case with engine oil. Do not overfill.

SPARK PLUG. Spark plug should be removed, cleaned and inspected after every 100 hours of use.

Recommended spark plug is NGK BM4A for Models G100 and GV100, NGK BPR4HS-10 for Model G150, NGK BR4HS for Model G200, and NGK

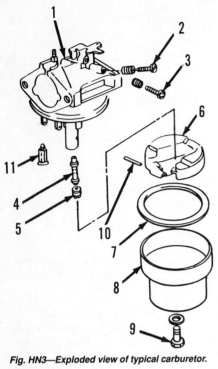

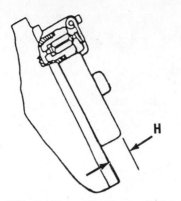

Fig. HN4—Position carburetor so float rests lightly against fuel inlet valve and measure float height (H).

Fig. HN3—Exploded view of typical carburetor.

1. Body	7. Gasket
2. Idle mixture screw	8. Fuel bowl
3. Idle speed screw	9. Screw
4. Nozzle	10. Float pin
5. Main jet	11. Fuel inlet valve
6. Float	

BPMR6A-10 for Model GV200. A resistor type spark plug may be required in some locations. Spark plug electrode gap should be 0.9-1.0 mm (0.035-0.039 in.) for all models.

When installing spark plug, manufacturer recommends installing spark plug finger tight, then for a new plug, tighten an additional ½ turn. For a used plug, tighten an additional ⅛ to ¼ turn.

CARBURETOR. All models are equipped with a Keihin float type carburetor with a fixed main fuel jet and an adjustable idle mixture screw.

Initial setting of idle mixture screw (2—Fig. HN3) from a lightly seated position is 1⅜ turns out on Models G100 and G150, 1⅞ turns out on Model GV100, 3¼ turns out on Model G200, and 1¾ turns out on Model GV200. Engine must be at normal operating temperature and running for final adjustment. Operate engine at idle speed and adjust low speed mixture screw (2) to obtain a smooth idle and satisfactory acceleration. Adjust idle speed by turning throttle stop screw (3) to obtain desired idle speed.

To check float height, remove fuel bowl and invert carburetor so float is vertical and resting lightly against fuel inlet valve as shown in Fig. HN4. Meas-

ure float height (H). Measurement should be 10.7-13.3 mm (0.42-0.52 in.) on Model GV100, and 6.7-9.7 mm (0.26-0.38 in.) on all other models. Renew float if float height is incorrect.

GOVERNOR. The mechanical flyweight type governor is located inside engine crankcase on the camshaft or on a separate shaft. To adjust external linkage, stop engine and make certain all linkage is in good condition and governor spring (2—Fig. HN5) is not stretched or damaged. Spring (4) must pull governor lever (3) and throttle pivot (6) toward each other. Loosen clamp bolt (8) and move governor lever (3) so throttle is completely open. Hold governor lever in this position and rotate governor shaft (1) in the same direction until it stops. Tighten clamp bolt.

Adjust maximum governed speed screw (9) so maximum engine speed is 3600 rpm on Model GV100, 3000-3200 rpm on Model GV200 and 3850-4000 rpm on Models G100, G150 and G200. Note that specified maximum speed may vary depending upon the equipment in which the engine is installed.

IGNITION SYSTEM. The breakerless ignition system requires no regular maintenance. The ignition coil unit is mounted outside the flywheel. Air gap between flywheel and ignition coil legs should be 0.2-0.6 mm (0.008-0.024 in.).

VALVE ADJUSTMENT. Valve tappet clearance (engine cold) should be as specified in following table:

G100
Intake . . . . 0.04-0.10 mm
(0.002-0.004 in.)

Exhaust . . . . 0.04-0.10 mm
(0.002-0.004 in.)

GV100
Intake . . . . 0.08-0.16 mm
(0.003-0.006 in.)

Exhaust . . . . 0.08-0.16 mm
(0.003-0.006 in.)

G150, G200
Intake . . . . 0.08-0.16 mm
(0.003-0.006 in.)

Exhaust . . . . 0.16-0.24 mm
(0.006-0.009 in.)

GV200
Intake . . . . 0.05-0.11 mm
(0.002-0.004 in.)

Exhaust . . . . 0.09-0.15 mm
(0.003-0.006 in.)

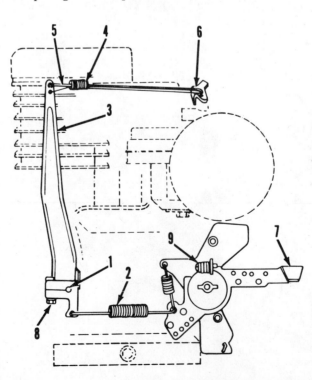

Fig. HN5—Drawing of external governor linkage used on horizontal crankshaft engine. Governor linkage on vertical crankshaft engines is similar.

1. Governor shaft
2. Tension spring
3. Governor lever
4. Spring
5. Governor rod
6. Throttle pivot
7. Throttle lever
8. Clamp bolt
9. Maximum speed screw

To check valve tappet clearance, remove spark plug and tappet chamber cover. Rotate crankshaft in normal direction of rotation until piston is at top dead center on compression stroke. Using a feeler gauge, measure clearance between valve tappets and intake and exhaust valve stem ends. Some models are equipped with valve stem caps that are available in varying thicknesses to adjust valve tappet clearance. On models not equipped with valve stem caps, adjust tappet clearance by grinding end of valve stem to increase clearance. To decrease clearance, grind valve seat deeper or renew valve and/or tappet.

CYLINDER HEAD AND COMBUSTION CHAMBER. Cylinder head, combustion chamber and top of piston should be cleaned and carbon and other deposits removed after every 300 hours of operation. Refer to REPAIRS section for service procedure.

REPAIRS

TIGHTENING TORQUES. Recommended tightening torque specifications are as follows:

Connecting rod:
  G100 . . . . 3 N•m
                            (27 in.-lbs.)
  GV100 . . . . 5 N•m
                            (44 in.-lbs.)
  All other models:
    6 mm . . . . 10 N•m
                           (89 in.-lbs.)
    7 mm . . . . 12 N•m
                       (106 in.-lbs.)
Crankcase cover/oil pan . . . . . . 10 N•m
                         (89 in.-lbs.)
Cylinder head:
  G100, GV100 . . . . 10 N•m
                         (89 in.-lbs.)
  All other models . . . . 25 N•m
                        (221 in.-lbs.)
Flywheel nut:
  G100 . . . . 4.8 N•m
                         (43 in.-lbs.)
  GV100 . . . . 50 N•m
                         (37 ft.-lbs.)

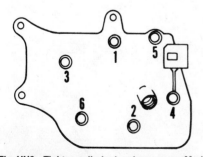

Fig. HN6—Tighten cylinder head screws on Models G100 and GV100 in sequence shown to 10 N•m (89 in.-lbs.).

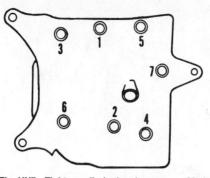

Fig. HN7—Tighten cylinder head screws on Models G150, G200 and GV200 in sequence shown to 25 N•m (221 in.-lbs.).

All other models . . . . 73 N•m
                         (54 ft.-lbs.)
Oil pump cover:
  GV200 . . . . 10 N•m
                         (89 in.-lbs.)

CYLINDER HEAD. To remove cylinder head, first remove cooling shrouds. Unscrew cylinder head screws in sequence shown in Fig. HN6 or HN7.

Remove carbon deposits from cylinder head, being careful not to scratch the gasket mating surface. Install cylinder head using a new gasket. Do not apply any sealant to the head gasket. Tighten cylinder head screws progressively in three steps following sequence shown in Fig. HN6 or HN7 to 10 N•m (89 in.-lbs.) on Models G100 and GV100 or to 25 N•m (221 in.-lbs.) on all other models.

VALVE SYSTEM. The valves can be removed from engine after first removing cylinder head and tappet chamber cover. Compress the valve spring and move slotted spring retainer sideways to disengage retainer from valve stem.

Valve face and seat angles are 45°. Standard valve seat width is 0.42-0.78 mm (0.016-0.031 in.) on Model G100 and 0.7 mm (0.028 in.) on all other models. Narrow seat if seat width is 1.0 mm (0.039 in.) or more on Models G100 and GV100 or 2.0 mm (0.079 in.) or more on all other models.

Standard valve stem diameter for Models G100 and GV100 is 5.480-5.490 mm (0.2157-0.2161 in.) for intake valve and 5.435-5.445 mm (0.2140-0.2144 in.) for exhaust valve. Renew intake valve if diameter is less than 5.450 mm (0.2146 in.). Renew exhaust valve if diameter is less than 5.400 mm (0.2126 in.).

Standard valve stem diameter for models except G100 and GV100 is 6.955-6.970 mm (0.2738-0.2744 in.) for intake valve and 6.910-6.925 mm (0.2720-0.2726 in.) for exhaust valve. Renew intake valve if diameter is less than 6.805 mm (0.2679 in.). Renew ex-

haust valve if diameter is less than 6.760 mm (0.2661 in.).

Standard valve guide inside diameter for Models G100 and GV100 is 5.50 mm (0.217 in.); renew guide if inside diameter is 5.56 mm (0.219 in.) or more. Standard valve guide inside diameter for all other models is 7.00 mm (0.276 in.); renew guide if inside diameter is 7.08 mm (0.279 in.) or more.

To remove and install valve guides on Model G100, use Honda valve guide tool 07969-8960000 to pull guide out of guide bore and press in new guide. New guide is pressed in to a depth of 18 mm (0.7 in.) measured from end of guide to cylinder head surface as shown in section (F—Fig. HN8) as (X). Finish ream guide after installation with reamer 07984-2000000.

To remove and install valve guides on all models except Model G100, use the following procedure and refer to the sequence of illustrations in Fig. HN8. Use driver 07942-8920000 on Model GV100 or driver 07942-8230000 on all other models and drive valve guide down into valve chamber slightly (A). Use a suitable cold chisel and sever guide adjacent to guide bore (B). Cover tappet opening to prevent fragments from entering crankcase. Drive remaining piece of guide into valve chamber (C) and remove from chamber. Place new guide on driver and start guide into guide bore (D). Alternate between driving guide into bore and measuring guide depth below cylinder head surface (E). Guide should be driven in to a depth of 18.5 mm (0.73 in.) on Model GV100 or 27.5 mm (1.08 in.) on all other models measured from end of guide to cylinder head surface as shown in section (F) at (X). Finish ream guide using reamer 07984-2000000 on Model GV100 or reamer 07984-5900000 on all other models.

CAMSHAFT. Camshaft rides directly in crankcase and crankcase cover or oil pan. Some models may be equipped with a camshaft that has an extension for use as a pto. This type camshaft is supported by a ball bearing at the pto end.

To remove camshaft, first remove engine from equipment. Remove crankcase cover or oil pan. If camshaft is gear-driven, rotate crankshaft so piston is at top dead center on compression stroke and withdraw camshaft. If camshaft is chain-driven (used on models with camshaft pto), the crankshaft, camshaft and timing chain must be removed as a unit. Refer to CRANKSHAFT section for crankshaft removal procedure.

Standard camshaft lobe height for Model G100 is 18.1-18.5 mm (0.71-0.73

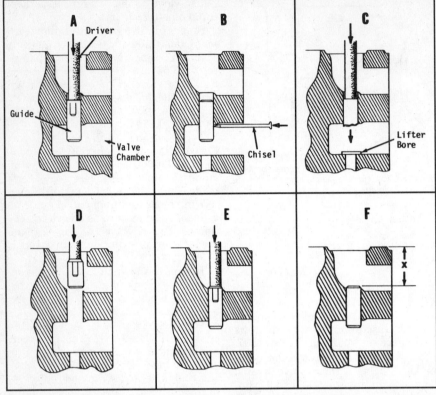

Fig. HN8—Refer to text and remove valve guide on all models except Model G100 in sequence shown.

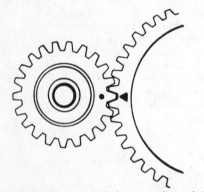

Fig. HN9—Align crankshaft gear and camshaft gear timing marks during installation.

in.) for both lobes. If intake or exhaust lobe measures 17.94 mm (0.706 in.) or less, renew camshaft.

Standard camshaft lobe height for Model GV100 is 20.82 mm (0.820 in.) for both lobes. If intake or exhaust lobe measures 20.45 mm (0.805 in.) or less, renew camshaft.

Standard camshaft lobe height for all models except Models G100 and GV100 is 33.5 mm (1.32 in.) for intake lobe and 33.8 mm (1.33 in.) for exhaust lobe. If intake lobe measures 33.25 mm (1.309 in.) or less, or exhaust lobe measures 33.55 mm (1.321 in.) or less, renew camshaft.

Service limit clearance between camshaft bearing journal and bearing hole

in crankcase, crankcase cover or oil pan is 0.1 mm (0.004 in.).

On models with extended camshaft, ball bearing should be a light press fit on camshaft journal and in crankcase cover. Camshaft seal should be pressed into crankcase cover 1.0 mm (0.04 in.).

On models with a gear-driven camshaft, timing marks on camshaft and crankshaft gears must be aligned as shown in Fig. HN9 during assembly.

On models with a chain-driven camshaft, timing marks on sprockets must be aligned as shown in Fig. HN10 during assembly.

PISTON, PIN AND RINGS. Piston and connecting rod are removed as an assembly. To remove piston and connecting rod, remove cylinder head,

Fig. HN10—On models with an extended camshaft for pto drive shaft, engines are equipped with timing sprockets and timing chain. Timing sprockets must be aligned with timing dots (T) aligned as shown during assembly.

crankcase cover or oil pan, and camshaft. Remove connecting rod cap screws and cap. Push connecting rod and piston assembly out of cylinder. Remove piston pin retaining rings and separate piston from connecting rod.

Standard piston diameter measured 90° from piston pin is listed in the following table:

Model	Piston Diameter
G100	45.98-46.00 mm (1.810-1.811 in.)
Wear limit	45.92 mm (1.808 in.)
GV100	49.995 mm (1.9683 in.)
Wear limit	49.92 mm (1.965 in.)
G150	63.98-64.00 mm (2.518-2.520 in.)
Wear limit	63.88 mm (2.515 in.)
G200, GV200	66.98-67.00 mm (2.637-2.638 in.)
Wear limit	66.88 mm (2.633 in.)

Check piston-to-cylinder clearance and refer to following table:

Model	Piston Clearance
G100	0.030 mm (0.0012 in.)
Wear limit	0.13 mm (0.005 in.)
GV100	0.005-0.035 mm (0.0002-0.0014 in.)
Wear limit	0.13 mm (0.005 in.)
G150	0.060 mm (0.0024 in.)
Wear limit	0.285 mm (0.0112 in.)
G200, GV200	0.045 mm (0.0018 in.)
Wear limit	0.285 mm (0.0112 in.)

Piston oversizes are available for some models.

Standard piston pin bore diameter in piston is 10.002 mm (0.3938 in.) for

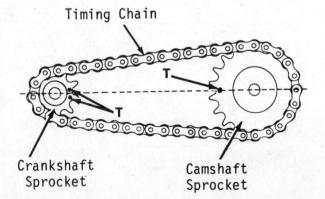

Timing Chain

T

Crankshaft Sprocket

Camshaft Sprocket

Models G100 and GV100, and 15.000-15.006 mm (0.5906-0.5908 in.) for remaining models. Renew piston if diameter exceeds 10.050 mm (0.3957 in.) on Models G100 and GV100, or 15.046 mm (0.5924 in.) on remaining models.

Standard piston pin diameter is 10.00 mm (0.394 in.) for Models G100 and GV100, and 15.00 mm (0.591 in.) for remaining models. Renew piston pin if diameter is less than 9.950 mm (0.3917 in.) on Models G100 and GV100, or 14.954 mm (0.5887 in.) on remaining models.

Ring side clearance in piston ring groove on Model G100 should be 0.025-0.055 mm (0.0009-0.0022 in.) for top ring and 0.010-0.040 mm (0.0004-0.0016 in.) for all remaining rings. Ring side clearance on Model GV100 for compression rings should be 0.055-0.085 mm (0.0022-0.0033 in.). Ring side clearance on all other models for compression rings should be 0.010-0.050 mm (0.0004-0.0020 in.). Maximum allowable side clearance is 0.10 mm (0.004 in.) on Models G100 and GV100 and 0.15 mm (0.006 in.) on all other models.

Ring end gap for compression rings on Models G100 and GV100 should be 0.15-0.35 mm (0.006-0.014 in.), while end gap on all other models should be 0.2-0.4 mm (0.008-0.016 in.). If ring end gap for any ring is 1.0 mm (0.039 in.) or more, renew ring and/or cylinder. Oversize piston rings are available for some models.

Install marked piston rings with marked side toward piston crown and stagger ring end gaps equally around piston. Lubricate piston and rings with engine oil prior to installation.

When assembling piston and rod or installing piston and rod in engine, refer to appropriate following paragraph.

Model G100. Assemble piston and rod so "896" stamped on piston crown (see Fig. HN11) is on same side as long side of connecting rod. Install piston and rod so "896" is toward valve side of engine. Align match marks on rod and rod cap.

Model GV100. Assemble piston and rod so "ZG1" stamped on piston crown

Fig. HN11—Note numbers on piston crown of Models G100 and GV100 when assembling piston and rod. See text.

(similar to Fig. HN11) is on same side as long side of connecting rod. Install piston and rod so "ZG1" is toward valve side of engine. Install rod cap so oil dipper points toward camshaft.

Models G150 And G200. Piston may be installed on rod in either direction. Install piston and rod so arrowhead on piston crown is toward valve side of engine. Align ribs on rod and rod cap and install oil dipper.

Model GV200. Assemble piston and rod so mark stamped on piston crown is toward ribbed side of connecting rod. Install piston and rod so ribbed side of rod is toward oil pan. Align match marks on rod and rod cap.

CONNECTING ROD. The aluminum alloy connecting rod rides directly on crankpin journal on all models. Refer to previous PISTON, PIN AND RINGS section for removal and installation procedure.

Standard piston pin bore diameter in connecting rod is 10.006 mm (0.3939 in.) for Models G100 and GV100, and 15.005-15.020 mm (0.5907-0.5913 in.) for remaining models. Renew connecting rod if diameter exceeds 10.050 mm (0.3957 in.) on Models G100 and GV100, or 15.070 mm (0.5933 in.) on remaining models.

Standard clearance between connecting rod and crankpin is 0.016-0.038 mm (0.0006-0.0015 in.) for Models G100 and GV100, and 0.040-0.066 mm (0.0016-0.0026 in.) for all other models. If clearance exceeds 0.1 mm (0.004 in.) on Models G100 or GV100, or 0.120 mm (0.0047 in.) on all other models, renew connecting rod and/or crankshaft.

Standard connecting rod side play on crankpin is 0.2-0.9 mm (0.008-0.035 in.) for Models G100 and GV100, and 0.10-0.80 mm (0.004-0.031 in.) for all other models. Maximum allowable rod side play is 1.1 mm (0.43 in.) on Models G100 and GV100, and 1.20 mm (0.47 in.) on all other models.

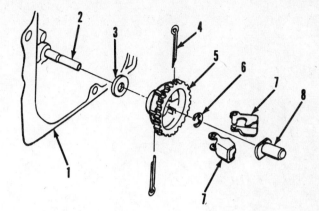

Fig. HN12—Exploded view of governor assembly used on Models G100 and GV100.

1. Crankcase cover
2. Governor stud
3. Thrust washer
4. Pin
5. Gear
6. "E" clip
7. Weights
8. Sleeve

GOVERNOR. Centrifugal flyweight type governor controls engine rpm via external linkage. The flyweight assembly on Models G100 and GV100 is located inside engine crankcase on a separate shaft and driven by the camshaft gear. A snap ring (6—Fig. HN12) secures the governor assembly on the shaft. A thrust washer is located under the governor gear.

On all other models, the flyweight assembly is located on the camshaft. Governor components on the camshaft cannot be serviced separately. The notch on the governor sleeve must index with a tab on the flyweight plate on the camshaft.

Adjust external linkage as outlined under GOVERNOR in MAINTENANCE section.

CRANKSHAFT, MAIN BEARINGS AND SEALS. Crankshaft is supported either by a ball bearing or a plain bearing.

To remove crankshaft, remove cooling shrouds, flywheel, cylinder head and crankcase cover or oil pan. Remove piston and connecting rod. Carefully remove camshaft and crankshaft. If necessary, remove main bearings and oil seals.

Standard crankpin diameter is 17.973-17.984 mm (0.7076-0.7080 in.) for Models G100 and GV100 and 25.967-25.980 mm (1.0223-1.0228 in.) for all other models. If crankpin diameter is less than 17.940 mm (0.7063 in.) for Models G100 and GV100 and 25.917 mm (1.0204 in.) for all other models, renew crankshaft.

On some models, crankshaft gear is a press fit on crankshaft. Mark position of gear on shaft before removal so gear can be installed on shaft in original position.

Except on Model GV100, install oil seal in crankcase cover or oil pan so seal is 4.5 mm (0.18 in.) below flush, and install oil seal in crankcase so seal is 2.0 mm (0.08 in.) below flush.

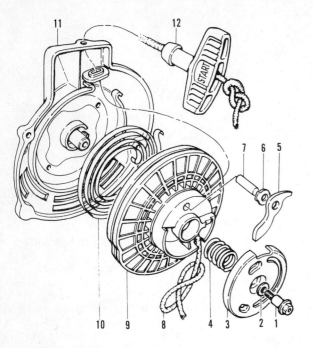

Fig. HN13—Exploded view of
rewind starter used on Model
GV100.

1. Screw
2. Washer
3. Cover
4. Friction spring
5. Pawl
6. Spring
7. Shaft
8. Rope
9. Pulley
10. Rewind spring
11. Starter housing
12. Rope handle

On models with a gear-driven camshaft, timing marks on camshaft and crankshaft gears must be aligned as shown in Fig. HN9 during assembly.

On models with a chain-driven camshaft, timing marks on sprockets must be aligned as shown in Fig. HN10 during assembly.

CYLINDER AND CRANKCASE.
Cylinder and crankcase are an integral casting. Refer to following table for standard cylinder bore size and wear limit. Rebore or renew cylinder if cylinder bore exceeds wear limit.

Model	Cylinder Bore
G100	46.00-46.01 mm
	(1.811-1.812 in.)
Wear limit	46.05 mm
	(1.813 in.)
GV100	50.00 mm
	(1.969 in.)
Wear limit	50.05 mm
	(1.970 in.)
G150	64.00-64.02 mm
	(2.519-2.520 in.)
Wear limit	64.17 mm
	(2.526 in.)
G200, GV200	67.00-67.02 mm
	(2.638-2.639 in.)
Wear limit	67.17 mm
	(2.644 in.)

Recondition cylinder bore if bore diameter exceeds wear limit. Oversize pistons are available for some models.

OIL PUMP. Model GV200 is equipped with an internal oil pump located inside crankcase. All other models are splash lubricated.

Oil pump rotors may be removed and checked without disassembling the engine. Remove oil pump cover and withdraw outer and inner rotors. Remove "O" ring. Make certain oil passages are clear and pump body is thoroughly clean. Inner-to-outer rotor clearance should be 0.15 mm (0.006 in.). If clearance is 0.20 mm (0.008 in.) or more, renew rotors. Clearance between oil pump body and outer rotor diameter should be 0.15 mm (0.006 in.). If clearance is 0.26 mm (0.010 in.) or more, renew rotors and/or oil pan. Outside diameter of outer rotor should be 23.15-23.28 mm (0.911-0.917 in.). If diameter is 23.23 mm (0.915 in.) or less, renew rotors.

Reverse disassembly procedure for reassembly and tighten oil pump cover screws to 10 N•m (89 in.-lbs.).

REWIND STARTER. Model GV100 may be equipped with the rewind starter shown in Fig. HN13 while the starter shown in HN14 may be used on other models. To disassemble starter, remove rope handle and allow rope to wind into starter. Remove pulley cover, if so equipped. Unscrew center retaining screw or nut. Wear appropriate safety eyewear and gloves before disengaging pulley from starter as spring may uncoil uncontrolled. Place shop towel around pulley and lift pulley out of housing. Use caution when detaching rewind spring from pulley or housing. If spring must be removed from pulley or housing, position pulley or housing so spring side is down and against floor. Tap pulley or housing to dislodge spring.

When assembling starter used on Model GV100 (Fig. HN13), install rewind spring on pulley so coil direction is counterclockwise from outer end. Wrap rope around pulley in counterclockwise direction viewed from flywheel side of pulley. Apply grease to side of rewind spring, then install pulley in starter

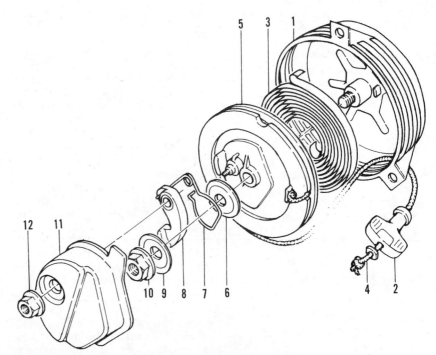

Fig. HN14—Exploded view of rewind starter used on some engines.

1. Starter housing
2. Rope handle
3. Rewind spring
4. Rope
5. Pulley
6. Friction plate
7. Spring
8. Pawl
9. Friction plate
10. Nut
11. Cover
12. Nut

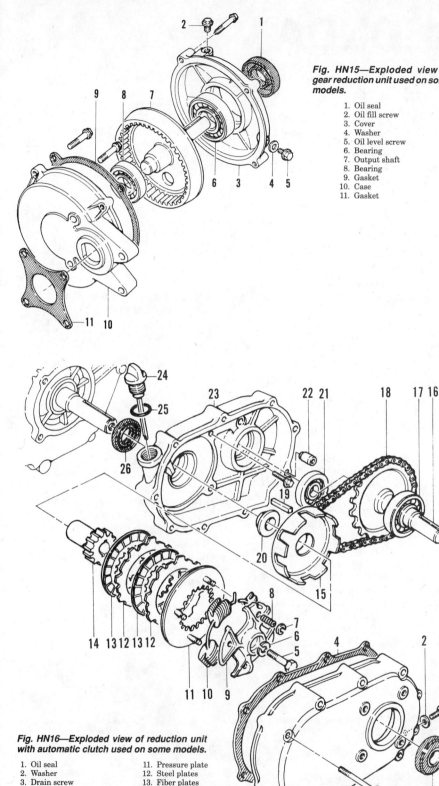

Fig. HN15—Exploded view of gear reduction unit used on some models.

1. Oil seal
2. Oil fill screw
3. Cover
4. Washer
5. Oil level screw
6. Bearing
7. Output shaft
8. Bearing
9. Gasket
10. Case
11. Gasket

Fig. HN16—Exploded view of reduction unit with automatic clutch used on some models.

1. Oil seal
2. Washer
3. Drain screw
4. Gasket
5. Screw
6. Spring washer
7. Spring holder
8. Clutch spring
9. Weight holder
10. Weights
11. Pressure plate
12. Steel plates
13. Fiber plates
14. Input shaft
15. Clutch drum
16. Output shaft
17. Bearing
18. Chain
19. Key
20. Spacer
21. Bearing
22. Dowel pin
23. Case
24. Dipstick
25. "O" ring
26. Oil seal

housing so end of spring engages notch on pulley and route rope through rope outlet in housing. Apply grease to pawl shaft and install remainder of starter components. To place tension on starter rope, pull rope out of housing until notch in pulley is aligned with rope outlet, then hold pulley to prevent pulley rotation. Pull rope back into housing while positioning rope in pulley notch. Turn rope pulley counterclockwise until spring is tight, allow pulley to turn clockwise until notch aligns with rope outlet, and disengage rope from notch. Check starter operation.

When assembling starter used on models other than Model GV100 (Fig. HN13), install rewind spring on pulley so coil direction is counterclockwise from outer end. Wrap rope around pulley in counterclockwise direction viewed from flywheel side of pulley. Apply grease to side of rewind spring and pulley shaft, then install pulley in starter housing so end of spring engages notch on pulley and route rope through rope outlet in housing. Attach rope handle. Apply grease to pawl shaft and friction plates (6 and 9). Install remainder of starter components. Convex sides of friction plates should be facing. Tighten center nut to 20-28 N·m (177-248 in.-lbs.) To place tension on starter rope, pull rope out of housing until notch in pulley is aligned with rope outlet, then hold pulley to prevent pulley rotation. Pull rope back into housing while positioning rope in pulley notch. Turn rope pulley counterclockwise until spring is tight, allow pulley to turn clockwise until notch aligns with rope outlet, and disengage rope from notch. Check starter operation.

REDUCTION GEAR. Engines equipped with a horizontal crankshaft may be equipped with a reduction gear unit that attaches to the crankcase cover. Power is transferred via gears or a chain, while some units may also contain a plate-type automatic clutch. Refer to Fig. HN15 and HN16 for an exploded view of unit.

Overhaul of unit is apparent after inspection of unit. On unit with an automatic clutch (Fig. HN16), wear limit thickness for friction plates is 3.0 mm (0.118 in.) and maximum allowable warpage of steel plates is 0.10 mm (0.004 in.). When assembling clutch pack in clutch drum, alternately install fiber and steel plates with a fiber plate first and a steel plate last.

HONDA

Model	Bore	Stroke	Displacement	Power Rating
GX110	57 mm	42 mm	107 cc	2.6 kW
	(2.2 in.)	(1.7 in.)	(6.6 cu.in.)	(3.5 hp)
GX120, GXV120	60 mm	42 mm	118 cc	2.9 kW
	(2.4 in.)	(1.7 in.)	(7.2 cu.in.)	(4.0 hp)
GX140, GXV140	64 mm	45 mm	144 cc	3.8 kW
	(2.5 in.)	(1.8 in.)	(8.8 cu.in.)	(5.0 hp)
GX160, GXV160	68 mm	45 mm	163 cc	4.0 kW
	(2.7 in.)	(1.8 in.)	(10.0 cu.in.)	(5.5 hp)
GX240	73 mm	58 mm	242 cc	6.0 kW
	(2.9 in.)	(2.3 in.)	(14.8 cu.in.)	(8 hp)

NOTE: Metric fasteners are used throughout engine.

ENGINE INFORMATION

All models are four-stroke, overhead valve, single-cylinder, air-cooled engines. Models GXV120, GXV140 and GXV160 are vertical crankshaft engines. All other models are horizontal crankshaft engines with the cylinder inclined 25°.

Engine model number is cast into side of crankcase (Fig. HN31). Engine serial number is stamped into crankcase (Fig. HN32). Always furnish engine model and serial number when ordering parts.

MAINTENANCE

LUBRICATION. Engine oil level should be checked prior to operating engine. Maintain oil level at top of reference marks (Fig. HN33) when checked with cap not screwed in, but just touching first threads.

Oil should be changed after the first 20 hours of engine operation and after every 100 hours thereafter.

Engine oil should meet or exceed latest API service classification. Use SAE 30 oil for temperatures above 50° F (10° C); use SAE 10W-30 oil for temperatures between 0° F (−18° C) and 85° F (30° C); below 32° F (0° C) SAE 5W-30 may be used.

Crankcase capacity is 0.6 L (0.63 qt.) for all models except Model GX240. Crankcase capacity for Model GX240 is 1.1 L (1.16 qt.).

Some engines are equipped with a low-oil warning system. If engine oil is low, an indicator lamp will light and the engine will stop or not run. Refer to OIL WARNING SYSTEM in REPAIRS section.

The reduction gear unit on horizontal crankshaft engines may be lubricated by oil from the engine or by a separate oil supply. If there is no oil level screw on the reduction case then the unit is lubricated by oil from the engine. The reduction gear unit with an automatic clutch is equipped with an oil dipstick to check oil level in the case. Oil level is checked similar to checking engine oil by unscrewing the oil fill plug. Maintain oil level at top of notch on gage when checked with plug not screwed in, but just touching first threads. Fill with same oil recommended for engine. Capacity is 0.3 L (0.32 qt.) on Model GX240 and 0.5 L (0.53 qt.) on all other models.

To check oil level on reduction units equipped with an oil level screw plug but no dipstick, add oil so level reaches oil level screw hole.

AIR CLEANER. Engine air filter should be cleaned and inspected after every 50 hours of operation, or more often if operating in extremely dusty conditions.

Remove foam and paper air filter elements from air filter housing. Foam element should be washed in a mild detergent and water solution, rinsed in clean water and allowed to air dry. Soak foam element in clean engine oil. Squeeze out excess oil.

Paper element may be cleaned by directing low pressure compressed air stream from inside filter toward the outside. Reinstall elements.

Some engines may be equipped with an oil bath type filter system in addition to the foam type element. Fill the oil compartment to the level indicated on the side of the case with engine oil. Do not overfill.

SPARK PLUG. Spark plug should be removed, cleaned and inspected after every 100 hours of use.

Fig. HN31—Engine model number (MN) is cast into side of engine crankcase.

Fig. HN32—Engine serial number (SN) is stamped on raised portion of crankcase.

Fig. HN33—Do not screw in oil plug and gage when checking oil level. Maintain oil level at top edge of reference marks (R) on dipstick.

Fig. HN34—View of Keihin float type carburetor used on all models showing location of low speed mixture screw (LS) and throttle stop screw (TS).

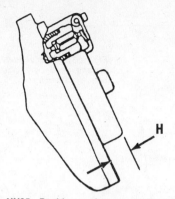

Fig. HN35—Position carburetor so float rests lightly against fuel inlet valve and measure float height (H).

Model		
GX140	#65	—
GX160	#72	#68
GX240	#85	#92
GXV120	#65	—
GXV140	#65	—
GXV160	#70	—

NOTE: Jet sizes may be different than listed above for some engine applications.

Recommended spark plug for Models GX120 and GX160 is NGK BPR6ES or ND W20EPR-U. Recommended spark plug for all other models is a NGK BP5ES or ND W16EP-U. A resistor type spark plug may be required in some locations. Spark plug electrode gap should be 0.7-0.8 mm (0.028-0.031 in.) for all models.

When installing spark plug, manufacturer recommends installing spark plug finger tight, then for a new plug, tighten an additional 1/2 turn. For a used plug, tighten an additional 1/8 to 1/4 turn.

CARBURETOR. All models are equipped with a Keihin float type carburetor with a fixed main fuel jet and an adjustable low speed fuel mixture screw.

Carburetor adjustment settings and jet sizes are different on some models due to external or internal venting of carburetor. Engines with type designation "QXC", stamped near the serial number (see Fig. HN32), have an internally vented carburetor. Note initial adjustment setting for low speed mixture screw (LS—Fig. HN34) in following table (turns out from a lightly seated position):

Model	Except Type QXC	Type QXC
GX110	3 turns	—
GX120	2 turns	2 3/8 turns
GX140	1 5/8 turns	—
GX160	3 turns	2 1/8 turns
GX240	2 turns	—
GXV120	3 turns	—
GXV140	1 1/4 turns*	—
GXV160	2 turns	—

*Some California models require 1 3/4 turns.

Adjust effective throttle cable length by loosening jam nuts and turning adjuster at handlebar end of cable. The choke should be fully closed when throttle lever is in "CHOKE" position, and engine should stop when throttle lever is in "STOP" position.

For final adjustment engine must be at normal operating temperature and running. Operate engine at idle speed and adjust low speed mixture screw (LS) to obtain a smooth idle and satisfactory acceleration. Adjust idle speed by turning throttle stop screw (TS) to obtain desired idle speed.

To check float height, remove fuel bowl and invert carburetor so float is vertical and resting lightly against fuel inlet valve as shown in Fig. HN35. Measure float height (H). Measurement should be 13.7 mm (0.54 in.) on Models GX120, GXV140 and GX160, 13.2 mm (0.52 in.) on Models GX240, and 12.2-15.2 mm (0.48-0.60 in.) for all other models. Renew float if float height is incorrect.

Refer to following table for standard main jet size. Optional jet sizes are available.

Model	Except Type QXC	Type QXC
GX110	#62	—
GX120	#60	#62

GOVERNOR. The mechanical flyweight type governor is located inside engine crankcase. To adjust external linkage, stop engine and make certain all linkage is in good condition and governor spring (5—Fig. HN36 or HN37) is not stretched or damaged. Spring (2) must pull governor lever (3) and throttle pivot toward each other. Loosen clamp bolt (7) and move governor lever (3) so throttle is completely open. Hold governor lever in this position and rotate governor shaft (6) in the same direction until it stops. Tighten clamp bolt.

With engine at normal operating temperature, adjust throttle stop screw (8) so engine operates at recommended maximum engine rpm. Some typical maximum recommended engine speeds are: lawn mower, 3100 rpm; generator, 3750 rpm; industrial, 3900 rpm.

IGNITION SYSTEM. The breakerless ignition system requires no regular maintenance. The ignition coil unit is mounted outside the flywheel. Air gap between flywheel and ignition coil legs should be 0.2-0.6 mm (0.008-0.024 in.).

To check ignition coil primary side, connect one ohmmeter lead to primary (black) coil lead and touch iron coil laminations with remaining lead. Ohmmeter should register 1.0-1.2 ohms on GXV120, 0.8-1.0 on Models GX120 and

Fig. HN36—View of governor linkage used on vertical crankshaft engines.

1. Governor-to-carburetor rod
2. Spring
3. Governor lever
4. Choke rod
5. Tension spring (behind plate & lever)
6. Governor shaft
7. Clamp bolt
8. Throttle stop screw

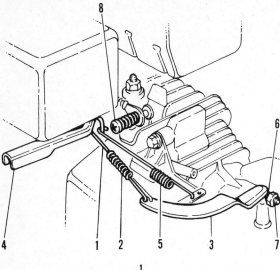

Fig. HN37—View of governor linkage used on horizontal crankshaft engines.

1. Governor rod
2. Tension spring
3. Governor lever
4. Throttle lever
5. Governor spring
6. Governor shaft
7. Clamp bolt
8. Throttle stop screw

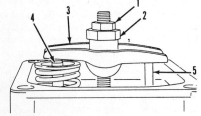

Fig. HN38—View of rocker arm and related parts.

1. Jam nut
2. Adjustment nut
3. Rocker arm
4. Valve stem clearance
5. Push rod

GX160, and 0.7-0.9 ohm on other models.

To check ignition coil secondary side, connect one ohmmeter lead to the spark plug lead wire and remaining lead to the iron core laminations. Ohmmeter should read 10k-14k ohms on GXV120, 5.9k-7.1k ohms on Models GX120 and GX160, and 6.3k-7.7k ohms on other models. If ohmmeter readings are not as specified, renew ignition coil.

VALVE ADJUSTMENT. Valve-to-rocker arm clearance should be checked and adjusted after every 300 hours of operation.

To adjust valve clearance, remove rocker arm cover. Rotate crankshaft in normal direction so piston is at top dead center (TDC) on compression stroke. Insert a feeler gauge between rocker arm (3—Fig. HN38) and end of valve stem (4). Loosen rocker arm jam nut (1) and turn adjusting nut (2) to obtain desired clearance. Specified clearance is 0.15 mm (0.006 in.) for intake and 0.20 mm (0.008 in.) for exhaust. Tighten jam nut and recheck clearance. Install rocker arm cover.

COMPRESSION PRESSURE. Specified compression pressure reading is 588-834 kPa (85-121 psi) for all models except GXV120. Compression read-

ing on Model GXV120 may be checked with compression release mechanism operational or disengaged. With compression release operating, compression reading should be 294-686 kPa (43-99 psi). With compression release disengaged (remove exhaust push rod), compression reading should be 784-1176 kPa (113-170 psi).

FLYWHEEL BRAKE. Some models may be equipped with a flywheel brake that should stop the engine within three seconds after the safety handle is released. When the flywheel brake is activated, an engine stop switch grounds the ignition.

Be sure the flywheel brake operates properly. Check free play of operating handle at top of handle. Free play should be 5-10 mm (3/16 to 3/8 in.) for lawn mower Models HR194, HR214 and HRA214, and 20-25 mm (3/4 to 1 in.) for Models HR195, HR215, HRA215, HRC215 and HRC216. Loosen jam nuts and turn adjuster at handlebar end of control cable to adjust free play.

CYLINDER HEAD AND COMBUSTION CHAMBER. Manufacturer recommends removal of carbon and lead deposits from cylinder head combustion chamber, valves and valve seats after every 300 hours of operation. Refer to CYLINDER HEAD paragraph in REPAIRS section for service procedure.

REPAIRS

TIGHTENING TORQUES. Recommended tightening torque specifications are as follows:
Connecting rod:
GX110, GX120, GX140,
GX160, GXV120,
GXV140, GXV160. . . . . 12 N•m
(106 in.-lbs.)

GX240 (serial number
1000001-1020232). . . . . . 32-34 N•m
(16-19 ft.-lbs.)
GX240 (after serial
number 1020232). . . . . 14 N•m
(124 in.-lbs.)
Crankcase cover/oil pan:
GX110, GX120,
GXV120, GXV140 . . . . 12 N•m
(106 in.-lbs.)
All other models . . . . 24 N•m
(18 ft.-lbs.)
Cylinder head:
GX110, GX120, GX140,
GX160, GXV120,
GXV140, GXV160 . . . . 24 N•m
(18 ft.-lbs.)
GX240. . . . . 35 N•m
(26 ft.-lbs.)
Flywheel nut:
GX110, GX120, GX140,
GX160, GXV120,
GXV140, GXV160 . . . . 75 N•m
(55 ft.-lbs.)
GX240. . . . . 115 N•m
(85 ft.-lbs.)
Oil drain plug. . . . . 18 N•m
(160 in.-lbs.)
Rocker arm cover. . . . . 8-12 N•m
(71-106 in.-lbs.)
Rocker arm jam nut. . . . . 8-12 N•m
(71-106 in.-lbs.)

CYLINDER HEAD. To remove cylinder head, remove blower shroud. Disconnect carburetor linkage and remove carburetor. Remove heat shield and muffler. Remove rocker arm cover and the four head screws. Remove cylinder head. Use care not to lose push rods.

When installing cylinder head on all models, tighten head screws to specified torque following sequence shown in Fig. HN39. Adjust valves as outlined in VALVE ADJUSTMENT paragraph.

VALVE SYSTEM. Remove rocker arms, compress valve springs and remove valve retainers. Note exhaust valve on Models GX110, GX120, GX140, GX160 and GX240 is equipped with a valve rotator on valve stem. Remove valves and springs. Remove push rod guide plate if necessary.

Valve face and seat angles are 45°. Standard valve seat width is 0.8 mm (0.032 in.). Narrow seat if seat width is 2.0 mm (0.079 in.) or more.

Standard valve spring free length for Model GX240 is 39.0 mm (1.54 in.). Renew valve spring if free length is 37.5 mm (1.48 in.) or less. Standard valve spring free length for all other models is 34.0 mm (1.339 in.). Renew valve spring if free length is 32.5 mm (1.280 in.) or less.

Standard valve guide inside diameter for Model GX240 is 6.60 mm (0.260 in.).

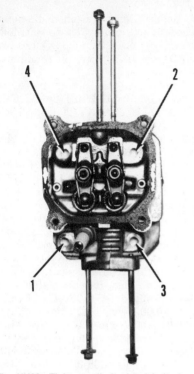

Fig. HN39—Tighten cylinder head bolts in sequence shown.

Renew guide if inside diameter is 6.66 mm (0.262 in.) or more. Standard valve guide inside diameter for all other models is 5.50 mm (0.2165 in.). Renew guide if inside diameter is 5.572 mm (0.2194 in.) or more.

Valve stem-to-guide clearance for all models except Model GX240 should be 0.020-0.044 mm (0.0008-0.0017 in.) for intake valve and 0.060-0.087 mm (0.0024-0.0034 in.) for exhaust valve. Renew valve and/or guide if clearance is 0.10 mm (0.004 in.) or more for intake valve or 0.12 mm (0.005 in.) or more for exhaust valve.

Valve stem-to-guide clearance for Model GX240 should be 0.010-0.037 mm (0.0004-0.0015 in.) for intake valve and 0.050-0.077 mm (0.002-0.003 in.) for exhaust valve. Renew valve and/or guide if clearance is 0.10 mm (0.004 in.) or more for intake valve or 0.12 mm (0.005 in.) or more for exhaust valve.

To renew valve guide on all models except Model GX240, heat entire cylinder head to 150° C (300° F) and use valve guide driver 07942-8920000 to remove and install guides. DO NOT heat head above recommended temperature as valve seats may loosen. Drive guides out toward rocker arm end of head. Note if exhaust valve guide has a locating clip around the top. If there is NO clip, drive guides into cylinder head until distance from end of guide to cylinder head mounting surface is 23.0 mm (0.905 in.) for Models GX110 and GXV120 or 25.5 mm (1.004 in.) for Models GX140, GXV140 and GXV160. If there is a clip around the exhaust valve guide, drive in the exhaust valve guide so the clip (C—Fig. HN40) is bottomed in the head. Drive in the intake valve guide so the top of the guide (S) stands 3.0 mm (0.12 in.) above the cylinder head boss. On all models, new valve guides must be reamed after installation to obtain specified valve stem clearance.

To renew valve guide on Model GX240, heat entire cylinder head to 150° C (300° F) and use valve guide driver 07742-0010200 to remove and install guides. DO NOT heat head above recommended temperature as valve seats may loosen. Drive guides out toward rocker arm end of head. Note if exhaust valve guide has a locating clip around the top. If there is NO clip, drive guides into cylinder head until distance from end of guide to cylinder head mounting surface is 7.0 mm (0.28 in.). If there is a clip around the exhaust valve guide, drive in the exhaust valve guide so the clip (C—Fig. HN40) is bottomed in the head. Drive in the intake valve guide so the top of the guide (S) stands 3.0 mm (0.12 in.) above the cylinder head boss on later models, or 9.0 mm (0.35 in.) on early models. New valve guides must be reamed after installation.

CAMSHAFT. Camshaft and camshaft gear are an integral casting equipped with a compression release mechanism (Fig. HN41). To remove camshaft, first remove engine from equipment. Remove crankcase cover or oil pan. Rotate crankshaft so piston is at top dead center on compression stroke. Withdraw camshaft from crankcase.

Standard camshaft bearing journal diameter for Models GX110, GX120, GX140, GX160, GXV120, GXV140 and GXV160 is 13.984 mm (0.5506 in.). Renew camshaft if journal diameter is 13.916 mm (0.5479 in.) or less. Standard camshaft bearing journal diameter for Model GX240 is 15.984 mm (0.6293 in.). Renew camshaft if journal diameter is 15.92 mm (0.627 in.) or less.

Standard camshaft lobe height for all models except Model GX240 is 27.7 mm (1.091 in.) for intake lobe and 27.75 mm (1.093 in.) for exhaust lobe. If intake lobe measures 27.45 mm (1.081 in.) or less, or exhaust lobe measures 27.50 mm (1.083 in.) or less, renew camshaft. Standard camshaft lobe height for Model GX240 is 31.52-31.92 mm (1.241-1.257 in.) for intake lobe and 31.56-31.96 mm (1.242-1.258 in.) for exhaust lobe. If intake or exhaust lobe measures 31.35 mm (1.234 in.) or less, renew camshaft.

Inspect compression release mechanism (Fig. HN42) for damage. Spring must pull weight tightly against camshaft so decompressor lobe holds exhaust valve slightly open. Weight overcomes spring tension at 1000 rpm and moves decompressor lobe away from cam lobe to release exhaust valve.

When installing camshaft, make certain camshaft and crankshaft gear tim-

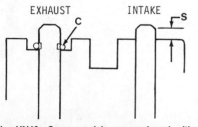

Fig. HN40—Some models are equipped with a locating clip (C) around the top of the exhaust valve guide. Install intake valve guide so standout (S) above cylinder head boss is as specified in text.

Fig. HN41—Camshaft and gear are an integral casting equipped with a compression release mechanism. Camshaft shown is for models with an auxiliary drive shaft.

Fig. HN42—Compression release mechanism spring (1) and weight (2) installed on camshaft gear.

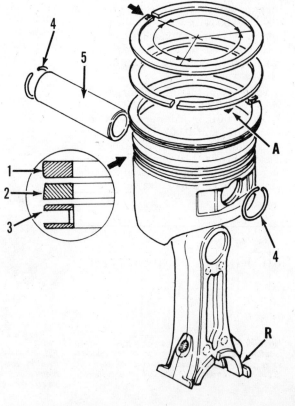

Fig. HN43—Align crankshaft gear and camshaft gear timing marks (F) during installation.

ing marks are aligned as shown in Fig. HN43.

PISTON, PIN AND RINGS. Piston and connecting rod are removed as an assembly. To remove piston and connecting rod, remove cylinder head, crankcase cover or oil pan, and camshaft. Remove connecting rod cap screws and cap. Push connecting rod and piston assembly out of cylinder. Remove piston pin retaining rings and separate piston from connecting rod.

Standard piston diameter measured 10 mm (0.4 in.) from lower edge of skirt and 90° from piston pin is listed in the following table:

Model	Piston Diameter
GX110	56.965-56.985 mm
	(2.2427-2.2435 in.)
Wear limit	56.55 mm
	(2.226 in.)
GX120	59.985 mm
	(2.3616 in.)
Wear limit	59.845 mm
	(2.3561 in.)
GX140	63.985 mm
	(2.5191 in.)
Wear limit	63.815 mm
	(2.5124 in.)
GX160	67.985 mm
	(2.6766 in.)
Wear limit	67.845 mm
	(2.6711 in.)
GX240	72.985 mm
	(2.8734 in.)
Wear limit	72.62 mm
	(2.859 in.)
GXV120	59.985 mm
	(2.3616 in.)
Wear limit	59.845 mm
	(2.3561 in.)
GXV140	63.985 mm
	(2.5191 in.)

Wear limit	63.815 mm
	(2.5124 in.)
GXV160	67.985 mm
	(2.6766 in.)
Wear limit	67.845 mm
	(2.6711 in.)

Piston oversizes are available for some models.

Standard piston pin bore diameter in piston is 13.002 mm (0.5119 in.) for Models GX110, GX120 and GXV120, and 18.002 mm (0.7087 in.) for remain-

Fig. HN44—View of rod and piston assembly. Long side of connecting rod (R) and arrowhead (A) on piston crown must be on the same side. Install piston rings with marked side toward piston crown.

1. Top ring (chrome plated)
2. Second ring (tapered face)
3. Oil control ring
4. Retaining rings
5. Piston pin

ing models. Renew piston if diameter exceeds 13.048 mm (0.5137 in.) on Models GX110, GX120 and GXV120, or 18.048 mm (0.7105 in.) on remaining models.

Standard piston pin diameter is 13.000 mm (0.5118 in.) for Models GX110, GX120 and GXV120, and 18.000 mm (0.7087 in.) for remaining models. Renew piston pin on Models GX110, GX120 and GXV120 if diameter is 12.954 mm (0.510 in.) or less. Renew piston pin on Models GX140, GX160, GXV140, GXV160 and GX240 if piston pin diameter is 17.954 mm (0.7068 in.) or less. Standard clearance between piston pin and pin bore in piston is 0.002-0.014 mm (0.0001-0.0006 in.) for all models. If clearance is 0.08 mm (0.003 in.) or greater, renew piston and/or pin.

Ring side clearance in piston ring groove should be 0.030-0.065 mm (0.0012-0.0026 in.) on Model GXV160 and 0.015-0.045 mm (0.0006-0.0018 in.) on all other models. Ring end gap for compression rings on all models should be 0.2-0.4 mm (0.008-0.016 in.). Ring end gap for oil control ring on all models should be 0.15-0.35 mm (0.006-0.014 in.). If ring end gap for any ring is 1.0 mm (0.039 in.) or more, renew ring and/or cylinder. Oversize piston rings are available for some models.

Refer to Fig. HN44 for correct installation of piston rings. Note that top ring is chrome plated and that second ring has a tapered face. Install marked pis-

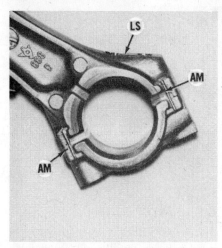

Fig. HN45—View of connecting rod used on Models GXV120, GXV140 and GXV160. Models GX110, GX120, GX140 and GX240 are equipped with an oil dipper on connecting rod cap.

Fig. HN46—Governor assembly on Models GXV120, GXV140 and GXV160 is mounted in oil pan. Refer to Fig. HN48 also. Governor assembly for horizontal crankshaft engines is shown in Fig. HN47.

G. Auxiliary drive gear
P. Retaining hair pin
W. Thrust washers
1. Bolt
2. Governor gear
3. Weight
4. Governor gear shaft
5. Auxiliary drive shaft

ton rings with marked side toward piston crown and stagger ring end gaps equally around piston as shown in Fig. HN44. Lubricate piston and rings with engine oil prior to installation.

When reassembling piston on connecting rod, long side of connecting rod and arrowhead on piston crown must be on the same side as shown in Fig. HN44. When installing piston pin retaining rings, do not align the end gap of the ring with the cutout in the piston pin bore.

When reinstalling piston and connecting rod assembly in cylinder, arrowhead on piston crown must be on camshaft side of engine. Align connecting rod cap and connecting rod match marks (AM—Fig. HN45), install connecting rod bolts and tighten to specified torque.

CONNECTING ROD. The aluminum alloy connecting rod rides directly on crankpin journal on all models. Connecting rod cap for all horizontal crankshaft models is equipped with an oil dipper. There is no dipper on vertical crankshaft models (Fig. HN45).

Refer to previous PISTON, PIN AND RINGS section for removal and installation procedure.

Standard piston pin bore diameter in connecting rod is 13.005 mm (0.512 in.) for Models GX110, GX120 and GXV120, 18.002 mm (0.7087 in.) for Models GX140, GX160 and GXV140, and 18.005 mm (0.7089 in.) for Models GXV160 and GX240. Renew connecting rod if diameter exceeds 13.07 mm (0.515 in.) on Models GX110, GX120 and GXV120, or 18.07 mm (0.711 in.) on Models GX140, GX160, GXV140, GXV160 or GX240.

Standard connecting rod bearing bore-to-crankpin clearance is 0.040-0.063 mm (0.0015-0.0025 in.) for all models. Renew connecting rod and/or crankshaft if clearance is 0.12 mm (0.0047 in.) or more. An undersize connecting rod is available for some models.

Connecting rod side play on crankpin should be 0.1-0.7 mm (0.004-0.028 in.) for all models. Renew connecting rod if side play is 1.1 mm (0.043 in.) or more.

Standard connecting rod big end diameter is 26.02 mm (1.0244 in.) for Models GX110, GX120 and GXV120, 30.02 mm (1.1819 in.) for Models GX140, GX160, GXV140 and GXV160, and 33.025 mm (1.3002 in.) for Model GX240. Renew connecting rod if big end diameter exceeds 26.066 mm (1.0262 in.) on Models GX110, GX120 and GXV120, 30.066 mm (1.1837 in.) on Models GX140, GX160, GXV140 and GXV160, and 33.07 mm (1.302 in.) on Model GX240.

GOVERNOR. Centrifugal flyweight type governor controls engine rpm via external linkage. Governor is located in oil pan on Models GXV120, GXV140 and GXV160, in crankcase cover on Model GX240, and on flywheel side of crankcase on Models GX110, GX120, GX140 and GX160. Refer to GOVERNOR paragraphs in MAINTENANCE section for adjustment procedure.

To remove governor assembly on Models GXV120, GXV140 and GXV160, remove oil pan. Remove governor assembly retaining bolt (1—Fig. HN46) and remove governor gear and weight assembly. Governor sleeve, thrust

Fig. HN47—Exploded view of typical governor flyweight assembly used on horizontal crankshaft engines.

1. Sleeve
2. Thrust washer
3. Clip
4. Weight
5. Pin
6. Gear
7. Thrust washer

washer, retaining clip and gear may be removed from shaft.

To remove governor assembly on Models GX110, GX120, GX140 and GX160, the crankshaft must be withdrawn. The governor assembly on Model GX240 is accessible after removing the crankcase cover. A typical governor assembly used on horizontal crankshaft engines is shown in Fig. HN47. Remove governor sleeve and washer. Remove retaining clip from governor gear shaft, then remove gear and weight assembly and remaining thrust washer.

Reinstall governor assemblies by reversing removal procedure. Adjust external linkage as outlined under GOVERNOR in MAINTENANCE section.

CRANKSHAFT, MAIN BEARINGS AND SEALS.

Crankshaft for Models GX110, GX120, GX140, GX160 and GX240 is supported at each end in ball bearing type main bearings. Crankshaft for Models GXV120, GXV140 and GXV160 is supported at flywheel end in a ball bearing main bearing and at pto end in a bushing type main bearing that is an integral part of the oil pan casting. To remove crankshaft, first remove engine from equipment. Remove blower housing, flywheel, cylinder head, crankcase cover or oil pan, camshaft, connecting rod and piston assembly. Withdraw crankshaft from crankcase.

Standard crankpin journal diameter is 25.98 mm (1.023 in.) for Models GX110, GX120 and GXV120, 29.980 mm (1.1803 in.) for Models GX140, GXV140, GXV160 and GXV160, and 32.985 mm (1.2986 in) for Model GX240. If crankpin diameter is 25.92 mm (1.020 in.) or less on Models GX110, GX120 and GXV120, renew crankshaft. If crankpin diameter is 29.92 mm (1.178 in.) or less on Models GX140, GX160, GXV140 or GXV160, renew crankshaft. If crankpin diameter is 32.92 mm (1.296 in.) or less on Model GX240, renew crankshaft. On some models an undersize connecting rod is available to fit a reground crankshaft.

On some models, the timing gear is a press fit on the crankshaft. Prior to removal of timing gear, mark position of gear on crankshaft using the timing mark on the gear as a reference point. Transfer marks to new timing gear so it can be installed in same position as old gear.

On engines with a horizontal crankshaft, the governor gear is driven by a gear on the crankshaft. The governor drive gear on the crankshaft is a press fit.

Ball bearing type main bearings are a press fit on crankshaft journals and in bearing bores of crankcase and cover. Renew bearings if loose, rough or fit loosely on crankshaft or in bearing bores.

Bushing type bearing in crankcase cover of vertical crankshaft models is an integral part of oil pan. Renew crankcase cover if bearing is worn, scored or damaged.

Seals should be pressed into seal bores until outer edge of seal is flush with seal bore.

When installing crankshaft, make certain crankshaft gear and camshaft gear timing marks are aligned as shown in Fig. HN43.

CYLINDER AND CRANKCASE.

Cylinder and crankcase are an integral casting. Refer to following table for standard cylinder bore size and wear limit. Rebore or renew cylinder if cylinder bore exceeds wear limit.

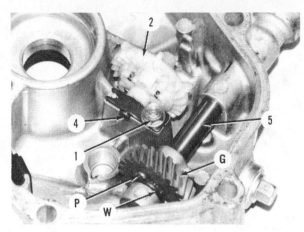

Fig. HN48—Auxiliary drive shaft is mounted in crankcase cover and driven by a gear which is an integral part of camshaft. See Fig. HN41.

1. Bolt
2. Governor gear
6. Governor gear shaft
7. Auxiliary drive shaft
8. Auxiliary drive gear
9. Retaining hair pin
10. Thrust washer

Model	Cylinder Diameter
GX110	57.000-57.015 mm (2.2441-2.2447 in.)
Wear limit	57.165 mm (2.2506 in.)
GX120	60.000 mm (2.3622 in.)
Wear limit	60.165 mm (2.3687 in.)
GX140	64.000 mm (2.5197 in.)
Wear limit	64.165 mm (2.5162 in.)
GX160	68.000 mm (2.6772 in.)
Wear limit	68.165 mm (2.6837 in.)
GX240	73.000 mm (2.8740 in.)
Wear limit	73.170 mm (2.8807 in.)
GXV120	60.000 mm (2.3622 in.)
Wear limit	60.165 mm (2.3687 in.)
GXV140	64.000 mm (2.5197 in.)
Wear limit	64.165 mm (2.5262 in.)
GXV160	68.000 mm (2.6772 in.)
Wear limit	68.165 mm (2.6837 in.)

Oversize pistons are available for some models.

AUXILIARY DRIVE.

Models GXV120, GXV140 and GXV160 may be equipped with an auxiliary drive. The auxiliary drive shaft (Fig. HN46 and HN48) is mounted in crankcase cover and driven by a gear which is an integral part of the camshaft.

Drive shaft is retained in crankcase cover by retaining pin (P—Fig. HN48). When reassembling, carefully slide drive shaft through cover oil seal, thrust washer, gear and remaining thrust washer. Insert retaining hairpin in hole in drive shaft so it is located between gear (G) and outer thrust washer (W).

OIL WARNING SYSTEM.

Some engines with a horizontal crankshaft may be equipped with a low-oil warning system that grounds the ignition system and lights a warning lamp if the oil level is low. To test circuit, run engine then disconnect yellow switch wire. Grounding the yellow wire to the engine should cause the warning light to flash and the engine should stop. Stop engine. With oil level correct and both switch leads disconnected, use an ohmmeter or continuity tester and check continuity between switch leads. Tester should indicate no continuity. With no oil in crankcase, tester should indicate continuity.

The oil level switch is connected to a module that contains the oil warning lamp and ignition switch. To check lamp, connect a 6-volt battery to yellow and black leads of module with positive battery terminal connected to black module lead. If lamp does not light, renew module or replace with a good unit.

The crankcase cover must be removed from the engine so the oil level switch can be removed for testing. Connect an ohmmeter or continuity tester and check continuity between switch leads. With switch in normal position, there should be continuity. With switch inverted, there should be no continuity. When switch is progressively inserted in a fluid, resistance should increase from zero to infinity.

REWIND STARTER.

The engine may be equipped with the rewind starter shown in Fig. HN49 or HN51. To disassemble starter remove rope handle and allow rope to wind into starter. Unscrew center retaining screw. Wear appropriate safety eyewear and gloves before disengaging pulley from starter as spring may uncoil uncontrolled.

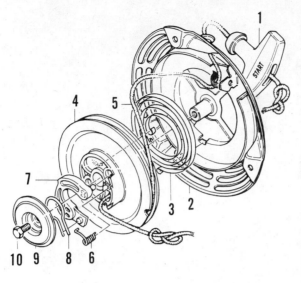

Fig. HN49—Exploded view of single-pawl rewind starter used on some engines.

1. Rope handle
2. Starter housing
3. Rewind spring
4. Pulley
5. Rope
6. Spring
7. Pawl
8. Friction spring
9. Friction plate
10. Screw

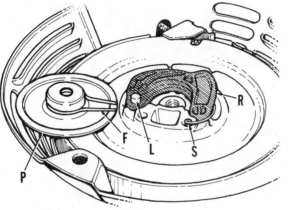

Fig. HN50—View of rewind starter showing correct installation of pawl (R) and spring (S). Install spring (F) in groove of friction plate (P). Ends of spring must fit around lug (L) on pawl.

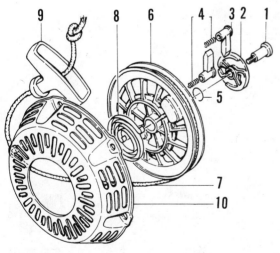

Fig. HN51—Exploded view of dual-pawl rewind starter used on some engines.

1. Screw
2. Retainer
3. Pawl
4. Friction spring
5. Snap ring
6. Pulley
7. Rope
8. Rewind spring
9. Rope handle
10. Starter housing

Place shop towel around pulley and lift pulley out of housing. Use caution when detaching rewind spring from pulley or housing. If spring must be removed from housing, position housing so spring side is down and against floor, then tap housing to dislodge spring.

On both types of starter, apply light grease to pulley shaft and sliding surfaces of rewind spring, pawls and friction spring.

On single-pawl starter (Fig. HN49), install rewind spring (3) so outer end engages starter housing notch and coils are in counterclockwise direction from outer end. Wrap rope around pulley in counterclockwise direction viewed from flywheel side of pulley. Install pulley in starter housing so inner end of spring engages notch on pulley and route rope through rope outlet in housing. Attach rope handle. Install pawl as shown in

Fig. HN50 and hook outer end of pawl spring (S) under pawl. Apply a light coat of grease to friction spring (F) and install spring in groove of friction plate (P). Install friction plate so the two ends of friction spring fit around lug (L) on pawl. Install center screw. Pretension rewind spring by turning pulley two turns counterclockwise before passing rope through rope outlet of housing and attaching handle.

On dual-pawl starter (Fig. HN51), wrap rope around pulley in counterclockwise direction viewed from flywheel side of pulley. Install rewind spring (8) in spring cavity in pulley in a counterclockwise direction from outer end. Install pulley in starter housing so inner end of spring engages tab on starter housing and route rope through rope outlet in housing. Attach rope handle. Install snap ring (5) on shaft. Install springs (4), pawls (3), retainer (2) and screw (1). To place tension on starter rope, pull rope out of housing until notch in pulley is aligned with rope outlet, then hold pulley to prevent pulley rotation. Pull rope back into housing while positioning rope in pulley notch. Turn rope pulley two turns counterclockwise, disengage rope from notch and allow rope to wind onto pulley. Check starter operation.

ELECTRIC STARTER. Some engines may be equipped with a 12 volt DC starter.

Test specifications for electric starter on Model GXV120 are: under load at more than 320 rpm and with a cranking voltage of 10.8 volts DC the current draw should be less than 30 amps; under no load with a cranking voltage of 11.7 volts DC the current draw should be less than 15 amps.

Test specifications for electric starter on Models GX120 and GX160 are: under load at more than 474 rpm and with a cranking voltage of 10.24 volts DC the current draw should be less than 50 amps; under no load with a cranking voltage of 11.4 volts DC the current draw should be less than 25 amps. Minimum brush length is 6 mm (0.24 in.).

Test specifications for electric starter on Models GXV120 and GXV160 are: under load at more than 367 rpm and with a cranking voltage of 9.8 volts DC the current draw should be less than 150 amps; under no load with a cranking voltage of 11.0 volts DC the current draw should be less than 18 amps. Minimum brush length is 6 mm (0.24 in.).

Test specifications for electric starter on Model GX240 are: under load at more than 393 rpm and with a cranking volt-

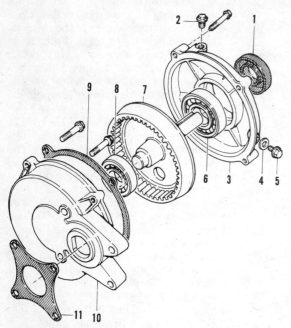

Fig. HN52—Exploded view of gear reduction unit used on some GX110 and GX140 models.

1. Oil seal
2. Oil fill screw
3. Cover
4. Washer
5. Oil level screw
6. Bearing
7. Output shaft
8. Bearing
9. Gasket
10. Case
11. Gasket

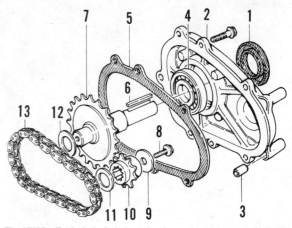

Fig. HN53—Exploded of chain reduction unit used on some GX110 and GX140 models.

1. Oil seal
2. Cover
3. Dowel pin
4. Bearing
5. Gasket
6. Key
7. Output shaft
8. Screw
9. Washer
10. Sprocket
11. Spacer
12. Spring washer
13. Chain

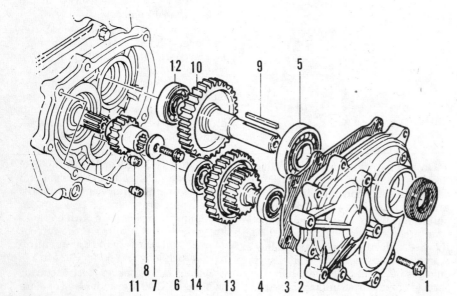

Fig. HN54—Exploded view of gear reduction unit used on some GX240 models. A similar unit equipped with a chain is also used.

1. Oil seal
2. Cover
3. Gasket
4. Bearing
5. Bearing
6. Screw
7. Washer
8. Input gear
9. Key
10. Output shaft
11. Dowel pin
12. Bearing
13. Countershaft
14. Bearing

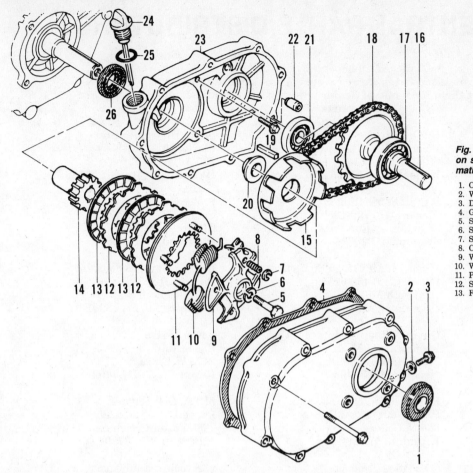

Fig. HN55—Exploded view of reduction unit used on some GX240 models equipped with an automatic clutch.

1. Oil seal
2. Washer
3. Drain screw
4. Gasket
5. Screw
6. Spring washer
7. Spring holder
8. Clutch spring
9. Weight holder
10. Weights
11. Pressure plate
12. Steel plates
13. Fiber plates
14. Input shaft
15. Clutch drum
16. Output shaft
17. Bearing
18. Chain
19. Key
20. Spacer
21. Bearing
22. Dowel pin
23. Case
24. Dipstick
25. "O" ring
26. Oil seal

age of 9.7 volts DC the current draw should be less than 80 amps; under no load with a cranking voltage of 11.5 volts DC the current draw should be less than 31 amps. Minimum brush length is 3.5 mm (0.138 in.).

REDUCTION GEAR. Engines equipped with a horizontal crankshaft may be equipped with a reduction gear unit that attaches to the crankcase cover. Power is transferred via gears or a chain, while some units may also contain a plate-type automatic clutch. Refer to Fig. HN52, HN53, HN54 and HN55 for an exploded view of unit.

Overhaul of unit is apparent after inspection of unit. On unit shown in Fig. HN54, tighten input gear or sprocket retaining screw to 24 N•m (212 in.-lbs.) and cover screws to 24 N•m (212 in.-lbs.). On unit with an automatic clutch (Fig. HN55), wear limit thickness for friction plates is 3.0 mm (0.118 in.) and maximum allowable warpage of steel plates is 0.10 mm (0.004 in.). When assembling clutch pack in clutch drum, alternately install fiber and steel plates with a fiber plate first and a steel plate last.

HONDA CENTRAL PARTS DISTRIBUTORS

(Arranged Alphabetically by States)
**These franchised firms carry extensive stocks of repair parts.
Contact them for name of dealer who will have replacement
parts.**

Totem Equipment & Supply, Inc.
Phone: (907) 276-2858
2536 Commercial Drive
Anchorage, AK 99501

Arizona Tru Power
Phone: (602) 470-0522
3652 E. Chipman Rd.
Phoenix, AZ 85282

Trimmer of Fresno, Inc.
Phone: (209) 266-0582
2531 East McKinley
Fresno, CA 93703

Tru-Cut, Inc.
Phone: (213) 258-4131
3221 San Fernando Rd.
Los Angeles, CA 90065

Bliss Power Lawn Equipment Co.
Phone: (916) 925-6936
101 Commerce Circle
Sacramento, CA 95815

Scotsco Pro Power Products
Phone: (916) 383-3511
8806 Fruitridge Road
Sacramento, California 95826

Red Ram
Phone: (303) 762-1515
2915 S. Tejon St.
Englewood, CO 80110

Roberts Supply, Inc.
Phone: (407) 657-5555
4203 Metric Dr.
Winter Park, FL 32792

Aloha Power Equipment Distributors
Phone: (808) 848-5534
330 Sand Island Access Road
Honolulu, HI 96819

Power Equipment Co.
Phone: (708) 834-8700
645 S. Route 83
Elmhurst, IL 60126

Iowa Power Products
Phone: (515) 648-2507
520 Brooks Rd.
Iowa Falls, IA 50126

Kansas City Power Products
Phone: (913) 321-7040
80 S. James St.
Kansas City, KS 66118

R.C.S. Distributing
Phone: (410) 799-1850
8019 Dorsey Run Road
Jessup, MD 20794

Plymouth Air Cooled Equipment
Phone: (313) 453-6258
739 South Mill
Plymouth, MI 48170

Great Northern Equipment Dist.
Phone: (218) 963-2921
218 N. Hazelwood Dr.
Nisswa, MN 56468

R.W. Distributors, Inc.
Phone: (601) 825-5878
806 N. College St.
Brandon, MS 39042

C.K. Power Products
Phone: (314) 868-8620
9290 W. Florissant
St. Louis, MO 63136

Anderson Industrial Engines Co.
Phone: (402) 558-8700
5532 Center Street
Omaha, NE 68106

Eastern Equipment, Inc.
Phone: (603) 437-0407
23 Londonderry Rd.
Londonderry, NH 03053

Brooks Gravely
Phone: (716) 424-1660
2425 Brighton Henrietta
Rochester, NY 14623

Engine Distribution Center
Phone: (919) 272-3857
103 E. Creekridge Rd.
Greensboro, NC 27406

Hayward Distributing Co.
Phone: (614) 272-5953
4061 Perimeter Drive
Columbus, OH 43228

Smith Distributing Co.
Phone: (405) 787-8304
220 Alliance Ct.
Oklahoma City, OK 73128

Scotsco
Phone: (503) 777-4726
9160 S.E. 74th Ave.
Portland, OR 97206

Paul B. Moyer & Sons, Inc.
Phone: (215) 348-1270
190 S. Clinton St.
Doylestown, PA 18901

Power House
Phone: (605) 348-8983
2425 W. Chicago
Rapid City, SD 57702

M.T.A. Distributors
Phone: (615) 726-2225
2940 Foster Creighton
Nashville, TN 37204

Lightbourn Equipment
Phone: (214) 233-5151
13649 Beta Rd.
Dallas, TX 75244

Lightbourn Equipment
Phone: (713) 741-2003
8272 El Rio, Suite 110
Houston, TX 77054

Red Ram
Phone: (801) 255-6326
6980 South 400 West, #2
Midvale, UT 84047

Tidewater Power Equipment Co.
Phone: (804) 497-4800
5321 Cleveland St.
Virginia Beach, VA 23462

Norwest Engine Distributor
Phone: (509) 534-8599
North 1403 Green #6
Spokane, WA 99202

Engine Power Inc.
Phone: (414) 544-1226
1923 MacArthur Rd.
Waukesha, WI 53188

KAWASAKI

KAWASAKI MOTORS CORP. USA
P.O. Box 504
Shakopee, Minnesota 55379

Model	Bore	Stroke	Displacement	Power Rating
FA76D	52 mm	36 mm	76 cc	1.25 kW
	(2.05 in.)	(1.42 in.)	(4.64 cu.in.)	(1.7 hp)
FA130D	62 mm	43 mm	129 cc	2.3 kW
	(2.44 in.)	(1.69 in.)	(7.9 cu.in.)	(3.2 hp)
FA210D	72 mm	51 mm	207 cc	3.9 kW
	(2.83 in.)	(2.01 in.)	(12.7 cu.in.)	(5.2 hp)

NOTE: Metric fasteners are used throughout engine.

ENGINE IDENTIFICATION

Kawasaki FA series engines are four-stroke, single-cylinder, air-cooled, horizontal crankshaft engines. Engine model number is located on the cooling shroud just above the rewind starter. Engine model number and serial number are both stamped on crankcase cover.

MAINTENANCE

LUBRICATION. The engine is splash lubricated by an oil slinger mounted on bottom of connecting rod cap. Engine oil level should be checked prior to each operating interval. To check oil level, insert dipstick into oil filler hole until plug touches the first threads. Do not screw dipstick plug in to check oil level. Oil level should be maintained between the reference marks on oil dipstick.

Engine oil should be changed after the first 20 hours of operation and every 50 hours thereafter. Crankcase capacity is 0.32 L (0.34 qt.) for Model FA76D, 0.50 L (0.53 qt.) for Model FA130D and 0.60 L (0.63 qt.) for Model FA210D.

Engine oil should meet or exceed latest API service classification. Normally, SAE 30 oil is recommended for use when temperatures are above 32° F (0° C). SAE 10W-30 oil is recommended when temperatures are below 32° F (0° C).

AIR FILTER. The engine is equipped with a foam type air filter element. The air filter element should be removed and cleaned after every 50 hours of operation, or more often if operating in extremely dusty conditions. To clean element, wash in a nonflammable cleaning solvent and gently squeeze element dry. Oil element with clean engine oil and gently squeeze out the excess oil.

SPARK PLUG. Recommended spark plug is NGK BM6A or BMR6A or equivalent. Spark plug should be removed, checked and cleaned after every 50 hours of operation. Renew spark plug if electrode is corroded or damaged.

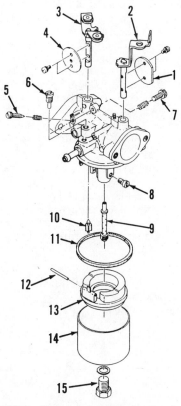

Fig.KW101—Exploded view of float type carburetor.

1. Choke valve
2. Choke shaft
3. Throttle shaft
4. Throttle valve
5. Idle mixture adjustment screw
6. Pilot jet
7. Idle speed adjustment screw
8. Main jet
9. Main nozzle
10. Fuel inlet valve
11. Gasket
12. Pin
13. Float
14. Float bowl
15. Bolt

Spark plug electrode gap should be set at 0.7-0.8 mm (0.028-0.031 in.).

FUEL VALVE. A filter screen and sediment bowl are mounted on bottom of fuel shut-off valve. Sediment bowl and screen should be removed and cleaned after every 50 hours of operation, or more often if water or dirt is visible in the sediment bowl.

CARBURETOR. Engine is equipped with a float type carburetor (Fig. KW101). Main fuel mixture is controlled by fixed main jet (8).

Initial adjustment of idle mixture needle (5) is 1⅛ turns out from a lightly seated position. Recommended engine idle speed is 1500-1700 rpm. To adjust idle speed, run engine until it reaches normal operating temperature and turn idle speed screw to obtained specified engine speed. Adjust idle mixture screw (7) so engine runs at maximum idle speed, then turn mixture screw out an additional ¼ turn.

To adjust engine maximum operating speed, refer to GOVERNOR paragraphs.

To check carburetor float level, remove the carburetor from the engine. Remove the float bowl and invert carburetor body. Surface of float should be parallel to carburetor body (Fig. KW102). The plastic float used on Model FA76D engine is nonadjustable and must be renewed if float level is incorrect.

Overhaul. To disassemble carburetor, remove fuel bowl retaining screw (15—Fig. KW101) and fuel bowl (14). Remove float pin (12), float (13) and fuel inlet needle (10). Remove throttle and choke shaft assemblies after unscrewing throttle and choke plate retaining screws. Remove idle mixture screw (5), idle mixture jet (6), main jet (8) and main fuel nozzle (9). Note that main jet must be removed before the nozzle as the jet blocks the nozzle.

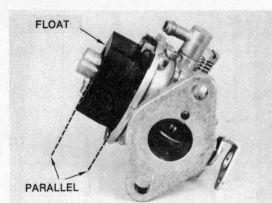

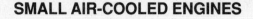

Fig. KW102—To check float level, hold carburetor so tab of float touches fuel inlet needle valve. Float should be parallel with surface of carburetor body.

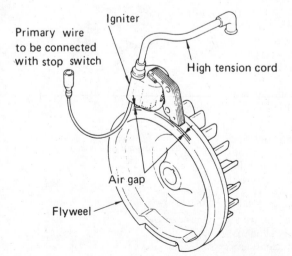

Fig. KW103—View of external governor linkage. Refer to text for adjustment procedure.

Clean carburetor using suitable carburetor cleaner. When assembling the carburetor note the following. Place a small drop of nonhardening sealant such as Permatex #2 or equivalent on throttle and choke plate retaining screws.

GOVERNOR. A gear-driven flyweight governor assembly is located inside the engine crankcase. To adjust external linkage, place engine throttle control in idle position. Move governor lever (Fig. KW103) to fully open carburetor throttle valve. Loosen governor lever clamp bolt and rotate governor shaft clockwise as far as possible. While holding governor shaft and lever in this position, tighten clamp bolt.

Fast idle engine speed (no load) may vary depending on the equipment on which the engine is used. Refer to the equipment specification. To adjust speed, turn the throttle cable adjusting nuts as necessary. Maximum engine speed must not exceed 4000 rpm.

IGNITION SYSTEM. All engines are equipped with a solid-state ignition system which has no moving parts except the flywheel. No adjustment is required on this ignition system. Correct engine timing will be obtained when ignition coil unit (Fig. KW104) is installed so that air gap between coil laminations and outer surface of flywheel is 0.5 mm (0.020 in.).

VALVE ADJUSTMENT. Clearance between tappet and end of valve stem should be checked after every 200 hours of operation. To check clearance, remove muffler cover and valve spring chamber cover. Turn crankshaft to position piston at top dead center on compression stroke. Measure valve clearance with engine cold using a feeler gauge. Specified clearance for FA76D engine is 0.10-0.20 mm (0.004-0.008 in.) for intake and 0.10-0.30 mm (0.004-0.012 in.) for exhaust. Specified clearance for FA130D and FA210D engines is 0.12-0.18 mm (0.005-0.007 in.) for intake and 0.10-0.24 mm (0.004-0.009 in.) for exhaust.

To adjust valve clearance, valves must be removed from engine. If clearance is too large, grind valve seat or renew valve to reduce clearance. If clearance is too small, grind end of valve stem to increase clearance.

CYLINDER HEAD AND COMBUSTION CHAMBER. Cylinder head, combustion chamber and top of piston should be cleaned and carbon and other deposits removed after every 200 hours of operation. Refer to REPAIRS section for service procedure.

GENERAL MAINTENANCE. Check and tighten all loose bolts, nuts or clamps prior to each period of operation. Check for fuel or oil leakage and repair if necessary.

Clean dirt, dust, grease or any foreign material from cylinder head and cylinder block cooling fins after every 100 hours of operation, or more often if operating in extremely dirty conditions.

REPAIRS

TIGHTENING TORQUES. Recommended tightening torques are as follows:

Connecting rod:
FA76D . . . . 7 N•m
(60 in.-lbs.)
FA130D, FA210D . . . . 12 N•m
(105 in.-lbs.)
Crankcase cover . . . . 6 N•m
(50 in.-lbs.)
Cylinder head:
FA76D . . . . 7 N•m
(60 in.-lbs.)
FA130D, FA210D . . . . 20 N•m
(175 in.-lbs.)
Flywheel nut:
FA76D . . . . 34 N•m
(25 ft.-lbs.)
FA130D, FA210D . . . . 62 N•m
(46 ft.-lbs.)

CYLINDER HEAD. To remove cylinder head, first remove the air cleaner, muffler cover, control panel and fuel tank. Remove rewind starter, flywheel blower fan housing and cylinder head

Fig. KW104—View of ignition unit. Air gap between flywheel and igniter coil laminations should be 0.5 mm (0.020 in.).

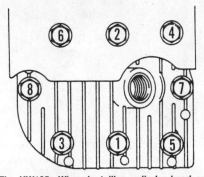

Fig. KW105—When installing cylinder head on Models FA76D and FA210D, tighten head bolts gradually following sequence shown.

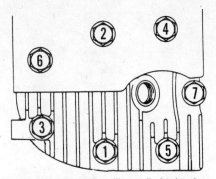

Fig. KW106—When installing cylinder head on Model FA130D, tighten head bolts gradually following sequence shown.

shroud. Loosen cylinder head screws gradually in ¼-turn increments until all screws are loose enough to be removed by hand. Remove cylinder head and gasket from the engine.

Remove spark plug and clean carbon and other deposits from cylinder head. Place cylinder head on a flat surface and check gasket sealing surface for distortion using a feeler gauge. Renew cylinder head if head is warped more than 0.15 mm (0.006 in.) on Model FA76D, 0.25 mm (0.010 in.) on Model FA130D, or 0.30 mm (0.012 in.) on Model FA210D.

Reinstall cylinder head using a new head gasket. Tighten cylinder head screws evenly in three steps to specified torque following tightening sequence shown in Fig. KW105 or KW106.

VALVE SYSTEM. The valves are accessible after removing cylinder head and valve spring chamber cover (23—Fig. KW107).

Minimum allowable valve margin is 0.55 mm (0.022 in.). Valve face and seat angles are 45°. Grinding the valve face is not recommended. Recommended valve seating width is 0.5-1.1 mm (0.020-0.043 in.) on Model FA76D or 1.0-1.6 mm (0.039-0.063 in.) for Models FA130D and FA210D.

Minimum valve stem diameter for Model FA76D is 5.462 mm (0.2150 in.) for intake valve and 5.435 mm (0.2140 in.) for exhaust valve. Minimum valve stem diameter for Models FA130D and FA210D is 5.960 mm (0.2346 in.) for intake valve and 5.940 mm (0.2339 in.) for exhaust valve.

Valve guides are integral with cylinder block. Specified valve guide maximum inside diameter for Model FA76D is 5.617 mm (0.2211 in.) for intake valve guide and 5.595 mm (0.2203 in.) for exhaust valve guide. Valve guide maxi-

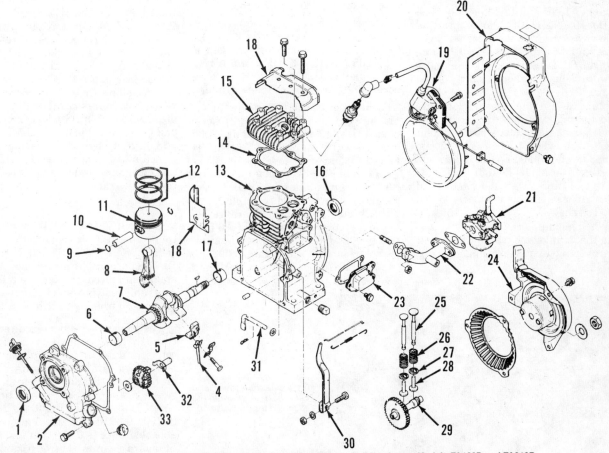

Fig. KW107—Exploded view of typical engine assembly. Note that main bearing (17) is a ball bearing on Models FA130D and FA210D.

1. Oil seal	8. Connecting rod	
2. Crankcase cover	9. Retaining ring	14. Head gasket
4. Oil slinger	10. Piston pin	15. Cylinder head
5. Rod cap	11. Piston	16. Oil seal
6. Main bearing	12. Piston rings	17. Main bearing
7. Crankshaft	13. Crankcase & cylinder assy.	18. Shrouds

19. Ignition coil assy.	24. Recoil starter assy.	29. Camshaft
20. Blower housing	25. Valve	30. Governor lever
21. Carburetor	26. Valve spring	31. Governor shaft
22. Intake tube	27. Retainer	32. Sleeve
23. Tappet chamber cover	28. Tappet	33. Governor assy.

mum inside diameter for Models FA130D and FA210D is 6.117 mm (0.2408 in.) for intake valve guide and 6.095 mm (0.2400 in.) for exhaust valve guide. Cylinder block should be renewed if valve guides are worn beyond specified limits.

Specified valve spring minimum free length for Model FA76D is 20.5 mm (0.81 in.) for both valve springs. Valve spring minimum free length for Models FA130D and FA210D is 23.5 mm (0.93 in.) for both valve springs.

CAMSHAFT. Camshaft (29—Fig. KW107) is supported directly in bores of the aluminum crankcase and crankcase cover. To remove camshaft, first remove crankcase cover. Turn crankshaft so timing marks on crankshaft gear and timing gear are aligned. Turn cylinder block upside down so tappets fall away from cam lobes, then remove camshaft and gear from crankcase. Identify tappets so they can be reinstalled in their original positions, then remove tappets from crankcase.

Minimum camshaft journal diameter is 9.945 mm (0.3915 in.) for Model FA76D, 11.937 mm (0.4670 in.) for Model FA130D and 12.935 mm (0.5093 in.) for Model FA210D. Maximum allowable clearance between camshaft journals and bearing bores is 0.10 mm (0.004 in.). If clearance is excessive, renew camshaft and/or crankcase and cover.

Make certain that camshaft lobes are smooth and free of scoring, pitting and other damage. Minimum lobe height for intake and exhaust lobes is 17.35 mm (0.683 in.) for Model FA76D, 23.25 mm (0.915 in.) for Model FA130D and 26.35 mm (1.037 in.) for Model FA210D. If camshaft is renewed, the tappets should also be renewed.

When installing camshaft, make certain that crankshaft gear and camshaft gear timing marks are aligned (Fig. KW108).

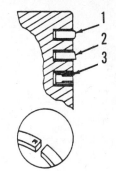

Fig. KW109—Piston rings must be installed with manufacturer's mark on end of ring facing upward. Note location and shape of top compression ring (1), second compression ring (2) and oil control ring (3).

PISTON, PIN AND RINGS. Piston and connecting rod are removed as an assembly after cylinder head and crankcase cover have been removed. Carbon deposits and ring ridge, if present, should be removed from top of cylinder prior to removing piston. Remove connecting rod cap and oil slinger, then push piston and connecting rod assembly out top of cylinder. Remove retaining rings and push pin out of piston to separate piston from connecting rod.

Renew piston if top or second piston ring side clearance exceeds 0.15 mm (0.006 in.). Specified ring end gap is 0.15-0.35 mm (0.006-0.014 in.). If ring end gap exceeds 1.0 mm (0.040 in.), rebore cylinder oversize or renew cylinder.

Measure piston diameter at bottom of piston skirt and 90° from pin bore. Specified standard diameter is 51.90 mm (2.0433 in.) for Model FA76D, 61.94 mm (2.4385 in.) for Model FA130D and 71.94 mm (2.8323 in.) for Model FA210D. On Model FA76D, piston should be renewed if chrome plating is worn through. Standard clearance between piston and cylinder is 0.06-0.09 mm (0.0024-0.0035 in.). Maximum allowable clearance is 0.20 mm (0.008 in.) on Model FA76D or 0.25 mm (0.010 in.)

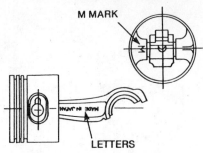

Fig. KW110—Piston and connecting rod should be assembled so that lettering on rod and "M" mark on underside of piston are on the same side.

on Models FA130D and FA210D. Piston and rings are available in standard size and 0.25 mm (0.010 in.) and 0.50 mm (0.020 in.) oversizes.

When installing rings on piston, make sure that manufacturer's mark on face of ring is toward piston crown (Fig. KW109). Note that top piston ring has a barrel face and second piston ring has a taper face.

Install piston on connecting rod so the side of connecting rod which is marked "MADE IN JAPAN" is toward the "M" stamped on piston pin boss (Fig. KW110). Secure piston pin with new retaining rings. Position the retaining rings in piston so the "tail" points straight up or down (Fig. KW111), otherwise inertia produced at top and bottom of piston stroke will cause retaining rings to compress and possibly fall out of piston pin bore. Stagger piston ring end gaps around piston so that top ring and oil control ring end gaps are positioned away from valves (Fig. KW112) and end gap of second ring is positioned toward the valves.

CONNECTING ROD. Piston and connecting rod are removed as an assembly. Refer to PISTON, PIN AND RINGS paragraphs for removal and installation procedure. Remove retaining rings and push pin out of piston to separate piston from connecting rod.

The connecting rod rides directly on crankshaft crankpin journal. Inspect connecting rod and renew if piston pin bore or crankpin bearing bore are scored or damaged. Standard clearance between piston pin and connecting rod small end bore is 0.006-0.025 mm (0.0002-0.0010 in.); maximum allowable clearance is 0.050 mm (0.002 in.). Standard clearance between connecting rod big end bore and crankshaft crankpin is 0.030-0.056 mm (0.0012-0.0022 in.). Renew connecting rod and/or crankshaft if clearance exceeds 0.070 mm (0.0028 in.).

Fig. KW108—Timing marks on crankshaft gear and camshaft gear must be aligned during installation.

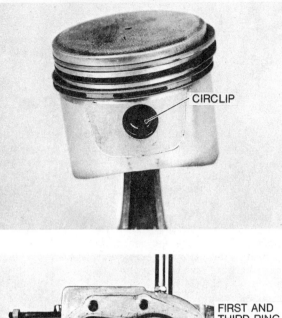

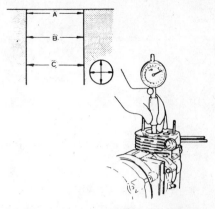

Fig. KW111—Install piston pin retaining rings with "tail" pointing either straight up or down.

Fig. KW113—To check cylinder bore for wear or out-of-round, measure bore in six different locations as shown.

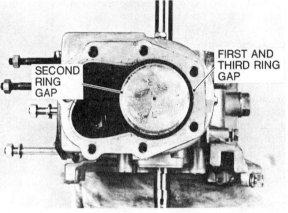

Fig. KW112—Stagger piston rings on piston so end gaps are positioned as shown.

Refer to following table for connecting rod small and big end maximum allowable diameters:

Small end diameter (max.):
FA76D . . . . 11.035 mm
(0.4344 in.)
FA130D . . . . 13.035 mm
(0.5134 in.)
FA210D . . . . 15.040 mm
(0.5912 in.)

Big end diameter (max.):
FA76D . . . . 20.057 mm
(0.7897 in.)
FA130D . . . . 24.072 mm
(0.9478 in.)
FA210D . . . . 27.072 mm
(1.0658 in.)

Connecting rod side play on crankpin should not exceed 0.5 mm (0.012 in.).

When installing connecting rod cap, making sure identification marks on side of cap are aligned with marks on connecting rod. Install oil slinger and new lock plate making sure that slinger arm and bendable tabs of lock plate are toward open side of crankcase. Tighten connecting rod screws to 7 N·m (60 in.-lbs.) on Model FA76D and to 12 N·m (105 in.-lbs.) on Models FA130D and FA210D. Bend tabs of lock plate.

GOVERNOR. The internal centrifugal flyweight governor (33—Fig. KW107) is located in the crankcase cover and is driven by the camshaft gear. Governor assembly should not be removed from crankcase cover unless renewal is necessary. The governor gear will be damaged as it is pulled off the cover stub shaft, and must be replaced with a new governor flyweight assembly.

Adjust governor external linkage as outlined in MAINTENANCE section.

CRANKSHAFT, MAIN BEARINGS AND SEALS. The crankshaft (7—Fig. KW107) is supported at each end in bushing type main bearings (6 and 17) which are pressed into crankcase and crankcase cover.

To remove crankshaft, remove flywheel, cylinder head, crankcase cover, camshaft, tappets, connecting rod and piston assembly,. Withdraw crankshaft from crankcase. Drive oil seals (1 and 16) out of crankcase and crankcase cover.

Minimum crankpin diameter is 19.95 mm (0.7854 in.) for Model FA76D, 23.92 mm (0.9417 in.) for Model FA130D and 26.92 mm (1.0598 in.) for Model FA210D. Maximum allowable clearance between crankpin and connecting rod big end bore is 0.070 mm (0.003 in.) on Model FA76D or 0.10 mm (0.004 in.) for all other models.

Maximum allowable clearance between crankshaft main journals and main bearing bushings is 0.13 mm (0.005 in.). When renewing main bearing bushings on Models FA76D and FA210D, make sure that oil hole in bushing is aligned with oil hole in crankcase and crankcase cover.

Crankshaft oil seals should be pressed into crankcase and crankcase cover until face of seal is flush with outer surface of crankcase and cover bores.

Lubricate oil seals, bushings and crankshaft journals with engine oil prior to reassembly. Be sure that crankshaft gear and camshaft gear timing marks are aligned during installation as shown in Fig. KW108.

CYLINDER AND CRANKCASE. Cylinder and crankcase are an integral casting. On Model FA76D, standard cylinder bore diameter is 52.00 mm (2.047 in.) and wear limit is 52.07 mm (2.050 in.). On Model FA130D, standard cylinder bore diameter is 62.00 mm (2.441 in.) and wear limit is 62.07 mm (2.444 in.). On Model FA210D, standard cylinder bore diameter is 72.00 mm (2.835 in.) and wear limit is 72.07 mm (2.838 in.). Cylinder bore should be measured at six different locations as shown in Fig. KW113. If cylinder is out-of-round more than 0.05 mm (0.0002 in.), rebore cylinder to appropriate oversize or renew cylinder.

NOTE: The cylinder block is made of an aluminum-silicon alloy and resizing requires special honing process. Special honing stones are available from Sunnen Products Company. Follow manufacturer's instruction for proper operation.

KAWASAKI

Model	Bore	Stroke	Displacement	Power Rating
FC150V	65 mm	46 mm	153 cc	3.7 kW
	(2.56 in.)	(1.81 in.)	(9.3 cu.in.)	(5.0 hp)

NOTE: Metric fasteners are used throughout engine.

ENGINE INFORMATION

Model FC150V is an air-cooled, four-stroke, single-cylinder engine. The engine has a vertical crankshaft and utilizes an overhead valve system.

MAINTENANCE

LUBRICATION. All models are lubricated by an oil slinger that is gear-driven by the governor gear.

Change oil after first five hours of operation and after every 50 hours of operation or at least once each operating season. Change oil weekly or after every 25 hours of operation if equipment undergoes severe usage.

Engine oil should meet or exceed latest API service classification. Use SAE 40 oil for temperatures above 68° F (20° C); use SAE 30 oil for temperatures between 32° F (0° C) and 95° F (35° C); use SAE 10W-30 or 10W-40 oil for temperatures between –4° F (–20° C) and 95° F (35° C); below 32° F (0° C) SAE 5W-20 may be used.

Crankcase capacity is 0.55 liters (1.16 U.S. pints).

FUEL FILTER. The fuel tank is equipped with a filter at the outlet. Check filter annually and periodically during operating season.

SPARK PLUG. Recommended spark plug is a NGK BP6ES; install a comparative resistor plug if required. Specified spark plug electrode gap is 0.7-0.8 mm (0.028-0.031 in.).

CARBURETOR. Adjustment. Idle speed at normal operating temperature should be 1400-1600 rpm. Adjust idle speed by turning idle speed screw (10—Fig. KW201). Idle mixture is controlled by idle jet (1) and idle mixture screw (8). Initial setting of idle mixture screw is one turn out. Turning screw clockwise will lean idle mixture. Adjust idle mixture screw so engine runs at maximum idle speed, then turn screw out an additional ¼ turn. Readjust idle speed screw. Idle mixture jet is not adjustable.

High speed operation is controlled by fixed main jet (12).

Overhaul. To disassemble carburetor, remove fuel bowl retaining screw (22), gasket (21) and fuel bowl (18). Remove float pin (16) by pushing against small end of pin toward the large end of pin. Remove float (15) and fuel inlet needle (13). Remove throttle and choke shaft assemblies after unscrewing throttle and choke plate retaining screws. Remove idle mixture screw (8), idle mixture jet (1), main jet (12) and main fuel nozzle (14). Note that main jet must be removed before the nozzle as the jet blocks the nozzle.

When assembling the carburetor note the following. Place a small drop of nonhardening sealant such as Permatex #2 or equivalent on throttle and

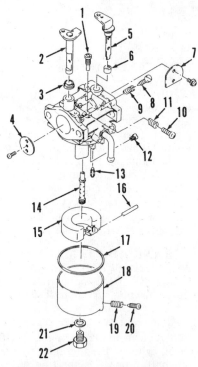

Fig. KW201—Exploded view of carburetor.

1. Idle jet	12. Main jet
2. Throttle shaft	13. Fuel inlet valve
3. Collar	14. Nozzle
4. Throttle plate	15. Float
5. Choke shaft	16. Float pin
6. Ring	17. Gasket
7. Choke plate	18. Fuel bowl
8. Idle mixture screw	19. Spring
9. Spring	20. Drain screw
10. Idle speed screw	21. Washer
11. Spring	22. Screw

choke plate retaining screws. Float height is not adjustable; replace any components which are damaged or excessively worn and adversely affect float position.

If removed, install carburetor insulator spacer and gaskets so gasket with square-shaped opening is toward cylinder and gasket with round opening is toward carburetor. Tighten carburetor mounting nuts to 7 N•m (62 in.-lbs.) torque.

CARBURETOR CONTROLS. Operate engine until normal operating temperature is reached and stop engine. Move throttle control to "FAST" position and loosen throttle cable clamp screw (C—Fig. KW202). Rotate speed control lever (T) so holes (H) in lever and plate (P) are aligned and insert a 6 mm (0.24 in.) rod or bolt into holes to maintain alignment. Start and run engine at fast, no-load idle speed specified by equipment manufacturer. Loosen screws (S) and move plate (P) to adjust engine speed. Stop engine and tighten screws. With throttle control in "FAST" position, tighten throttle cable clamp screw (C).

To adjust choke, insert a 6 mm (0.24 in.) diameter rod through holes (H) in speed control lever and control plate. Back out choke screw (K) so it does not touch lever (L), then turn screw in so it just touches lever. Remove the 6 mm (0.24 in.) rod from control lever. With throttle control in "CHOKE" position, the carburetor choke plate should be closed.

GOVERNOR. The engine is equipped with a mechanical, flyweight type governor. To adjust governor linkage, proceed as follows: Loosen governor lever clamp nut (N—Fig. KW203), rotate governor lever (L) so throttle plate is fully open and hold lever in place. Turn governor shaft (S) counterclockwise as far as possible, then tighten nut (N).

If internal governor assembly must be serviced, refer to REPAIRS section.

IGNITION SYSTEM. The engine is equipped with an electronic ignition system and periodic maintenance is not required. All components are located

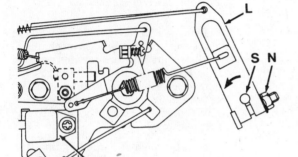

Fig. KW202—View of carburetor and governor control linkage. Refer to text for adjustment.

Fig. KW203—View showing location of ignition control unit (ECU) and governor linkage.

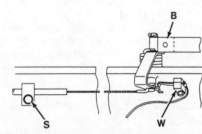

Fig. KW204—Control unit wire (2) and stop switch wire (3) must be connected to stud terminal (8) as shown. Note 45° angle of wires.

1. Nuts
2. Control unit wire
3. Stop switch wire
5. Washer
6. Insulator
7. Insulator
8. Stud

outside the flywheel. Ignition timing is not adjustable.

Ignition coil armature leg air gap should be 0.3 mm (0.012 in.). Adjust air gap by loosening ignition coil mounting screws and repositioning ignition coil. Primary side ignition coil resistance should be 0.67-1.10 ohms; secondary side resistance should be 6000-10000 ohms.

The ignition control unit (ECU) is located adjacent to the carburetor and governor controls shown in Fig. KW203. The only positive way of determining if the control unit is faulty is to install a good control unit and check ignition performance. Note in Fig. KW204 the correct attachment of wires from the stop

Fig. KW205—Stop switch (W) should contact brake lever when safety handle is released.

switch and control unit to the stud terminal.

An ignition stop switch is located adjacent to the flywheel band brake actuating lever on the engine. Be sure stop switch operates properly and stops ignition when the equipment safety handle is released.

FLYWHEEL BRAKE. Mower engines may be equipped with a brake band (B—Fig. KW205) that contacts the flywheel when the mower safety handle is released. Stop switch (W) contacts the brake lever and grounds the ignition when the brake is activated. The brake should stop the engine within three seconds after the safety handle is released.

Adjust brake cable by releasing the equipment safety handle, loosening cable clamp screw (S) and pulling slack out of outer cable housing. Retighten clamp screw.

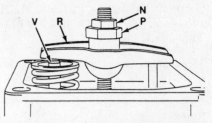

Fig. KW206—Loosen lock nut (N) and turn pivot nut (P) to adjust clearance between rocker arm (R) and valve stem end (V) which should be 0.12 mm (0.005 in.).

VALVE ADJUSTMENT. Engine must be cold for valve adjustment. Rotate crankshaft so piston is at top dead center on compression stroke. Remove rocker arm cover. Clearance between rocker arm (R—Fig. KW206) and valve stem end (V) should be 0.12 mm (0.005 in.) for both the intake and exhaust valves. Loosen lock nut (N) and turn pivot nut (P) to obtain desired clearance. Tighten lock screw to 7.0 N•m (62 in.-lbs.) torque.

COMPRESSION PRESSURE. Minimum allowable compression pressure is 343 kPa (50 psi).

CYLINDER HEAD. Manufacturer recommends that after every 100 hours of operation the cylinder head is removed and cleaned of deposits.

REPAIRS

TIGHTENING TORQUES. Recommended tightening torques specifications are as follows:

Blade brake mounting
  bolt (M8).............. 18 N•m
            (160 in.-lbs.)
Carburetor mounting screw ... 7 N•m
            (62 in.-lbs.)
Connecting rod............ 12 N•m
            (106 in.-lbs.)
Crankcase cover............ 7 N•m
            (62 in.-lbs.)
Crankshaft (pto end)........ 38 N•m
            (28 ft.-lbs.)
Cylinder head.............. 23 N•m
            (204 in.-lbs.)
Flywheel.................. 45 N•m
            (33 ft.-lbs.)
Muffler .... 7 N•m
            (62 in.-lbs.)
Oil drain plug.............. 21 N•m
            (186 in.-lbs.)
Rocker arm stud .... 7 N•m
            (62 in.-lbs.)
Standard screws:
  M5..................... 3.5 N•m
            (31 in.-lbs.)
  M6..................... 6 N•m
            (53 in.-lbs.)
  M8..................... 15 N•m
            (133 in.-lbs.)

CRANKCASE BREATHER.
Crankcase pressure is vented to the cylinder head where two reed valves are located. One reed valve is found in the rocker arm chamber while the other reed valve is situated in a breather chamber on the top side of the cylinder head. Renew reed valve if tip of reed stands up more than 0.2 mm (0.008 in.), or if reed is damaged or worn excessively.

CYLINDER HEAD AND VALVE SYSTEM. To remove cylinder head, remove rewind starter, blower housing, exhaust and intake components, and rocker arm cover. Unscrew cylinder head screws and remove head. Push rods, rocker arms and pivot nuts should be marked so they can be reinstalled in their original position.

Clean deposits from cylinder head and inspect for cracks or other damage. Check flatness of cylinder head gasket surface using a straightedge and feeler gauge. Renew head if warped more than 0.07 mm (0.003 in.).

Valve face and seat angles are 45°. Specified seat width is 0.53-1.16 mm (0.021-0.046 in.). Minimum allowable valve margin is 0.5 mm (0.020 in.). Minimum allowable valve stem diameter is 5.435 mm (0.2319 in.) for intake valve and 5.420 mm (0.2133 in.) for exhaust valve. Renew valve if runout exceeds 0.03 mm (0.0012 in.) measured at midpoint of stem. Renew valve if valve stem end is worn so length from stem end to groove in valve stem is less than 3.8 mm (0.15 in.).

Valve guides are not renewable. If valve guide inner diameter is larger than 5.550 mm (0.2185 in.) for intake or 5.560 mm (0.2189 in.) for exhaust, renew cylinder head.

Renew push rod if runout exceeds 0.6 mm (0.024 in.) measured at midpoint of push rod.

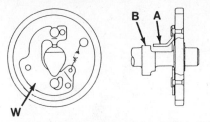
Fig. KW208—Compression release arm (A) should protrude above cam lobe base (B) when weight (W) is in innermost (starting) position.

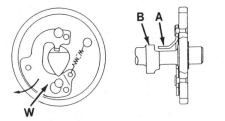
Fig. KW209—Compression release arm (A) should be beneath surface of cam lobe base (B) when weight (W) is extended to running position.

Rocker arm studs are threaded into cylinder head. When installing studs, apply Loctite to threads and tighten to 7 N•m (62 in.-lbs.) torque.

Note the following when reinstalling cylinder head: Do not apply sealer to cylinder head gasket. Piston should be at top dead center on compression stroke. Tighten cylinder head bolts evenly in three steps using sequence shown in Fig. KW207 until final torque reading of 23 N•m (204 in.-lbs.) is obtained.

CAMSHAFT. Camshaft and camshaft gear are an integral casting which is equipped with a compression release mechanism. The compression release arm (A—Fig. KW208) extends at cranking speed to hold the exhaust valve open slightly thereby reducing compression pressure.

To remove camshaft proceed as follows: Drain crankcase oil and remove engine from equipment. Clean pto end of crankshaft and remove any burrs or rust. Remove rocker arms and push rods and mark them so they can be returned to original position. Unscrew fasteners and remove crankcase cover (oil sump). Rotate crankshaft so timing marks on crankshaft and camshaft gears are aligned (this will position valve tappets out of way). Withdraw camshaft and remove tappets.

Minimum allowable camshaft bearing journal diameter is 13.920 mm (0.5480 in.) for both ends. Renew camshaft if either cam lobe height is less than 22.80 mm (0.898 in.). Minimum allowable inside diameter of bearing bore in crankcase or crankcase cover (oil sump) is 14.070 mm (0.554 in.).

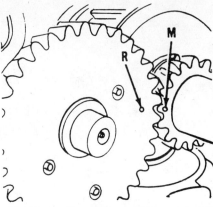
Fig. KW210—Align timing mark (R) on camshaft with timing mark (M) on crankshaft.

Refer to Figs. KW208 and KW209 to check compression release mechanism. With weight (W) in starting position (Fig. KW208), arm (A) should protrude above cam lobe base. With weight (W) extended in running position (Fig. KW209), arm (A) should be beneath surface of cam lobe base.

Install camshaft while aligning timing marks (Fig. KW210) on crankshaft and camshaft gears. Be sure governor weights are closed and governor gear will align with camshaft gear. Mate crankcase cover with crankcase, but do not force. Tighten crankcase screws evenly in steps using sequence shown in Fig. KW211 to 7 N•m (62 in.-lbs.) torque. Reassemble remainder of components.

PISTON, PIN, RINGS AND CONNECTING ROD. To remove piston and rod assembly, drain engine oil and remove engine from equipment. Remove cylinder head as previously outlined. Clean pto end of crankshaft and remove any burrs or rust. Unscrew fasteners and remove crankcase cover (oil sump). Rotate crankshaft so timing marks on crankshaft and camshaft gears are aligned (this will position valve tappets out of the way). Withdraw camshaft from cylinder block. Remove carbon or ring ridge, if present, from top of cylinder prior to removing piston. Unscrew connecting rod screws, remove rod cap and push piston and rod out of cylinder block.

Renew piston if top or second ring side clearance in piston ring groove exceeds 0.10 mm (0.004 in.). Maximum piston ring end gap is 1.0 mm (0.040 in.) for top or second ring and 1.5 mm (0.060 in.) for oil ring. Maximum allowable piston pin hole diameter is 15.050 mm (0.5925 in.). Minimum allowable piston pin diameter is 14.975 mm (0.5896 in.). Pistons are available in oversizes of 0.25, 0.50 and 0.75 mm.

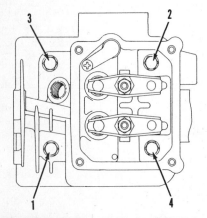
Fig. KW207—Tighten cylinder head screws to a torque of 23 N•m (204 in.-lbs.) in sequence shown above.

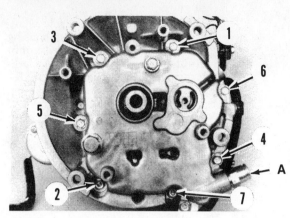

Fig. KW211—Tighten crankcase screws to 7 N•m (62 in.-lbs.) torque and follow sequence shown above.

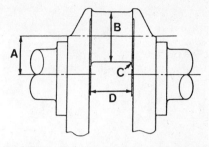

Fig. KW214—Refer to diagram and following dimensions to grind crankpin for 0.5 mm undersize connecting rod.

A. 22.950-23.000 mm (0.9035-0.9055 in.)	C. 1.5-1.8 mm (0.059-0.071 in.)
B. 27.467-27.480 mm (1.0814-1.0819 in.)	D. 24.00-24.10 mm (0.945-0.949 in.)

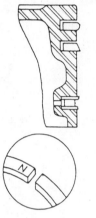

Fig. KW212—Install piston rings as shown above with "N" on ring toward piston crown.

Renew connecting rod if small end inner diameter is greater than 15.050 mm (0.5925 in.) or big end inner diameter exceeds 28.070 mm (1.1051 in.). A connecting rod with 0.5 mm (0.020 in.) undersize big end hole is available to fit a crankshaft with a reground crankpin.

When assembling piston and rod, arrow on piston crown must be on same side as "MADE IN JAPAN" on rod. Renew piston pin retaining rings whenever they are removed. Install piston rings on piston as shown in Fig. KW212, making sure that "N" mark on top and second ring faces up. Assemble piston and rod in cylinder block with arrow mark on piston crown toward flywheel side of engine. Install rod cap so index grooves on rod and cap align. Tighten connecting rod screws to 12 N•m (106 in.-lbs.) torque.

Install camshaft while aligning timing marks (Fig. KW210) on crankshaft and camshaft gears. Be sure governor weights are closed and governor gear will align with camshaft gear. Mate crankcase cover with crankcase, but do not force. Tighten crankcase screws evenly in steps using sequence shown in Fig. KW211 to 7 N•m (62 in.-lbs.) torque. Reassemble remainder of components.

GOVERNOR. The governor gear and flyweight assembly is located on the inside of the crankcase cover (oil sump). The plunger in the gear assembly contacts the governor arm and shaft in the crankcase. The governor shaft and arm transfer governor action to the external governor linkage.

To gain access to the governor gear assembly, drain crankcase oil and remove engine from equipment. Clean pto end of crankshaft and remove any burrs or rust. Unscrew fasteners and remove crankcase cover.

Governor assembly rides on a stud pressed into the crankcase cover. The governor assembly must be pried loose from the stud which damages the assembly so it must be discarded. When installing a new governor assembly, assemble plunger in governor gear before installing gear on crankcase cover stud. Place a thrust washer around the stud, then position governor gear assembly on stud so step in governor bore indexes on groove of stud. Check governor and flyweight movement which must be free without binding.

If removed, install governor shaft and arm in side of crankcase and attach cotter pin as shown in Fig. KW213.

To reassemble, position governor gear assembly in crankcase cover. Be sure governor arm is in proper position to contact governor, and camshaft end will properly engage oil pump drive. Install crankcase cover and tighten cover screws to 7 N•m (62 in.-lbs.) torque in

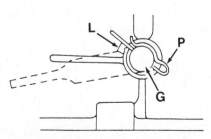

Fig. KW213—Install cotter pin (P) on governor shaft (G) as shown. Note location of lug (L).

sequence shown in Fig. KW211. Do not force mating of cover with crankcase. Reassemble remainder of components.

CRANKSHAFT AND MAIN BEARINGS. To remove crankshaft, first drain crankcase oil and remove engine from equipment. Clean pto end of crankshaft and remove any burrs or rust. Remove cylinder head, flywheel, crankcase cover and camshaft. Unscrew connecting rod cap and remove piston and rod. Withdraw crankshaft.

Renew crankshaft if main bearing journal diameter is less than 24.920 mm (0.9811 in.) for either end. With crankshaft supported at ends, runout measured at bearing journals must not exceed 0.20 mm (0.008 in.). Crankshaft must be renewed or reground if crankpin diameter is less than 27.920 mm (1.0992 in.). Refer to Fig. KW214 for crankpin regrinding dimensions to fit a 0.5 mm (0.020 in.) undersize connecting rod.

Crankshaft timing gear is removable. When installing gear be sure key does not protrude past outer face of timing gear.

Crankshaft rides in a ball bearing at the flywheel end. Use a suitable tool to remove and install bearing in crankcase. Install oil seal so flat side is out and flush with crankcase surface. Crankshaft rides directly in bore in crankcase cover (oil sump). Renew crankcase cover if bearing bore inside diameter exceeds 25.100 mm (0.9882 in.). Install oil seal so flat side is out and flush with crankcase surface.

CYLINDER. If cylinder bore exceeds 65.060 mm (2.5614 in.) or if out-of-round of bore exceeds 0.060 mm (0.0024 in.), cylinder should be bored to the next oversize of 0.25, 0.50 or 0.75 mm. Bore cylinder to following diameter:

0.25 mm	65.230-65.250 mm (2.5681-2.5689 in.)
0.50 mm	65.480-65.500 mm (2.5779-2.5787 in.)

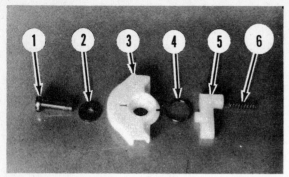

Fig. KW215—Note position
of recoil starter pawl (5) before removing it from
starter.

1. Screw
2. Washer
3. Retainer
4. Snap ring
5. Pawl
6. Spring

rection as viewed from pawl side of pulley. Position pulley in starter housing and turn pulley counterclockwise until spring tension is felt. Apply thread locking compound to threads of screw (1—Fig. KW215), then install pawl (5) and retainer (3). Tighten retainer screw to 3.5 N•m (30 in.-lbs.) torque; do not overtighten. After installing retainer, hold rope in notch and preload spring by turning pulley three turns counterclockwise, then pass rope through rope outlet in housing and install handle.

0.75 mm . . . . 65.730-65.750 mm
(2.5878-2.5886 in.)

REWIND STARTER. To disassemble starter, remove starter from engine. Remove rope handle and allow rope to wind slowly into starter. Note position of pawl (5—Fig. KW215), then unscrew center screw and remove retainer and pawl assembly. Wear appropriate safety eyewear and gloves before disengaging pulley from starter as spring may uncoil uncontrolled. Place shop cloth around pulley and lift pulley out of housing; spring should remain with pulley. If spring must be removed from pulley, position pulley so spring side is down and against floor, then tap pulley to dislodge spring.

Reassemble starter by reversing disassembly procedure while noting the following. Lightly grease sides of rewind spring and pulley. Install spring in pulley so coil direction is counterclockwise from outer spring end. Wind rope around pulley in counterclockwise di-

ELECTRICAL SYSTEM. Some engines may be equipped with a 12-volt electrical system that is separate from the ignition system. The electrical system comprises an electric starter, alternator and battery. Starter and alternator are considered unit assemblies. No service specifications are available for starter. The alternator is located under the flywheel and should be renewed if output is less than 8 volts DC.

KAWASAKI

Model	Bore	Stroke	Displacement	Power Rating
FE120	60 mm	44 mm	124 cc	2.8 kW
	(2.36 in.)	(1.73 in.)	(7.6 cu.in.)	(3.7 hp)
FE170	66 mm	50 mm	171 cc	4.0 kW
	(2.60 in.)	(1.97 in.)	(10.4 cu.in.)	(5.3 hp)

NOTE: Metric fasteners are used throughout engine.

ENGINE IDENTIFICATION

All models are four-stroke, single-cylinder, air-cooled engines with a horizontal crankshaft. Splash lubrication is used on all models. Engine serial number plate is located on the flywheel blower housing. Models with a "D" suffix are direct-drive engines, while models with a "G" suffix are equipped with an extended camshaft that provides a 2:1 pto. Always furnish engine model and serial numbers when ordering parts or service information.

MAINTENANCE

LUBRICATION. Engine oil level should be checked prior to each operating interval. Oil should be maintained between reference marks on dipstick with dipstick just touching first threads. Do not screw dipstick in to check oil level.

Engine oil should meet or exceed latest API service classification. Use SAE 40 oil for temperatures above 68° F (20° C); use SAE 30 oil for temperatures between 32° F (0° C) and 95° F (35° C); SAE 10W-30 or 10W-40 oil may be used for temperatures between −4° F (−20° C) and 95° F (35° C); below 32° F (0° C) SAE 5W-20 should be used.

Oil should be changed after first five hours of operation and then after every 50 hours of operation. Oil should be drained while engine is warm. Crankcase capacity is approximately 0.6 L (0.63 quart). Do not overfill.

The engine may be equipped with a low-oil warning system. If engine oil is low, an indicator lamp will light and the engine will stop or not run. Refer to OIL WARNING SYSTEM in REPAIRS section.

AIR FILTER. The air filter element should be removed and cleaned after every 50 hours of operation, or more often if operating in extremely dusty conditions. Paper element should be renewed after every 300 hours of operation. Element should also be renewed if it is very dirty or if it is damaged in any way.

Clean foam element in solution of warm water and liquid detergent, then squeeze out excess water and allow to air dry. DO NOT wash paper element. Apply light coat of engine oil to foam element and squeeze out excess oil. Clean paper element by tapping gently to remove dust. DO NOT use compressed air to clean element. Inspect paper element for holes or other damage. Reinstall by reversing removal procedure.

CRANKCASE BREATHER. Crankcase pressure is vented to the cylinder head. A reed valve is located on the top of the cylinder head in the rocker arm chamber. Renew reed valve if tip of reed stands up excessively or if reed is damaged or worn excessively. Reed valve seat must be smooth.

SPARK PLUG. Recommended spark plug is NGK BPR5ES. Spark plug should be removed, cleaned and electrode gap set after every 100 hours of operation. Specified spark plug gap is 0.7-0.8 mm (0.028-0.032 in.). Renew spark plug if electrode is severely burnt or damaged. Tighten spark plug to 25 N•m (18 ft.-lbs.).

CARBURETOR. All models are equipped with a float type side draft carburetor. Keihin Model BBK 18-12 carburetor is used on FE120 engine and Keihin Model BBK 18-15 carburetor is used on FE170 engine. Idle mixture is adjustable while high speed mixture is determined by a fixed main jet.

Initial setting for idle mixture screw (IM—Fig. KW401) is 1½ turns out from a lightly seated position on Model FE120 and ⅞ turn on Model FE170. Make final adjustment with engine at operating temperature and running. Push governor lever (L) toward carburetor so carburetor throttle is closed, then adjust idle speed screw (IS) on carburetor so engine idles at 1200 rpm. Continue to hold carburetor throttle closed and adjust idle mixture screw to obtain maximum engine idle speed, then turn idle mixture screw out an additional ¼ turn. If necessary, readjust idle speed screw so engine idles at 1200 rpm. Release governor lever and adjust idle stop screw (W) on speed control bracket so engine idles at 1600 rpm on Model FE120 or 1300 rpm on Model FE170. On some equipment, governed idle speed may be different.

Disassembly of carburetor is self-evident upon examination of unit and reference to Fig. KW402. Do not clean plastic carburetor parts in carburetor cleaner.

With carburetor inverted, float should be parallel with fuel bowl mating surface. Float height is not adjustable; replace any components which are damaged or excessively worn and adversely affect float position.

GOVERNOR. A gear-driven flyweight type governor is located inside engine crankcase. Before adjusting governor linkage, make certain all linkage is in good condition and that governor spring (G—Fig. KW403) is not stretched. Spring (R) around governor-to-carburetor rod must pull governor lever (L) and throttle lever toward each other.

To adjust external linkage, place engine throttle control in idle position. Move governor lever (L) so throttle is in wide open position. Loosen governor lever clamp bolt nut (N) and turn governor shaft (S) clockwise as far as possible. Tighten clamp bolt nut. Maximum no-

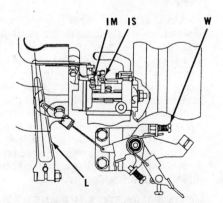

Fig.KW401—Drawing showing location of idle mixture screw (IM), idle speed screw (IS), governor lever (L) and throttle stop screw (W).

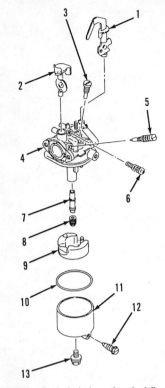

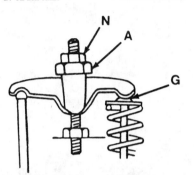

+Tester Lead

		Black	Brown	Yellow
−Tester Lead	Black		B	D
	Brown	A		B
	Yellow	C	C	

Fig. KW404—Connect ohmmeter leads to ignition control unit leads as indicated in chart and note desired readings in legend.

A. 1k-10k ohms　　　　C. 50k-200k ohms
B. 5k-20k ohms　　　　D. 10k-30k ohms

Fig. KW402—Exploded view of typical float type carburetor used on all models.

1. Choke shaft assy.	7. Main nozzle
2. Throttle shaft assy.	8. Main jet
3. Pilot jet	9. Float
4. Body	10. Gasket
5. Idle mixture screw	11. Fuel bowl
6. Idle speed screw	12. Drain screw

Fig. KW405—Valve clearance gap (G) between valve stem and rocker arm is adjusted by loosening nut (N) and rotating adjusting pivot nut (A). Valve clearance should be 0.12 mm (0.005 in.).

load engine speed should be 4000 rpm and is adjusted by turning high speed governed speed screw (M).

IGNITION SYSTEM. The engine is equipped with a solid-state ignition system that does not require regular maintenance. Ignition timing is not adjustable. Ignition coil is located outside flywheel. Air gap between ignition coil and flywheel should be 0.30 mm (0.012 in.).

To test ignition coil, remove cooling shrouds and disconnect spark plug ca-

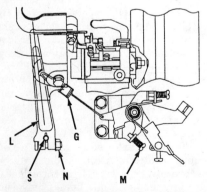

Fig. KW403—Drawing showing location of governor spring (G), governor lever (L), maximum governed speed screw (M), clamp nut (N), throttle rod and spring (R), and governor shaft (S).

ble and primary lead wire. Connect ohmmeter test leads between coil core (ground) and high tension (spark plug) lead. Secondary coil resistance should be 7.0k-9.0k ohms. Remove test lead connected to high tension lead and connect tester lead to coil primary terminal. Primary coil resistance should be 0.8-1.0 ohm. If readings vary significantly from specifications, renew ignition coil.

To test control unit, disconnect leads to control unit and connect an ohmmeter to control unit leads as shown in chart shown in Fig. KW404. Renew or replace with a good control unit if ohmmeter readings are not as specified in chart.

VALVE ADJUSTMENT. Clearance between valve stem ends and rocker arms should be checked and adjusted after every 300 hours of operation. Engine must be cold for valve adjustment. Rotate crankshaft so piston is at top dead center on compression stroke. Remove rocker arm cover. Valve clearance gap (G—Fig. KW405) for both valves should be 0.12 mm (0.005 in.). Loosen nut (N) and turn adjusting screw or nut (A) to obtain desired clearance. Tighten nut (N) and recheck adjustment.

COMPRESSION PRESSURE. Standard compression reading should be 390 kPa (57 psi).

TIGHTENING TORQUES. Recommended tightening torques are as follows:

Connecting rod. . . . . 12 N·m
　　　　　　　　　　　　(105 in.-lbs.)
Crankcase cover:
　FE120. . . . . 8.8 N·m
　　　　　　　　　　　　　(78 in.-lbs.)
　FE170. . . . . 25 N·m
　　　　　　　　　　　　(221 in.-lbs.)
Cylinder head:
　FE120. . . . . 23 N·m
　　　　　　　　　　　　(204 in.-lbs.)
　FE170. . . . . 25 N·m
　　　　　　　　　　　　(221 in.-lbs.)
Flywheel:
　FE120. . . . . 62 N·m
　　　　　　　　　　　　　(46 ft.-lbs.)
　FE170. . . . . 64 N·m
　　　　　　　　　　　　　(47 ft.-lbs.)
Oil drain plug. . . . . 20 N·m
　　　　　　　　　　　　(177 in.-lbs.)
Spark plug . . . . 25 N·m
　　　　　　　　　　　　　(18 ft.-lbs.)

CYLINDER HEAD. To remove cylinder head, remove cylinder head shroud and blower housing. Remove carburetor and muffler. Remove rocker arm cover, loosen cylinder head mounting bolts evenly and remove cylinder head and gasket.

Remove carbon deposits from combustion chamber being careful not to damage gasket sealing surface. Inspect cylinder head for cracks, nicks or other damage. Place cylinder head on a flat surface and check entire sealing surface for distortion using a feeler gauge. Renew cylinder head if sealing surface is warped more than 0.05 mm (0.002 in.).

To reinstall cylinder head, reverse removal procedure. Surfaces of cylinder head gasket are coated with a sealant and do not require additional sealant. Push rods should be installed in their original positions. Tighten cylinder head screws in three steps in sequence shown in Fig. KW406 to final torque of 23 N·m (204 in.-lbs.) on Model FE120 and 25 N·m (221 in.-lbs.) on Model FE170. Adjust valve clearance as outlined in MAINTENANCE section.

VALVE SYSTEM. Valves are actuated by rocker arms mounted on a stud threaded into the cylinder head. Valve face and seat angles are 45° for intake and exhaust. Valve seat width should be 0.5-0.9 mm (0.020-0.040 in.) and service limit is 1.7 mm (0.070 in.). A seat width of 0.70 mm (0.028 in.) is desirable. Seats can be narrowed using a 30° stone or cutter.

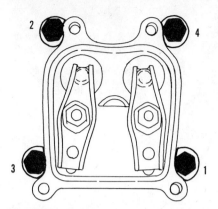

Fig. KW406—Tighten cylinder head bolts in sequence shown.

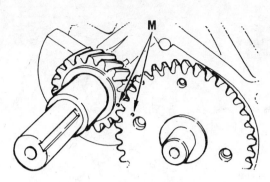

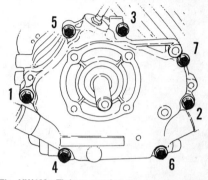

Fig. KW408—View of timing marks (M) on crankshaft and camshaft gears.

Minimum allowable valve head thickness (margin) is 0.5 mm (0.020 in.). Renew valve if valve margin is less than service limit. Valve stem runnout should not exceed 0.05 mm (0.020 in.).

Minimum allowable valve stem diameter is 5.430 mm (0.2138 in.) for intake valve and 5.415 mm (0.2132 in.) for exhaust valve. Maximum allowable valve guide inside diameter is 5.568 mm (0.2192 in.).

Valve guides can be renewed using suitable valve guide driver. Press or drive guide out towards rocker arm side of head after heating cylinder head to 350° F (175° C). Press or drive guide into top of cylinder head after heating so valve guide is flush with boss. Ream guide with Kawasaki reamer 57001-1079 or other suitably sized reamer. Valve guide finished inside diameter should be 5.500-5.512 mm (0.2165-0.2170 in.).

Minimum valve spring free length is 31.75 mm (1.250 in.). Maximum push rod runnout is 0.3 mm (0.012 in.).

CAMSHAFT. Camshaft and camshaft gear are an integral casting which is equipped with a compression release mechanism. The compression release arm (A—Fig. KW407) extends at cranking speed to hold the exhaust valve open slightly thereby reducing compression pressure. Camshaft is supported at each end in bushings which are integral part of crankcase or crankcase cover. On

engines with a pto shaft at the end of the camshaft ("G" models), a ball bearing in the crankcase cover supports the pto end of the camshaft.

To remove camshaft, remove rocker arm push rods and mark them so they can be returned to original position. Remove crankcase cover. Rotate crankshaft so piston is at top dead center on compression stroke and invert engine so tappets will not interfere with camshaft removal and remove camshaft. If tappets are removed, mark tappets so they can be installed in their original guides.

Camshaft minimum lobe height for intake and exhaust lobes is 26.157 mm (1.0298 in.) for FE120 engine and 30.024 mm (1.1820 in.) for FE170 engine.

Bearing journal minimum diameter for both ends is 14.910 mm (0.5870 in.) for Model FE120 and 15.910 mm (0.6264 in.) for Model FE170. If camshaft has an extended pto shaft ("G" models), the minimum journal diameter at the pto end is 19.930 mm (0.7846 in.) on Model FE120 and 24.930 mm (0.9815 in.) on Model FE170.

Camshaft bearing bore maximum inside diameter for both ends is 15.063 mm (0.5930 in.) for Model FE120 and 16.063 mm (0.6324 in.) for Model FE170.

Refer to Fig. KW407 to check compression release mechanism. With weight (W) in starting position, arm (A) should protrude (H) above cam lobe base at least 0.6 mm (0.024 in.) on Model FE120 and 1.1 mm (0.043 in.) on Model FE170. With weight extended in running position, arm should be beneath surface of cam lobe base.

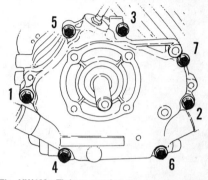

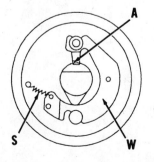

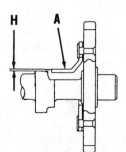

Fig. KW407—Drawing showing location of compression release arm (A), spring (S) and weight (W). Arm (A) should protrude (H) above cam lobe base at least 0.6 mm (0.024 in.) on Model FE120 and 1.1 mm (0.043 in.) on Model FE170.

Fig. KW409—Tighten crankcase cover screws in sequence shown.

When reinstalling camshaft and tappets, be sure that tappets are installed in their original position. If camshaft is renewed, tappets should also be renewed. Make sure that timing marks (M—Fig. KW408) on camshaft gear and crankshaft gear are aligned. Tighten crankcase cover screws evenly in sequence shown in Fig. KW409 to 8.8 N•m (76 in.-lbs.) on Model FE120 and to 25 N•m (221 in.-lbs.) on Model FE170.

PISTON, PIN AND RINGS. Piston and connecting rod are removed as an assembly after removing cylinder head, crankcase cover, camshaft and connecting rod cap.

Measure piston diameter 9 mm (0.35 in.) above bottom of piston skirt and perpendicular to piston pin hole. Minimum allowable standard diameter is 59.905 mm (2.3584 in.) on Model FE120 and 65.885 mm (2.5939 in.) on Model FE170.

Maximum inside diameter of pin bore in piston is 14.042 mm (0.5528 in.) for Model FE120 and 16.033 mm (0.6312 in.) for Model FE170. Minimum piston pin outside diameter is 13.975 mm (0.5502 in.) for Model FE120 and 15.975 mm (0.6289 in.) for Model FE170.

Maximum piston ring side clearance is 0.15 mm (0.006 in.) for top ring and 0.12 mm (0.005 in.) for second ring. Minimum piston ring thickness for either compression ring is 1.42 mm (0.056 in.).

Maximum allowable end gap is 1.00 mm (0.039 in.) for compression rings and 1.20 mm (0.047 in.) for oil control ring. If piston ring gap is greater than specified, check cylinder bore for wear.

When assembling piston and rod, note that piston crown is marked with "R" and "L". On direct drive engines ("D" models), install piston on connecting rod so "R" on top of piston is on same side of "MADE IN JAPAN" side of connecting rod. On reduction drive engines ("G" models), install piston on connecting rod so "L" on top of piston is on same side of "MADE IN JAPAN" side of connecting rod.

When installing piston and rod in direct drive engines ("D" models), "R" mark on top of piston must be toward flywheel side of engine. When installing piston and rod in reduction drive engines ("G" models), "L" mark on top of piston must be toward flywheel side of engine. Tighten connecting rod screws to 12 N•m (105 in.-lbs.). Install cylinder head and camshaft as previously outlined.

CONNECTING ROD. Connecting rod and piston are removed as an assembly as outlined in previous section.

Connecting rod rides directly on crankshaft journal. Maximum allowable inside diameter for connecting rod big end is 26.052 mm (1.0257 in.) for Model FE120 and 29.052 mm (1.1438 in.) for Model FE170. An undersize connecting rod is available for use with undersize crankshaft crankpin. Refer to CRANKSHAFT AND BALANCER section.

Maximum inside diameter of connecting rod small end is 14.042 mm (0.5528 in.) for Model FE120 and 16.047 mm (0.6318 in.) for Model FE170.

Minimum connecting rod width at big end is 23.0 mm (0.91 in.).

Install connecting rod as outlined in previous section.

GOVERNOR. The internal centrifugal flyweight governor is mounted in the crankcase cover. The governor is gear-driven by a gear on the crankshaft.

To remove governor assembly, remove crankcase cover from crankcase. Use two screwdrivers to snap governor gear and flyweight assembly off governor stub shaft. Governor unit will be damaged when removed and must be renewed if removed. A thrust washer is located under the governor. Install governor by pushing down until it snaps onto the locating groove.

Refer to MAINTENANCE section for external governor linkage adjustment.

CRANKSHAFT. The crankshaft is supported at both ends by ball type main bearings in the crankcase and crankcase cover.

Crankshaft main journal minimum diameter for both ends is 19.930 mm (0.7846 in.) for Model FE120 and 24.930 mm (0.9815 in.) for Model FE170. Main journals cannot be resized. Refer to CYLINDER, CRANKCASE, MAIN BEARINGS AND SEALS section for main bearing dimensions. Measure crankshaft runout at the main journals. Crankshaft should be renewed if runout exceeds 0.05 mm (0.002 in.). Crankshaft cannot be straightened.

Crankpin journal minimum diameter is 25.944 mm (1.0214 in.) for Model FE120 and 28.944 mm (1.1395 in.) for Model FE170. Undersize connecting rods are not available.

CYLINDER, CRANKCASE, MAIN BEARINGS AND SEALS. Cylinder and crankcase are an integral casting. Standard cylinder bore diameter is 59.980-60.000 mm (2.3614-2.3622 in.) for Model FE120 and 65.980-66.000 mm (2.5976-2.5984 in.) for Model FE170. Cylinder bore wear limit is 60.067 mm (2.3648 in.) for Model FE120 and 66.067 mm (2.6011 in.) for Model FE170. Maximum allowable cylinder bore out-of-round is 0.056 mm (0.0022 in.). Cylinder can be bored or honed to fit oversize pistons. Bore the cylinder to the following dimensions to obtain proper piston clearance:

Model	Cyl. Bore Diameter
0.25 mm (0.010 in.) Oversize:	
FE120	60.230-60.250 mm (2.3712-2.3720 in.)
FE170	66.230-66.250 mm (2.6075-2.6083 in.)
0.50 mm (0.020 in.) Oversize:	
FE120	60.480-60.500 mm (2.3811-2.3819 in.)
FE170	66.480-66.500 mm (2.6173-2.6181 in.)
0.75 mm (0.030 in.) Oversize:	
FE120	60.730-60.750 mm (2.3909-2.3917 in.)
FE170	66.730-66.750 mm (2.6272-2.6279 in.)

The flywheel end of the crankshaft rides in an integral bushing in the crankcase. Renew crankcase if bearing bore exceeds 30.075 mm (1.1840 in.).

Renew crankshaft seals if worn or damaged. Install seals with lip facing inside of engine. Press in oil seal in crankcase until flush with surface. Press in oil seal in crankcase cover until 4.0 mm (0.16 in.) below surface.

Tighten crankcase cover screws in sequence shown in Fig. KW409 to 8.8 N•m (76 in.-lbs.) on Model FE120 and to 25 N•m (221 in.-lbs.) on Model FE170.

BALANCER. The engine may be equipped with a balancer shaft that is driven by a gear on the crankshaft. The balancer shaft rides in a nonrenewable bushing in the crankcase and a ball bearing in the crankcase cover.

Minimum balancer shaft journal diameter is 14.933 mm (0.5879 in.) at flywheel end and 14.943 mm (0.5883 in.) at gear end. Maximum bushing bore in crankcase is 15.063 mm (0.5930 in.).

Timing marks (M—Fig. KW410) on crankshaft and balancer gears must be aligned.

OIL WARNING SYSTEM. The engine may be equipped with a low-oil warning system that grounds the ignition system and lights a warning lamp if the oil level is low. To test circuit, run engine then disconnect yellow switch wire. Grounding the yellow wire to the engine should cause the warning light to flash and the engine should stop. Stop engine. With oil level correct and both switch leads disconnected, use an ohmmeter or continuity tester and check continuity between switch leads. Tester should indicate no continuity. With no oil in crankcase, tester should indicate continuity.

The crankcase cover must be removed from the engine so the oil level switch can be removed for testing. Con-

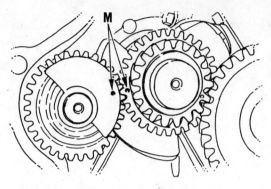

Fig. KW410—View of timing marks (M) on balancer gear and crankshaft gear.

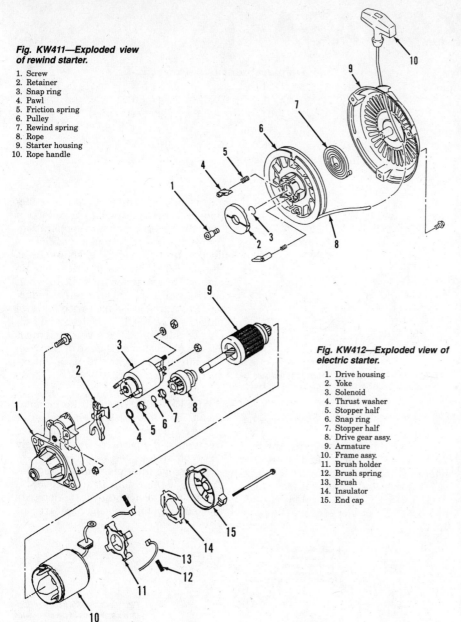

Fig. KW411—Exploded view of rewind starter.

1. Screw
2. Retainer
3. Snap ring
4. Pawl
5. Friction spring
6. Pulley
7. Rewind spring
8. Rope
9. Starter housing
10. Rope handle

Fig. KW412—Exploded view of electric starter.

1. Drive housing
2. Yoke
3. Solenoid
4. Thrust washer
5. Stopper half
6. Snap ring
7. Stopper half
8. Drive gear assy.
9. Armature
10. Frame assy.
11. Brush holder
12. Brush spring
13. Brush
14. Insulator
15. End cap

in a fluid, resistance should increase from zero to infinity.

REWIND STARTER. Refer to Fig. KW411 for an exploded view of rewind starter. To disassemble starter, remove starter from engine. Remove rope handle and allow rope to wind slowly into starter. Unscrew center retaining screw (1). Wear appropriate safety eyewear and gloves before disengaging pulley from starter as spring may uncoil uncontrolled. Place shop towel around pulley and lift pulley out of housing. Use caution when detaching rewind spring from pulley.

Apply light grease to pulley shaft and sliding surfaces of rewind spring. Wrap rope around pulley in counterclockwise direction viewed from flywheel side of pulley (on "G" models, wrap rope in clockwise direction). Install rewind spring (7) in spring cavity in pulley in a counterclockwise direction from outer end (on "G" models, install spring in clockwise direction). Install pulley in starter housing so inner end of spring engages tab on starter housing and route rope through rope outlet in housing. Attach rope handle. Install snap ring (3) on shaft. Install springs (5), pawls (4), retainer (2) and screw (1). To place tension on starter rope, pull rope out of housing until notch in pulley is aligned with rope outlet, then hold pulley to prevent pulley rotation. Pull rope back into housing while positioning rope in pulley notch. Turn rope pulley two turns counterclockwise, disengage rope from notch and allow rope to wind onto pulley. Check starter operation.

ELECTRIC STARTER. The engine may be equipped with a 12 volt DC starter. Under no load with a cranking voltage of 11.5 volts DC the current draw should be less than 22 amps.

Refer to Fig. KW412 for an exploded view of electric starter. Renew brushes if length is less than 10 mm (0.40 in.). Minimum commutator diameter is 27.0 mm (1.06 in.).

nect an ohmmeter or continuity tester and check continuity between switch leads. With switch in normal position, there should be continuity. With switch inverted, there should be no continuity. When switch is progressively inserted

KAWASAKI

Model	Bore	Stroke	Displacement	Power Rating
FG150	64 mm	47 mm	151 cc	2.7 kW
	(2.51 in.)	(1.85 in.)	(9.2 cu.in.)	(3.6 hp)
FG200	71 mm	51 mm	201 cc	3.7 kW
	(2.79 in.)	(2.01 in.)	(12.3 cu.in.)	(5.0 hp)
FG300	78 mm	62 mm	296 cc	10.0 kW
	(3.07 in.)	(2.44 in.)	(18.07 cu.in.)	(7.5 hp)

NOTE: Metric fasteners are used throughout engine.

ENGINE IDENTIFICATION

Kawasaki FG series engines are four-stroke, single-cylinder, air-cooled gasoline engines. Engine model number is located on the shroud adjacent to rewind starter, and engine model number and serial number are located on crankcase cover (Fig. KW301). Models with a "D" suffix are direct-drive engines, while models with a "R" suffix are equipped with a reduction unit. Always furnish engine model and serial numbers when ordering parts or service information.

MAINTENANCE

LUBRICATION. The engine is lubricated by oil thrown by an oil dipper attached to the connecting rod. Some engines are equipped with a sensor that detects a low oil level in the crankcase.

Engine oil level should be checked prior to each operating interval. Oil level should be maintained between reference marks on dipstick with dipstick just touching first threads. Do not screw dipstick in when checking oil level (Fig. KW302).

Engine oil should meet or exceed latest API service classification. Use SAE 40 oil for temperatures above 95° F (35° C); use SAE 30 oil for temperatures between 50° F (10° C) and 95° F (35° C);

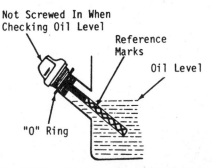

Fig. KW302—Oil plug and gauge should not be screwed into crankcase when checking oil level.

use SAE 10W-30 for temperatures below 50° F (10° C).

Oil should be changed after the first 20 hours of operation and every 100 hours of operation thereafter. Crankcase capacity is 0.5 L (1.0 pt.) for Model FG150, 0.7 L (1.5 pt.) for Model FG200 and 0.8 L (1.7 pt.) for Model FG300.

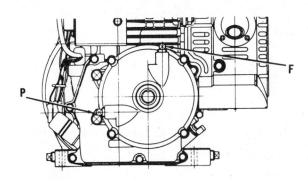

Fig. KW303—Oil in 6:1 reduction case should be even with oil level plug (P). Fill plug (F) also functions as a vent and must be at top of case.

On engines equipped with a reduction unit, check oil level in the reduction case after every 100 hours of operation. An oil dipstick is provided on 2.7:1 reduction unit. The oil level should be at the "H" mark on the dipstick, but no higher. Do not screw dipstick in when checking oil level. Oil level on 6:1 reduction units should be even with oil level plug (P—Fig. KW303). Note that oil fill plug (F) also serves as a vent and must always be located at the top of the case. Recommended oil for both types of reduction units is SAE 10W-30.

AIR FILTER. The air filter element should be removed and cleaned after every 50 hours of operation, or more frequently if operating in extremely dirty conditions.

To remove filter elements (2 and 3—Fig. KW304), unsnap air filter cover (1) and pull out elements. Clean elements in nonflammable solvent, squeeze out

Fig.KW301—View showing location of engine model number and serial number.

Engine Model And
Serial Number Location

excess solvent and allow to air dry. Soak elements in SAE 30 engine oil and squeeze out excess oil. Reinstall by reversing removal procedure.

FUEL FILTER. A fuel filter screen is located in the sediment bowl below fuel shut-off valve (Fig. KW305). Sediment bowl and screen should be cleaned after every 50 hours of operation. To remove sediment bowl, shut off fuel valve and unscrew sediment bowl from valve body. Clean bowl and filter screen

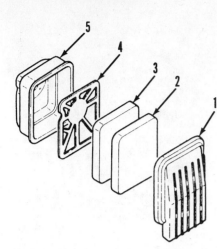

Fig. KW304—Exploded view of air cleaner assembly.

1. Cover
2. Element
3. Element
4. Plate
5. Housing (case)

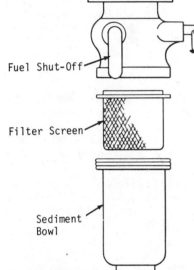

Fuel Shut-Off

Filter Screen

Sediment Bowl

Fig. KW305—Exploded view of fuel filter and sediment bowl assembly used on most models.

using suitable solvent. When reinstalling, make certain that gasket is in position before installing sediment bowl.

SPARK PLUG. Recommended spark plug for all models is NGK BPR4HS or equivalent.

Spark plug should be removed, cleaned and inspected after every 100 hours of operation. Renew spark plug if burned or damaged. Electrode gap should be set at 0.6-0.7 mm (0.024-0.027 in.) on all models. Tighten spark plug to 28 N·m (20 ft.-lbs.) torque.

CARBURETOR. All models are equipped with a float type carburetor (Fig. KW306 or KW307). Low speed fuel:air mixture is controlled by the pilot air jet and high speed fuel:air mix-

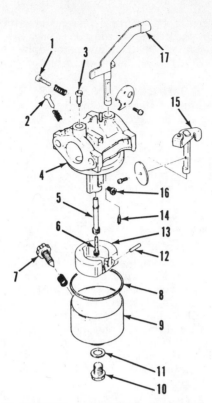

Fig. KW306—Exploded view of float type carburetor used on Model FG150 engine.

1. Idle speed screw
2. Idle mixture screw
3. Pilot jet
4. Carburetor body
5. Nozzle
6. Bleed tube
7. Drain screw
8. Gasket
9. Float bowl
10. Plug
11. Gasket
12. Float pin
13. Float
14. Fuel inlet needle
15. Throttle shaft
16. Main jet
17. Choke shaft

ture is controlled by a fixed main jet. Carburetor adjustment should be checked whenever poor or erratic performance is noted.

Engine idle speed is adjusted by turning throttle stop screw (1) clockwise to increase idle speed or counterclockwise to decrease idle speed. Initial adjustment of idle mixture screw (2) from a lightly seated position is 7/8 turn open for Model FG150, 1 turn open for Model FG200 and 1¼ turn open for Model FG300. Final adjustment should be made with engine at operating temperature and running. Adjust idle mixture screw to obtain the smoothest idle and acceleration.

Overhaul. To disassemble carburetor, refer to Fig. KA306 or Fig. KA307 and remove fuel bowl retaining screw and fuel bowl. Remove float pin, float and fuel inlet needle. Remove throttle and choke shaft assemblies after unscrewing throttle and choke plate retaining screws. Remove idle mixture screw, pilot jet, main jet and main fuel nozzle. Note that main jet for Model FG150 carburetor must be removed be-

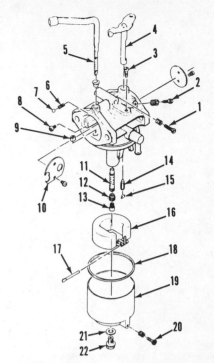

Fig. KW307—Exploded view of float type carburetor used on Models FG200 and FG300.

1. Idle speed screw
2. Pilot screw
3. Pilot air jet
4. Throttle shaft
5. Choke shaft
6. Spring
7. Ball
8. Air jet
9. Collar
10. Choke valve
11. Nozzle
12. Main jet holder
13. Main jet
14. Fuel inlet needle
15. Clip
16. Float
17. Float pin
18. Gasket
19. Float bowl
20. Drain screw
21. Gasket
22. Bolt

fore the nozzle as the jet blocks the nozzle.

When assembling the carburetor, note the following: Place a small drop of nonhardening sealant such as Permatex #2 or equivalent on throttle and choke plate retaining screws. Pilot jet size is #37.5 for Model FG150 and #45 for Models FG200 and FG300. Standard main jet size is #72.5 for Model FG150, #87.5 for Model FG200 and #91.3 for Model FG300.

To check float level, carburetor must be removed from engine. Float should be parallel to carburetor body when carburetor is held so that float tab just touches fuel inlet valve needle. To adjust, carefully bend float tab.

GOVERNOR. A gear-driven flyweight governor assembly is located inside engine crankcase. To adjust external governor linkage, place engine throttle control in idle position. Make certain that all linkage is in good condition and that tension spring (4—Fig. KW308) is not stretched. Loosen clamp bolt (5) and rotate governor shaft (6) clockwise as far as possible and move governor lever (3) to fully open carbure-

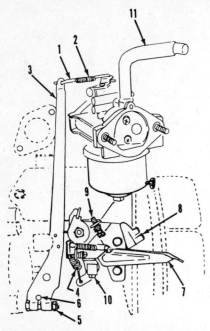

Fig. KW308—View of external governor linkage. Refer to text for adjustment procedure.

1. Governor-to-carburetor rod	7. Throttle lever
2. Spring	8. Throttle plate
3. Governor lever	9. Maximum speed adjusting screw
4. Governor spring	10. Idle speed adjusting screw
5. Clamp bolt	11. Throttle pivot
6. Governor shaft	

tor throttle. Then tighten clamp bolt while holding governor lever and shaft in these positions.

Start engine and use a tachometer to check maximum engine speed. Adjust maximum speed set screw (9) to obtain fast idle speed as specified by the equipment manufacturer.

IGNITION SYSTEM. All models are equipped with a solid-state ignition system and no regular maintenance is required. Ignition timing is not adjustable. Ignition coil is located outside the flywheel. Air gap between ignition coil and flywheel magnets should be 0.3 mm (0.010 in.).

VALVE ADJUSTMENT. Clearance between valve stem ends and tappets should be checked after every 200 hours of operation. To check clearance, remove tappet chamber cover and rotate crankshaft to position piston at top dead center on compression stroke. Measure valve clearance with engine cold using a feeler gauge. Specified clearance for FG150 engine is 0.12-0.18 mm (0.005-0.007 in.) for intake and 0.21-0.23 mm (0.008-0.009 in.) for exhaust. Specified clearance for FG200 and FG300 engines is 0.12-0.18 mm (0.005-0.007 in.) for intake and 0.19-0.25 mm (0.008-0.010 in.) for exhaust.

If clearance is not as specified, valves must be removed and end of stems

ground off to increase clearance or seats ground deeper to reduce clearance. Refer to VALVE SYSTEM paragraphs in REPAIRS section for service procedure.

COMPRESSION PRESSURE. Minimum allowable compression pressure is 297 kPa (43 psi).

CYLINDER HEAD AND COMBUSTION CHAMBER. Cylinder head, combustion chamber and piston should be cleaned and carbon and other deposits removed after every 100-200 hours of operation. Refer to REPAIRS section for service procedure.

GENERAL MAINTENANCE. Check and tighten all loose bolts, nuts or clamps prior to each interval of operation. Check for fuel or oil leakage and repair as necessary.

Clean dust, dirt, grease or any foreign material from cylinder head and cylinder block cooling fins after every 100 hours of operation, or more frequently if operating in extremely dirty conditions.

REPAIRS

TIGHTENING TORQUES. Recommended tightening torques are as follows:

Spark plug . . . . 28 N·m
(20 ft.-lbs.)
Cylinder head . . . . 23.5-24.5 N·m
(17.5-18 ft.-lbs.)
Connecting rod
FG150 & FG200 . . . . 20-21 N·m
(177-185 in.-lbs.)
FG300 . . . . 21.5-22.5 N·m
(190-200 in.-lbs.)
Crankcase cover
FG150 & FG200 . . . . 9-11 N·m
(80-97 in.-lbs.)
FG300 . . . . 14 N·m
(10.5 ft.-lbs.)
Flywheel . . . . 59-64 N·m
(44-47 ft.-lbs.)

CYLINDER HEAD. To remove cylinder head, first remove cylinder head shroud. Clean engine to prevent entrance of foreign material. Loosen cylinder head bolts in ¼-turn increments following sequence shown in Fig. KW309 until all bolts are loose enough to remove by hand. Note the position of each cylinder head bolt so they can be reinstalled in original location; Model FG300 has high tension bolts on exhaust valve side of cylinder head.

Remove spark plug and clean carbon and other deposits from cylinder head. Place cylinder head on a flat surface and check entire sealing surface for distortion. Renew cylinder head if head is warped more than 0.3 mm (0.012 in.).

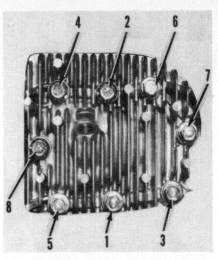

Fig. KW309—Tighten and loosen cylinder head bolts following sequence shown.

Reinstall cylinder head using a new head gasket. DO NOT use any sealant on the new head gasket. Lubricate threads of cylinder head bolts with light oil, then tighten evenly in three steps to 23.5-24.5 N·m (17.5-18 ft.-lbs.) torque in sequence shown in Fig. KW309.

VALVE SYSTEM. Valve face and seat angle is 45° for intake and exhaust valves. Valve seat width should be 1.0-1.6 mm (0.039-0.063 in.). Valve should be renewed if valve head margin is less than 0.6 mm (0.024 in.).

Refer to the following table for valve stem and valve guide wear limit specifications:

Model FG150
Valve stem OD (min.):
Intake . . . . 5.948 mm
(0.2342 in.)
Exhaust . . . . 5.935 mm
(0.2336 in.)
Valve guide ID (max.):
Intake . . . . 6.08 mm
(0.239 in.)
Exhaust . . . . 6.08 mm
(0.239 in.)

Model FG200
Valve stem OD (min.):
Intake . . . . 6.942 mm
(0.2733 in.)
Exhaust . . . . 6.952 mm
(0.2737 in.)
Valve guide ID (max.):
Intake . . . . 7.09 mm
(0.279 in.)
Exhaust . . . . 7.09 mm
(0.279 in.)

Model FG300
Valve stem OD (min.):
Intake . . . . 7.442 mm
(0.2929 in.)
Exhaust . . . . 7.425 mm
(0.2923 in.)

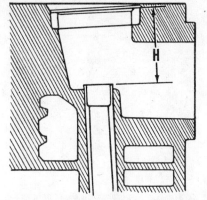

Fig. DW310—Install new valve guide so distance (H) between top of valve seat and top of guide is 23 mm (0.910 in.) for Models FG150 and FG200 or 26 mm (1.02 in.) for Model FG300.

Valve guide ID (Max.):

Intake .... 7.59 mm
(0.299 in.)
Exhaust .... 7.59 mm
(0.299 in.)

Valve guides are renewable. Use suitable tools to remove old guide and press in new guide. Lubricate outer surface of new valve guide with oil, then press into cylinder block until top of guide is 23 mm (0.91 in.) below top of valve seat for Models FG150 and FG200 or 26 mm (1.02 in.) for Model FG300 (see Fig. KW310). New guides must be reamed to final size after installation.

CAMSHAFT AND BEARINGS.

Camshaft is supported at crankcase end in a bearing which is an integral part of crankcase casting and is supported at crankcase cover end in a ball bearing. Refer to CRANKCASE AND BEARINGS paragraphs for camshaft removal.

Inspect camshaft and tappets for excessive wear, scoring, pitting or other damage and renew as necessary. Camshaft and tappets should be renewed as a set. Refer to the following table for wear limit specifications:

Model FG150

Cam lobe height:

Intake .... 27.3 mm
(1.075 in.)
Exhaust .... 27.1 mm
(1.067 in.)

Bearing journal diameter

Pto end .... 14.940 mm
(0.5882 in.)
Flywheel end......... 14.942 mm
(0.5883 in.)

Model FG200

Cam lobe height

Intake & exhaust .... 31.7 mm
(1.248 in.)

Bearing journal diameter

Pto end.............. 14.966 mm
(0.5892 in.)

Flywheel end .... 16.942 mm
(0.6670 in.)

Model FG300

Cam lobe height

Intake & exhaust .... 32.67 mm
(1.286 in.)

Bearing journal diameter

Pto end .... 19.960 mm
(0.7858 in.)
Flywheel end .... 19.955 mm
(0.7856 in.)

Camshaft bearing bore in crankcase wear limit is 15.043 mm (0.5922 in.) for Model FG150, 17.043 mm (0.6710 in.) for Model FG200 and 20.056 mm (0.7869 in.) for Model FG300.

Camshaft ball bearing should spin smoothly with no rough spots or looseness. Bearing should be a light press fit on camshaft and in crankcase cover bore. It may be necessary to heat crankcase cover slightly to install ball bearing.

Lubricate tappets and camshaft with engine oil prior to installation. When installing camshaft, make certain camshaft gear and crankshaft gear timing marks are aligned as shown in Fig. KW311. On Model FG300, install balancer shaft in crankcase, aligning timing mark on balancer gear with timing mark on crankshaft gear.

PISTON, PIN AND RINGS. Piston and connecting rod are removed as an assembly after removing cylinder head and splitting crankcase. Remove connecting rod cap bolts and remove rod cap (8—Fig. KW312). Remove carbon and ring ridge, if present, from top of cylinder before removing piston. Push piston and connecting rod assembly out through top of cylinder block. Remove snap rings (4), push piston pin (5) out and separate piston from connecting rod if necessary.

Measure piston diameter at point 6 mm (0.24 in.) above bottom of skirt and 90° from piston pin bore center line. Minimum allowable piston diameter is 63.810 mm (2.2526 in.) for Model FG150, 70.810 mm (2.7878 in.) for

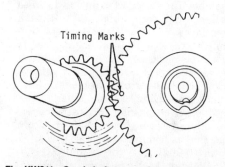

Fig. KW311—Crankshaft gear and camshaft gear timing marks "T" must be aligned during crankshaft or camshaft installation.

Model FG200 and 77.810 mm (3.0634 in.) for Model FG300. Clearance between piston and cylinder bore should not exceed 0.25 mm (0.010 in.). Renew piston if excessively worn, scored or damaged.

Maximum allowable piston pin bore inside diameter is 13.037 mm (0.5133 in.) for Model FG150, 15.026 mm (0.5916 in.) for Model FG200 and 16.038 mm (0.6314 in.) for Model FG300. Minimum allowable piston pin outside diameter is 12.986 mm (0.5113 in.) for Model FG150, 14.977 mm (0.5896 in.) and 15.987 mm (0.6294 in.) for Model FG300.

Maximum allowable piston ring groove width on Model FG150 is 2.120 mm (0.0835 in.) for top groove, 2.115 mm (0.0833 in.) for second groove and 3.06 mm (0.120 in.) for oil control ring groove. Maximum allowable piston ring groove width on Model FG200 is 2.120 mm (0.0835 in.) for top groove, 2.100 mm (0.0827 in.) for second groove and 3.06 mm (0.120 in.) for oil control ring groove. Model FG300 uses a tapered top ring and top ring groove width is difficult to measure. Maximum allowable width of second ring groove is 2.10 mm (0.0827 in.) and 4.06 mm (0.1598 in.) for oil control ring groove. Renew piston if width of any ring groove exceeds maximum specification.

Minimum allowable piston ring width on Model FG150 is 1.950 mm (0.0768 in.) for top compression ring and 1.955 mm (0.0770 in.) for second compression ring. Minimum allowable piston ring width on Model FG200 is 1.950 mm (0.0768 in.) for both top and second compression rings. Minimum allowable piston ring width for second compression ring on Model FG300 is 1.950 mm (0.0768 in.).

Maximum allowable piston ring end gap is 0.8 mm (0.031 in.) for Model FG150 or 1.0 mm (0.004 in.) for Models FG200 and FG300. If end gap is excessive, recondition cylinder bore for installation of oversize piston and rings.

When installing rings on piston, make sure that manufacturer's mark on face of ring is toward piston crown (Fig. KW313). On Models FG150 and FG200, note that top piston ring has a barrel face and second piston ring has a taper face.

Install piston on connecting rod so the "R" mark on top of piston is on side of connecting rod with Japanese characters. Install rod and piston assembly in cylinder block with numbered side (EC—Fig. KW312) of connecting rod toward pto end of crankshaft. Match marks (AM) on rod and cap must be aligned. Tighten connecting rod screws

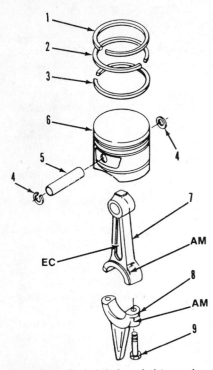

Fig. KW312—Exploded view of piston and connecting rod assembly. When assembling, be sure match marks (AM) on connecting rod and cap are aligned.

1. Compression ring
2. Compression ring
3. Oil control ring
4. Retaining ring
5. Piston pin
6. Piston
7. Connecting rod
8. Connecting rod cap
9. Connecting rod bolts

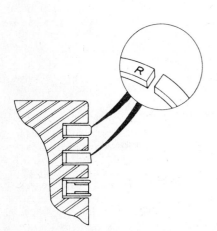

Fig. KW313—Top compression ring has a barrel face while second compression ring is tapered. Note that side toward piston crown is marked with a "R".

evenly to 23-24 N•m (17-18 ft.-lbs.) torque.

CONNECTING ROD. Piston and connecting rod are removed as an assembly as outlined in PISTON, PIN AND RINGS paragraphs.

The connecting rod rides directly on crankshaft crankpin journal. Inspect connecting rod and renew if piston pin

bore or crankpin bearing bore are scored or damaged. Refer to the following connecting rod wear limits.

Small end diameter (max.):
FG150 . . . . 13.043 mm
(0.5135 in.)
FG200 . . . . 15.027 mm
(0.5916 in.)
FG300 . . . . 16.046 mm
(0.6317 in.)

Big end diameter (max.):
FG150 . . . . 24.553 mm
(0.9667 in.)
FG200 . . . . 27.050 mm
(1.0650 in.)
FG300 . . . . 30.050 mm
(1.1831 in.)

Big end width (min.)
FG150 . . . . 23.30 mm
(0.917 in.)
FG200 . . . . 24.30 mm
(0.957 in.)
FG300 . . . . 26.10 mm
(1.0275 in.)

Assemble piston and connecting rod as outlined in PISTON, PIN AND RINGS paragraphs. Match marks (AM—Fig. KW312) on rod and cap must be aligned. Tighten connecting rod screws evenly to 23-24 N•m (17-18 ft.-lbs.) torque.

GOVERNOR. The internal centrifugal flyweight governor is driven by the camshaft gear. Refer to GOVERNOR paragraph in MAINTENANCE section for external governor adjustments.

To remove governor assembly, remove external linkage, shrouds and crankcase cover. Remove governor gear cover mounting screws and remove cover. Lift gear assembly from governor shaft.

To reinstall governor assembly, reverse removal procedure.

CRANKSHAFT AND MAIN BEARINGS. Crankshaft is support by ball bearings at each end. To remove crankshaft, remove fan housing, flywheel, cylinder head and crankcase cover. Remove connecting rod cap and remove piston and connecting rod assembly. Turn crankcase upside down so tappets fall away from camshaft, then withdraw camshaft and tappets. Remove crankshaft from crankcase. It may be necessary to heat crankcase cover and crankcase slightly to remove crankshaft bearings. Remove crankshaft seals using suitable puller.

Inspect crankshaft for wear, scoring or other damage. Crankshaft crankpin journal minimum diameter is 24.447 mm (0.9625 in.) for Model FG150, 26.950 mm (1.0610 in.) for Model FG200 and 29.950 mm (1.1791 in.) for Model FG300. On all models, crankshaft

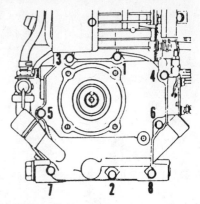

Fig. KW314—Tighten crankcase cover screws to 9-11 N•m (80-97 in.-lbs.) torque in sequence shown on Models FG150 and FG200.

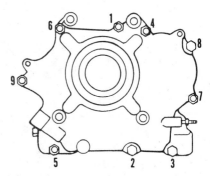

Fig. KW315—Tighten crankcase cover screws to 14 N•m (10.5 ft.-lbs.) torque in sequence shown on Model FG300.

runout should not exceed 0.05 mm (0.002 in.).

Ball bearing main bearings should be a light press fit on crankshaft and in bearing bores in crankcase and crankcase cover. Bearings should spin smoothly and have no rough spots. It may be necessary to heat crankcase or crankcase cover slightly to install bearings.

During assembly, crankshaft gear and camshaft gear timing marks must be aligned as shown in Fig. KW311. On Model FG300, timing mark on balancer shaft gear must be aligned with mark on crankshaft gear. When installing crankcase cover, tighten bolts evenly to specified torque following sequence shown in Fig. KW314 or Fig. KW315.

CYLINDER AND CRANKCASE. Cylinder and crankcase are an integral casting. Standard cylinder bore diameter is 63.98-64.00 mm (2.519-2.520 in.) for Model FG150, 70.98-71.00 mm (2.794-2.795 in.) for Model FG200 and 77.98-78.00 mm (3.070-3.071 in.). If cylinder diameter exceeds specified limit or if bore is out-of-round more than 0.05 mm (0.002 in.), recondition or renew cylinder. Bore cylinder to the next over-

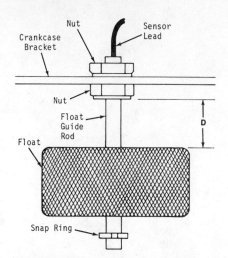

Fig. KW316—On models equipped with low oil sensor, sensor float and switch assembly is mounted on crankcase cover. Float gap (D) must be 9.5-15.5 mm (0.37-0.61 in.). Refer to text.

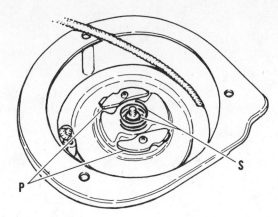

Fig. KW317—Install springs (S) and pawls (P) as shown.

size of 0.25, 0.50 or 0.75 mm. Bore cylinder to following diameter:

Model FG150

Piston Oversize	Bore Diameter
0.25 mm	64.23-64.25 mm (2.52875-2.5295 in.)
0.50 mm	64.48-64.50 mm (2.5386-2.5394 in.)
0.75 mm	64.73-64.75 mm (2.5485-2.5492 in.)

Model FG200

Piston Oversize	Bore Diameter
0.25 mm	71.23-71.25 mm (2.8043-2.8051 in.)
0.50 mm	71.48-71.50 mm (2.8142-2.8149 in.)
0.75 mm	71.73-71.75 mm (2.8240-2.8248 in.)

Model FG300

Piston Oversize	Bore Diameter
0.25 mm	78.23-78.25 mm (3.0799-3.0807 in.)
0.50 mm	78.48-78.50 mm (3.0898-3.0905 in.)
0.75 mm	78.73-78.75 mm (3.0996-3.1004 in.)

On Model FG200 equipped with a 2.7:1 reduction unit, note that two oil seals are used in the crankcase cover. The seals must be installed with faces abutting. The lip of the inner seal will be toward the inside of the crankcase cover while the lip of the outer seal will be out toward the reduction case.

When installing crankcase cover, tighten screws evenly to specified torque following sequence shown in Fig. KW314 or Fig. KW315.

LOW OIL SENSOR. Float type low oil sensor is located inside the crankcase (Fig. KW316). To remove or check float

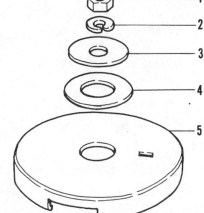

Fig. KW318—Exploded view of rewind starter nut (1), lockwasher (2), metal washer (3), plastic washer (4) and retainer (5).

gap (D), it is necessary to remove crankcase cover.

To check oil level sensor switch, remove bearing plate and float cover. Disconnect electrical leads and connect ohmmeter test leads to switch leads. Slide float to top of shaft. Ohmmeter should indicate infinite resistance. Slide float slowly down the shaft until ohmmeter just deflects. Gap between top of float and lower nut (Fig. KW316) should be 9.5-15.5 mm (0.37-0.61 in.). If not, renew switch.

REWIND STARTER. To disassemble starter, remove starter from engine. Remove rope handle and allow rope to wind slowly into starter. Unscrew retaining nut on spindle and remove cover. Remove pawls (P—Fig. KW317) and springs (S). Wear appropriate safety eyewear and gloves before disengaging pulley from starter as spring may uncoil uncontrolled. Place shop cloth around pulley and lift pulley out of housing; spring should remain in housing. If spring must be removed from housing, position housing so spring side

is down and against floor, then tap housing to dislodge spring.

Reassemble starter by reversing disassembly procedure while noting the following. Lightly grease sides of rewind spring and pulley. Install spring in housing so coil direction is counterclockwise from outer spring end. Wind rope around pulley in counterclockwise direction as viewed from pawl side of pulley. Position pulley in starter housing and install pawls and springs as shown in Fig. KW317. Install retainer, nut and washers as shown in Fig. KW318. Rotate pulley counterclockwise until spring tension is felt. Hold rope in notch and preload spring by turning pulley two turns counterclockwise, then pass rope through rope outlet in housing and install handle. Check starter operation.

6:1 GEAR REDUCTION. Models FG150 and FG200 may be equipped with the 6:1 gear reduction unit. Disassembly and reassembly are evident after inspection of unit and referral to exploded view in Fig. KW319. Clean dirt and rust from pto shaft and remove any burrs from shaft before removing outer case half (3). Heat cases to no more than 475° F (250° C) to ease removal and installation of bearings. Apply small amount of grease to lip of oil seal (1) before installing outer case, and be careful not to damage seal lip when case is installed over pto shaft.

2.7:1 CHAIN REDUCTION. Model FG200 may be equipped with a 2.7:1 chain reduction unit that uses a shoe type clutch to engage the pto shaft. Disassembly and reassembly are evident after inspection of unit and referral to exploded view in Fig. KW320. Clean dirt and rust from pto shaft and remove any burrs from shaft before removing outer case half (4).

Renew chain if chain slack measured midway between sprockets exceeds 9

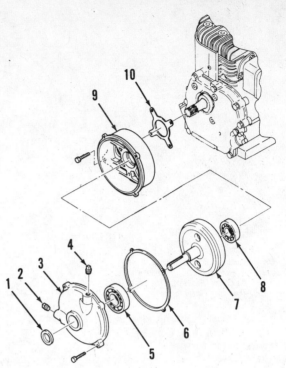

Fig. KW319—Exploded view of 6:1 reduction unit.

1. Oil seal
2. Oil level plug
3. Case half
4. Fill plug
5. Ball bearing
6. Gasket
7. Pto shaft/gear
8. Ball bearing
9. Case half
10. Gasket

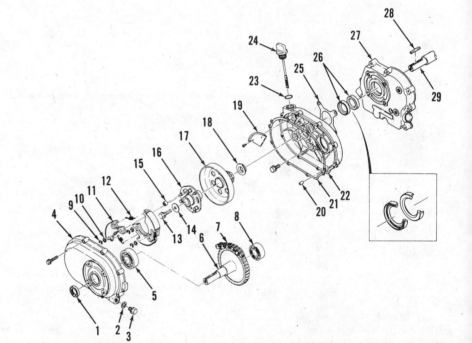

Fig. KW320—Exploded view of 2.7:1 reduction unit.

1. Oil seal
2. Gasket
3. Drain plug
4. Case half
5. Ball bearing
6. Pto shaft/sprocket
7. Chain
8. Ball bearing
9. Snap ring
10. Washer
11. Clutch shoe
12. Clutch spring
13. Screw
14. Washer
15. Damper
16. Clutch hub
17. Clutch drum
18. Collar
19. Cover
20. Dowel pin
21. Gasket
22. Case half
23. "O" ring
24. Dipstick
25. Gasket
26. Oil seals
27. Crankcase cover
28. Key
29. Crankshaft

mm (0.35 in.), measured midway between the sprockets. Inspect clutch components for excessive wear and damage. Clutch shoe contact surface can be reshaped with a file so surfaces of clutch shoe and drum conform. Renew clutch hub (16—Fig. KW318) if outer diameter is less than 25.955 mm (1.0218 in.). Renew clutch drum (17) if hub inside diameter is greater than 25.056 mm (0.9864 in.).

Heat cases to no more than 475° F (250° C) to ease removal and installation of bearings. Note that two oil seals (26) are used in the crankcase cover. The seals must be installed with faces abutting. The lip of the inner seal will be toward the inside of the crankcase cover while the lip of the outer seal will be out toward the reduction case. Apply small amount of grease to lip of oil seals (1 and 26) before installing case halves, and be careful not to damage seal lip when case is installed over the shafts.

KOHLER

KOHLER COMPANY
444 Highland Drive
Kohler, Wisconsin 53044

Model	Bore	Stroke	Displacement	Power Rating
K91	2.375 in.	2.0 in.	8.9 cu.in.	4 hp
	(60.3 mm)	(50.8 mm)	(145 cc)	(3.0 kW)

ENGINE IDENTIFICATION

Kohler engine identification and serial number decals are located on engine shrouding (Fig. KO1). The model number designates displacement (in cubic inches), digit after displacement on late models indicates number of cylinders and the letter suffix indicates the specific version. No suffix indicates a rope start model. Therefore, K91 would indicate a Kohler (K), 9 cubic inch (9), single-cylinder (1) engine with rope start (no suffix). Suffix interpretation is as follows:

A   Special oil pan
C   Clutch model
EP   Electric plant
G   Generator application
P   Pump model
Q   Quiet model
R   Gear reduction
S   Electric start
ST   Electric & retractable start
T   Retractable start

Engine model, specification and serial number are usually required when ordering parts.

MAINTENANCE

LUBRICATION. Periodically check oil level; do not overfill. If oil dipstick is threaded, rest dipstick plug on top of hole to check oil level; do not screw in dipstick. If dipstick is bayonet type, insert dipstick fully to check oil level. Change oil after first 5 hours of operation. Thereafter change oil after every 25 hours of operation. Oil should be drained while engine is warm. Oil required for oil change is approximately 1 pint (0.47 L). Dry oil capacity is approximately 1.5 pints (0.7 L). Check oil level on dipstick and do not overfill.

Engine oil should meet or exceed latest API service classification. Use SAE 30 oil for temperatures above 32°F (0°C). When operating in temperatures below 32°F (0°C), SAE 5W-30 may be used. Manufacturer recommends use of SAE 30 oil for first 5 hours of operation of overhauled engines or new short blocks, then change oil according to ambient temperature requirements.

If engine is equipped with a reduction unit, fill housing with engine oil to height of oil level plug on side of gearcase.

AIR FILTER. The engine is equipped with a paper type air filter (4—Fig. KO2) and may be equipped with a foam precleaner element (3). Service the air filter after every 100 hours of operation and the precleaner after every 25 hours of operation. Service more frequently if engine is operated in severe conditions.

The air filter should be renewed rather than cleaned. Do not wash or direct pressurized air at filter.

Clean precleaner element by washing in soapy water. Allow to dry then apply clean engine oil. Squeeze out excess oil.

FUEL FILTER. The engine may be equipped with a fuel filter below the fuel shut-off valve or an inline fuel filter. Periodically inspect fuel filter and clean or renew as needed.

CRANKCASE BREATHER. A crankcase breather is located in the valve cover on the side of the engine. A filter is located in the cover and a reed valve maintains vacuum in the crankcase. The breather filter should be cleaned periodically as needed. Also check the reed for proper operation.

SPARK PLUG. Recommended spark plug is Champion RC12YC or equivalent. Specified electrode gap is 0.025 inch (0.7 mm). Tighten spark plug to 18-22 ft.-lbs. (24.5- Nm).

NOTE: Manufacturer does not recommend spark plug cleaning using abrasive grit as grit may enter engine.

CARBURETOR. Initial adjustment of idle mixture screw (7—Fig. KO3) is 1½ turns out from a lightly seated position. Initial setting for high speed mixture screw (1) is 2 turns out from a lightly seated position. Perform carburetor adjustments with engine at normal operating temperature. Turn high speed mixture screw clockwise until engine begins to stumble and note screw position. Turn high speed mixture screw counterclockwise until engine begins to stumble and note screw position. Turn the high speed mixture screw clockwise to a position that is midway from the clockwise (lean) and counterclockwise (rich) positions. Run engine at idle and adjust idle speed screw so engine speed is 1200 rpm. Turn idle mixture screw clockwise until engine begins to stumble and note screw position. Turn idle

KOHLERengine
HP

Spec. no.

Model no.

Refer to owners manual for operation and maintenance instructions.

KOHLER COMPANY
KOHLER WISCONSIN USA

Serial No.

Fig. KO1—View showing name plate and serial number plate on Kohler engine. Refer to text.

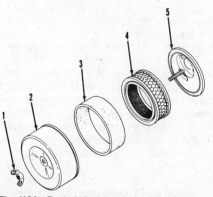

Fig. KO2—Exploded view of air cleaner. Foam precleaner (3) is optional.

1. Wing nut
2. Cover
3. Precleaner
4. Filter
5. Adapter

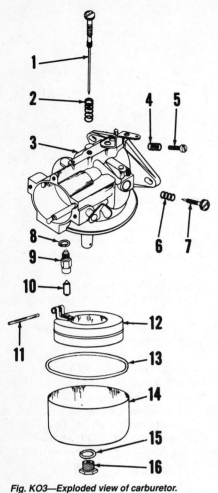

Fig. KO3—Exploded view of carburetor.

1. High speed mixture screw
2. Spring
3. Body assy.
4. Spring
5. Idle speed screw
6. Spring
7. Idle mixture screw
8. Sealing washer
9. Fuel inlet valve seat
10. Fuel inlet valve
11. Pin
12. Float
13. Gasket
14. Fuel bowl
15. Sealing washer
16. Retainer

mixture screw counterclockwise until engine begins to stumble and note screw position. Turn the idle mixture screw clockwise to a position that is midway from the clockwise (lean) and counterclockwise (rich) positions. With engine running at idle, rapidly move speed control to full throttle position. If engine stumbles or hesitates, slightly turn idle mixture screw counterclockwise and repeat test. Recheck idle speed and, if necessary, readjust idle speed screw.

To check float level, invert carburetor body and float assembly. There should be $^{11}\!/_{64}$ inch (4 mm) clearance between machined surface of body casting and free end of float as shown in upper view of Fig. KO4. Adjust as necessary by bending float lever tang. Measure float drop between machined surface of body and bottom of float free end as shown in middle view of Fig. KO4. Float drop should be $1^{1}\!/_{32}$ inches (26 mm). Float-to-

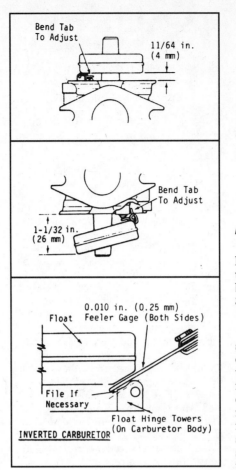

Fig. KO4—View of float adjustment procedure. Refer to text.

float hinge tower clearance should be 0.010 inch (0.25 mm) as shown in lower view of Fig. KO4. File float hinge tower as needed to obtain recommended clearance.

Throttle and choke plates and shafts should not be removed unless renewal is required. Note that a detent ball and spring are located behind the choke shaft and will be loose when the shaft is withdrawn.

A bushing kit is available to repair excessively worn shaft bores in carburetor body. Proceed as follows to install bushings. Accurately position carburetor in a drill press so drill is aligned with centerline of throttle or choke shaft bore. Run a $^{7}\!/_{32}$-inch drill through shaft bore, then ream bore with a $^{5}\!/_{16}$-inch reamer. Clean bore. If installing choke shaft bushings, thread two screws until lightly seated in threaded holes on face of carburetor (the screws act as stops for the bushings). Apply Loctite in repair kit to outside diameter of either throttle or choke shaft bushings. Push choke shaft bushings into carburetor body until bushings bottom against screws. Install throttle shaft bushings by pushing bushings in with the round driver in the

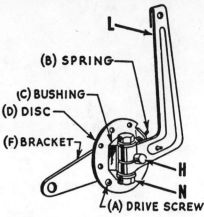

Fig. KO5—Points of governor adjustment.

kit until the driver bottoms. Insert throttle or choke shaft and check for binding. Allow Loctite to set for at least 30 minutes.

GOVERNOR. To adjust governed speed, first synchronize linkage by loosening clamp bolt nut (N—Fig. KO5), turn shaft (H) counterclockwise until internal resistance is felt. Pull arm (L) completely to the left (away from carburetor) and tighten the clamp bolt nut. To increase or decrease maximum engine speed, vary the tension of governor spring (B). On engine with remote throttle control, spring (B) tension is varied by moving bracket (F) up or down. On engines without remote throttle control, rotate disc (D) after loosening bushing (C) to vary spring (B) tension.

IGNITION SYSTEM. The engine is equipped with a magneto type ignition system. Breaker points are contained under a cover on the side of the engine. Breaker point gap should be 0.020 inch (0.51 mm).

Ignition timing may be checked with a timing light or continuity checker. Timing marks on the flywheel are visible through a port in the bearing plate. Ignition (breaker points open) should occur when the S or SP mark on flywheel is aligned with mark on bearing plate. If checking ignition timing with engine running, engine speed should be 1200-1800 rpm. Vary point gap to adjust ignition timing.

VALVE ADJUSTMENT. Valve stem-to-tappet clearance should be checked after every 500 hours of engine operation. The intake valve tappet gap should be 0.005-0.009 inch (0.13-0.23 mm). The exhaust valve tappet gap should be 0.011-0.015 inch (0.28-0.38 mm). Valve tappet clearance is adjusted by carefully grinding end of valve stem to increase clearance or by grinding valve seats deeper and/or renewing valve or lifter to decrease clearance.

CYLINDER HEAD. After 500 hours of engine operation, the cylinder head should be removed and any carbon or deposits should be removed.

REPAIRS

TIGHTENING TORQUES. Recommended tightening torques are as follows:

Bearing plate............ 115 in.-lbs.
(13 N•m).
Connecting rod .... 140 in.-lbs.*
(15.8 N•m).
Cylinder head .... 200 in.-lbs.*
(22.6 N•m).
Flywheel nut .... 40-50 ft.-lbs.
(54-68 N•m)
Oil pan .... 250 in.-lbs.
(28.2 N•m)
Spark plug............ 18-22 ft.-lbs.
(24-30 N•m)
*With threads lubricated lightly.

FLYWHEEL. A suitable puller should be used to remove flywheel. Striking the flywheel may damage the crankshaft. Tighten flywheel nut to 40-50 ft.-lbs. (54-68 N•m).

WARNING: Do not attempt to remove ignition magnet on flywheel. Unscrewing magnet retaining screws may allow magnet to be thrown from flywheel during operation.

CYLINDER HEAD. To remove cylinder head, first remove all necessary metal shrouds. Clean engine to prevent entrance of foreign material and remove cylinder head retaining screws.

Renew cylinder head if warpage exceeds 0.003 inch (0.08 mm). Always use a new head gasket when installing cylinder head. Tighten cylinder head screws evenly and in graduated steps using the sequence shown in Fig. KO6.

VALVE SYSTEM. The exhaust valve seats on a renewable seat insert and the intake valve seat is machined directly into cylinder block surface. Valve face and valve seat angles are 45°. Valve seat width for both valves should be 0.037-0.045 inch (0.94-1.14 mm).

Fig. KO6—Tighten cylinder head screws to 200 in.-lbs. (22.6 N•m) in sequence shown.

Specified valve stem diameter is 0.2480-0.2485 inch (6.299-6.312 mm) for intake valve and 0.2460-0.2465 inch (6.248-6.261 mm) for exhaust valve. Maximum allowable valve stem clearance is 0.005 inch (0.13 mm) for intake valve and 0.007 inch (0.18 mm) for exhaust valve.

PISTON, PIN AND RINGS. The piston and connecting rod are removed as an assembly after cylinder head and oil pan have been removed. Unscrew connecting rod cap screws and push piston and rod out top of cylinder. Detach snap rings (9—Fig. KO7), push out piston pin and separate piston and rod.

Measure piston diameter just below oil control ring groove and perpendicular to piston pin. Standard piston diameter is 2.369-2.371 inches (60.173-60.223 mm). Minimum allowable standard piston diameter is 2.366 inches (60.096 mm). Piston clearance should be 0.0035-0.0060 inch (0.089-0.152 mm). Oversize pistons and rings are available.

NOTE: Some factory engines are equipped with a 0.003 inch oversize piston. These engines are stamped ".003" on the cylinder head mating surface. Do not install a standard size piston. Standard size piston rings are used on 0.003 inch oversize pistons.

Piston pin fit in piston pin bore should be 0.0002 inch (0.005 mm) interference to 0.0002 inch (0.005 mm) loose. Piston pin diameter should be 0.5623-0.5625 inch (14.282-14.288 mm). Heating piston will aid removal and installation of piston pin.

Maximum allowable piston ring side clearance is 0.006 inch (0.15 mm). Piston ring end gap should be 0.007-0.017

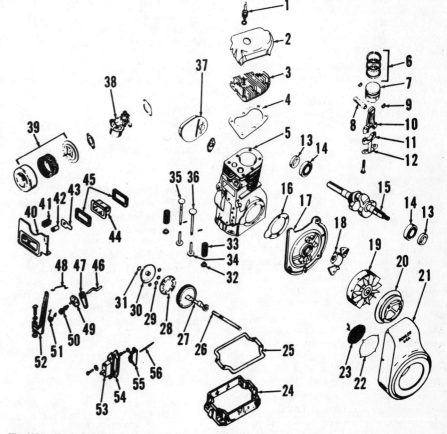

Fig. KO7—Exploded view of engine. Breaker points (55) are actuated by cam on right end of camshaft (27) through push rod (56).

1. Spark plug	15. Crankshaft	29. Steel balls	43. Breather reed
2. Air baffle	16. Gasket	30. Thrust cone	44. Breather plate
3. Cylinder head	17. Bearing plate	31. Snap ring	45. Gaskets
4. Head gasket	18. Magneto	32. Spring retainer	46. Governor shaft
5. Cylinder block	19. Flywheel	33. Valve spring	47. Bracket
6. Piston rings	20. Pulley	34. Valve tappets	48. Link
7. Piston	21. Shroud	35. Exhaust valve	49. Speed disc
8. Piston pin	22. Screen retainer	36. Intake valve	50. Bushing
9. Retaining rings	23. Screen	37. Muffler	51. Governor spring
10. Connecting rod	24. Oil pan	38. Carburetor	52. Governor lever
11. Rod cap	25. Gasket	39. Air cleaner assy.	53. Breaker cover
12. Rod bolt lock	26. Camshaft pin	40. Valve cover	54. Gasket
13. Oil seal	27. Camshaft	41. Filter	55. Breaker points
14. Ball bearing	28. Flyball retainer	42. Breather seal	56. Push rod

inch (0.18-0.43 mm) in a new bore. Maximum allowable ring end gap is 0.027 inch (0.68 mm).

Install piston rings with marked side toward piston crown. Follow instructions included with piston rings. Stagger ring end gaps evenly around piston circumference.

Piston may be installed on connecting rod either way. Install piston and connecting rod in engine so match marks on rod and cap are aligned and marks are toward flywheel end of crankshaft. Tighten connecting rod screws to 140 in.-lbs. (15.8 N·m).

CONNECTING ROD. The aluminum alloy connecting rod rides directly on the crankpin journal. Connecting rod and piston are removed as a unit as outlined in PISTON, PIN AND RINGS section.

Connecting rod small end diameter should be 0.5630-0.5633 inch (14.300-14.308 mm). Piston pin clearance in rod should be 0.0005-0.0010 inch (0.013-0.025 mm).

Clearance between rod and crankpin should be 0.0010-0.0025 inch (0.025-0.063 mm). Maximum allowable clearance is 0.003 inch (0.076 mm).

CRANKSHAFT, MAIN BEARINGS AND SEALS. The crankshaft is supported at each end by a ball bearing type main bearing (14—Fig. KO7). Main bearings should be a light press fit in crankcase and bearing plate bores. Renew bearings if rough or loose.

Specified crankpin journal diameter is 0.9355-0.9360 inch (23.762-23.774 mm) and minimum diameter is 0.9350 inch (23.75 mm). Maximum crankpin journal out-of-round is 0.0005 inch (0.013 mm) and maximum journal taper is 0.001 inch (0.025 mm).

Before installing crankshaft, camshaft must be installed. Use a press to install crankshaft in bearing in crankcase. Make certain crankshaft gear and camshaft gear timing marks are aligned (Fig. KO8). Fabricate alignment studs when installing bearing plate. Install required shim gaskets (16—Fig. KO7). Tighten bearing plate screws evenly to pull bearing plate onto crankshaft. Tighten bearing plate screws to 115 in.-lbs. (13 N·m). Crankshaft end play should be 0.004-0.023 inch (0.10-0.58 mm). Adjust end play by varying number and thickness of shim gaskets (16).

Front and rear crankshaft oil seals should be pressed into seal bores so outside edge of seal is $\frac{1}{32}$ inch (0.80 mm) below seal bore surface.

CAMSHAFT AND BEARINGS. The hollow camshaft and integral cam gear (27—Fig. KO7) rotate on pin (26).

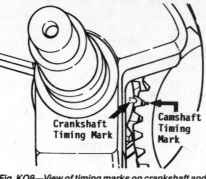

Fig. KO8—View of timing marks on crankshaft and camshaft gears.

Camshaft can be removed after removing bearing plate (17) and crankshaft, then drive pin (26) out toward bearing plate side of crankcase. When reinstalling camshaft, make certain camshaft gear and crankshaft gear timing marks are aligned (Fig. KO8).

Camshaft pin (26) is a press fit in closed (crankcase) side and a slip fit with 0.0005-0.0012 inch (0.013-0.030 mm) clearance in bearing plate side.

Camshaft-to-camshaft pin clearance should be 0.001-0.0025 inch (0.025-0.063 mm). Camshaft end play should be 0.005-0.020 inch (0.13-0.50 mm) and is controlled by varying number and thickness of shim washers between camshaft and bearing plate side of crankcase.

Governor flyball retainer (28), flyballs (29), thrust cone (30) and snap ring (31) are attached to, and rotate with, the camshaft assembly.

CYLINDER. Standard cylinder bore diameter is 2.3745-2.3755 inches (60.312-60.338 mm). Maximum allowable cylinder bore diameter is 2.378 inches (60.40 mm). Maximum allowable cylinder taper or out-of-round is 0.003 inch (0.076 mm). Oversize pistons are available.

REWIND STARTER. To disassemble starter, remove rope handle and allow rope to wind into starter housing. Detach "E" ring (1—Fig. KO9) then remove engagement components (2 thru 7) shown in Fig. KO9. Wear appropriate safety eyewear and gloves before disengaging pulley from starter as spring may uncoil uncontrolled. Place shop cloth around pulley and lift pulley out of housing; spring should remain in housing. If spring must be removed from housing, position housing so spring side is down and against floor, then tap housing to dislodge spring.

When reassembling unit, lubricate rewind spring, cover shaft and its bore in rotor with Lubriplate or equivalent. Install rewind spring in housing so spring is coiled in a counterclockwise direction from outer end. Install rope on pulley in a counterclockwise direction as viewed from engagement side of pulley. Check friction shoe ends and renew if necessary. Install friction washers, friction shoe assembly, brake spring, retainer washer and retainer ring. Make certain friction shoe assembly is installed properly for correct starter rotation. If properly installed, sharp ends of friction shoe plates will extend when rope is pulled. Pass rope through rope outlet and install rope handle. To apply spring tension to pulley, pull a loop of rope into notch in pulley and rotate pulley counterclockwise a couple of turns, then pull rope out of notch and allow rope to wind into starter housing. Rope handle should be snug against housing, if not, increase spring tension by rotating pulley another turn. Check for excessive spring tension by pulling rope to fully extended length. With rope fully extended it should be possible to rotate pulley at least ½ turn counterclockwise, if not reduce spring tension one turn.

REDUCTION UNIT. Some engines may be equipped with a reduction unit attached to the crankcase. The drive gear is an integral part of the engine

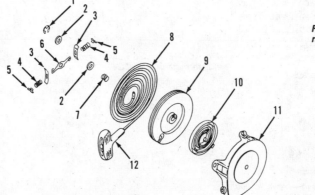

Fig. KO9—Exploded view of rewind starter.

1. "E" ring
2. Friction washers
3. Friction shoes
4. Retainers
5. Springs
6. Brake lever
7. Spring
8. Rope
9. Pulley
10. Rewind spring
11. Starter housing
12. Rope handle

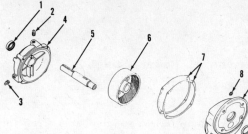

Fig. KO10—Exploded view of reduction unit.

1. Oil seal
2. Fill plug
3. Oil level plug
4. Cover
5. Output shaft
6. Ring gear
7. Gaskets
8. Copper washers
9. Housing
10. Expansion plug
11. Oil seal

crankshaft. The driven gear (6—Fig. KO10) is pressed on the output shaft (5) which is supported by bushings in the housing (9) and cover (4). Oil seals are located in the housing and cover.

Disassembly and reassembly is evident after inspection and referral to Fig. KO10. Note that copper washers are used on the two internal cap screws. Adjust output shaft end play to 0.001-0.006 inch (0.02-0.15 mm) by varying number of cover gaskets (7).

Fill housing with engine oil to height of oil level plug (3).

KOHLER

Model	Bore	Stroke	Displacement	Power Rating
C5	67 mm	51 mm	180cc	3.7 kW
	(2.64 in.)	(2.01 in.)	10.98 cu. in.)	(5 hp)

NOTE: Metric fasteners are used throughout engine.

The Kohler C5 engine is a 5 horsepower, four-stroke, air-cooled engine using an overhead valve system. Refer to preceding Kohler section for engine identification information.

MAINTENANCE

LUBRICATION. Periodically check oil level; do not overfill. Oil dipstick should be resting on tube to check oil level; do not screw in dipstick. Change oil after first 5 hours of operation. Thereafter change oil after every 100 hours. Oil capacity is 0.66 liter (0.7 qt.).

Manufacturer recommends using oil with an API service classification of SF or SG. Use 10W-30 or 10W-40 oil for temperatures above 0° F (–18° C); below 32° F (0° C) SAE 5W-20 or 5W-30 may be used. Manufacturer recommends use of SAE 10W-30 API SF oil for first 5 hours of operation of overhauled engines, then change oil according to ambient temperature requirements.

The engine may be equipped with a low-oil sensor. The sensor circuit may be designed to stop engine or trigger a warning device if oil level is low.

AIR FILTER. The engine is equipped with a foam precleaner element and paper type air filter. Service the precleaner after every 25 hours of operation and the air filter after every 100 hours of operation. Service more frequently if engine is operated in severe conditions.

Clean precleaner element by washing in soapy water. Allow to dry then apply clean engine oil. Squeeze out excess oil.

The air filter should be renewed rather than cleaned. Do not wash or direct pressurized air at filter.

FUEL FILTER. If so equipped, periodically inspect fuel filter. If dirty or damaged, renew filter.

BREATHER. On some engines, a breather is attached to the top of the cylinder head under the rocker cover. The breather tube is connected to the air cleaner base. Breather must be renewed as a unit assembly. Clean unit as needed to prevent or remove restrictions.

SPARK PLUG. Recommended spark plug is Champion RC12YC or equivalent. Specified electrode gap is 0.76 mm (0.030 in.). Tighten spark plug to 25-30 N•m (18-22 ft.-lbs.).

NOTE: Manufacturer does not recommend spark plug cleaning using abrasive grit as grit may enter engine.

CARBURETOR. Initial setting of idle mixture screw (M—Fig. KO40) is one turn out from lightly seated. Final adjustment of idle mixture screw should be made with engine at normal operating temperature. Adjust idle speed screw (S) so engine idles at 1200 rpm. Turn idle mixture screw (M) counterclockwise until engine rpm decreases and note screw position. Turn screw clockwise until engine rpm decreases again and note screw position. Turn screw to midpoint between the two noted positions. Reset idle speed screw (M) if necessary to obtain idle speed of 1200 rpm, or to equipment manufacturer's specification.

High speed mixture is controlled by a main jet. No optional jets are offered, although a high altitude kit is available.

To disassemble carburetor, refer to Fig. KO41. The edges of throttle and choke plates (4 and 8) are beveled and must be reinstalled in their original positions. Mark choke and throttle plates before removal to ensure correct reassembly. Use a suitably sized screw to pull out the fuel inlet seat (10) and discard seat. Use a sharp punch to pierce Welch plug and pry plug from carburetor body.

Clean all parts in suitable carburetor cleaner and blow out all passages with compressed air. Be careful not to enlarge any fuel passages or jets as calibration of carburetor may be altered.

Press new fuel inlet seat into carburetor body so seat is bottomed. Apply Loctite to throttle plate retaining screw. Be sure throttle plate is properly seated against carburetor bore before tightening screw. Be sure choke shaft properly

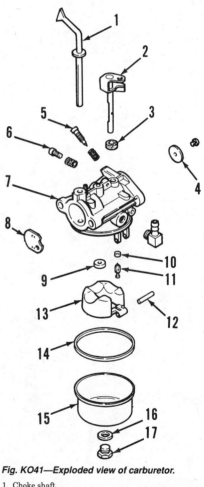

Fig. KO41—Exploded view of carburetor.

1. Choke shaft
2. Throttle shaft
3. Seal
4. Throttle plate
5. Idle mixture screw
6. Idle speed screw
7. Carburetor body
8. Choke plate
9. Main jet
10. Fuel inlet valve seat
11. Fuel inlet valve
12. Pin
13. Float
14. Gasket
15. Fuel bowl
16. Gasket
17. Screw

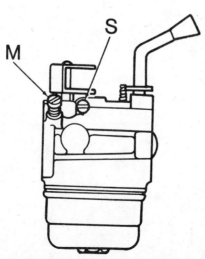

Fig. KO40—Drawing showing location of carburetor idle mixture screw (M) and idle speed screw (S). Refer to text for adjustment.

Illustrations courtesy Kohler Co.

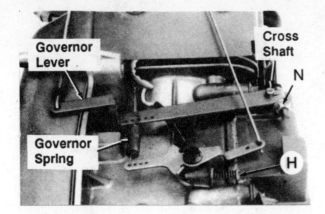

Fig. KO42—View of governor linkage. Refer to text for adjustment.

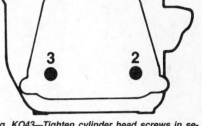

Fig. KO43—Tighten cylinder head screws in sequence shown above.

engages detent spring on carburetor. Locking tabs on choke plate must straddle choke shaft. Use a suitable sealant on Welch plug.

IGNITION. The engine is equipped with a breakerless, electronic magneto ignition system. The electronic ignition module is mounted outside the flywheel. The ignition switch grounds the module to stop the engine. There is no periodic maintenance or adjustment required with this ignition system.

Air gap between module and flywheel should be 0.20-0.30 mm (0.008-0.012 in.). Loosen module retaining screws and position module to obtain desired gap. Tighten screws to 4 N•m (35 in.-lbs.) for used engines or to 6.2 N•m (55 in.-lbs.) on a new engine cylinder block.

If ignition module fails to produce a spark, check for faulty kill switch or grounded wires. Measure resistance of ignition module secondary using suitable ohmmeter. Connect one test lead to spark plug terminal of high tension wire and other test lead to module core laminations. Resistance should be 7900-10850 ohms. If resistance is low or infinite, renew module.

GOVERNOR. A flyweight type governor is located in the crankcase. The governor gear is driven by the camshaft gear. Refer to REPAIRS section for overhaul information.

To adjust governor linkage, proceed as follows: Loosen clamp nut (N—Fig. KO42) and push governor lever so throttle is wide open. Turn governor cross shaft counterclockwise as far as possible and tighten clamp nut. Turn high speed stop screw (H) so maximum no-load speed is 3600 rpm, or as specified by equipment manufacturer.

Governor sensitivity is determined by location of governor spring in one of five holes in governor lever. Attaching the spring to a hole farther from cross shaft increases governor sensitivity.

VALVE ADJUSTMENT. To adjust rocker arm-to-valve stem clearance, engine must be cold. Remove rocker arm cover and breather assembly. Rotate flywheel so piston is at top dead center on compression stroke. Turn rocker arm adjusting nut so clearance is 0.038-0.051 mm (0.0015-0.0020 in.) for either valve.

REPAIRS

TIGHTENING TORQUES. Recommended tightening torques are as follows:

Connecting rod	9.0 N•m
	(80 in.-lbs.)
Crankcase cover	22.6 N•m
	(200 in.-lbs.)
Cylinder head	22.6 N•m
	(200 in.-lbs.)
Flywheel	68 N•m
	(50 ft.-lbs.)
Rocker arm cover	3.4 N•m
	(30 in.-lbs.)
Spark plug	25-30 N•m
	(18-22 ft.-lbs.)

FUEL PUMP. Some engines may be equipped with a pulse-operated diaphragm type fuel pump. Individual components are not available; pump must be renewed as a unit assembly.

CYLINDER HEAD. To remove cylinder head, remove air cleaner assembly and base. Remove fuel tank and detach speed control rod. Remove carburetor and muffler. Remove air baffles and shields. Remove rocker arm cover and breather assembly. Rotate crankshaft so piston is at top dead center on compression stroke. Push rods, rocker arms and pivot nuts should be marked so they can be reinstalled in their original position. Loosen rocker arm nuts and remove push rods. Unscrew cylinder head bolts and remove cylinder head and gasket.

Clean combustion deposits from cylinder head and inspect for cracks or other damage. Maximum allowable

warpage of head surface is 0.076 mm (0.003 in.).

Reverse removal procedure to reinstall head. Tighten cylinder head screws to 22.6 N•m (200 in.-lbs.) using sequence shown in Fig. KO43. Adjust valve clearance as previously outlined. Tighten rocker arm cover screws to 3.4 N•m (30 in.-lbs.).

VALVE SYSTEM. See Fig. KO44 for exploded view of valve components. Valve face and seat angles are 45°. Minimum allowable valve margin is 1.0 mm (0.040 inch). Specified valve stem-to-guide clearance is 0.039-0.075 mm (0.0015-0.0029 in.) for intake valve and 0.061-0.099 mm (0.0024-0.0039 in.) for exhaust valve. Specified new valve guide diameter for either valve is 4.999-5.010 mm (0.1968-0.1972 in.). Minimum allowable valve stem diameter is 4.934 mm (0.1943 in.) for intake valve and 4.911 mm (0.1933 in.) for exhaust valve. Maximum allowable valve guide inside diameter is 5.08 mm (0.200 in.) for either valve. Valve guides are not renewable, however, guides may be reamed to accept valves with 0.25 mm oversize stem.

PISTON, PIN, RINGS AND ROD. To remove piston and rod, remove cylinder head as previously outlined. Remove crankcase cover screws. Use a suitable tool and pry off crankcase cover by inserting tool between pry tabs (Fig. KO45). Rotate crankshaft so timing marks (Fig. KO46) on crankshaft and camshaft gears are aligned. Remove camshaft. Either secure tappets in their bores or remove and mark them so they can be returned to original position. Detach rod cap and remove piston and connecting rod. Remove piston pin retaining rings and separate piston and rod.

To determine piston clearance in cylinder, measure piston diameter perpendicular to piston pin and 6 mm (0.24 in.) from bottom of skirt. Measure cylinder

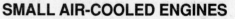

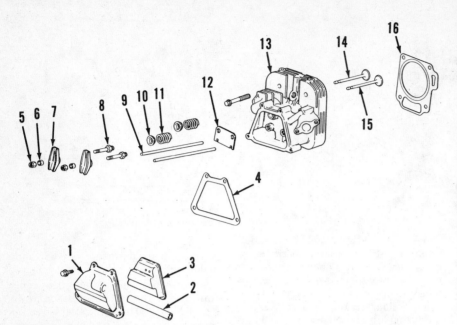

Fig. KO44—Exploded view of cylinder head and valve system.

1. Rocker arm cover	5. Locknut
2. Breather hose	6. Pivot ball
3. Breather assy.	7. Rocker arm
4. Gasket	8. Stud

9. Push rod	13. Cylinder head
10. Valve spring retainer	14. Exhaust valve
11. Valve spring	15. Intake valve
12. Push rod guide plate	16. Gasket

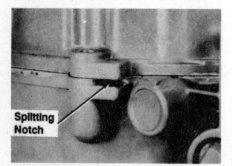

Fig. KO45—Insert a suitable prying tool between tabs on crankcase and cover to dislodge cover.

bore inside diameter at point of greatest wear, approximately 40 mm (1.6 in.) below top of cylinder and perpendicular to piston pin. Piston clearance should be 0.016-0.059 mm (0.0006-0.0023 in.). Piston and rings are available in standard size and oversizes of 0.35 and 0.50 mm (0.014 and 0.020 in.).

Specified piston pin bore is 14.006-14.014 mm (0.5514-0.5517 in.). Specified piston pin diameter is 13.996-14.000 mm (0.5510-0.5512 in.). Piston-to-piston pin clearance should be 0.005-0.018 mm (0.0002-0.0007 in.). Specified connecting rod small end diameter is 14.015-14.023 mm (0.5518-0.5520 in.) with a wear limit of 14.036 mm (0.5526 in.).

Connecting rod bearing clearance should be 0.030-0.056 mm (0.0011-0.0022 in.). Maximum allowable clearance is 0.0635 mm (0.0025 in.). A connecting rod with 0.25 mm undersize big end is available. Specified rod side clearance on crankpin is 0.431-0.661 mm (0.017-0.026 in.).

Insert new rings in piston ring grooves and measure ring side clearance using a feeler gauge. Piston ring side clearance should be 0.040-0.085 mm (0.0016-0.0033 in.) for top compression ring; 0.040-0.072 mm (0.0016-0.0028 in.) for second compression ring; 0.140-0.275 mm (0.0055-0.0108 in.) for oil ring. Renew piston if side clearance is excessive.

Specified piston ring end gap for compression rings is 0.25-0.45 mm (0.010-0.018 in.). Maximum allowable ring end gap in a used cylinder is 0.75 mm (0.030 in.).

When mating piston and rod, numbers on side of rod must be opposite "FLY" embossed in piston crown as shown in Fig. KO48. Install piston rings so "TOP" side is towards piston crown. Top compression ring has a round face and a blue stripe on face. Second compression ring has a bevel face and a pink stripe on face. See Fig. KO49. Install piston and rod so "FLY" and arrow on piston crown are towards flywheel as shown in Fig. KO50. Match marks (Fig. KO51) on rod and rod cap must be aligned. Tighten rod cap screws to 9.0 N•m (80 in.-lbs.).

Reverse disassembly procedure to reinstall components. Align timing marks (Fig. KO46) on crankshaft and camshaft gears as shown. No gasket is used with crankcase cover; apply silicone gasket compound to mating surface of cover. Tighten crankcase cover screws to 22.6 N•m (200 in.-lbs.) using sequence shown in Fig. KO52. Install cylinder head as previously outlined.

Fig. KO46—View of timing marks on crankshaft and camshaft gears.

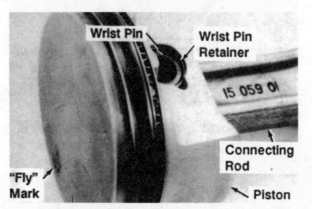

Fig. KO48—Assemble rod and piston so numbers on side of rod are opposite "FLY" on piston crown.

Illustrations courtesy Kohler Co.

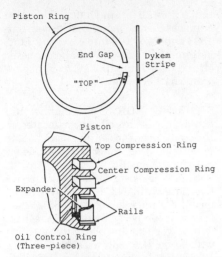

Fig. KO49—Drawing showing configuration and location of piston rings on piston.

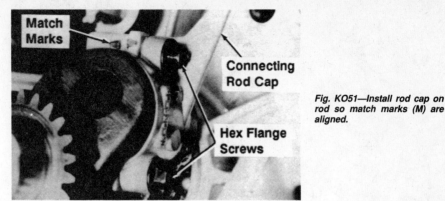

Fig. KO51—Install rod cap on rod so match marks (M) are aligned.

CAMSHAFT. To remove camshaft, rotate crankshaft so piston is at top dead center on compression stroke. Remove rocker cover and breather assembly. Loosen rocker arms, then remove push rods while marking them so they can be returned to original position. Detach governor lever from governor cross shaft. Remove oil fill tube. Remove crankcase cover screws. Use a suitable tool and pry off crankcase cover by inserting tool between pry tabs (Fig. KO45). Rotate crankshaft so timing marks (Fig. KO46) on crankshaft and camshaft gears are aligned. Remove camshaft. Either secure tappets in their bores or remove and mark them so they can be returned to original position.

Inspect camshaft and tappets for damage and excessive wear. Minimum cam lobe height for both lobes is 5.40 mm (0.213 in.). Bearing journal diameter should be 15.954 mm (0.6281 in.) at crankcase end and 25.350 mm (0.9980 in.) at cover end.

The camshaft is equipped with a compression reduction device to aid starting. The lever and weight mechanism on the camshaft gear moves a pin inside the exhaust cam lobe. During starting

Arrow Must Point Towards Flywheel

the pin protrudes above the cam lobe and forces the exhaust valve to stay open longer thereby reducing compression. At running speeds the pin remains below the surface of the cam lobe. Inspect mechanism for proper operation.

Reverse disassembly procedure to reinstall components. Align timing marks (Fig. KO46) on crankshaft and camshaft gears as shown. No gasket is used with crankcase cover; apply silicone gasket compound to mating surface of cover. Tighten crankcase cover screws to 22.6 N·m (200 in.-lbs.) using sequence shown in Fig. KO52. Adjust valve clearance as previously outlined. Adjust governor as previously outlined.

GOVERNOR. The engine is equipped with a flyweight mechanism mounted on governor gear (9--Fig. KO53). Remove crankcase cover (1) for access to governor gear. Inspect gear assembly for excess wear and damage. Gear and flyweight assembly are available only as a unit assembly. Detach snap ring (4) to remove governor shaft (6). On some engines the governor shaft rides in needle bearings (7).

No gasket is used with crankcase cover; apply silicone gasket compound to mating surface of cover. Tighten crankcase cover screws to 22.6 N·m (200 in.-lbs.) following sequence shown in Fig. KO52. Adjust governor as pre-

Fig. KO50—Install piston and rod assembly so "FLY" and arrow on piston crown are towards flywheel.

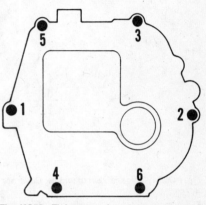

Fig. KO52—Tighten crankcase cover screws in sequence shown above.

viously outlined in MAINTENANCE section.

CRANKSHAFT. The crankshaft is supported by a ball bearing at both ends. To remove crankshaft, remove starter and flywheel. Remove crankcase cover, piston, connecting rod and camshaft as previously outlined. Remove crankshaft.

Specified crankpin diameter is 30.947-30.960 mm (1.2184-1.2189 in.). Minimum allowable crankpin diameter is 30.934 mm (1.2179 in.). Maximum allowable crankpin taper is 0.025 mm (0.0010 in.). Maximum allowable crankpin out-of-round is 0.013 mm (0.0005 in.). Crankpin may be ground to accept a connecting rod that is 0.25 mm (0.010 in.) undersize. Maximum allowable crankshaft runout is 0.10 mm (0.004 in.) measured at pto end.

CYLINDER/CRANKCASE. Cylinder bore standard diameter is 67.000-67.030 (2.6378-2.6390 in.). Maximum bore out-of-round is 0.15 mm (0.006 in.). Maximum bore taper is 0.10 mm (0.004 in.). Cylinder may be bored to accept an oversize piston.

Install crankshaft oil seal in crankcase using seal driver KO-1034 and force seal into crankcase until tool bot-

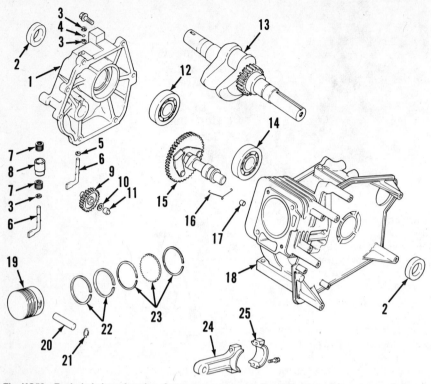

Fig. KO53—Exploded view of engine. Governor components (7 and 8) are used on some engines.

1. Crankcase cover
2. Crankshaft seals
3. Washer
4. Snap ring
5. Washer
6. Governor shaft
7. Needle bearings
8. Bushing
9. Governor gear
10. Spacer
11. Stub shaft
12. Main bearing
13. Crankshaft
14. Main bearing
15. Camshaft assy.
16. Compression release spring
17. Dowel
18. Crankcase
19. Piston
20. Piston pin
21. Snap ring
22. Compression rings
23. Oil ring
24. Connecting rod
25. Rod cap

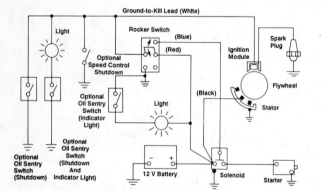

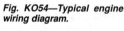

Fig. KO54—Typical engine wiring diagram.

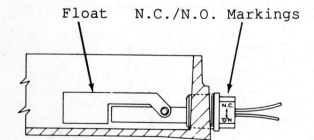

Fig. KO55—Drawing of oil sensor in installed position. Note location of "N.C./N.O." letters and flats on fitting.

toms. Install crankshaft oil seal in crankcase cover using seal driver KO-1043 and force seal into cover until tool bottoms. Oil seal in crankcase cover should be 5-7 mm (0.20-0.28 in.) below crankshaft bore lip.

OIL SENSOR. Some engines are equipped with an oil sensor located in the crankcase. The sensor uses a float to monitor oil level. Refer to Fig. KO54 for a wiring schematic.

Note that sensor threaded fitting has two flats with "N.C/N.O." lettering on one of the flats (see Fig. KO55). When sensor is properly installed, the flats will be vertical and flat with lettering will be on left side as shown in Fig. KO55.

Sensor can be removed without removing crankcase cover. Drain engine oil. Turn fitting ¼ turn so flats shown in Fig. KO55 are horizontal. Turn fitting counterclockwise ½ turn in a continuous motion and stop so flats are again horizontal. Continue rotating half turns and stopping when flats are horizontal until fitting is loose. Unscrew by hand to determine if float is striking crankcase. Do not pause when turning as float will strike crankcase. If float strikes crankcase, turn fitting clockwise so flats and letters are as shown in Fig. KO55, then resume removal procedure.

To install sensor, apply Teflon sealer to threads and lubricate float mechanism. Insert assembly so flats and letters on fitting are as shown in Fig. KO55, then turn fitting ¼ turn clockwise so flats are horizontal. Turn fitting clockwise ½ turn and stop so flats are again horizontal. Continue rotating half turns and stopping when flats are horizontal until fitting is tight. Tighten by hand as much as possible to determine if float strikes crankcase. Do not pause when turning fitting as float will strike crankcase. If float strikes crankcase, turn fitting counterclockwise so flats and letters are as shown in Fig. KO55, then resume installation procedure.

If installed properly, there should be no continuity at sensor leads when float is up (full oil level) and continuity when float is down (low oil level).

REWIND STARTER. To disassemble starter, remove rope handle and allow pulley to totally unwind. Unscrew pulley retaining screw (1—Fig. KO56). Remove retainer (2), pawls (4), springs (5), brake spring (3) and washer (6). Wear appropriate safety eyewear and gloves before disengaging pulley from starter as spring may uncoil uncontrolled. Place shop towel around pulley and lift pulley out of housing; spring and cup may remain in housing or stay in

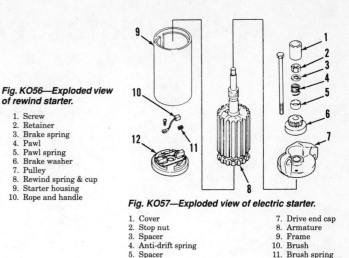

Fig. KO56—Exploded view of rewind starter.

1. Screw
2. Retainer
3. Brake spring
4. Pawl
5. Pawl spring
6. Brake washer
7. Pulley
8. Rewind spring & cup
9. Starter housing
10. Rope and handle

Fig. KO57—Exploded view of electric starter.

1. Cover	7. Drive end cap
2. Stop nut	8. Armature
3. Spacer	9. Frame
4. Anti-drift spring	10. Brush
5. Spacer	11. Brush spring
6. Pinion	12. End cap

pulley. Do not attempt to separate spring from cup as they are a unit assembly.

Inspect components for damage and excessive wear. Reverse disassembly procedure to install components. Apply a light coat of grease to rewind spring and spring contact area on inside of pulley. Note that spring cup (8) will only fit in pulley (7) if lugs are properly engaged. To assemble pulley, spring cup and housing, install spring cup with spring in pulley and hold pulley so spring cup is on top. Place starter housing on top of pulley. Rotate pulley counterclockwise until spring tension is felt indicating inner end of spring has engaged spring anchor. Install pawl springs (5) so long end is up and spring forces pawl towards center of starter.

Apply grease to ends of brake spring (3). Flanges of retainer (2) must be towards rewind spring. Tighten center screw (1) to 7.9 N•m (70 in.-lbs.). Rotate pulley counterclockwise until spring is wound tight. Let pulley unwind just until rope holes in pulley and starter housing are aligned. Insert rope through holes and tie knot at pulley end. Allow pulley to rotate to wind rope onto pulley and install rope handle.

When installing starter on engine, pull out rope so dogs engage starter cup and center starter before tightening starter mounting screws.

ELECTRIC STARTER. Manufacturer recommends lubricating pinion drive splines with Kohler lubricant 52

357 01 annually or after every 500 hours of operation, whichever occurs first.

Refer to Fig. KO57 for an exploded view of electric starter used on some engines. Disassembly is evident after inspection of unit and referral to Fig. KO57. Make alignment marks across end caps (7 and 12) and frame (9) before disassembly. Tighten nut (2) to 17.0-19.2 N•m (150-170 in.-lbs.).

ALTERNATOR. Some engines are equipped with a 0.5 amp alternator. Alternator stator is located outside flywheel. An inline diode rectifies the alternating current to direct current.

To check alternator output, connect a voltmeter capable of RMS reading to black alternator stator lead. With engine running, dc voltage reading should be 4.20-5.57 volts at 2000 rpm; 9.68-11.90 volts at 2800 rpm; 13.29-14.97 volts at 3600 rpm.

KOHLER CENTRAL PARTS DISTRIBUTORS

(Arranged Alphabetically by States)
**These franchised firms carry extensive stocks of repair parts.
Contact them for name of dealer in their area who will have re-
placement parts.**

Auto Electric & Carburetor Company
Phone: (205) 323-7155
2625 4th Avenue South
Birmingham, Alabama 35233

Loftin Equipment Company
Phone: (602) 272-9466
12 N. 45th Avenue
Phoenix, Arizona 85043

H.G. Makelim Company
Phone: (714) 978-7515
1520 South Harris Ct.
Anaheim, California 92806

H.G. Makelim Company
Phone: (415) 873-4757
219 Shaw Road
South San Francisco, California
94083-2827

Spitzer Industrial Products Company
Phone: (303) 287-3414
6601 North Washington Street
Thornton, Colorado 80229

Spencer Engine, Inc.
Phone: (904) 262-1661
5200 Sunbeam Road
Jacksonville, Florida 32257

Power Systems
Phone: (208) 342-6541
4499 Market Street
Boise, Idaho 83705

Medart Engines of Kansas
Phone: (913) 888-8828
15500 West 109th Street
Lenexa, Kansas 66219

The Grayson Company of Louisiana
Phone: (318) 222-3871
215 Spring Street
Shreveport, Louisiana 71162

W.J. Connell Company
Phone: (508) 543-3600
65 Green Street
Foxboro, Massachusetts 02035

Central Power Distributors, Inc.
Phone: (612) 633-5179
2976 North Cleveland Avenue
St. Paul, Minnesota 55113

Medart Engines of St. Louis
Phone: (314) 343-0505
100 Larkin Williams Industrial Ct.
Fenton, Missouri 63026

Original Equipment, Inc.
Phone: (406) 245-3081
905 Second Avenue, North
Billings, Montana 59103

Power Distributors, Inc.
Phone: (908) 225-4922
71 Northfield Avenue
Edison, New Jersey 08837

Gardner, Inc.
Phone: (614) 488-7951
1150 Chesapeake Avenue
Columbus, Ohio 43212

MICO, Inc.
Phone: (918) 627-1448
7450 East 46th Place
Tulsa, Oklahoma 74145

E.C. Distributing Co.
Phone: (503) 224-3623
1835 NW 21st Avenue
Portland, Oregon 97210

Pitt Auto Electric Company
Phone: (412) 766-9112
2900 Stayton Street
Pittsburgh, Pennsylvania 15212

Medart Engines of Memphis
Phone: (901) 795-4365
4365 Old Lamar
Memphis, Tennessee 38118

Tri-State Equipment Company
Phone: (915) 532-6931
410 South Cotton Street
El Paso, Texas 79901

Waukesha-Pearce Industries, Inc.
Phone: (713) 723-1050
12320 South Main Street
Houston, Texas 77235

Chesapeake Engine Distributors
Division of RBI Corporation
Phone: (804) 550-2231
103 Sycamore Drive
Ashland, Virginia 23005

E.C. Distributing of Washington
Phone: (206) 872-7011
6410 S. 196th Street
Kent, Washington 98032

Kohler Company
Engine Division
Phone: (414) 457-4441
Kohler, Wisconsin 53044

CANADIAN DISTRIBUTORS

Lotus Equipment Sales, Ltd.
Phone: (403) 253-0822
5726 Burleigh Circle
Calgary, Alberta T2H 1Z8

Yetman's Ltd.
Phone: (204) 586-8046
949 Jarvis Avenue
Winnipeg, Manitoba R2X 0A1

W.N. White Company, Ltd.
Phone: (902) 443-5000
2-213-215 Bedford Highway
Halifax, Nova Scotia B3M 2J9

CPT Canada Power Technologies, Ltd.
Phone: (416) 890-6900
161 Watline Avenue
Mississauga, Ontario L4Z 1P2

LAWN BOY

LAWN BOY, INC.
P.O. Box 152
Plymouth, Wisconsin 53073

Model	Bore	Stroke	Displacement	Power Rating
F-Series	2.38 in.	1.75 in.	7.78 cu.in.	4.5 hp
	(60.4 mm)	(44.4 mm)	(127 cc)	(3.3 kW)

ENGINE IDENTIFICATION

Lawn Boy F-Series engines are used on Gold Series lawnmowers. The engine is an air-cooled, single-cylinder, two-stroke engine with a vertical crankshaft.

MAINTENANCE

LUBRICATION. Engine is lubricated by mixing gasoline and oil at a ratio of 32:1. Gasoline and oil should be mixed in a separate container before filling engine fuel tank. Lawn Boy recommends using Lawn Boy 2-CYCLE ASHLESS OIL and advises against mixing different oil brands if another oil is used. DO NOT use automotive (4-cycle) engine oil.

Manufacturer recommends unleaded gasoline with an 85 octane rating or higher. Gasoline containing methanol is not recommended, although gasoline containing 10 percent or less of ethanol may be used. If gasoline containing ethanol is used, it must be drained from the fuel system before storing the engine.

Lawn Boy 2+4 Fuel Conditioner is recommended if fuel is stored for an extended period.

AIR FILTER. The foam type filter element should be cleaned, inspected and re-oiled after every 25 hours of engine operation or annually, whichever occurs first. To remove filter element, unsnap filter cover latch (2—Fig. LB71) and swing filter cover (1) to the side. Clean the filter element in soapy water then squeeze the filter until dry (don't twist the filter).

Inspect the filter for tears and holes or any other opening. Discard the filter if it cannot be cleaned satisfactorily or if the filter is torn or otherwise damaged.

Pour approximately one tablespoon of clean SAE 30 engine oil into the filter, then squeeze the filter to remove the excess oil and distribute oil throughout the filter. Be sure filter fits properly in filter box.

SPARK PLUG. Recommended spark plug is a Champion CJ14 or equivalent. Electrode gap should be 0.035 inch (0.9 mm). Tighten spark plug to 15 ft.-lbs. (20.4 N•m) torque.

CARBURETOR. The engine is equipped with a float type carburetor. The governor air vane is attached to the carburetor throttle shaft.

The carburetor may be removed after removing air filter element, then unscrewing air filter base and carburetor mounting screws (S—Fig. LB72).

To adjust carburetor, turn idle mixture needle (8—Fig. LB73) in (clockwise) until lightly seated, then turn needle out ¾ turn. Start engine and run until normal operating temperature is reached. Set speed control lever to "Low Speed" position, turn fuel flow needle in slowly until engine starts surging or slowing down, then turn needle slowly out until engine runs smoothly. Allow engine to run a few minutes at this setting to make certain fuel mixture is not too lean. With engine running at idle, rapidly move speed control to full throttle position. If engine stumbles or hesitates, slightly turn idle mixture screw ⅛ turn counterclockwise and repeat test.

To disassemble carburetor, refer to Fig. LB73 and remove bowl screw (19) and float bowl (18). Remove float pin (11), float (16) and fuel inlet valve (15). If necessary to remove inlet valve seat (14), use a stiff wire with a hook on one end to pull seat from carburetor body. Fuel inlet valve and seat should be serviced as a set. To remove the Welch plugs (7 and 10), pierce the plug with a sharp pin punch, then pry out the plug, but do not damage underlying metal. Remove idle mixture needle (8). Remove retaining screw from throttle plate (12) and withdraw throttle shaft (5). Pull choke plate (9) out of slot in choke shaft (6), then withdraw shaft.

Inspect carburetor and renew any damaged or excessively worn components. The body must be replaced if there is excessive throttle or choke shaft play as bushings are not available. Use a ³⁄₁₆ inch (5 mm) diameter rod to install the fuel inlet valve seat. The groove on the seat must be down (towards carburetor bore). Push in the seat until it bottoms. Install Welch plugs while being careful not to indent the plugs; the plugs should be flat after installation. Apply a nonhardening sealant around the plug. The float should be approximately parallel with the body when the carburetor is inverted. Bend float hinge tang to adjust float level, or renew float and/or fuel inlet valve.

GOVERNOR. Governor is an air vane (pneumatic) type that is attached to the carburetor throttle shaft. Air vane (1—Fig. LB73) extends through cooling shroud base and responds to air flow created by the flywheel fan. Fluctuation in engine speed, due to change in engine

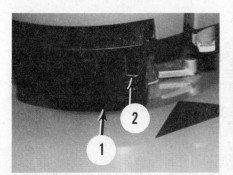

Fig. LB71—To service air filter element, unsnap latch (2) and swing filter cover (1) to the left.

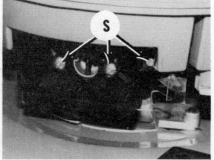

Fig. LB72—View of air filter housing and carburetor mounting screws (S).

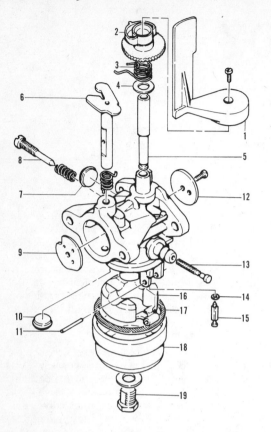

Fig. LB73—Exploded view of carburetor.

1. Governor air vane
2. Collar
3. Spring
4. Washer
5. Throttle shaft
6. Choke shaft
7. Welch plug
8. Idle mixture screw
9. Choke plate
10. Welch plug
11. Float shaft
12. Throttle plate
13. Fuel filter
14. Valve seat
15. Fuel inlet valve
16. Float
17. Gasket
18. Float bowl
19. Screw

break plate loose from crankshaft. Remove three screws securing muffler. Refer to Fig. LB74 for view of muffler plate (1) and engine exhaust ports (2). Rotate crankshaft so piston blocks exhaust ports, then use a 3/8-inch (10 mm) diameter wooden dowel to break carbon away from port openings. Clean muffler openings.

CAUTION: Particular care must be used when cleaning exhaust ports so piston is not damaged. Metal tools must not be used.

When reassembling, tighten muffler screws to 155 in.-lbs. (17.5 N•m). Tighten blade nut to 47 ft.-lbs. (64 N•m).

REPAIRS

TIGHTENING TORQUES. Recommended tightening torques are as follows:

Blade nut . . . . 47 ft.-lbs. (64 N•m)
Carburetor mounting screw   65 in.-lbs. (7.3 N•m)
Connecting rod . . . . 55-65 in.-lbs. (6.2-7.3 N•m)
Crankcase cover . . . . . . 100-120 in.-lbs. (11-13 N•m)
Flywheel nut . . . . 33 ft.-lbs. (44.8 N•m)
Muffler . . . . 155 in.-lbs. (17.5 N•m)
Spark plug . . . . 15 ft.-lbs. (20.4 N•m)

CRANKCASE PRESSURE TEST. An improperly sealed crankcase can cause the engine to be hard to start, run rough, have low power and overheat. Refer to ENGINE SERVICE in the FUNDAMENTALS SECTION of this manual for crankcase pressure test procedure. If crankcase leakage is indicated, pressurize crankcase and use a soap and water solution to check gaskets, seals, carburetor pulse line and casting for leakage.

load, opens or closes throttle plate to maintain desired engine rpm. Recommended governor high speed is 3100-3300 rpm.

To adjust governor, hold base of air vane and turn collar (2—Fig. LB73) to obtain desired engine speed. Each click represents approximately 50 rpm. Clockwise rotation of collar increases engine speed, while counterclockwise rotation decreases engine speed.

IGNITION SYSTEM. A solid-state ignition system is used on all models. All ignition components, including the ignition coil, are contained in a module located outside the flywheel. Ignition timing is not adjustable. The air gap between the ignition module and flywheel should be 0.010 inch (0.25 mm). Insert Lawn Boy gauge 604659 or correct thickness plastic strip between module and flywheel magnets to check air gap.

If an ignition malfunction is suspected, first be sure spark plug and high tension wire lead are in good condition. The high tension lead is available separately. There are no test procedures for ignition module. If module is suspected, install a module that is known to be good and recheck ignition performance.

CARBON. If ignition, fuel supply to combustion chamber and compression are satisfactory, but engine will not run

or runs poorly, it is likely that there is heavy carbon buildup in exhaust ports and muffler. A common symptom is "four-cycling" or firing every other power stroke.

Manufacturer recommends cleaning out carbon after every 50 hours of operation. Before undertaking carbon cleaning procedure, drain fuel from fuel tank and, if so equipped, remove battery. Tip mower on its side and secure in position by blocking.

WARNING: Be sure spark plug wire is disconnected from spark plug and wire is properly grounded.

Remove mower blade nut, blade and blade mounting plate. Strike mounting plate in a counterclockwise direction to

Fig. LB74—View showing muffler plate (1) and engine exhaust ports (2).

FLYWHEEL. Remove starter assembly and fuel tank and shroud assembly for access to flywheel. Use suitable puller to remove flywheel from tapered crankshaft.

When installing flywheel, be sure flywheel key is installed correctly as shown in Fig. LB75. Tighten flywheel nut to 33 ft.-lbs. (44.8 N·m) torque.

PISTON, PIN AND RINGS. Engine must be removed from the equipment to service piston (6—Fig. LB76), pin (4) and rings (3). Piston and rod assembly can be removed after removing flywheel (1), shroud (2), carburetor and crankcase cover half (17). Do not lose the loose bearing rollers (10) when disconnecting rod and cap from crankshaft.

Piston, pin and rings are available in standard size only. Recommended piston-to-cylinder clearance is 0.003-0.004 inch (0.08-0.10 mm).

Piston pin is retained by snap rings at each end of pin bore in piston. Use Lawn Boy tool 602884 or a suitable wooden dowel to drive piston pin out. Piston pin diameter is 0.4998-0.5000 inch (12.695-12.700 mm).

Top piston ring end gap should be 0.007-0.017 inch (0.18-0.43 mm) and bottom ring end gap should be 0.015-0.025 inch (0.38-0.64 mm). Renew rings if end gap is excessive. Note that a semi-keystone ring is used in top ring groove. When installing rings, stagger ring gaps at least 30° apart. Reassemble connecting rod to piston and install piston pin retaining rings. Position piston pin retaining rings (5) so beveled side is toward pin and opening faces upward toward piston crown to prevent ring from popping out during operation.

Before installing piston and rod assembly, place Lawn Boy stop tool 677389 or a suitable equivalent in spark plug hole.

NOTE: Without piston stop, piston could enter cylinder too far allowing top ring to become stuck in cylinder.

Install piston and connecting rod in cylinder so "BTM" letters on piston skirt are facing toward exhaust ports in cylinder. Attach connecting rod to crankshaft as outlined in following section.

CONNECTING ROD. The aluminum connecting rod rides on renewable steel inserts and 33 needle roller bearings surrounding the crankpin as shown in Fig. LB77. Piston and rod assembly can be removed after separating crankcase cover half from cylinder. Do not lose the loose bearing rollers when disconnecting rod from crankshaft.

Fig. LB75—When installing flywheel, be sure key is installed correctly in crankshaft groove as shown.

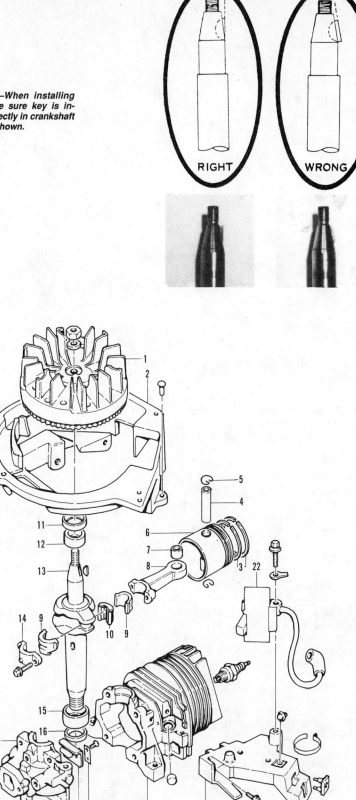

RIGHT WRONG

Fig. LB76—Exploded view of engine assembly.

1. Flywheel	7. Bearing	
2. Base shroud	8. Connecting rod	13. Crankshaft
3. Piston rings	9. Split liner	14. Rod cap
4. Piston pin	10. Needle bearings	15. Lower main bearing
5. Retaining ring	11. Oil seal	16. Oil seal
6. Piston	12. Upper main bearing	17. Crankcase cover

18. Reed
19. Reed plate
20. Cylinder & crankcase assy.
21. Ignition bracket
22. Ignition module

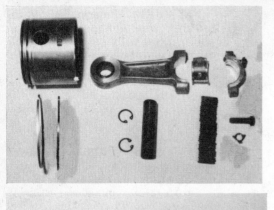

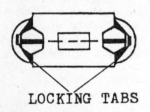

Fig. LB78—View showing connecting rod and piston assembled on crankshaft. Alignment marks on connecting rod and rod cap must be aligned.

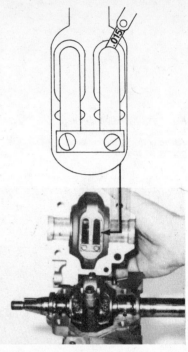

Fig. LB80—View showing positioning of reed valves in crankcase cover. Maximum allowable clearance between reed tip and machined surface is 0.015 inch (0.38 mm).

LOCKING TABS

Fig. LB79—Bend locking tabs snugly up against connecting rod cap screws.

When reassembling connecting rod to crankpin, install steel inserts in connecting rod end and cap making sure they are centered and dovetail guides will mate correctly when assembled. If bearing rollers are being reused, be sure that none are damaged. Coat insert surfaces with Lawn Boy needle bearing grease 378642 or a suitable equivalent and stick rollers to connecting rod and cap. Fit 17 rollers to rod cap and 16 rollers to rod.

If installing a new bearing roller set, lay strip of 33 rollers on forefinger and carefully peel backing off rollers. Curl forefinger around crankpin to transfer rollers from finger to journal. Grease on roller will hold them together and to crankpin journal.

Install connecting rod cap; BE SURE rod and cap mating marks (Fig. LB78) are aligned. Tighten rod cap screws to initial torque of 15-20 in.-lbs. (1.7-2.3 N•m), then alternately tighten screws in two steps to a final torque of 55-65 in.-lbs. (6.2-7.3 N•m). Bend locking tabs (Fig. LB79) snugly against flats on head of screws. Check to be sure that rod assembly is free on crankpin and that

none of the rollers dropped out during assembly.

REED VALVE. Reed valves permit air:fuel mixture to enter crankcase on the compression stroke and seal the air:fuel mixture in crankcase on the power stroke. Reed valves are mounted in crankcase cover as shown in Fig. LB80. Reeds can be cleaned with solvent or carburetor cleaner. DO NOT use compressed air on reed assemblies or distortion may occur resulting in hard starting or loss of power. Bent or distorted reeds cannot be repaired. Rough edge of reed must be positioned away from machined surface of mounting plate. Check reed-to-plate clearance as shown Fig. LB80. If clearance between reed tip and machined surface exceeds 0.015 inch (0.38 mm), renew reed. Coat reed mounting screws with Loctite and tighten to 10-13 in.-lbs. (1.12-1.47 N•m) torque.

CRANKCASE, CRANKSHAFT, BEARINGS AND SEALS. Crankcase, cylinder and cylinder head are an integral assembly (20—Fig. LB76). Cylinder bore cannot be bored for oversize piston. Crankcase cover (17) houses reed valve assembly. The crankshaft (13) is supported by caged roller bearings that contain 20 bearing rollers each. Top main bearing is shorter than lower bearing.

Crankshaft main bearing journal diameter at either end should be 0.8773-0.8778 inch (22.283-22.296 mm).

Diametral clearance between crankshaft journal and roller bearing should be 0.002-0.003 inch (0.05-0.08 mm). Maximum allowable clearance is 0.005 inch (0.13 mm).

Crankpin diameter should be 0.742-0.743 inch (18.85-18.87 mm) with a diametral clearance of 0.0025-0.0035 inch (0.064-0.089 mm) between pin and connecting rod assembly. Renew crankshaft if crankpin is out-of-round more than 0.0015 inch (0.38 mm). Crankshaft end play should be 0.006-0.016 inch (0.15-0.41 mm).

Install crankshaft seals (11 and 16—Fig. LB76) with lips (grooved side of seal) to inside of crankcase. When installing crankshaft seals, use a protective sleeve or wrap tape over keyway and sharp shoulders on crankshaft to prevent damage to seals.

Before assembly, remove any gasket material on mating surfaces of crankcase and cover, then apply a thin coat of Lawn Boy Gasket Maker 682302 or a suitable equivalent to sealing surface of crankcase cover. Before installing crankcase cover, be sure crankshaft main bearing dowels align in crankcase locator slots as shown in Fig. LB81. Install crankcase cover and securing screws, then using a crossing pattern, tighten screws in small increments until a final torque of 100-120 in.-lbs. (11.3-13.6 N•m) is obtained. Check crankshaft for free rotation.

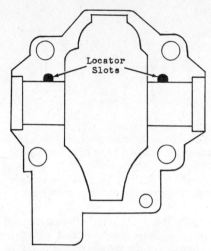

Fig. LB81—Drawing showing crankcase locator slots used in aligning crankshaft main bearings. Be sure main bearing dowels align in slots.

Fig. LB83—Install dog spring (8) so long end is up and outside the dog as shown. Spring should force dog toward center of starter.

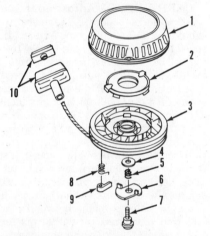

Fig. LB82—Exploded view of rewind starter.

1. Starter housing
2. Spring cup
3. Pulley
4. Washer
5. Brake spring
6. Retainer
7. Screw
8. Spring
9. Dog
10. Handle & insert

REWIND STARTER. To disassemble starter, first remove starter assembly from engine. Pull rope partially out of starter and remove rope handle, then allow pulley to unwind slowly. Unscrew pulley retaining screw (7—Fig. LB82). Remove retainer (6), dogs (9), springs (8), brake spring (5) and washer (4). Wear appropriate safety eyewear and gloves before disengaging pulley from starter as spring may uncoil uncontrolled. Place shop towel around pulley (3) and lift pulley out of housing (1); spring and cup may remain in housing or stay in pulley. Do not attempt to separate spring from cup as they are a unit assembly.

Inspect components for damage and excessive wear. Reverse disassembly procedure to install components. Apply a light coat of grease to rewind spring and spring contact area on inside of pulley. Note that spring cup (2) will only fit in pulley if lugs are properly engaged. To assemble pulley, spring cup and housing, install spring cup with spring in pulley and hold pulley so spring cup is on top. Place starter housing on top of pulley. Rotate pulley counterclockwise until spring tension is felt indicating inner end of spring has engaged spring anchor. Install dog springs (8) so long end is up and spring forces dog toward center of starter as shown in Fig. LB83. Apply grease to ends of brake spring. Flanges of retainer must be toward rewind spring. Tighten center screw to 70 in.-lbs. (7.9 N·m) torque. Rotate pulley counterclockwise until spring is wound tight. Let pulley unwind just until rope holes in pulley and starter housing are aligned. Insert rope through holes and tie knot at pulley end (Fig. LB84). In-

Fig. LB84—After rotating pulley to place tension on rewind spring, insert rope through housing outlet (O) and pulley hole and tie a knot in pulley end of rope.

stall rope handle on outer end of rope, then allow pulley to unwind to wind rope onto pulley.

When installing starter on engine, pull out rope so dogs engage starter cup and center starter before tightening starter mounting screws. Tighten starter mounting screws to 25 in.-lbs. (2.8 N·m) torque.

LAWN BOY

Model	Bore	Stroke	Displacement	Power Rating
M25	2.380 in.	2.00 in.	8.9 cu.in.	5 hp
	(60.4 mm)	(50.8 mm)	(145.6 cc)	(3.7 kW)

ENGINE INFORMATION

The Lawn Boy Model M25 engine is a two-stroke, single-cylinder engine used on "M" Series lawnmowers. The engine is air cooled and has a vertical crankshaft.

MAINTENANCE

LUBRICATION. Some models are equipped with an oil pump (known as "FreshLube") that injects oil into the incoming air-fuel mixture through a passage in spacer (11—Fig. LB201). Lawn Boy recommends using Lawn Boy 2-CYCLE ASHLESS OIL and advises against mixing different oil brands if another oil is used.

On engines equipped with an oil pump, oil is contained in an oil tank; do not add oil to fuel. Later models are equipped with an oil level sensor in the oil tank that stops the engine if the oil level is low. Early models do not have an oil level sensor and the engine will be damaged if all oil is consumed while the engine is running.

NOTE: Tipping the mower over may allow air to enter the oil line if the oil tank filter is exposed. If bubbles in the oil line are more than ¾ inch (19 mm) long, empty fuel tank, fill fuel tank with a fuel:oil mix of 64:1, be sure oil tank contains oil, and start and run mower. This will insure that there is sufficient lubrication while air bubbles are purged.

If oil pump control linkage has been disturbed, refer to OIL PUMP section for adjustment.

On models not equipped with an oil pump, the oil must be mixed with the fuel. Manufacturer recommends using Lawn Boy oil. Fuel:oil ratio when using Lawn Boy oil is 64:1. DO NOT use automotive (4-cycle) engine oil.

Manufacturer advises that using gasoline with alcohol (gasohol) may adversely affect some engine components. If gasoline containing ethanol is used, it must be drained from the fuel system before storing the engine. The use of premium gasoline or gasoline containing methanol is not recommended.

SPARK PLUG. Recommended spark plug is a Champion CJ14 or equivalent. Electrode gap should be 0.035 inch (0.9 mm). Tighten spark plug to 15-18 ft.-lbs. (20-24 N•m).

CARBURETOR. Adjustment. Idle speed at normal operating temperature should be 2100-2300 rpm. Adjust idle speed by turning idle speed screw (16—Fig. LB202). Idle mixture is controlled by idle jet (6) and idle mixture screw (14). Initial setting of idle mixture screw is one turn out. Make final adjustment with engine running and at operating temperature. To adjust idle mixture, turn idle mixture screw clockwise and lean mixture until engine speed just starts to slow, then turn screw counterclockwise and enrich mixture just until engine speed begins to slow. Turn idle screw to halfway point between lean and rich positions. Turn idle mixture screw counterclockwise in small increments to enrich mixture if engine will not accelerate without stumbling. Idle mixture jet is not adjustable. High speed operation is controlled by main jet (19—Fig. LB201) and is not adjustable. Standard size main jet only is available.

Overhaul. To disassemble carburetor, remove fuel bowl retaining screw

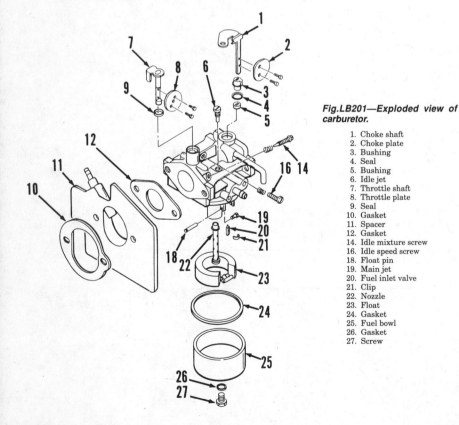

Fig.LB201—Exploded view of carburetor.

1. Choke shaft
2. Choke plate
3. Bushing
4. Seal
5. Bushing
6. Idle jet
7. Throttle shaft
8. Throttle plate
9. Seal
10. Gasket
11. Spacer
12. Gasket
14. Idle mixture screw
16. Idle speed screw
18. Float pin
19. Main jet
20. Fuel inlet valve
21. Clip
22. Nozzle
23. Float
24. Gasket
25. Fuel bowl
26. Gasket
27. Screw

Fig. LB202—View showing location of idle jet (6), idle mixture screw (14) and idle speed screw (16).

Fig. LB203—View showing location of air jets (AJ). Jets are not available except with carburetor body.

Fig. LB204—Install choke plate with flat side toward fuel inlet.

Fig. LB205—Install throttle plate so numbers are out and on side opposite to fuel inlet.

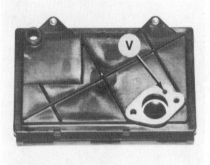

Fig. LB206—Air filter base gasket must be installed so vent hole (V) matches holes in carburetor and filter base.

properly matches opening in spacer. When installing air filter base gasket, be sure vent hole (V—Fig. LB206) matches holes in filter base and carburetor.

GOVERNOR. All models are equipped with a ball type mechanical governor located internally on the crankshaft. As engine speed increases, the internal governor assembly rotates governor shaft (S—Fig. LB207) which in turn rotates governor lever (L) and closes the throttle. The movement of governor lever is counterbalanced by tension of governor spring (G) which tends to open the throttle. When the two forces balance, governed engine speed is obtained.

To adjust governor, loosen governor shaft clamp bolt (B—Fig. LB207). Move throttle control so carburetor throttle plate is in full open position. Turn governor shaft (S) clockwise as far as possible and retighten clamp bolt (be sure throttle is in full open position). With engine warm, maximum engine speed at full throttle should be 3100-3300 rpm. Adjust maximum engine speed by bending governor spring arm (A). Bending the arm up will decrease spring tension and reduce engine speed. DO NOT exceed specified maximum engine speed.

IGNITION SYSTEM. A solid-state ignition system is used on all models. All ignition components including the ignition coil are contained in a module located outside the flywheel. Ignition timing is not adjustable. The air gap (G—Fig. LB208) between the ignition module and flywheel should be 0.010 inch (0.25 mm). Insert Lawn Boy gauge 604659 or correct thickness plastic strip between module and flywheel magnets and tighten module mounting screws to 100 in.-lbs. (11.3 N•m).

(27—Fig. LB201), gasket (26) and fuel bowl (25). Remove float pin (18) by pushing against round end of pin toward the square end of pin. Remove float (23) and fuel inlet needle (20). Remove throttle and choke shaft assemblies after unscrewing throttle and choke plate retaining screws. Remove idle mixture screw (14), idle mixture jet (6), main jet (19) and main fuel nozzle (22). Air jets shown in Fig. LB203 are only available as part of body unit assembly.

When assembling the carburetor note the following. Place a small drop of nonhardening sealant such as Permatex #2 or equivalent on throttle and choke plate retaining screws. Flat side of choke plate must be on fuel inlet side of carburetor (Fig. LB204). Numbers on throttle plate must face out and be on side of carburetor opposite of fuel inlet (Fig. LB205). Tighten nozzle (22—Fig.

LB201) to 7 in.-lbs. (0.8 N•m) and tighten idle jet (6) and main jet (19) to 6 in.-lbs. (0.7 N•m). Retaining clip (21) for fuel inlet valve must engage groove of fuel inlet valve and fit around tab of float. Float height is not adjustable.

If removed, install spacer (11—Fig. LB201) so "D" shaped opening is toward cylinder and oil fitting (if so equipped) is up. Be sure opening in gasket (10)

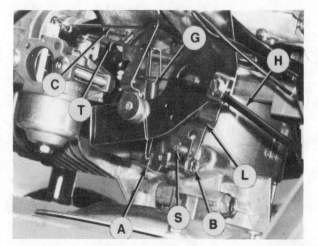

Fig. LB207—View of governor linkage.

A. Spring arm
B. Clamp bolt
C. Choke rod
G. Governor spring
H. Throttle/choke control cable
L. Governor lever
S. Governor shaft
T. Throttle rod

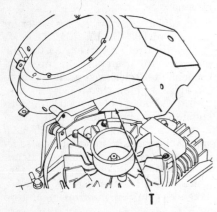

Fig. LB208—Air gap (G) between ignition module and flywheel should be 0.010 inch (0.25 mm).

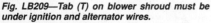

Fig. LB209—Tab (T) on blower shroud must be under ignition and alternator wires.

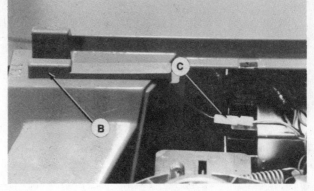

Fig. LB210—View showing location of chute interlock switch (B) and connector (C).

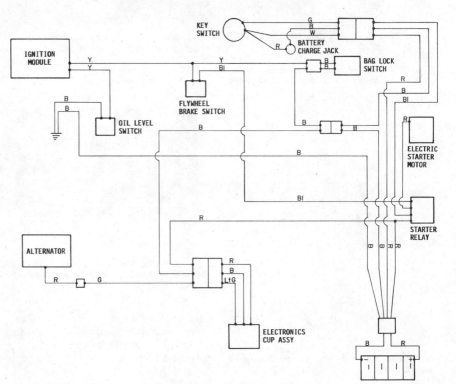

Fig. LB211—Wiring schematic for M25 engine on Model M21EMR lawn mower.

B. Black	G. Green	W. White
Bl. Blue	R. Red	Y. Yellow

If an ignition malfunction is suspected, first be sure spark plug and high tension wire lead are in good condition. The high tension lead is available separately. There are no test procedures for ignition module. If module is suspected to be faulty, install a module that is known to be good and recheck ignition performance.

Note that when installing blower shroud, the tab (T—Fig. LB209) must be under ignition and alternator wires.

An interlock system is used to prevent unsafe mower operation. Various devices are used to ground the ignition system, either stopping the engine while running or preventing starting. Depending on the components installed, a chute interlock switch (LB210), blade brake clutch switch or throttle switch may be present. A wiring diagram for electric start models is shown in Fig. LB211.

CAUTION: Be sure all safety related devices work properly; DO NOT run engine otherwise.

REPAIRS

TIGHTENING TORQUES. Recommended tightening torques are as follows:

Alternator . . . . 75 in.-lbs.
(8.5 N·m)
Carburetor mounting screw 65 in.-lbs.
(7.3 N·m)
Crankcase screws . . . . 170 in.-lbs.
(19.2 N·m)
Cylinder head . . . . 190 in.-lbs.
(21.5 N·m)
Electric starter . . . . 170 in.-lbs.
(19.2 N·m)
Engine plate . . . . 155 in.-lbs.
(17.5 N·m)
Flywheel nut. . . . . 33 ft.-lbs.
(44.8 N·m)
Ignition module . . . . 100 in.-lbs.
(11.3 N·m)

Muffler mounting screw . . 170 in.-lbs.
(19.2 N·m)
Oil pump mounting screw . . 75 in.-lbs.
(8.5 N·m)
Spark plug . . . . 15 ft.-lbs.
(20.4 N·m)

CRANKCASE PRESSURE TEST. An improperly sealed crankcase can cause the engine to be hard to start, run rough, have low power and overheat. Refer to SERVICE SECTION TROUBLE-SHOOTING section of this manual for crankcase pressure test procedure. If crankcase leakage is indicated, pressurize crankcase and use a soap and water solution to check gaskets, seals, carburetor pulse line and casting for leakage.

OIL TANK. An oil tank is attached to engines equipped with an oil pump.

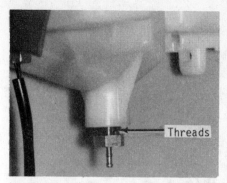

Fig. LB212—To prevent damage to "O" ring, there should be one to two threads exposed on oil tank fitting.

Fig. LB214—Marks (M) on oil pump lever (L) and oil pump (P) must be aligned when throttle is full open.

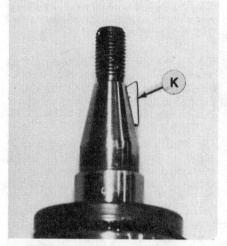

Fig. LB213—Straight edge of crankshaft key (K) must be parallel to crankshaft.

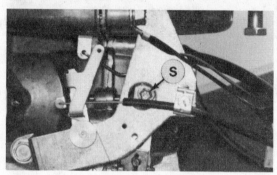

Fig. LB215—Loosen screw (S) and relocate oil pump link end to align oil pump marks shown in Fig. LB214.

If oil tank fitting is removed, do not screw fitting all the way into tank but leave one or two threads exposed as shown in Fig. LB212. Screwing fitting in completely may damage "O" ring at upper end of fitting.

FLYWHEEL. To remove flywheel, first drain fuel and oil tanks. On later models equipped with oil injection, disconnect oil sensor lead located beneath tanks. Unscrew mounting screws, detach retaining clip on underside and remove fuel and oil tank module. Remove air filter and unscrew top screw securing air filter base to blower shroud. Unscrew top screw securing muffler guard to blower shroud. Detach flywheel brake spring. Remove rewind starter. Remove blower shroud. Use tool 613593 or another suitable tool to hold flywheel and unscrew flywheel nut. Use tool 611600 or other suitable tool and pull flywheel off crankshaft. If holes in flywheel are not threaded, use a $^5/_{16}$-18 tap and cut threads in flywheel puller holes.

Discard flywheel if damaged. Note that flywheel surface contacting fly-

wheel brake must be smooth and free of grease and oil. Be sure flywheel key is parallel with shaft as shown in Fig. LB213, and NOT parallel with taper. Note that when installing blower shroud, the tab (T—Fig. LB209) must be under ignition and alternator wires. Tighten flywheel nut to 31-33 ft.-lbs. (42-45 N•m). On engines equipped with oil injection system, refer to LUBRICATION section and purge air from oil line as noted.

OIL PUMP. The oil pump (P—Fig. LB214) is located under the flywheel and driven by a worm on the crankshaft.

Adjust. Maximum oil pump output must occur at full throttle. To adjust pump output, the flywheel must be removed. Push governor lever so carburetor is fully open. Alignment marks (M) on oil pump and control lever should match as shown in Fig. LB214. To align marks, loosen screw (S—Fig. LB215) and relocate end of control link. Tighten screw and recheck alignment.

Overhaul. Remove flywheel for access to oil pump. Detach oil lines from tank and injection fitting; do not attempt to pull lines from oil pump.

Oil pump and worm are available only as a unit assembly. Individual oil pump components are not available.

Install worm on crankshaft so flat (F—Fig. LB216) on inside corresponds with flat on crankshaft. Install oil pump

and tighten mounting screws to 75 in.-lbs. (8.5 N•m).

Note that holes in crankcase for oil lines are marked "IN" and "OUT" (see Fig. LB217). Oil line passing through "IN" hole connects to oil tank fitting while oil line passing through "OUT" hole connects with injection fitting (oil lines will cross).

After assembly, refer to LUBRICATION section and purge air from oil line as noted.

OIL INJECTION CHECK VALVE. On models equipped with the FreshLube oil injection system, a check

Fig. LB216—Flats (F) on crankshaft and oil pump worm must index when installing worm.

Fig. LB217—Crankcase has "IN" and "OUT" cast on side to identify oil lines.

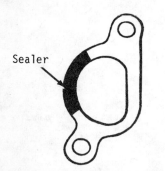

Fig. LB218—Apply sealer to exhaust port side of muffler gasket on rounded portion of gasket indicated.

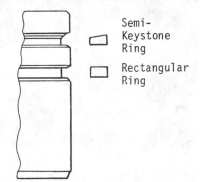

Fig. LB219—Cylinder head must be installed so bosses (B) are toward flywheel. Tighten cylinder head bolts in sequence shown.

Fig. LB221—Piston must be installed on rod so "EXH" on piston crown will be toward exhaust port side of engine. Piston ring end gaps (G) must align with locating pin (P).

Fig. LB220—Install rectangular piston ring in second piston ring groove and semi-keystone piston ring in top piston ring groove with bevel toward piston crown.

valve is located in the oil injection fitting. The fitting is attached to the spacer plate between the carburetor and engine. The check valve prevents entry of oil into the engine when it is not running. To test the check valve, connect a pressure tester with an oil-filled line to the fitting (there must be oil in the test line or the check valve will not operate properly). The check valve must hold at least 1 psi (6.9 kPa) but require no more than 3-4 psi (20.7-27.6 kPa) before flowing.

MUFFLER. When installing muffler, apply gasket sealer to muffler gasket in area shown in Fig. LB218 on side that contacts muffler. Tighten muffler mounting screws to 170 in.-lbs. (19.2 N•m).

CYLINDER HEAD. The blower shroud must be removed for access to cylinder head. To remove shroud, first drain fuel and oil tanks. On later models equipped with oil injection, disconnect oil sensor lead located beneath tanks. Unscrew mounting screws, detach retaining clip on underside and remove fuel and oil tank module. Remove air filter and unscrew top screw securing air filter base to blower shroud. Unscrew top screw securing muffler guard to blower shroud. Remove rewind starter. Remove blower shroud. Un-

screw cylinder head screws and remove head.

Inspect cylinder head and discard if cracked or otherwise damaged. Remove carbon using a tool that will not damage cylinder head surface.

Install cylinder head so blower shroud screw bosses (B—Fig. LB219) are up. Tighten cylinder head screws in three steps to final torque of 190 in.-lbs. (21.5 N•m) using crossing pattern shown in Fig. LB219. On models equipped with oil pump, refer to LUBRICATION section and purge air from oil line as noted.

PISTON, PIN AND RINGS. To remove piston it is necessary to separate crankcase halves as outlined in CRANKSHAFT AND CRANKCASE section. Detach snap ring and use a suitable puller to extract piston pin so piston can be separated from connecting rod.

Piston diameter at bottom of skirt at right angle to piston pin is 2.3782-2.3792 inch (60.406-60.432 mm). Piston diameter at bottom of skirt inline with piston pin is 2.3742-2.3752 inch

(60.305-60.330 mm). Specified piston pin hole diameter in piston is 0.5631-0.5635 inch (14.303-14.313 mm). Piston pin is available in standard size only. Piston ring end gap should be 0.005-0.015 inch (0.13-0.38 mm).

Piston and rings are available in standard size and 0.020 inch (0.51 mm) oversize. Piston ring nearer the piston crown is a semi-keystone type and the bevel must be toward the piston crown (see Fig. LB220). Second piston ring is rectangular and may be installed either direction.

Install piston on rod so "EXH" on piston crown (see Fig. LB221) will be toward exhaust port. Be sure piston ring end gaps (G) are aligned with locating pin (P) in piston ring grooves when installing piston in cylinder.

CRANKSHAFT AND CRANKCASE. To disassemble crankcase, remove blade and blade clutch if so equipped. Disconnect flywheel brake cable if so equipped. Disconnect throttle cable. On electric start model, disconnect battery. Remove flywheel as previously outlined. Remove air filter, carburetor, muffler, governor linkage and ignition module. On models so equipped, remove alternator and electric starter. Withdraw wiring from engine. On self-propelled models, remove belt cover behind engine. Remove oil pump and worm. Unscrew fasteners securing engine plate to lawn mower deck and remove engine.

Remove engine plate from engine. Remove taper pin (P—Fig. LB222) by driving pin away from cylinder as shown in Fig. LB222. Unscrew crankcase screws and separate crankcase by carefully prying between halves. Do not damage crankcase mating surfaces. When removing crankshaft assembly

Fig. LB222—Tapered alignment pin (P) is removed by driving away from cylinder (direction of arrow).

Fig. LB225—Governor arm (A) must be toward pto end of crankcase. Bearing retaining rings must fit into grooves (G) in both crankcase halves.

from crankcase, note that there are four governor balls that will be loose when governor sleeve (S—Fig. LB223) slides away from crankpin.

Specified main bearing journal diameter for both ends is 0.9835-0.9840 inch (24.981-24.994 mm). Main bearing bore in crankcase should be 2.0462-2.0469 inches (51.973-51.991 mm). Connecting rod small end diameter should be 0.7870-0.7874 inch (19.990-20.000 mm).

Crankshaft and connecting rod are available only as a unit assembly. Bearings (B—Fig. LB223) and seals (L) will fit either end. Top retainer ring (TR) at flywheel end is 0.030 inch (0.76 mm) thick while retainer ring (BR) at blade end is 0.050 inch (1.27 mm) thick.

When assembling crankshaft, install governor so balls are next to crankpin and pin (P—Fig. LB224) engages hole in crank throw. Do not install seals until crankcase is fastened together. Position governor arm (A—Fig. LB225) so arm is toward blade end of crankcase. Use a suitable gasket removing solvent and be sure crankcase mating surfaces are clean, flat and free of burrs or other damage. Apply Loctite 515 to mating surface of crankcase and spread evenly. Install crankshaft and piston assembly being sure piston ring gaps properly engage pin in ring grooves. Bearing re-

taining rings (BR and TR—Fig. LB223) must engage grooves (G—Fig. LB225) in crankcase and ring gaps must be toward cylinder. Install crankcase half and loosely install crankcase screws. Install tapered alignment pin (P—Fig. LB222). Tighten crankcase screws

Fig. LB224—Pin (P) of governor sleeve (S) must index in hole in crankpin.

evenly to 170 in.-lbs. (19.2 N·m). Rotate crankshaft and check for binding. Install oil seals with flat side out using protector 613598 and driver 613594. On self-propelled models, install belt pulley so set screw end of pulley is toward blade and belt groove is toward engine. Pin end of pulley set screw must index in hole in crankshaft. Apply Loctite to set screw threads. Complete reassembly while referring to appropriate component sections.

REWIND STARTER. To disassemble starter, first remove starter assembly from engine. Pull rope partially out of starter, remove rope handle and allow pulley to unwind slowly. Unscrew pulley retaining screw (7—Fig. LB226). Remove retainer (6), dogs (9), springs (8), brake spring (5) and washer (4). Wear appropriate safety eyewear and gloves

Fig. LB223—View of crankshaft assembly after removal of crankcase half.

   B. Bearing
BR. Retaining ring
   C. Crankshaft
   L. Seal
TR. Retaining ring
   S. Governor sleeve

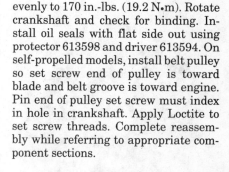

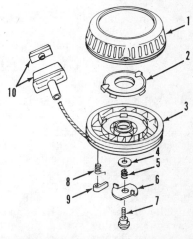

Fig. LB226—Exploded view of rewind starter.

1. Starter housing
2. Spring cup
3. Pulley
4. Washer
5. Brake spring
6. Retainer
7. Screw
8. Spring
9. Dog
10. Rope handle

Fig. LB227—Install dog spring (8) so long end is up and outside the dog as shown. Spring should force dog toward center of starter.

Fig. LB228—After rotating pulley to place tension on rewind spring, insert rope through housing outlet (O) and pulley hole and tie a knot in pulley end of rope.

inner end of spring has engaged spring anchor. Install dog springs (8) so long end is up and spring forces dog toward center of starter as shown in Fig. LB227. Apply grease to ends of brake spring. Flanges of retainer must be toward rewind spring. Tighten center screw to 70 in.-lbs. (7.9 N•m). Rotate pulley counterclockwise until spring is wound tight. Let pulley unwind just until rope holes in pulley and starter housing are aligned. Insert rope through holes and tie knot at pulley end (Fig. LB228). Attach handle to outer end of rope, then allow pulley to wind rope onto pulley.

When installing starter on engine, pull out rope so dogs engage starter cup and center starter before tightening starter mounting screws. Tighten

before disengaging pulley from starter as spring may uncoil uncontrolled. Place shop towels around pulley (3) and lift pulley out of housing; spring and cup (2) may remain in housing or stay in pulley. Do not attempt to separate spring from cup as they are a unit assembly.

Inspect components for damage and excessive wear. Reverse disassembly procedure to install components. Apply a light coat of grease to rewind spring and spring contact area on inside of pulley. Note that spring cup (2) will only fit in pulley if lugs are properly engaged. To assemble pulley, spring cup and housing, install spring cup with spring in pulley and hold pulley so spring cup is on top. Place starter housing on top of pulley. Rotate pulley counterclockwise until spring tension is felt, indicating

starter mounting screws to 25 in.-lbs. (2.8 N•m).

ELECTRIC STARTER. Engine may be equipped with an electric starter. To remove starter, tilt rear access door back and disconnect battery connector. Drain fuel, then remove spring clip and screws retaining fuel and oil tank assembly to mounting bracket and position tank assembly on top of blower shroud. Disconnect wire from starter. Remove starter retaining screws and withdraw starter from engine.

To disassemble starter drive components, detach "E" ring (E—Fig. LB229) and lift off pinion assembly. Remove cover (C) for access to gears (Fig. LB230). The remainder of the starter is available only as a unit assembly.

To test starter windings, disconnect starter lead and connect an ohmmeter to starter lead. Ground other ohmmeter lead to starter frame. Slowly rotate starter shaft at least one complete revolution (rapid rotation will give false readings). Resistance should be less than 10 ohms. Renew starter if tester displays high resistance or infinity.

When assembling drive components, note the following: Coat spur gears with Lubriplate; do not lubricate any other drive components. Loop end of drag spring (DS—Fig. LB229) must be around stud. Large end of spring (S) must be next to plastic gear (G). Fingers on washer (W) must engage slots on helix (H).

After installation, rotate starter gear to fully engaged position. Check gap (G—Fig. LB231) between base of flywheel gear teeth and top of starter gear

Fig. LB229—View of electric starter.

C. Cover
DS. Drag spring
E. "E" ring
G. Pinion gear
H. Helix & clutch assy.
S. Spring
W. Flanged washer

Fig. LB230—View of electric starter drive gears.

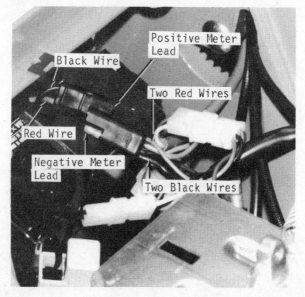

Fig. LB231—Gap (G) between base of flywheel and top of starter gear teeth should be 0.078 inch (1.98 mm).

teeth. Gap should be 0.078 inch (1.98 mm). Adjust starter position to obtain desired gap and tighten starter mounting screws to 170 in.-lbs. (19.2 N•m).

ALTERNATOR. Models with an electric starter are equipped with an alternator to charge the battery. The air gap (G—Fig. LB232) between the alternator (A) and flywheel should be 0.010 inch (0.25 mm). Insert Lawn Boy gauge 604659 or correct thickness plastic strip

between module and flywheel magnets and tighten module mounting screws to 75 in.-lbs. (8.5 N•m).

To check alternator output, first be sure air gap is correct. Run engine at 3100-3300 rpm, then disconnect battery at connector. Reconnect connector as shown in Fig. LB233 so red wires are connected but black wires are not. Connect negative ammeter lead to male end of two black wires and positive lead to female end of one black wire. Ammeter should read 190-450 mA.

NOTE: DO NOT start engine with electric starter when meter is connected as meter will be damaged.

With engine stopped, resistance reading between alternator green lead and ground should be 2.7-3.3 ohms.

FLYWHEEL BRAKE. A flywheel brake is used on many models. The brake will stop the engine within three seconds when the equipment safety handle is released. The ignition circuit is also grounded when the brake is actuated.

Detach spring (S—Fig. LB234) to deactivate brake. Inspect mechanism for excessive wear and damage. Renew brake if brake pad thickness is less than 0.031 inch (0.78 mm) at narrowest point. Brake pad must be clean and free of grease and oil. Flywheel surface must be clean and smooth. Tighten large retaining screw to 155 in.-lbs. (17.5 N•m) and smaller retaining screw and nut to 75 in.-lbs. (8.5 N•m).

To adjust flywheel brake cable, turn adjusting nuts (N—Fig. LB235) so distance (D) from bracket to brake lever is 2⅜ inches (60.3 mm) when the safety bail is pulled against the handlebar.

Fig. LB232—Air gap (G) between the alternator (A) and flywheel should be 0.010 inch (0.25 mm).

Fig. LB233—Reconnect battery connector so red wires are connected and black wires are disconnected, then check alternator using an ammeter as outlined in text.

Fig. LB234—Disable flywheel brake by detaching spring (S).

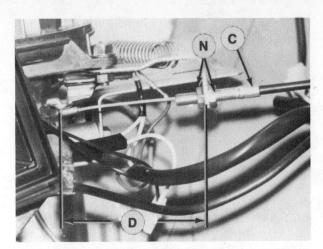

Fig. LB235—When safety bail is pulled against handlebar, distance (D) from bracket to brake lever must be 2-3/8 inches (60.3 mm). Turn nuts (N) to adjust distance.

SHINDAIWA

SHINDAIWA
P.O. Box 1090
Tualatin, Oregon 97062

Model	Bore	Stroke	Displacement
S25P	32 mm	30 mm	24.1 cc
	(1.26 in.)	(1.18 in.)	(1.47 cu. in.)

ENGINE INFORMATION

The S25P is an air-cooled, single-cylinder, two-stroke engine. The engine provides power for the Shindaiwa Model GP25 pump.

MAINTENANCE

LUBRICATION. The engine is lubricated by mixing oil with the fuel. Recommended oil is a good quality oil designed for use in air-cooled, two-stroke engines. Fuel:oil mixture ratio should be 25:1. Use a separate container to mix oil and gasoline.

AIR FILTER. The air filter is made of foam and may be cleaned using a suitable solvent. Dry filter, then apply a light oil and squeeze out excess oil.

SPARK PLUG. Recommended spark plug is Champion CJ8. Specified electrode gap is 0.6 mm (0.025 in.). Tighten spark plug to 16.7-18.6 N·m (148-164 in.-lbs.).

CARBURETOR. The engine is equipped with a TK carburetor. Refer to Fig. SH11 for an exploded view of carburetor.

Idle mixture is not adjustable. Idle speed is adjusted by rotating idle speed screw (17). High speed mixture is controlled by fixed main jet (22). Mid-range mixture is determined by position of clip (6) in grooves on jet needle (7). Installing clip in a lower groove on jet needle will provide a richer mixture. Normal clip position is in second groove from top if jet needle has four grooves, or in middle groove if jet needle has three grooves. See Fig. SH12.

Note that fuel inlet fitting (9—Fig. SH11) serves as fuel inlet seat and is renewable. Note that needle valve (18) may have either a solid tip or Viton tip. Hard tipped valve is used with fuel inlet fitting having a nonradiused seat. Viton tipped needle valve must be assembled with fuel inlet fitting having a radiused seat.

To determine float level, invert carburetor and measure height of float hinge (20) from bowl mating surface of carburetor as shown in Fig. SH13. Height (H) should be 3 mm (1/8 in.). Bend float hinge to obtain desired height.

When assembling carburetor, be sure groove in side of throttle slide (8—Fig. SH11) indexes with pin in throttle slide bore of carburetor.

IGNITION SYSTEM. The engine is equipped with an electronic breakerless ignition system. The ignition module is located adjacent to the carburetor (note insulators between module and engine shown in Fig. SH14).

Ignition occurs at 23° BTDC at 6000 rpm and is not adjustable. Air gap between ignition coil and flywheel should be 0.30-0.35 mm (0.012-0.014 in.).

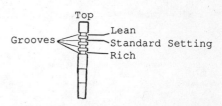

Fig. SH12—Installing jet needle clip in upper groove of jet needle will lean mid-range mixture and installing clip in lower grooves will provide richer mixture.

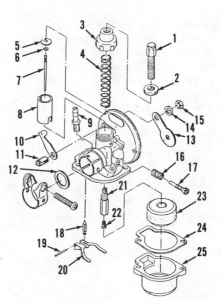

Fig. SH11—Exploded view of TK carburetor. A cable-actuated version of this carburetor is shown. Some engines may use a lever-actuated version.

1. Cable adjuster
2. Nut
3. Cap
4. Spring
5. Retainer
6. Jet needle clip
7. Jet needle
8. Throttle slide
9. Fuel inlet fitting
10. Choke lever
11. Shaft
12. Gasket
13. Choke plate
14. Lockwasher
15. Nut
16. Spring
17. Idle speed screw
18. Fuel inlet valve
19. Float pin
20. Float hinge
21. Nozzle
22. Main jet
23. Float
24. Gasket
25. Fuel bowl

Fig. SH13—Adjust float hinge to obtain height (H) of 3 mm (1/8 in.).

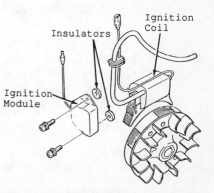

Fig. SH14—Insulator washers must be installed between ignition module and engine.

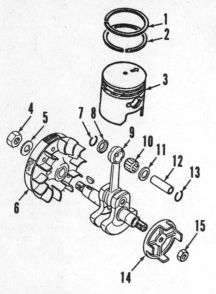

Fig. SH15—Exploded view of crankshaft, rod and piston assemblies.

1. Piston ring
2. Piston ring
3. Piston
4. Nut
5. Washer
6. Flyweight
7. Retainer
8. Thrust washer
9. Crankshaft & rod assy.
10. Bearing
11. Thrust washer
12. Piston pin
13. Retainer
14. Rewind pulley
15. Nut

Loosen ignition coil mounting screws and move ignition coil to adjust air gap.

Ignition coil primary windings resistance should be approximately 0.830 ohm, and secondary windings resistance should be approximately 5600 ohms.

REPAIRS

TIGHTENING TORQUES. Recommended tightening torque specifications are as follows:

Crankcase screws . . . . 4-5 N•m
(35-44 in.-lbs.)
Cylinder . . . . 4-5 N•m
(35-44 in.-lbs.)
Flywheel nut . . . . 11.8-13.8 N•m
(104-122 in.-lbs.)
Spark plug . . . . 16.7-18.6 N•m
(148-164 in.-lbs.)

PISTON, PIN, RING AND CYLINDER. To remove piston, first unbolt and remove pump from engine. Remove recoil starter assembly, lock engine starter pulley and remove impeller from pump shaft. Unbolt and remove end plate, adapter plate and pump shaft from engine. Remove fan housing, muffler and carburetor. Remove cylinder retaining screws and withdraw cylinder from piston. Remove piston pin retaining rings, push pin out of piston and remove piston from connecting rod.

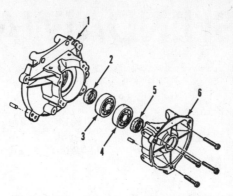

Fig. SH16—Exploded view of crankcase assembly.

1. Crankcase half
2. Seal
3. Bearing
4. Bearing
5. Seal
6. Crankcase half

The cylinder bore is chrome plated and should be inspected for excessive wear and damage to cylinder bore. Cylinder should be renewed if chrome plating is worn through. Inspect piston and discard if excessive wear, scoring or other damage is evident.

Specified piston diameter is 31.97 mm (1.2586 in.) with a wear limit of 31.90 mm (1.256 in.). Specified cylinder diameter is 32.00 mm (1.2598 in.) with a wear limit of 32.10 mm (1.264 in.). Maximum allowable cylinder out-of-round is 0.03 mm (0.0012 in.).

Specified piston pin bore in piston is 9.00 mm (0.354 in.) with a wear limit of 9.03 mm (0.3555 in.). Specified piston pin diameter is 9.00 (0.354 in.) with a wear limit of 8.98 mm (0.3535 in.).

Piston ring end gap should be 0.1-0.3 mm (0.004-0.012 in.); maximum allowable gap is 0.6 mm (0.024 in.). Specified piston ring groove width is 1.50 mm (0.059 in.). Specified piston ring side clearance is 0.04-0.08 mm (0.0016-0.0031 in.) with a wear limit of 0.20 mm (0.008 in.).

Piston must be installed so arrow on piston crown points towards exhaust port. Heating piston in hot water or oil

will make removal and installation of piston pin easier. When installing piston pin, do not apply side pressure to connecting rod as rod may be bent. Install new piston pin retaining rings, positioning open end of rings 90° from grooves cut in side of piston.

A locating pin in the piston ring groove prevents piston ring rotation. Be sure piston ring ends properly engage pin when installing cylinder. Lubricate cylinder and piston with oil. Install cylinder using a new gasket, apply thread locking compound to retaining screws and tighten screws evenly to 4-5 N•m (35-44 in.-lbs.).

CRANKSHAFT, CONNECTING ROD AND CRANKCASE. Refer to Fig. SH15 for an exploded view of crankshaft and rod assembly. Refer to Fig. SH16 for an exploded view of crankcase components.

Remove cylinder and piston as previously outlined. Use a suitable puller to remove flywheel from crankshaft. Do not strike end of crankshaft with a hammer as shock may damage flywheel or crankshaft. After removing screws from crankcase halves, tap crankcase with a soft-faced hammer to separate halves. Do not pry between crankcase halves as crankcase sealing surfaces may be damaged. Heat crankcase halves to release bearings.

Connecting rod and crankshaft are a unit assembly and not available separately. Do not attempt to separate rod and crankshaft.

Specified small end diameter of connecting rod is 12.00 mm (0.4724 in.) with a wear limit of 12.03 mm (0.4736 in.).

Maximum allowable crankshaft deflection measured at main bearing journals is 0.07 mm (0.0027 in.).

Heat crankcase halves to approximately 100° C (212° F) to install oil seals and main bearings. Apply lithium base grease to lip of oil seals. Apply thin coat

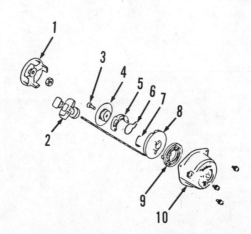

Fig. SH17—Exploded view of rewind starter.

1. Starter pulley
2. Handle
3. Screw
4. Friction plate
5. Ratchet
6. Friction spring
7. Return spring
8. Rope pulley
9. Rewind spring
10. Housing

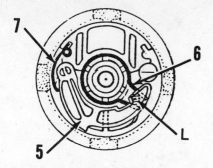

Fig. SH18—Install friction spring (6) so prongs straddle lug (L) on ratchet (5).

of liquid gasket maker to mating surfaces of both crankcase halves, then assemble crankshaft and crankcase halves. Make certain crankshaft turns freely. Apply Loctite thread locking compound to crankcase screws and tighten evenly to 4-5 N•m (35-44 in.-lbs.).

REWIND STARTER. To disassemble starter, remove rope handle (2—Fig. SH17) and allow rope to wind into starter. Unscrew retaining screw (3) and remove friction plate (4), ratchet (5), friction spring (6) and return spring (7). Wear appropriate safety eyewear and gloves before disengaging pulley (8) from starter housing (10) as rewind spring may uncoil uncontrolled. Turn pulley to disengage pulley from recoil spring and lift pulley out of housing; spring should remain in starter housing. If spring must be removed from housing, position housing so spring side is down and against floor, then tap housing to dislodge spring.

Install rewind spring so outer end engages starter housing notch and coils are in counterclockwise direction from outer end. Wrap rope around pulley in counterclockwise direction viewed from flywheel side of pulley. Position pulley in starter housing and engage recoil spring in pulley. Install ratchet (5) as shown in Fig. SH18 and hook outer end of ratchet spring (7) under ratchet. Apply a light coat of grease to friction spring (6) and install spring in groove of friction plate. Install friction plate so the two ends of friction spring (6) fit around lug (L) on ratchet. Install center screw. Pretension rewind spring by turning pulley two turns counterclockwise before passing rope through rope outlet of housing and attaching handle.

STIHL

STIHL INC.
536 Viking Drive
Virginia Beach, Virginia 23452

Model	Bore	Stroke	Displacement	Rated Power
08S	49 mm*	32 mm	60 cc*	3.0 kW
	(1.93 in.)	(1.26 in.)	(3.66 cu.in.)	(4.1 hp)

*On early models, bore diameter was 47 mm (1.85 in.) and displacement was 56 cc (3.39 cu. in.).

ENGINE IDENTIFICATION

The Model 08S covered here is an air-cooled, single-cylinder, two-stroke engine used on Stihl augers.

MAINTENANCE

LUBRICATION. The engine is lubricated by mixing oil with the fuel. Regular unleaded gasoline is recommended. Recommended oil is Stihl two-stroke engine oil mixed at a 40:1 or 50:1 ratio. If Stihl oil is not available, a good quality oil designed for use in air-cooled, two-stroke engines may be used when mixed at a 25:1 ratio. Use a separate container to mix oil and gasoline. Do not mix directly in engine fuel tank.

SPARK PLUG. Recommended spark plug is Bosch W175 for models with 47 mm (1.85 in.) bore and Bosch WSR6F for models with 49 mm (1.93 in.) bore. Specified electrode gap is 0.5 mm (0.020 in.). Tighten spark plug to 24 N·m (18 ft.-lbs.).

CARBURETOR. The engine may be equipped with a Tillotson HL or Zama LAS8 diaphragm type carburetor. Refer to Fig. SL1 for an exploded view of carburetor.

Initial setting of idle mixture screw (5) and high speed mixture screw (7) is one turn open from a lightly seated position. Adjust idle mixture screw and idle speed screw (13) so engine idles cleanly just below clutch engagement speed and engine will accelerate without hesitation. Adjust high speed mixture screw to obtain optimum performance with engine under load. Final adjustments must be made with engine at normal operating temperature. Do not set high speed mixture screw too lean as engine may be damaged.

The carburetor is equipped with a governor valve (9) that enriches the mixture at high speed to prevent overspeeding.

To overhaul the carburetor, refer to Fig. SL1 and remove fuel inlet (31) and screen (29). Remove fuel pump body (28), pump diaphragm (27), diaphragm cover (25) and metering diaphragm (24). Remove diaphragm lever (19), spring (18) and fuel inlet valve (22). Unscrew fuel inlet seat (21) from carburetor body. Remove idle mixture screw (5) and high speed mixture screw (7). The valve jet (17) can be removed by pushing it from the diaphragm chamber towards the venturi. Mark throttle and choke plates (11 and 14) so they can be returned to their original positions. Note that a detent ball (15) and spring (16) are located behind the choke shaft. Thoroughly clean all parts using suitable solvent. Clean all fuel passages with compressed air. Do not use wires or drills to clean orifices or passages.

Diaphragms (24 and 27) should be flat and free of holes, creases, tears and other imperfections. Be careful when handling spring (18) as stretching, compressing or otherwise damaging spring will affect carburetor operation. Ball in valve jet (17) must be free. Fuel inlet valve and seat should be renewed as a set.

Install Welch plug so convex side is out, then flatten plug using a 5/16-inch (8 mm) diameter flat-end punch. A correctly installed Welch plug is flat; a concave plug may leak. Install valve jet (17) so rear edge of the jet is flush with floor of diaphragm chamber. Install fuel inlet valve seat, valve, spring and metering lever. Be sure that spring (18) sits on dimple of diaphragm lever (19). The diaphragm lever must rotate freely on pin (12). The lever must be flush with the diaphragm chamber floor as shown in Fig. SL2. Three cast pins are located on

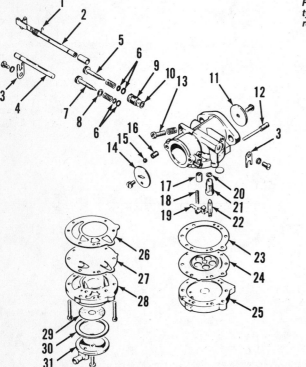

Fig. SL1—Exploded view of typical diaphragm type carburetor.

1. Spring
2. Throttle shaft
3. Retainer clips
4. Choke shaft
5. Idle mixture screw
6. "O" rings
7. High speed mixture screw
8. Retaining ring
9. Governor valve
10. Gasket
11. Throttle plate
12. Diaphragm lever pivot pin
13. Idle speed screw
14. Choke plate
15. Detent ball
16. Spring
17. Valve jet
18. Spring
19. Diaphragm lever
20. Gasket
21. Fuel inlet valve seat
22. Fuel inlet valve
23. Gasket
24. Diaphragm
25. Diaphragm cover
26. Gasket
27. Fuel pump diaphragm & valves
28. Pump body
29. Screen
30. Gasket
31. Fuel inlet

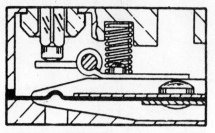

Fig. SL2—Diaphragm lever should be flush with diaphragm chamber floor.

carburetor body and fuel pump cover. Diaphragm gaskets, diaphragms and cover must be correctly positioned on the cast pins to insure correct installation and operation.

GOVERNOR. Maximum governed engine speed is determined by a governor valve (9—Fig. SL1) that causes an excessively rich air:fuel mixture at high speed to prevent overspeeding. The governor valve is not adjustable.

IGNITION SYSTEM. The engine is equipped with a solid-state breakerless

ignition system. Flywheel must be removed for access to ignition components.

The electronic ignition module is mounted behind the flywheel. The ignition switch grounds the module to stop the engine. There is no periodic maintenance or adjustment required with this ignition system. Specified ignition coil-to-flywheel air gap is 0.2-0.3 mm (0.008-0.012 in.).

If ignition module fails to produce a spark, check for faulty kill switch or grounded wires. Measure resistance of ignition module secondary using suitable ohmmeter. Connect one test lead to spark plug terminal of high tension wire and other test lead to module core laminations. If resistance is either very low or infinite, renew module.

REPAIRS

TIGHTENING TORQUES. Recommended tightening torque specifications are as follows:

Clutch nut................. 29 N•m
(22 ft.-lbs.)

Crankcase screws .... 5 N•m
(44 in.-lbs.)
Cylinder .... 10 N•m
(89 in.-lbs.)
Flywheel nut .... 29 N•m
(22 ft.-lbs.)
Spark plug .... 24 N•m
(18 ft.-lbs.)

CRANKCASE PRESSURE TEST. An improperly sealed crankcase can cause the engine to be hard to start, run rough, have low power and overheat. Refer to SERVICE SECTION TROUBLESHOOTING section of this manual for crankcase pressure test procedure. If crankcase leakage is indicated, pressurize crankcase and use a soap and water solution to check gaskets, seals, carburetor pulse line and casting for leakage.

CYLINDER, PISTON, PIN AND RINGS. The piston is equipped with two piston rings. The floating piston pin is retained in the piston by a snap ring at each end. The pin bore of the piston is unbushed. The connecting rod has a caged needle roller piston pin bearing.

To remove cylinder and piston, first remove fan housing and control handle shroud. Remove carburetor, muffler and spark plug. Remove cylinder retaining screws and withdraw cylinder from piston.

NOTE: If crankshaft and connecting rod are to be removed, insert a wood block between piston and crankcase and loosen flywheel and clutch retaining nuts before removing piston.

Remove piston pin snap rings (7—Fig. SL3) and push pin (6) out of piston and connecting rod.

The cylinder and head are one piece. Cylinder is available only with a fitted piston. On early models, cylinders and pistons are graded according to size and stamped with a letter from A to E, with A denoting the smallest diameter. New cylinder and piston should have same letter. Service parts are graded B. Replacement piston may be used in all cylinders. Cylinder bore on all models is chrome plated. Cylinder should be renewed if chrome plating is flaking, scored or worn away.

Install piston on connecting rod so arrow on piston crown points toward exhaust port side of crankcase. Piston rings are held in position by a pin located in each piston ring groove. Be sure that piston ring ends properly engage the pins when installing cylinder on piston. Tighten cylinder retaining screws evenly to 10 N•m (89 in.-lbs.) torque.

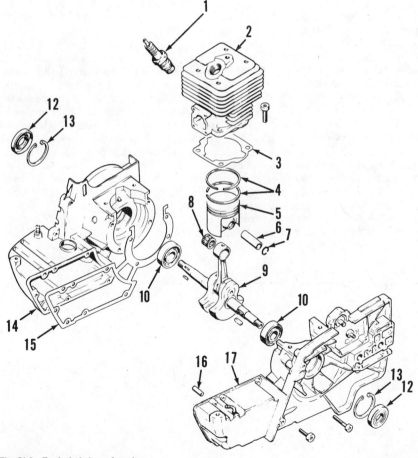

Fig. SL3—Exploded view of engine.

1. Spark plug
2. Cylinder
3. Gasket
4. Piston rings
5. Piston
6. Piston pin
7. Snap ring
8. Roller bearing
9. Crankshaft & connecting rod assy.
10. Ball bearings
12. Seal
13. Snap ring
14. Crankcase half
15. Gasket
16. Dowel pin
17. Crankcase half

Illustrations courtesy Stihl, Inc.

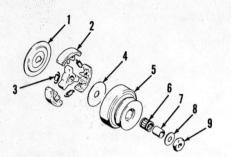

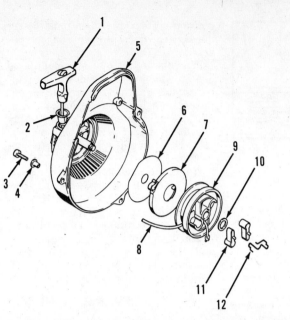

Fig. SL4—Exploded view of clutch used on some models.

1. Washer
2. Clutch shoe
3. Clutch spring
4. Washer
5. Clutch drum
6. Roller bearing
7. Inner race
8. Washer
9. Nut (L.H.)

Fig. SL5—Exploded view of rewind starter.

1. Rope handle
2. Bushing
3. Screw
4. Sleeve
5. Housing
6. Washer
7. Rewind spring
8. Rope
9. Pulley
10. Washer
11. Pawl
12. Spring clip

CONNECTING ROD, CRANK-SHAFT AND CRANKCASE. Clutch, flywheel, ignition unit, cylinder and piston must be removed and crankcase halves must be split for access to crankshaft (9—Fig. SL3). Remove both keys from crankshaft and drive crankcase dowel pins (16) into ignition side crankcase half (14) before splitting crankcase halves. Do not damage crankcase mating surfaces. During disassembly if main bearing is tight in crankcase half, heat crankcase half so bearing is released.

The crankshaft is supported by ball bearings at both ends. Connecting rod, crankpin and crankshaft (9) are a pressed-together assembly and available only as a unit. Connecting rod big end rides on a roller bearing and should be inspected for excessive wear and damage. If rod, bearing or crankshaft is damaged, complete crankshaft and connecting rod assembly must be renewed.

When installing main bearing (10—Fig. SL3), heat crankcase half and seat bearing against snap ring (13). It may be necessary to heat main bearing to ease insertion of crankshaft. Tighten crankcase screws evenly to 5 N•m (44 in.-lbs.). Install new crankshaft oil seals (12) being careful not to damage seal lip on sharp edge of crankshaft.

It is recommended that the crankcase be pressure tested for leakage before proceeding with remainder of engine reassembly.

CLUTCH. Refer to Fig. SL4 for an exploded view of clutch used on some models. Note that clutch retaining nut (9) has left-hand threads. Install washer (1) so concave side is towards crankcase and slot engages crankshaft key. Springs and shoes are available only as sets. Stihl tools 1107 890 4500 and 1108 893 4500 are available to remove and install clutch hub.

REWIND STARTER. Refer to Fig. SL5 for an exploded view of rewind starter. To disassemble, remove fan housing (5) with starter assembly. While holding rope pulley from turning, unwind two turns of starter rope from pulley. Release pulley and allow to unwind slowly to relieve spring tension. Disengage spring clip (12) and withdraw starter components from fan housing shaft.

Install rewind spring (7) in housing so spring is wound clockwise from outer end. Rope must be wound on pulley in clockwise direction when viewed from pawl side of pulley.

To place tension on starter rope, pull rope out of housing until notch in pulley is aligned with rope outlet, then hold pulley to prevent pulley rotation. Pull rope back into housing while positioning rope in pulley notch. Turn rope pulley clockwise until spring is tight, allow pulley to turn counterclockwise until notch aligns with rope outlet, and disengage rope from notch. Check starter operation. Rope handle should be held against housing by spring tension, but it must be possible to rotate pulley at least ½ turn clockwise when rope is pulled out fully.

TANAKA (TAS)

TANAKA KOGYO (USA) LTD.
7509 S. 228th St.
Kent, WA 98031

Model	Bore	Stroke	Displacement	Rated Power
Tanaka	30 mm	28 mm	20.0 cc	0.75 kW
	(1.18 in.)	(1.10 in.)	(1.22 cu.in.)	(1.0 hp)
Tanaka	31 mm	28 mm	21 cc	0.82 kW
	(1.22 in.)	(1.10 in.)	(1.28 cu.in.)	(1.1 hp)
Tanaka	33 mm	30 mm	26.0 cc	1.12 kW
	(1.30 in.)	(1.18 in.)	(1.59 cu.in.)	(1.5 hp)
Tanaka	36 mm	30 mm	31 cc	1.34 kW
	(1.42 in.)	(1.18 in.)	(1.86 cu.in.)	(1.8 hp)
Tanaka	38 mm	30 mm	34 cc	1.49 kW
	(1.50 in.)	(1.18 in.)	(2.07 cu.in.)	(2.0 hp)
Tanaka	40 mm	32 mm	40 cc	1.71 kW
	(1.57 in.)	(1.26 in.)	(2.44 cu.in.)	(2.3 hp)
Tanaka	38 mm	38 mm	43 cc	1.71 kW
	(1.50 in.)	(1.50 in.)	(2.62 cu.in.)	(2.3 hp)
Tanaka	41 mm	38 mm	50.2 cc	1.86 kW
	(1.61 in.)	(1.50 in.)	(3.06 cu.in.)	(2.5 hp)

These engines are used on Tanaka equipment, as well as equipment of some other manufacturers.

ENGINE INFORMATION

Tanaka two-stroke, air-cooled gasoline engines are used by several manufacturers of string trimmers, brush cutters and blowers.

MAINTENANCE

LUBRICATION. Engine lubrication is obtained by mixing gasoline with an good quality oil designed for two-stroke, air-cooled engines. Do not use automotive type oil. Manufacturer also does not recommend the use of gasoline containing alcohol. The recommended fuel:oil mixture ratio is 25 parts gasoline to 1 part oil.

SPARK PLUG. Recommended spark plug is NGK BM6A or equivalent. Electrode gap should be 0.6 mm (0.024 in.).

CARBURETOR. Various types of carburetors have been used. Refer to the appropriate following section for carburetor service.

Walbro HDA. The Walbro HDA is used on a blower engine. Initial setting of idle and high speed mixture screws is 1 1/8 turns out from a lightly seated position. Final adjustment is performed with engine running at normal operating temperature. All blower tubing must be attached. Adjust idle speed screw so engine idles at 2500 rpm. Adjust idle mixture screw to obtain highest idle speed, then turn screw 1/6 turn

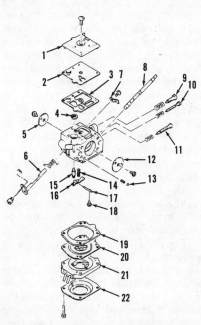

Fig. TA101—Exploded view of Walbro HDA carburetor.

1. Cover	12. Choke plate
2. Fuel pump diaphragm	13. Detent ball
3. Gasket	14. Spring
4. Screen	15. Fuel inlet valve
5. Throttle plate	16. Metering lever
6. Throttle shaft	17. Pin
7. Throttle stop	18. Screw
8. Choke shaft	19. Gasket
9. Idle speed screw	20. Metering diaphragm
10. Idle mixture screw	21. Plate
11. High speed mixture screw	22. Cover

counterclockwise and readjust idle speed screw to 2500 rpm. Adjust high speed mixture screw to obtain highest engine speed at full throttle, then turn screw 1/6 turn counterclockwise. Do not adjust high speed mixture screw too lean as engine may be damaged.

Carburetor disassembly and reassembly is evident after inspection of carburetor and referral to Fig. TA101. Clean and inspect all components. Wires or drill bits should not be used to clean passages as fuel flow may be altered. Inspect diaphragms (2 and 20) for defects which may affect operation. Examine fuel inlet valve and seat. Inlet valve (15) is renewable, but carburetor body must be renewed if seat is damaged or excessively worn. Discard carburetor body if mixture screw seats are damaged or excessively worn. Screens should be clean. Be sure throttle plate fits shaft and carburetor bore properly. Apply Loctite to throttle plate retaining screws. Adjust metering lever height so metering lever tip is flush with body as shown in Fig. TA102.

Walbro WA. Some engines are equipped with a Walbro WA diaphragm type carburetor. Initial adjustment of idle and high speed mixture screws is 1 1/4 turns out from a lightly seated position. Final adjustments are performed with trimmer line at recommended length or blade installed. Engine must be at operating temperature and running. Be sure engine air filter is clean before adjusting carburetor.

Fig. TA102—Tip of metering lever should be flush with body. Bend metering lever as needed.

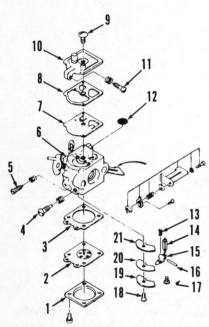

Fig. TA103—Exploded view of Walbro WA carburetor.

1. Cover	11. Idle speed screw
2. Diaphragm	12. Screen
3. Gasket	13. Spring
4. High speed	14. Fuel inlet valve
mixture screw	15. Metering lever
5. Idle mixture screw	16. Pin
6. Body	17. Screw
7. Fuel pump diaphragm	18. Screw
8. Gasket	19. Circuit plate
9. Screw	20. Diaphragm
10. Cover	21. Gasket

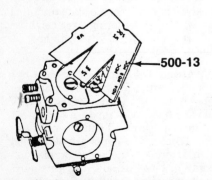

Fig. TA104—Metering lever should just touch leg of Walbro tool 500-13. Bend lever to obtain correct lever height.

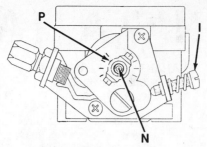

Fig. TA105—On Walbro WY or WYJ carburetor, idle speed screw is located at (I), idle mixture limiter plate is located at (P) and idle mixture needle is located at (N). A plug covers the idle mixture needle.

Adjust idle speed screw (11—Fig. TA103) so trimmer head or blade does not rotate. Adjust idle mixture screw (5) to obtain maximum idle speed possible, then turn idle mixture screw 1/6 turn counterclockwise. Readjust idle speed. Operate unit at full throttle and adjust high speed mixture screw (4) to obtain maximum engine rpm, then turn high speed mixture screw 1/6 turn counterclockwise.

When overhauling carburetor, refer to exploded view in Fig. TA103. Examine fuel inlet valve and seat. Inlet valve (14) is renewable, but carburetor body must be renewed if seat is excessively worn or damaged. Inspect mixture screws and seats. Renew carburetor body if seats are excessively worn or damaged. Clean fuel screen. Inspect diaphragms for tears and other damage.

Check metering lever height as shown in Fig. TA104 using Walbro tool 500-13. Metering lever should just touch leg on tool. Bend lever to obtain correct lever height.

Walbro WY and WYJ. Some engines may be equipped with a Walbro WY or WYJ carburetor. This is a diaphragm type carburetor that uses a barrel type throttle rather than a throttle plate.

Idle fuel for the carburetor flows up into the throttle barrel where it is fed into the air stream. On some models, the idle fuel flow can be adjusted by turning an idle mixture limiter plate (P—Fig. TA105). Initial setting is in center notch. Rotating the plate clockwise will lean the idle mixture. Inside the limiter plate is an idle mixture needle (N—Fig. TA106) that is preset at the factory. If removed, use the following procedure to determine correct position. Back out needle (N) until unscrewed. Screw in needle 5 turns on Model WY or 15 turns on Model WYJ. Rotate idle mixture plate (P—Fig. TA105) to center notch. Run engine until normal operating temperature is attained. Adjust idle speed screw so trimmer head or blade does not rotate. Rotate idle mixture needle (N—Fig. TA106) and obtain highest

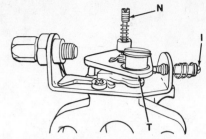

Fig. TA106—View of idle mixture needle (N) used on Walbro WY carburetor.

rpm (turning needle clockwise leans the mixture), then turn needle 1/4 turn counterclockwise. Readjust idle speed screw. Note that idle mixture plate and needle are available only as an assembly with throttle barrel (21—Fig. TA107).

The high speed mixture is controlled by a removable fixed jet (16—Fig. TA107).

To overhaul carburetor, refer to exploded view in Fig. TA107 and note the following: On models with a plastic body, clean only with solvents approved for use with plastic. Do not use wire or drill bits to clean fuel passages as fuel flow may be altered. Do not disassemble throttle barrel assembly (21). Examine fuel inlet valve and seat. Inlet valve (9) is renewable, but fuel pump body (11) must be renewed if seat is excessively worn or damaged. Clean fuel screen. Inspect diaphragms for tears and other damage. When installing plates and gaskets (12 through 15) note that tabs (T) on ends will "stairstep" when correctly installed. Adjust metering lever height to obtain 1.5 mm (0.059 in.) between carburetor body surface and lever as shown in Fig. TA108.

Walbro WZ. Some engines may be equipped with a Walbro WZ carburetor. This is a diaphragm type carburetor that uses a barrel type throttle rather than a throttle plate.

Idle fuel for the carburetor flows up into the throttle barrel where it is fed into the air stream. Idle fuel flow can be adjusted by turning idle mixture limiter plate (P—Fig. TA109). Initial setting is in center notch. Rotating the plate clockwise will lean the idle mixture. Inside the limiter plate is an idle mixture needle (N—Fig. TA106) that is preset at the factory (a plug covers the needle). If idle mixture needle is removed, use the following procedure to determine correct position. Back out needle (N) until unscrewed, then screw in needle 6 turns. Rotate idle mixture plate (P—Fig. TA109) to center notch. Run engine until normal operating temperature is attained. Adjust idle speed screw so trimmer head or blade does not rotate. Rotate idle mixture needle (N—Fig. TA106) and obtain highest rpm (turning

Illustrations courtesy Tanaka Ltd.

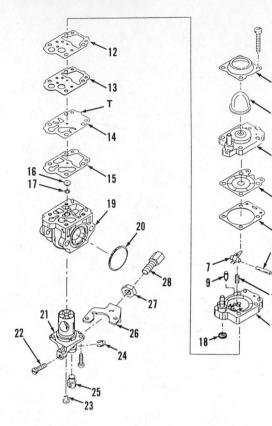

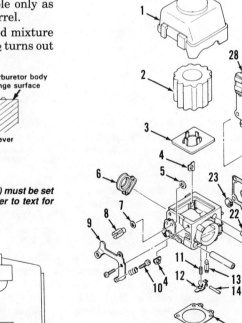

Illustrations courtesy Tanaka Ltd.

Fig. TA107—Exploded view of Walbro WYJ. Model WY is similar.

1. Cover
2. Primer bulb
4. Plate
5. Metering diaphragm
6. Gasket
7. Metering lever
8. Pin
9. Fuel inlet valve
10. Spring
11. Fuel pump body
12. Gasket
13. Fuel pump plate
14. Fuel pump diaphragm
15. Gasket
16. Main jet
17. "O" ring
18. Fuel screen
19. Body
20. "O" ring
21. Throttle barrel assy.
22. Idle speed screw
23. Plug
24. "E" ring
25. Swivel
26. Bracket
27. Nut
28. Adjuster

from a lightly seated position. Adjust high speed mixture screw to obtain highest engine speed, then turn screw ¼ turn counterclockwise. Do not adjust mixture too lean as engine may be damaged.

To overhaul carburetor, refer to exploded view in Fig. TA110 and note the following: Clean only with solvents approved for use with plastic. Do not disassemble throttle barrel assembly. Examine fuel inlet valve and seat. Inlet valve (13) is renewable, but carburetor body must be renewed if seat is excessively worn or damaged. Inspect high speed mixture screw and seat. Renew carburetor body if seat is excessively worn or damaged. Clean fuel screen. Inspect diaphragms for tears and other damage. When installing plates and gaskets (24, 25 and 26) note that tabs (T) on ends will "stairstep" when correctly installed. Adjust metering lever height to obtain 1.5 mm (0.059 in.) between carburetor body surface and lever as shown in Fig. TA108.

Float Type Carburetor. Two different float type carburetors have been used (see Fig. TA111 or TA112). Service

needle clockwise leans the mixture), then turn needle counterclockwise until rpm decreases 200-500 rpm. Readjust idle speed screw. Note that idle mixture plate and needle are available only as an assembly with throttle barrel.

Initial setting of high speed mixture screw (21—Fig. TA110) is 1 ½ turns out

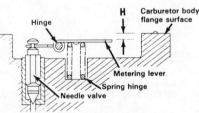

Fig. TA108—Metering lever height (H) must be set on diaphragm type carburetors. Refer to text for specified height.

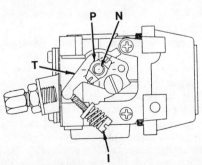

Fig. TA109—On Walbro WZ carburetor idle speed screw is located at (I), idle mixture limiter plate is located at (P) and idle mixture needle is located at (N). A plug covers the idle mixture needle.

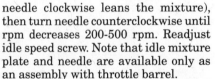

Fig. TA110—Exploded view of Walbro WZ carburetor.

1. Cover
2. Air cleaner element
3. Plate
4. Plate
5. Stop ring
6. Gasket
7. Thrust washer
8. Swivel
9. Bracket
10. Idle speed screw
11. Spring
12. Metering lever
13. Fuel inlet valve
14. Pin
15. Gasket
16. Metering diaphragm
17. Cover
18. Bracket
19. Nut
20. Cable adjuster
21. High speed mixture screw
22. Sleeve
23. Screen
24. Gasket
25. Fuel pump diaphragm
26. Plate
27. Gasket
28. Fuel pump cover
29. Gasket
30. Primer bulb
31. Retainer

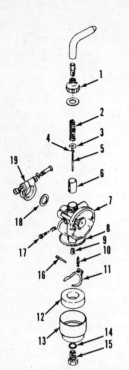

Fig. TA111—*Exploded view of float type carburetor used on some models.*

1. Cap	11. Float lever
2. Spring	12. Float
3. Seat	13. Float bowl
4. Clip	14. Gasket
5. Jet needle	15. Screw
6. Throttle slide	16. Pin
7. Body	17. Idle speed screw
8. Gasket	18. Gasket
9. Main jet	19. Clamp
10. Fuel inlet valve	

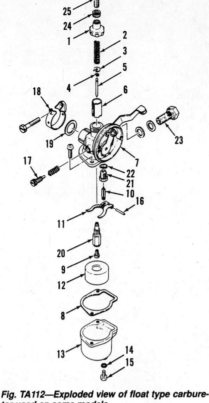

Fig. TA112—*Exploded view of float type carburetor used on some models.*

1. Cap	14. Gasket
2. Spring	15. Screw
3. Seat	16. Pin
4. Clip	17. Idle speed screw
5. Jet needle	18. Clamp
6. Throttle slide	19. Gasket
7. Body	20. Nozzle
8. Gasket	21. Fuel inlet valve seat
9. Main jet	22. "O" ring
10. Fuel inlet valve	23. Fuel inlet fitting
11. Float lever	24. Nut
12. Float	25. Cable adjuster
13. Float bowl	

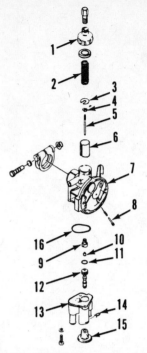

Fig. TA113—*Exploded view of rotary type carburetor used on some models.*

1. Cap	9. Main jet
2. Spring	10. Cap
3. Seat	11. "O" ring
4. Clip	12. Check valve
5. Jet needle	13. Reservoir
6. Throttle slide	14. Clip
7. Body	15. Rubber cap
8. Idle speed screw	16. Gasket

Fig. TA114—*On models with breaker points, align "M" mark on flywheel magneto with mark on crankcase as shown.*

procedure for both carburetors is similar.

Idle mixture is not adjustable. Idle speed may be adjusted by turning adjusting screw (17). High speed mixture is controlled by removable fixed jet (9). Midrange mixture is determined by the position of clip (4) on jet needle (5). There are three grooves in upper end of the jet needle and normal position of clip is in the middle groove. The mixture will be leaner if clip is installed in the top groove, or richer if clip is installed in the bottom groove.

Before removing carburetor from engine, unscrew cap (1) and withdraw throttle slide (6) assembly. When overhauling carburetor, refer to Fig. TA111 or TA112 and note the following: Examine fuel inlet valve and seat. Inlet valve (10) is renewable on all models. Inlet seat (21) is renewable on some models. On some models, the inlet seat is not renewable and carburetor body (7) must be renewed if seat is excessively worn or damaged. When installing throttle slide, be sure groove in side of throttle slide (6) indexes with pin in bore of carburetor body.

Rotary Type Carburetor. Refer to Fig. TA113 for an exploded view of the rotary type carburetor used on some models.

Idle mixture is not adjustable. High speed mixture is controlled by removable fixed jet (9). Midrange mixture is determined by the position of clip (4) on jet needle (5). There are three grooves in upper end of the jet needle and normal position of clip is in the middle groove. The mixture will be leaner if clip is installed in the top groove, or richer if clip is installed in the bottom groove.

Before removing carburetor from engine, unscrew cap (1) and withdraw throttle slide (6) assembly. When overhauling carburetor, refer to Fig. TA113 and note the following: Remove reservoir (13), clip (14) and rubber cap (15). Pull check valve (12) out of reservoir. Remove check valve cap (10) and "O" ring (11).

Inspect components and renew if damaged or excessively worn. When installing throttle slide, be sure groove in side of throttle slide (6) indexes with pin in bore of carburetor body.

IGNITION SYSTEM. The engine may be equipped with a breaker point type ignition system or a solid-state type ignition system. Refer to appropriate following paragraphs for service information.

Breaker-Point Ignition System. The ignition condenser and breaker points are located behind the flywheel. To adjust breaker point gap the flywheel must be removed. Breaker point gap should be 0.35 mm (0.014 in.).

Ignition timing is adjusted by moving magneto stator plate. Breaker points should open when "M" mark on flywheel is aligned with reference mark on crankcase (see Fig. TA114).

On models with external ignition coil, air gap between coil legs and flywheel

Illustrations courtesy Tanaka Ltd.

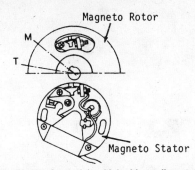

Fig. TA115—On models with ignition coil on stator plate, the metal core legs of the ignition coil should be flush with outer edge of the stator plate.

should be 0.35 mm (0.014 in.). On models with ignition coil on stator plate (Fig. TA115), the metal core legs of the ignition coil should be flush with outer edge of the stator plate.

Solid-State Ignition System. All later models are equipped with a solid-state ignition system. The engine may be equipped with a one-piece ignition module that includes the ignition coil, or a two-piece ignition system that consists of a separate ignition module and ignition coil. The ignition coil or module may be mounted on the blower housing, crankcase or cylinder.

The ignition system is considered satisfactory if a spark will jump across the 3 mm (1/8 in.) gap of a test spark plug. If no spark is produced, check on/off switch, wiring and ignition module/coil air gap. Air gap between flywheel magnet and ignition module/coil should be O.25-0.35 mm (0.010-0.014 in.).

REPAIRS

CRANKCASE PRESSURE TEST. An improperly sealed crankcase can cause the engine to be hard to start, run rough, have low power and overheat. Refer to SERVICE SECTION TROUBLE-SHOOTING section of this manual for crankcase pressure test procedure. If crankcase leakage is indicated, pressurize crankcase and use a soap and water solution to check gaskets, seals, carburetor pulse line and casting for leakage.

PISTON, PIN AND RINGS. The piston is equipped with two piston rings. Ring rotation is prevented by a locating pin in each piston ring groove. Piston and rings are available only in standard diameter.

To remove piston, remove engine cover, starter housing and fuel tank. Remove air cleaner, carburetor and muffler. Remove cylinder mounting screws and pull cylinder off piston. Remove retaining rings (5—Fig. TA116), then use a suitable pin puller to push piston pin (6) out of piston. Remove piston (4) and bearing (7) from connecting rod.

Lubricate piston, rings and cylinder wall with engine oil before installing piston. Install piston so arrow on piston crown points towards exhaust port. On 50.2 cc engine, install piston so shaped portion of crown is towards intake. Be sure piston ring gaps are correctly indexed with locating pins in piston ring grooves when installing cylinder.

CYLINDER. Some engines are equipped with a chrome plated cylinder. Renew cylinder if bore is excessively worn, scored or otherwise damaged. Cylinder is available only in standard size.

CRANKSHAFT, CONNECTING ROD AND CRANKCASE. Crankshaft, connecting rod and rod bearing are a unit assembly; individual components are not available separately. The crankshaft is supported by ball bearings at both ends.

To remove crankshaft and connecting rod assembly, separate engine from trimmer drive housing. Remove engine cover, starter housing, fuel tank, air cleaner, carburetor and muffler. Remove spark plug and install a lock screw or the end of a starter rope in spark plug hole to lock piston and crankshaft. Remove nuts attaching flywheel and starter pulley to crankshaft. Use suitable puller to remove flywheel. Remove cylinder mounting screws and pull cylinder off piston. Remove crankshaft keys and crankcase retaining bolts. Carefully tap on one end of crankshaft to separate crankcase halves. Remove crankshaft and connecting rod assembly from crankcase. Use suitable puller to push piston pin from piston and separate piston from crankshaft. If main bearings (13) remain in crankcase, heat crankcase halves slightly to aid bearing removal. Drive seals (12) out of crankcase halves.

A renewable needle bearing (7) is located in the small end of the connecting rod. Maximum allowable side clearance at rod big end is 0.5 mm (0.020 in.).

Heat crankcase if necessary to aid removal of main bearings. Install seals (12—Fig. TA116) with lip towards inside of crankcase. On some engines, shims (16) are available to adjust crankshaft end play. Maximum allowable crankshaft end play is 0.1 mm (0.004 in.).

CLUTCH. 50.2 cc Engine. Trimmers with a 50.2 engine are equipped

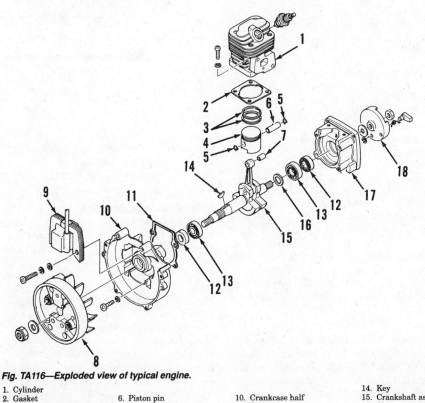

Fig. TA116—Exploded view of typical engine.

1. Cylinder
2. Gasket
3. Piston rings
4. Piston
5. Retaining rings
6. Piston pin
7. Needle bearing
8. Flywheel
9. Ignition coil
10. Crankcase half
11. Gasket
12. Seal
13. Bearings
14. Key
15. Crankshaft assy.
16. Shim
17. Crankcase half
18. Starter pulley

Illustrations courtesy Tanaka Ltd.

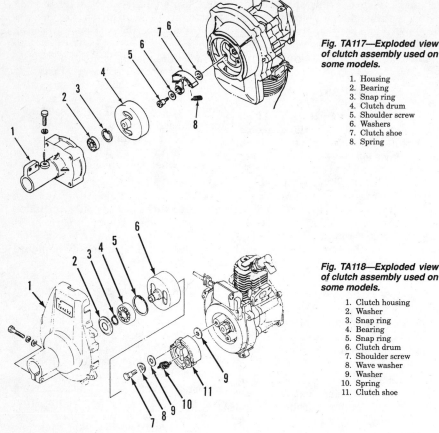

Fig. TA117—Exploded view
of clutch assembly used on
some models.

1. Housing
2. Bearing
3. Snap ring
4. Clutch drum
5. Shoulder screw
6. Washers
7. Clutch shoe
8. Spring

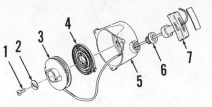

Fig. TA120—Exploded view of rewind starter used
on some models.

1. Screw
2. Washer
3. Pulley
4. Rewind spring
5. Housing
6. Rope guide
7. Handle

Fig. TA118—Exploded view
of clutch assembly used on
some models.

1. Clutch housing
2. Washer
3. Snap ring
4. Bearing
5. Snap ring
6. Clutch drum
7. Shoulder screw
8. Wave washer
9. Washer
10. Spring
11. Clutch shoe

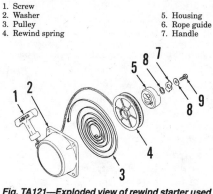

Fig. TA121—Exploded view of rewind starter used
on some models.

1. Handle
2. Housing
3. Rewind spring
4. Pulley
5. Pawl
6. Washer
7. Friction plate
8. Washer
9. Screw

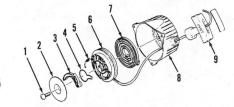

Fig. TA122—Exploded view of rewind starter used
on some models.

1. Screw
2. Plate
3. Pawl
4. Friction spring
5. Spring
6. Pulley
7. Rewind spring
8. Housing
9. Handle

Fig. TA119—Exploded view of clutch drum and
bearing assembly used on some models.

1. Housing
2. Isolator
3. Carrier
4. Snap ring
5. Bearings
6. Clutch drum

with the clutch shown in Fig. TA117.
The upper end of the drive shaft is
threaded into the clutch drum hub. To
service clutch, disconnect throttle cable
and stop switch wires from engine. Detach drive housing (1) from engine. To
service clutch drum (4), remove gear
head from drive shaft housing. Insert a
tool through slot of clutch drum so it
cannot rotate, then turn square end
(trimmer end) of drive shaft so drum
unscrews from drive shaft. Detach snap
ring (3) then force drive shaft and bearing out towards large end of drive housing.

Clutch shoes (7) are available only as
a set. Clutch springs are available only
as a set.

All Other Engines. The engine is
equipped with a two-shoe clutch (Fig.
TA118). The clutch shoe assembly (11)
is mounted on the flywheel. The clutch
drum (6) rides in the drive housing (1),
which may be the fan housing on some
engines.

To remove clutch drum, first disconnect throttle cable and stop switch wires
from engine. Separate trimmer drive
housing assembly from engine. On models with slots in clutch face, reach
through slot and detach snap ring (5—
Fig. TA118), then press clutch drum (6)
and bearing (4) out of housing. Heating

housing will ease removal. On other engines, detach carrier (3—Fig. TA119)
from drive housing (1) and press drum
and bearing assembly out of carrier. Detach snap ring (4) and press clutch drum
shaft out of bearings (5). Reverse removal procedure to install clutch drum.

Clutch shoes are available only as a
pair.

REWIND STARTER. Refer to Fig.
TA120, TA121 or TA122 for an exploded
view of starter. To disassemble starter,
detach starter housing from engine. Remove rope handle and allow rope to
wind into starter. Unscrew center screw
and remove rope pulley. Wear appropri-

ate safety eyewear and gloves before
detaching rewind spring from housing
as spring may uncoil uncontrolled.

To assemble starter, lubricate center
post of housing and spring side with
light grease. On engines with a plate
attached to the crankshaft that carries
a pawl, install pawl (P—Fig. TA123) in
hole marked "R". Install rewind spring
so coil windings are counterclockwise
from outer end (clockwise on Tanaka
Models TBC-160, TBC-202 and TBC-
210). Assemble starter while passing
rope through housing rope outlet and
attach rope handle to rope. To place
tension on starter rope, pull rope out of
housing. Engage rope in notch on pulley
and turn pulley counterclockwise
(clockwise on Tanaka Models TBC-160,
TBC-202 and TBC-210). Hold pulley
and disengage rope from pulley notch.

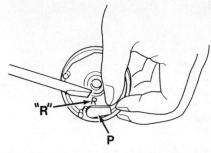

Fig. TA123—Install pawl (P) in hole marked "R".

Release pulley and allow rope to wind on pulley. Check starter operation. Rope handle should be held against housing by spring tension, but it must be possible to rotate pulley at least ¼ turn counterclockwise (clockwise on Tanaka Models TBC-160, TBC-202 and TBC-210) when rope is pulled out fully.

ELECTRIC STARTER. Engines with 20 cc or 21 cc displacement may be equipped with an electric starter. Refer to Fig. TA124 for a diagram of starter components. A battery pack of four or six 1.2v/1.2AH nicad batteries provides power to the starter motor. A relay is used to energize the starter when the start button is depressed. The batteries are charged by a circuit off the ignition coil. A rectifier converts alternating current to direct current for battery charging. The batteries can also be charged using 120v line current and a transformer as outlined in trimmer section.

To check motor operation, remove spark plug, disconnect four-wire connector to handle and use a jumper wire to connect terminals (1 and 4—Fig. TA125) of female connector. Starter motor should run; if not, check relay. If motor runs, but engine does not run, check drive gears. To check relay, remove cover (1—Fig. TA124) and connect voltmeter test leads to motor terminals. Connect jumper wire between terminals (1 and 4—Fig. TA125). If no voltage is indicated at motor terminals, renew relay. If battery voltage is indicated at

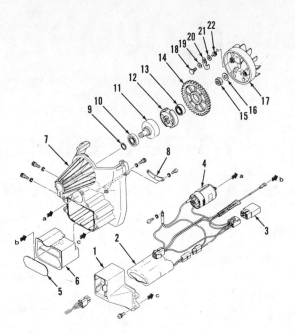

Fig. TA124—Exploded view of electric starter components on 21 cc engine. Components on 20 cc engine are similar.

1. Cover
2. Battery pack
3. Relay
4. Starter motor
5. Cushion
6. Case
7. Housing
8. Bracket
9. Snap ring
10. Bearing
11. Clutch drum
12. Clutch assy.
13. Bearing
14. Gear
15. Nut
16. Washer
17. Flywheel
18. Screw
19. Washer
20. Pawl
21. Washer
22. Spring

motor terminals but motor does not turn, motor is faulty.

To check engine charging system, run engine and disconnect two-wire connector to battery pack and four-wire connector to handlebar. Check voltage at terminals (1 and 2—Fig. TA125) of female connector by connecting negative tester lead to terminal (1) and positive lead to terminal (2). With engine running at 6000 rpm, tester should indicate more than 1.1 volts. If voltage is lower than 1.1 volts, check wiring harness for open circuit. Check air gap between flywheel magnet and ignition coil. If both items are satisfactory, renew ignition coil.

To check wiring harness, shut off engine and disconnect four-wire connector at engine fan cover. Connect voltmeter negative test lead to terminal (1—Fig. TA125) and positive test lead to terminal (4). Battery voltage should be indicated on tester. Connect negative test lead to terminal (1) and positive test lead to terminal (3). There should be no voltage indicated on tester. Renew wiring harness if either of these tests are failed.

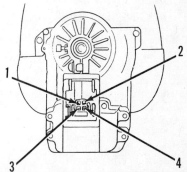

Fig. TA125—Tests may be performed on starter electrical system by attaching tester to connector terminals as outlined in text.

To check battery pack, recharge battery pack then disconnect four-wire connector at handlebar. With engine stopped, connect negative lead of a voltmeter to terminal (1) of female connector and positive tester lead to terminal (2). Battery voltage reading should be at least 4.8 volts on 20 cc engine or 7.0 volts on 21 cc engine.

Starter motor must be serviced as a unit assembly; components are not available.

TECUMSEH

TECUMSEH PRODUCTS COMPANY
900 North Street
Grafton, Wisconsin 53024

2-STROKE ENGINES

Model	Bore	Stroke	Displacement	Power Rating
AV520	2.09 in.	1.50 in.	5.2 cu.in.	3.0 hp
	(53 mm)	(38 mm)	(85 cc)	(2.2 kW)
AH600, AV600	2.09 in.	1.75 in.	6.00 cu.in.	3.8 hp
	(53 mm)	(44.5 mm)	(98 cc)	(2.8 kW)
HSK600	2.09 in.	1.75 in.	6.00 cu.in.	3.0 hp
	(53 mm)	(44.5 mm)	(98 cc)	(2.24 kW)
TVS600	2.09 in.	1.75 in.	6.00 cu.in.	3.25 hp
	(53 mm)	(44.5 mm)	(98 cc)	(2.4 kW)

ENGINE IDENTIFICATION

All models are two-stroke, single-cylinder, air-cooled engines with a unit block type construction.

Prefixes before type and serial number indicate the following information.

T-Tecumseh
A-Aluminum
V, VS-Vertical crankshaft
H-Horizontal crankshaft

Engine prefix, type number and serial numbers are located as shown in Fig. TP1. Figs. TP2 and TP3 show number interpretations. Always furnish engine model, type and serial numbers when ordering parts or service material.

MAINTENANCE

LUBRICATION. Engines are lubricated by mixing a good quality engine oil designed for two-stroke, air-cooled engines with unleaded, regular grade gasoline. Oil must be rated SAE 30 or

Fig.TP1—View showing location of engine model, type and serial number. Refer also to Figs. TP2 and TP3.

40. Automotive or multiviscosity type oils are not recommended. Manufacturer states that gasoline containing methanol must not be used, and if gasohol is used, it must not contain more

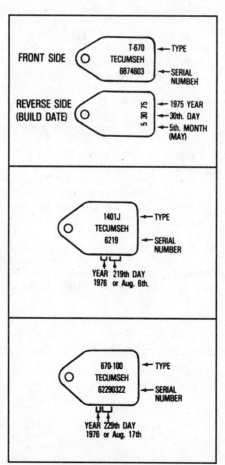

Fig. TP2—View showing identification tag from engine and interpretation of letters and numbers. Refer to text.

than 10 per cent ethanol. Recommended fuel:oil ratio is 24:1 for Models AV520 and AV600, 32:1 for Models AH600 and TVS600, and 50:1 for Model HSK600. Use a separate container to mix oil and gasoline. Do not mix directly in engine fuel tank.

SPARK PLUG. On Model AH600 engines, the spark plug used will depend on engine application and usage. Tecumseh recommends Champion CJ8, RCJ8Y or RJ17LM spark plugs for Model AH600 engines. Recommended spark plug for Model HSK600 is Champion RCJ8Y. Where applicable, follow equipment manufacturer's recommendation. On all other engines, Tecumseh recommends a Champion RJ17LM spark plug.

Electrode gap should be 0.030 inch (0.76 mm) for all models. Tighten spark plug to 216 in.-lbs. (24.4 N•m) torque.

CARBURETOR. Tecumseh diaphragm and float type carburetors are used. Refer to the appropriate following paragraphs for information on specific carburetors.

Diaphragm Type Carburetor. Carburetor identification numbers are

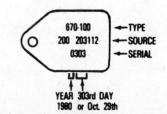

Fig. TP3—View of identification tag from replacement short block and interpretation of letters and numbers.

Illustrations courtesy Tecumseh Products Co.

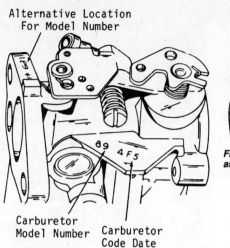

Alternative Location For Model Number

Carburetor Model Number

Carburetor Code Date

Fig. TP4—View showing location of carburetor identification number on Tecumseh diaphragm type carburetor.

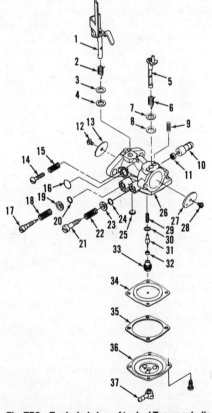

Fig. TP5—Exploded view of typical Tecumseh diaphragm type carburetor.

1. Throttle shaft	20. "O" ring
2. Spring	21. High speed
3. Washer	mixture screw
4. Felt washer	22. Spring
5. Choke shaft	23. Washer
6. Spring	24. "O" ring
7. Washer	25. Welch plug
8. Felt washer	26. Body
9. Choke retainer	27. Choke plate
10. Inlet fitting	28. Screw
11. Outlet check valve	29. Spring
12. Screw	30. Gasket
13. Throttle plate	31. Fuel inlet valve
14. Idle stop screw	32. Valve seat
15. Spring	33. Fitting
16. Welch plug	34. Diaphragm
17. Idle mixture screw	35. Gasket
18. Spring	36. Cover
19. Washer	37. Primer hose fitting

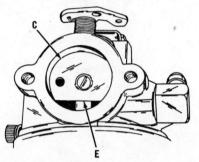

Fig. TP6—Install throttle plate (T) so lines (L) are as shown.

Fig. TP7—Install choke plate (C) so flat edge (E) is toward bottom of carburetor.

stamped on carburetor as shown in Fig. TP4. Refer to Fig. TP5 for identification and exploded view of Tecumseh diaphragm type carburetor.

Initial adjustment of idle mixture screw (17—Fig. TP5) and high speed mixture screw (21) from a lightly seated position is one turn open for each mixture screw. Final adjustments are made with engine at operating temperature and running. Operate engine at idle speed and adjust idle mixture screw to obtain smoothest engine idle. Operate engine at rated speed, under load and adjust main fuel mixture screw for smooth engine operation. If engine cannot be operated under load, adjust main fuel mixture needle so engine runs smoothly at rated rpm and accelerates properly when throttle is opened quickly.

To overhaul carburetor, refer to Fig. TP5 and remove cover (36) and diaphragm (34). Unscrew fuel inlet valve fitting (33) and remove seat (32), valve (31) and spring (29). Inlet valve needle and seat are replaced as an assembly. Remove idle and high speed mixture screws. Use a small chisel or scratch awl to pierce and remove Welch plug (25). Note the following: The fuel inlet fitting (10—Fig. TP5) is pressed into bore of carburetor body of some models. On these models, the fuel strainer (11) behind inlet fitting can be cleaned by reverse flushing with compressed air after inlet needle and seat are removed. Remove mounting screws from throttle plate and choke plate, and withdraw throttle shaft and choke shaft. The body must be replaced if there is excessive throttle or choke shaft play as bushings are not available.

NOTE: On TVS600 models, the governor air vane is mounted on the choke shaft. Refer to GOVERNOR section.

Clean all metallic parts in suitable solvent. Blow out all passages with compressed air in the opposite direction of normal fuel flow. Do not use wires or drills to clean orifices or passages. Diaphragm (34) should be flat and free of holes, creases, tears and other imperfections.

When assembling the carburetor note the following: Throttle plate (T—Fig. TP6) should be installed with lines (L) stamped on plate facing out and toward top of carburetor and 3 o'clock position as shown. Choke plate (C—Fig. TP7) should be installed with flat edge (E) toward bottom of carburetor as shown. Install Welch plug so convex side is out, then flatten plug using a $\frac{5}{16}$-inch (8 mm) diameter flat-end punch. A correctly installed Welch plug is flat; a concave plug may leak. When installing diaphragm (34—Fig. TP5), head of rivet (R—Fig. TP8) must be

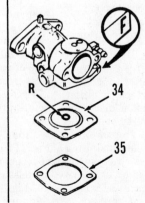

Fig. TP8—Install diaphragm so rivet head (R) is towards carburetor body. Note position of diaphragm (34) and gasket (35) on carburetor with and without "F" on carburetor flange.

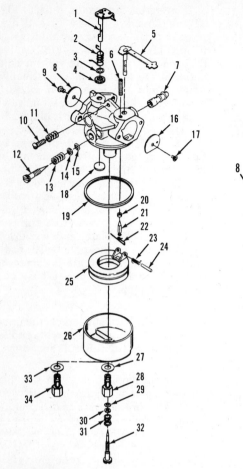

Fig. TP9—Exploded view of typical Tecumseh float type carburetor used on some models.

1. Throttle shaft	18. Welch plug
2. Spring	19. Gasket
3. Washer	20. Fuel inlet seat
4. Felt washer	21. Fuel inlet valve
5. Choke shaft	22. Clip
6. Choke retainer	23. Spring
7. Inlet fitting	24. Float pin
8. Screw	25. Float
9. Throttle plate	26. Fuel bowl
10. Idle stop screw	27. Washer
11. Spring	28. Bowl retaining nut
12. Idle mixture screw	29. "O" ring
13. Spring	30. Washer
14. Washer	31. Spring
15. "O" ring	32. High speed mixture screw
16. Choke plate	33. Washer
17. Screw	34. Bowl retaining nut

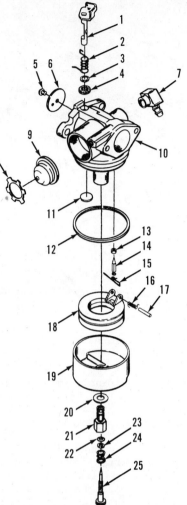

Fig. TP10—Exploded view of Tecumseh float type carburetor equipped with a primer bulb.

1. Throttle shaft	14. Fuel inlet valve
2. Spring	15. Clip
3. Washer	16. Spring
4. Felt washer	17. Float pin
5. Screw	18. Float
6. Throttle plate	19. Fuel bowl
7. Inlet fitting	20. Washer
8. Retainer	21. Bowl retaining nut
9. Primer bulb	22. "O" ring
10. Body	23. Washer
11. Welch plug	24. Spring
12. Gasket	25. High speed
13. Fuel inlet seat	mixture screw

against fuel inlet valve. On carburetor with an "F" embossed on carburetor body (Fig. TP8), diaphragm (34) must be installed next to the carburetor body. If no "F" is present on the carburetor body, gasket (35) must be next to the carburetor body.

Carburetors utilizing a remote primer bulb are equipped with a check valve (11—Fig. TP5) which is either brass or Teflon. The Teflon disc type valve is accessible after removing the fuel inlet fitting. Place a drop of oil on disc to hold it in position when installing fitting. To remove a brass check valve, carefully drill into outlet check valve (10) to a depth of 1/8 inch (3.18 mm) using a 9/64-inch drill. Take care not to

drill into carburetor body. Thread an 8-32 tap into the check valve. Using a proper size nut and flat washer, convert the tap into a puller to remove check valve from carburetor body. Press new check valve into carburetor body until face of valve is flush with surrounding base of fuel inlet chamber. Apply Loctite 271 sealer to shank of fuel inlet fitting before pressing fitting into carburetor body.

Float Type Carburetor. Refer to Fig. TP9 or Fig. TP10 for exploded view of Tecumseh float type carburetor. Some models are equipped with a primer bulb (9—Fig. TP10) on the side of the carburetor and the choke plate is absent.

Some carburetors are equipped with an idle or high speed mixture adjustment screw, while some carburetors are equipped with a fixed jet in the fuel bowl mounting nut (34—Fig. TP9). Initial adjustment of either the idle or high speed fuel mixture screw is one turn open from a lightly seated position. Final adjustments are made with engine at operating temperature and running. Operate engine at rated speed and adjust high speed mixture screw for smoothest engine operation. Operate engine at idle speed and adjust idle mixture screw for smoothest engine idle. Adjust idle speed screw so engine idles at desired idle speed. If engine fails to accelerate smoothly, slight adjustment of idle fuel mixture screw may be necessary.

To disassemble carburetor, remove fuel bowl mounting nut and fuel bowl. Remove float pin, float and fuel inlet valve. Compressed air can be used to dislodge the inlet valve seat by directing air through the fuel inlet fitting. Remove idle and high speed mixture screws. Use a small chisel or scratch awl to pierce and remove Welch plug. Remove mounting screws from throttle plate and choke plate, and withdraw throttle shaft and choke shaft. The carburetor body must be replaced if there is excessive throttle or choke shaft play as bushings are not available.

NOTE: On TVS600 models, the governor air vane is mounted on the choke shaft. Refer to GOVERNOR section.

Clean all metallic parts in suitable solvent. Blow out all passages with compressed air in the opposite direction of normal fuel flow. Do not use wires or drills to clean orifices or passages.

Install new Welch plug with the raised side facing up. Use a flat-end punch that is the same size as the plug to flatten the plug. Do not indent the plug. Install the fuel valve inlet seat with the flat side out and the grooved side towards the carburetor (Fig. TP11). Throttle plate (T—Fig. TP6) should be installed with lines (L) stamped on plate facing out and toward top of carburetor and 3 o'clock position as shown. Choke

Fig. TP11—The Viton fuel inlet valve seat must be installed so grooved side is toward carburetor body.

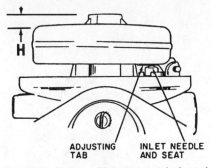

Fig. TP12—Distance (H) between end of nozzle stanchion and float with body inverted should be 0.162-0.215 inch (4.11-5.46 mm).

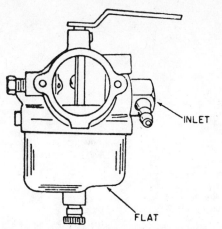

Fig. TP14—Flat part of fuel bowl should be located under the fuel inlet fitting.

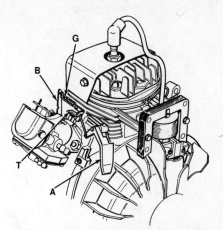

Fig. TP16—View of horizontal engine using a plastic air vane (A). The air vane is attached to the carburetor throttle shaft (T) and tension of governor spring (G) holds throttle (T) open. Bend tab (B) to adjust engine speed.

plate (C—Fig. TP7) should be installed with flat edge (E) toward bottom of carburetor as shown. Install the fuel inlet valve retaining clip on the float tab so the long end is toward the choke end of the carburetor.

With carburetor inverted, measure float height (H—Fig. TP12) at a point opposite the fuel inlet valve. The float level should be 0.162-0.215 inch (4.11-5.46 mm). Bend the float tab to adjust the float level. Tecumseh has a gauge (Tecumseh part 670253A) that can be used to determine the correct float level. With carburetor inverted, position gauge 670253A at a 90° angle to the carburetor bore and resting on the nozzle stanchion as shown in Fig. TP13. The toe of the float should not be higher than the first step on the gauge or lower than the second step.

If removed, reinstall fuel inlet fitting (7—Fig. TP9 or TP10) by pushing it into the carburetor so the inlet is in its original position. Apply Loctite sealant before installation. Install the fuel bowl so the indented portion is on the same side as the fuel inlet (Fig. TP14).

To install the primer bulb on carburetors so equipped, place the retainer ring around the bulb, then push the bulb and retainer into the carburetor using a ¾-inch deep well socket.

GOVERNOR. An air vane type governor attached to the carburetor throttle shaft is used on some models. On models with variable speed governor, the high speed (maximum) stop screw is located on the speed control lever as

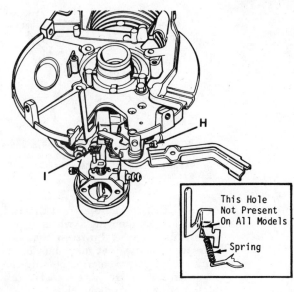

Fig. TP13—Use procedure described in text to measure float height using tool 670253A.

Fig. TP15—View of governor and carburetor linkage on vertical crankshaft engine with variable engine speed control. Turn screw (I) to adjust idle speed and screw (H) to adjust high speed.

shown in Fig. TP15. On models with fixed engine speed, rpm is adjusted by moving the governor spring bracket (B—Fig. TP16 or TP17) or relocating governor spring arm (R—Fig. TP18). To increase engine speed, bracket or arm must be moved to increase governor spring tension holding throttle open.

On some TVS series engines, an air vane (1—Fig. TP19) is mounted on the choke shaft, independent of the choke plate. The governor air vane is con-

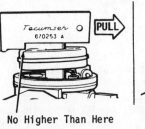

No Higher Than Here

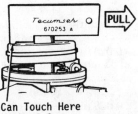

Can Touch Here Without Gap

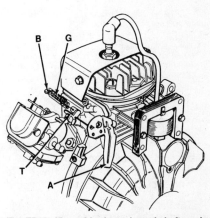

Fig. TP17—View of horizontal crankshaft engine using an aluminum air vane (A). The air vane is attached to the carburetor throttle shaft (T) and tension of governor spring (G) holds throttle (T) open. Reposition bracket (B) to adjust engine speed.

Illustrations courtesy Tecumseh Products Co.

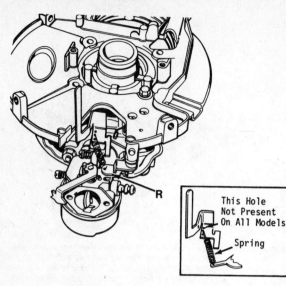

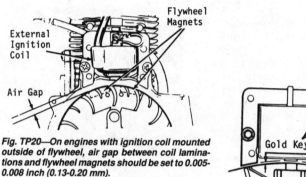

Fig. TP18—View of governor linkage on vertical crankshaft engine with fixed speed. Relocate governor arm (R) to adjust engine speed.

Fig. TP21—The solid-state ignition charging coil, triggering system and mounting plate (2 and 10) are available only as an assembly and cannot be serviced.

2. Charging coil
8. Pulse transformer
10. Trigger system
11. Low tension lead
12. High tension lead

Fig. TP20—On engines with ignition coil mounted outside of flywheel, air gap between coil laminations and flywheel magnets should be set to 0.005-0.008 inch (0.13-0.20 mm).

Fig. TP22—On solid-state ignition system, set coil lamination air gap at 0.0125 inch (0.32 mm) at locations shown. Ignition module will be marked (arrow) to indicate color of flywheel sleeve (S) or crankshaft key (K) that must be installed.

Fig. TP19—Exploded view of air vane governor used on some carburetors.

1. Air vane
2. Choke/governor shaft
3. Spring
4. Governor spring
5. Sleeve
6. Serrated disc
7. Washer
8. Spring
9. Choke plate
10. Screw

nected to the throttle arm by governor spring (4). The governor spring can be adjusted to vary the tension, causing the engine to run faster or slower as desired. One increment movement of sleeve (5) on serrated disc (6) will vary engine speed approximately 100 rpm.

IGNITION SYSTEM. Engines may be equipped with either a magneto type (breaker points) or a solid-state ignition system. Refer to the appropriate paragraph for model being serviced. Refer to

FLYWHEEL paragraph in REPAIRS section when removing flywheel.

Breaker-Point Ignition System. Breaker-point gap at maximum opening should be set before adjusting the ignition timing. On some models, ignition timing is not adjustable.

Some models may be equipped with external coil and magnet laminations. Air gap (Fig. TP20) between external ignition coil laminations and flywheel magnet should be 0.005-0.008 inch (0.13-0.20 mm).

Ignition points and condenser are located under flywheel and flywheel must be removed for service. Refer to the following specifications for point gap and piston position before top dead center (BTDC) when breaker points just begin to open to correctly set ignition timing.

AV520:
Point gap. . . . . 0.020 in.
(0.51 mm)
BTDC . . . . 0.070 in.*
(1.78 mm)
AH600:
Point gap. . . . . 0.020 in.
(0.51 mm)
BTDC . . . . *
AV600:
Point gap. . . . . 0.020 in.
(0.51 mm)

BTDC . . . . 0.070 in.
(1.78 mm)
TVS600:
Point gap . . . . 0.020 in.
(0.51 mm)
BTDC . . . . *

*On models with external coil, timing is not adjustable. Set coil air gap to specified dimension.

Solid-State Ignition System. The Tecumseh solid-state ignition system does not use ignition points. The only moving part of the system is the rotating flywheel with the charging magnets. Early systems utilize individual components contained under the flywheel as shown in Fig. TP21, while later systems are modular with a one-piece ignition coil and ignition module externally mounted as shown in Fig. TP22. Note that module is square as opposed to a breaker point ignition coil which is round.

On early systems, the ignition charging coil, electronic triggering system and mounting plate are available only as an assembly. If necessary to renew this assembly, place the unit in position on the engine. Start the retaining screws, turn the mounting plate counterclockwise as far as possible, then tighten retaining screws to 5-7 ft.-lbs. (7-10 N•m) torque.

The ignition coil/module on later models is a one-piece unit. The module is marked with "GOLD KEY" on the module and a gold colored crankshaft key and a gold colored flywheel sleeve must be used. See Fig. TP22. The correct air gap setting between the flywheel magnets and the laminations on ignition module is 0.0125 inch (0.32 mm). Use Tecumseh gauge 670297 or equivalent thickness plastic strip to set gap as shown in Fig. TP22. Tighten mounting screws to 30-40 in.-lbs. (3.4-4.5 N•m) torque.

CARBON. Muffler and exhaust ports should be cleaned after every 50 to 75 hours of operation. The cylinder head, piston and cylinder wall should be cleaned of carbon if excessive carbon buildup is noted.

REPAIRS

TIGHTENING TORQUES. Recommended tightening torque specifications are as follows:

Connecting rod	40-50 in.-lbs. (5-6 N•m)
Crankcase cover	80-100 in.-lbs. (9.0-11.3 N•m)
Cylinder head	80-100 in.-lbs. (9.0-11.3 N•m)
Flywheel nut	22-27 ft.-lbs. (30-36 N•m)
Spark plug	216 in.-lbs. (24.4 N•m)

CRANKCASE PRESSURE TEST. An improperly sealed crankcase can

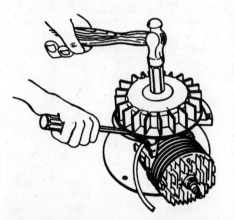

Fig. TP23—View showing use of a knock-off tool (nut) to separate flywheel from crankshaft end.

Illustrations courtesy Tecumseh Products Co.

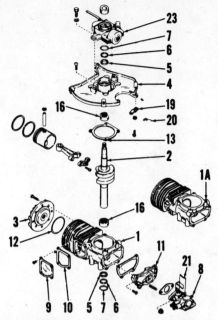

Fig. TP24—Exploded view typical of AV520 and AV600 engines.

1.	Cylinder & crankcase	11.	Reed valve
2.	Crankshaft	12.	Seal ring
3.	Cylinder head	13.	Gasket
4.	Crankcase cover	16.	Main bearings (roller)
5.	Seal	19.	Spring bracket
6.	Retainer	20.	Governor spring
7.	Snap ring	21.	Air vane (on carburetor)
8.	Carburetor	23.	Magneto

cause the engine to be hard to start, run rough, have low power and overheat. Refer to SERVICE SECTION TROUBLE-SHOOTING section of this manual for crankcase pressure test procedure. If crankcase leakage is indicated, pressurize crankcase and use a soap and water solution to check gaskets, seals, carburetor pulse line and casting for leakage.

FLYWHEEL. Disengage flywheel brake as outlined in FLYWHEEL BRAKE section. If flywheel has tapped holes, use a suitable puller to remove flywheel. If no holes are present, screw a knock-off nut onto crankshaft as shown in Fig. TP23 so there is a small gap between nut and flywheel. Gently pry against bottom of flywheel while tapping sharply on nut. After installing flywheel, tighten flywheel nut to 22-27 ft.-lbs.(30-36 N•m) torque.

On engines with a ball bearing at flywheel end of crankshaft, using a knock-off nut may result in bearing and crankshaft moving toward pto end. This will reduce clearance between crankshaft and crankcase. Rap sharply on pto end of crankshaft with a rawhide mallet to reseat ball bearing.

Some engines are equipped with a flywheel sleeve similar to the sleeve shown in Fig. TP22. Gold colored sleeves are used with solid-state igni-

tions while grey colored sleeves are used with breaker-point ignitions. The sleeve must be discarded if it is sheared or damaged. Install sleeve so it is flush or just below inside surface of flywheel.

CYLINDER HEAD. To remove cylinder head, remove all interfering shrouds and brackets. Clean area around cylinder head to prevent entrance of foreign material. Unscrew cylinder head screws and remove cylinder head.

A new head gasket should be installed when installing cylinder head. Tighten cylinder head retaining screws evenly to a torque of 80-100 in.-lbs. (9.0-11.3 N•m) torque.

DISASSEMBLY. Refer to Figs. TP24, TP25 and TP26 for exploded view of engine. When disassembling engine, note that piston and connecting rod must be removed before crankcase cover is removed. The crankshaft is removed with the crankcase cover. Remove reed valve or crankcase end cover for access to connecting rod. Remove connecting rod cap and push the rod and piston unit out through top of cylinder. Do not lose bearing rollers which will fall out when rod cap is loosened. To prevent damage to piston and rings, it may be necessary to remove the ridge from top of cylinder bore before removing piston and connecting rod assembly.

To separate crankshaft and crankcase cover, carefully heat bearing area so crankshaft and bearing can be separated from crankcase cover; do not use excessive heat.

PISTON, PIN, RINGS AND CYLINDER. The piston and connecting rod can be removed after removing the cylinder head and end cover or reed valve. The piston pin rides in a roller bearing in the small end of the connecting rod. The piston pin can be extracted after removing the retaining clips and using a suitable piston pin puller or press.

Standard cylinder bore diameter for all models is 2.093-2.094 inches (53.162-53.188 mm). The bore can be bored to accept a 0.010 inch (0.25 mm) oversize piston on some models. Check parts availability to determine applicable models.

Standard piston diameter is listed in the following table.

Type Number

653-01 thru 653-05, 661-01 thru 661-29	2.0865-2.0875 in. (52.997-53.022 mm)
653-07 thru 653-10, 660-39, 660-39A, 660-40, 661-30 thru 661-45	2.0880-2.0885 in. (53.035-53.048 mm)

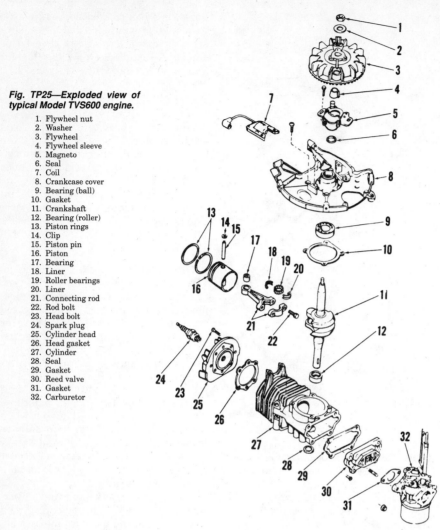

Fig. TP25—Exploded view of typical Model TVS600 engine.

1. Flywheel nut
2. Washer
3. Flywheel
4. Flywheel sleeve
5. Magneto
6. Seal
7. Coil
8. Crankcase cover
9. Bearing (ball)
10. Gasket
11. Crankshaft
12. Bearing (roller)
13. Piston rings
14. Clip
15. Piston pin
16. Piston
17. Bearing
18. Liner
19. Roller bearings
20. Liner
21. Connecting rod
22. Rod bolt
23. Head bolt
24. Spark plug
25. Cylinder head
26. Head gasket
27. Cylinder
28. Seal
29. Gasket
30. Reed valve
31. Gasket
32. Carburetor

660-11 thru 660-38,
  670-01 thru 670-109 2.0870-2.0880 in.
    (53.010-53.035 mm)
1620 . . . . 2.0875-2.0885 in.
    (53.022-53.048 mm)
1624 thru 1642 . . . 2.0882-2.0887 in.
    (53.040-53.053 mm)

Piston-to-cylinder bore clearance should be as listed in the following table.

Type Number

653-01 thru 653-05,
  661-01 thru 661-29 0.0055-0.0075 in.
    (0.140-0.190 mm)
653-07 thru 653-10,
  660-39, 660-39A, 660-40,
  661-30 thru 661-45 . 0.0045-0.006 in.
    (0.114-0.152 mm)
660-11 thru 660-38,
  670-01 thru 670-109 . 0.005-0.007 in.
    (0.13-0.18 mm)
1620 . . . . 0.0045-0.0065 in.
    (0.114-0.165 mm)
1624 thru 1642 . . . 0.0043-0.0058 in.
    (0.109-0.147 mm)

The piston pin on most models is available only as an assembly with the piston. Piston pin diameter should be 0.4997-0.4999 inch (12.692-12.697 mm).

The piston is equipped with two piston rings. Top piston ring groove width should be 0.0655-0.0665 inch (1.664-1.689 mm). Bottom piston ring groove width should be 0.0645-0.0655 inch (1.638-1.664 mm). Piston ring end gap should be 0.007-0.017 inch (0.18-0.43 mm) for type number 1620 and 0.006-0.016 inch (0.15-0.41 mm) for all other models.

If connecting rod on vertical crankshaft engine is equipped with a lubrication hole in side of rod, when assembling piston and connecting rod, hole must be toward top of engine. On engines with a piston marked with a "V" on the piston crown, install the piston in the engine so the "V" mark is toward the right side as shown in Fig. TP27.

When installing piston and rod in engine, stagger piston ring end gaps and use a ring compressor. Lubricate piston and rings with engine oil. Make certain rings do not catch in recess at top of cylinder.

CONNECTING ROD. The engine is equipped with an aluminum connecting rod. The piston pin rides in a cartridge type needle bearing that is pressed into connecting rod small end. A steel insert (liner) is used on inside of connecting rod and needle bearing rollers are used at the crankpin end. Connecting rod may contain 30 or 37 bearing rollers, depending on engine model. Standard journal diameter is either 0.8113-0.8118 inch (20.607-20.620 mm) or 0.8442-0.8450 inch (21.443-21.463 mm), depending on engine model. Measure unworn portion of crankpin to determine standard diameter.

Inspect connecting rod, crankpin, bearing rollers and bearing liners for damage and excessive wear. Bearing rollers should be renewed only as a set. Renew bearing set if any roller is damaged.

When installing rod on crankshaft, make certain match marks on connecting rod and cap are aligned. Ends of bearing liners must correctly engage when match marks on connecting rod and cap are aligned. New rollers are serviced in a strip and can be installed by wrapping the strip around crankpin. Old bearing rollers can be held in place using grease. After new needle rollers and connecting rod cap are installed, force lacquer thinner into needles to remove the beeswax or grease, then lubricate bearing with oil.

Tighten connecting rod cap retaining screws to 40-50 in.-lbs. (5-6 N·m) torque.

CRANKSHAFT AND CRANKCASE. The crankshaft can be removed after the piston, connecting rod, flywheel and magneto end bearing plate are removed.

Crankshaft main bearings may be either ball type or cartridge needle roller. If ball type main bearings are used, it should be necessary to bump the crankshaft out of the bearing inner races. Ball bearing may be retained on crankshaft with a retainer ring. If a retainer ring is not used, apply Loctite 609 into grooves on crankshaft where ball bearing contacts shaft. Ball and roller bearing outer races should be a tight fit in bearing bores. If new ball bearings are to be installed, heat the crankcase when removing or installing bearings.

On all models, bearings should be installed with printed face on race toward center of engine. If the crankshaft is equipped with thrust washers at ends, make certain they are installed

Illustrations courtesy Tecumseh Products Co.

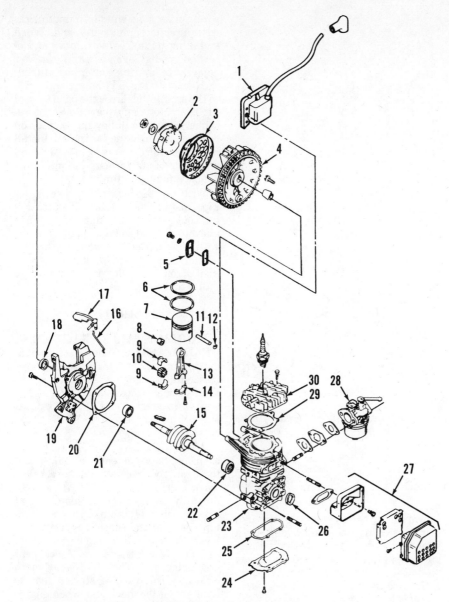

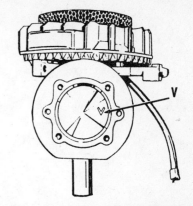

Fig. TP27—The "V" mark stamped on top of piston must be toward side shown. Lubrication hole in side of connecting rod must be toward top on all vertical shaft models.

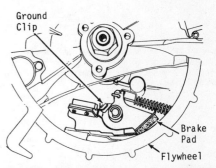

Fig. TP28—Some engines are equipped with flywheel band brake system.

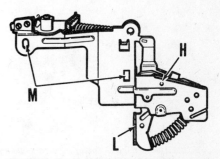

Fig. TP29—Push against lever (L) and insert a pin through holes (H) to hold brake pad away from flywheel.

Fig. TP26—Exploded view of typical Model AH600 engine.

1. External ignition	8. Bearing	16. Governor link
2. Starter cup	9. Liner	17. Governor air vane
3. Screen	10. Roller needles	18. Oil seal
4. Flywheel	11. Piston pin	19. Crankcase
5. Compression	12. Retaining ring	end plate
release cover	13. Connecting rod	20. Gasket
6. Piston rings	14. Rod cap	21. Bearing
7. Piston	15. Crankshaft	22. Bearing

23. Crankcase	
24. Crankcase cover	
25. Gasket	
26. Oil seal	
27. Muffler assy.	
28. Carburetor	
29. Head gasket	
30. Cylinder head	

when assembling. Crankshaft end play for all models is zero. Install crankshaft seals so lip is toward inside of engine. It is recommended that the crankcase be pressure tested for leakage before proceeding with remainder of engine reassembly.

REED VALVES. Vertical crankshaft engines are equipped with a reed type inlet valve located in the crankcase end cover. Reed petals should not stand out more than 0.010 inch (0.25 mm) from the reed plate and must not be bent, distorted or cracked. The reed plate must be smooth and flat. Reed petals

are available only as part of the crankcase end cover.

FLYWHEEL BRAKE. A flywheel brake is used on some engines that will stop the engine within three seconds when the equipment safety handle is released. The ignition circuit is grounded also when the brake is actuated. Refer to Fig. TP28.

The brake shown in Fig. TP28 contacts the inside of the flywheel. Before the flywheel can be removed, the brake must be disengaged from the flywheel. Push lever (L—Fig. TP29) toward spark plug so brake pad moves away from

flywheel, then insert Tecumseh tool 670298 or a suitable pin in hole (H) to hold lever.

Inspect mechanism for excessive wear and damage. Minimum allowable thickness of brake pad at narrowest point is 0.060 inch (1.52 mm). Flywheel surface must be clean and undamaged. Install brake mechanism and push up on bracket so bracket mounting screws are at bottom of slotted holes (M) in bracket. Tighten mounting screws to 90 in.-lbs. (10.2 N·m) torque.

REWIND STARTER

Dog Type Starter

TEARDROP HOUSING. To disassemble starter (Fig. TP30), release ten-

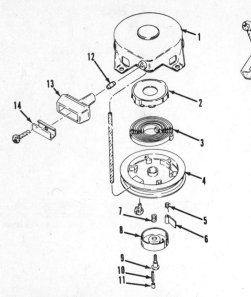

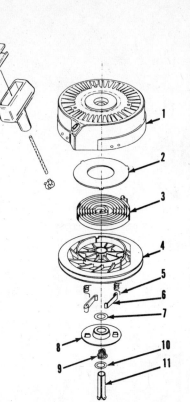

Fig. TP30—Exploded view of typical pawl type rewind starter with teardrop shaped housing (1) using retainer screw (9). Some starters may have three starter pawls (6).

1. Housing
2. Spring keeper
3. Rewind spring
4. Pulley
5. Spring
6. Dog
7. Brake spring
8. Retainer
9. Screw
10. Centering pin
11. Nylon bushing
12. Rope coupler
13. Handle
14. Insert

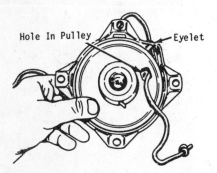

Fig. TP31—Insert rope through starter housing eyelet and hole in pulley, then tie knot in rope end.

sion of rewind spring by removing rope handle and allowing rope to wind slowly into starter. Remove retainer screw (9), retainer (8) and spring (7). Remove dog (6) and spring (5). Remove pulley with spring. Wear appropriate safety eyewear and gloves before disengaging keeper (2) and rewind spring (3) from pulley as spring may uncoil uncontrolled.

To reassemble, reverse the disassembly procedure. Standard rope length is 69 inches (175 cm) and diameter is 5/32 inch (4 mm). Spring (3) should be lightly greased. Install the pawl (6) and spring (5) so the spring end forces the pawl toward the center of the pulley. Assemble starter but install rope last as follows: Turn pulley counterclockwise until tight, then allow to unwind so hole in pulley aligns with rope outlet as

Fig. TP32—Exploded view of "stylized" rewind starter.

1. Starter housing
2. Cover
3. Rewind spring
4. Pulley
5. Springs (2)
6. Dogs (2)
7. Plastic washers (2)
8. Retainer pawl
9. Brake spring
10. Metal washer
11. Pin

shown in Fig. TP31. Insert rope through starter housing and pulley hole, tie a knot in rope end, allow rope to wind onto pulley and install rope handle. Some models use centering pin (10—Fig. TP30) to align starter with starter cup. Place nylon bushing (11) on pin (10), then bottom pin in hole in retainer screw (9). Pin and bushing should index in end of crankshaft when installing starter on engine.

STYLIZED STARTER. The "stylized" starter is shown in Fig. TP32. To disassemble starter, remove rope handle and allow rope to wind slowly into starter. Position a suitable sleeve support under retainer pawl (8) and using a 5/16 inch (8 mm) diameter punch, drive pin (11) free of starter. Remove brake spring (9), retainer (8), dogs (6) and springs (5). Wear appropriate safety eyewear and gloves before disengaging pulley from starter as spring may uncoil uncontrolled. Place shop towel around pulley and lift pulley out of housing; spring should remain with pulley.

Inspect components for damage and excessive wear. Reverse disassembly procedure to install components. Re-

wind spring coils wind in counterclockwise direction from outer end. When installing pulley, be sure inner end of rewind spring engages spring retainer adjacent to housing center post. Install the pawl (6—Fig. TP32) and spring (5) so the spring end forces the pawl toward the center of the pulley. Place the retainer (8) on the pulley hub so tabs on retainer are inside ends of pawls. Install pin (11) so top of pin is 1/8 inch (3.2 mm) below top of starter. Driving pin in too far may damage retainer pawl.

With starter assembled, except for rope, install rope as follows: Rotate pulley counterclockwise until tight, then allow to unwind so hole in pulley aligns with rope outlet. Insert rope through starter housing and pulley hole, tie a knot in rope end, allow rope to wind onto pulley and install rope handle.

Side-Mounted Starter

HORIZONTAL ENGAGEMENT GEAR STARTER. When the rope handle is pulled, the starter gear moves horizontally to engage the gear teeth on the flywheel. An exploded view of the starter is shown in Fig. TP33. Most starters use a rope that is 61 inches (155 cm) long.

After installing starter assembly, clearance between teeth on gear (6) and teeth on flywheel should be checked when starter is operated. When teeth are fully engaged, there should be at least 1/16 inch (1.6 mm) clearance from top of gear tooth to base of opposite gear teeth. Remove spark plug wire, operate starter several times and check gear engagement. Insufficient gear tooth clearance could cause starter gear to hang up on flywheel gear when engine starts, which could damage starter.

To disassemble starter, proceed as follows: Detach rope from insert (8—Fig. TP33) and handle (9), then allow the rope to wind slowly onto the pulley to relieve spring tension. Unscrew screws securing spring cover (13) and carefully remove spring cover without disturbing rewind spring. Remove rewind spring. Safety eyewear and gloves should be worn when working on or around the rewind spring. Remove hub screw (15) and spring hub (14), then withdraw pulley and gear assembly. Remove snap ring (3), washer (4), brake spring (5) and gear (6) from pulley.

Reassemble starter using the following procedure: If removed, install rope guide (2) on starter bracket so dimple on guide fits in depression on bracket. Install rope on pulley. Wrap rope around pulley in a counterclockwise direction when viewing pulley from rewind spring side of pulley. Place gear on pul-

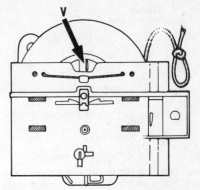

Fig. TP35—Drawing of vertical engagement gear starter. Note location of "V" notch.

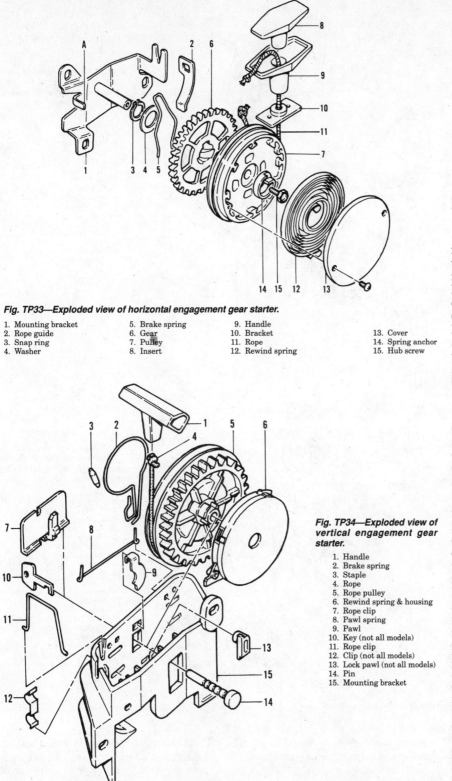

Fig. TP33—Exploded view of horizontal engagement gear starter.

1. Mounting bracket
2. Rope guide
3. Snap ring
4. Washer
5. Brake spring
6. Gear
7. Pulley
8. Insert
9. Handle
10. Bracket
11. Rope
12. Rewind spring
13. Cover
14. Spring anchor
15. Hub screw

Fig. TP34—Exploded view of vertical engagement gear starter.

1. Handle
2. Brake spring
3. Staple
4. Rope
5. Rope pulley
6. Rewind spring & housing
7. Rope clip
8. Pawl spring
9. Pawl
10. Key (not all models)
11. Rope clip
12. Clip (not all models)
13. Lock pawl (not all models)
14. Pin
15. Mounting bracket

from outer end. A new spring is contained in a holder that allows spring installation by pushing spring from holder into spring cavity on pulley. Pass outer end of rope through rope bracket (10) and install rope handle and insert. Pull a portion of rope out past rope guide and wrap any excess rope around pulley, then turn pulley 2-2½ turns against spring tension to preload rewind spring. Check starter operation.

VERTICAL ENGAGEMENT GEAR STARTER. When the rope handle is pulled, the starter gear moves vertically to engage the gear teeth on the flywheel. An exploded view of the starter is shown in Fig. TP34.

To replace the starter rope, proceed as follows: If the starter bracket has a "V" notch (V—Fig. TP35), then the inner end of the rope is accessible and the rope can be replaced without disassembling the starter. Typical rope lengths are 65 inches (165 cm) and 98 inches (249 cm). If rope is unbroken, remove rope handle and let rope wind onto rope pulley. Note that inner end of rope was originally retained by a staple, while inner end on replacement ropes is inserted through a hole in pulley and knotted. Rotate pulley so that either the stapled rope end or knotted rope end is visible in the "V" notch. Pry out staple or untie knot and pull out rope. Rotate pulley counterclockwise until rewind spring is tight, then allow pulley to turn clockwise until rope hole in pulley is visible in "V" notch. Route new rope through pulley hole and tie a knot in rope end. Pull the knot into pulley cavity so rope end does not protrude. Allow pulley to wind rope onto pulley. Attach rope handle to rope end.

To disassemble starter, proceed as follows: If rope is unbroken, remove rope handle and let rope wind onto rope pulley. Drive or press out pulley spindle (14—Fig. TP34) by placing starter over a deep-well socket and forcing spindle

ley. Do not lubricate helix in gear or on pulley shaft. Install brake spring (5) in groove of gear (6). The bent end of brake spring should point away from gear. The brake spring should fit snugly in groove. Do not lubricate brake spring. Install washer (4) and snap ring (3) on pulley shaft. Lightly lubricate shaft on starter

bracket then install pulley and gear assembly on bracket shaft. The closed end of brake spring (5) must fit around tab (A) on bracket. Install spring anchor (14) and screw (15). Tighten screw to 44-55 in.-lb. (5.08-6.21 N·m) torque. Install rewind spring. The spring coils should wind in a clockwise direction

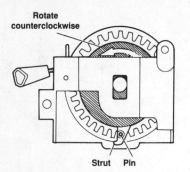

Rotate counterclockwise

Strut   Pin

Fig. TP36—Housing rotation may be prevented by inserting a pin or rod through strut hole and into gear teeth.

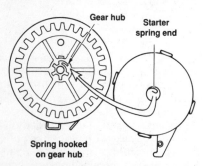

Gear hub   Starter spring end

Spring hooked on gear hub

Fig. TP37—If removed, place spring housing on rope pulley while being sure inner spring end engages anchor on pulley.

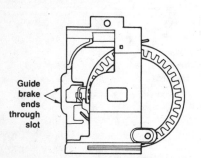

Guide brake ends through slot

Fig. TP38—Install rope pulley assembly in starter bracket while inserting brake spring legs into slots in starter bracket.

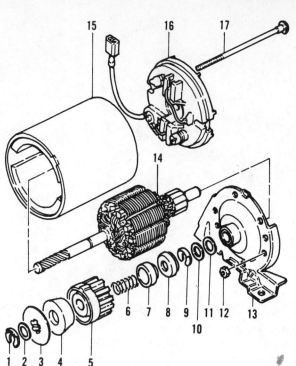

Fig. TP39—Exploded view of typical round frame starter motor with inward-acting starter drive.

1. "E" ring
2. Plastic washer
3. Drive hub
4. Rubber driver
5. Pinion gear
6. Spring
7. Plastic spring cup
8. Metal cup
9. "E" ring
10. Metal washer
11. Plastic washer
12. Nut
13. Drive plate
14. Armature
15. Frame
16. End cap assy.
17. Through-bolt

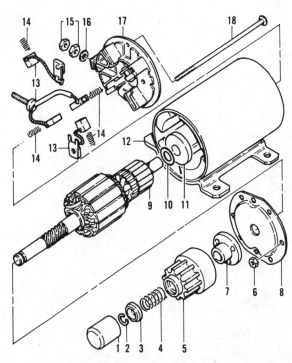

Fig. TP40—Exploded view of typical round frame starter motor with outward-acting starter drive.

1. Boot
2. Snap ring
3. Spring cup
4. Spring
5. Pinion gear
6. Nut
7. Drive hub
8. Drive plate
9. Armature
10. Washer
11. Washer
12. Frame
13. Brushes
14. Brush springs
15. Nuts
16. Washer
17. End cap
18. Through-bolt

into socket. Turn spring housing (6) so strut aligns with legs on brake spring (2). Prevent housing rotation by inserting a pin or rod through strut hole and into gear teeth as shown in Fig. TP36. Remove pulley assembly from starter bracket.

CAUTION: Do not allow spring housing to separate from rope pulley until any spring tension has been relieved.

Hold spring housing (6—Fig. TP34) against rope pulley so spring housing cannot rotate. Withdraw pin or rod in spring housing strut (Fig. TP36) and allow spring housing to rotate thereby relieving rewind spring tension. If necessary, separate spring housing from rope pulley. Do not attempt to remove

rewind spring from spring housing. The spring and housing are available only as a unit assembly.

If rope replacement is necessary, note that inner end of rope was originally retained by a staple, while inner end on replacement ropes is inserted through a hole in pulley and knotted. Pry out staple, if so equipped, to release rope. Route new rope through pulley hole and tie a knot in rope end. The knot must not protrude above the pulley.

To assemble starter, proceed as follows: Do not lubricate any starter components. When viewed from gear side of rope pulley, wind rope onto pulley in a clockwise direction. Install brake spring (2—Fig. TP34) on rope pulley while being careful not to distort spring legs. If removed, place spring housing on rope pulley while being sure inner spring end engages anchor on pulley. See Fig. TP37. Rotate spring housing four turns counterclockwise, then align legs on brake spring with hole in strut. Prevent housing rotation by inserting a pin or rod through strut hole and into gear teeth. If so equipped, install clip (7—Fig. TP34), key (10) and pawl (9). Install rope pulley assembly in starter bracket while inserting brake spring legs into slots in starter bracket (Fig. T89). Route outer rope end past rope guide and install rope handle. Withdraw pin or rod in strut hole. The strut will rotate until it contacts starter bracket. Press or drive in a new pulley spindle (14—Fig. TP34).

ELECTRIC STARTER. While several electric starter motors have been used on Tecumseh engines, the motors are basically divided into those with a round frame or a square frame. Several variations of each type have been produced, but service is basically similar except as noted in following paragraphs.

Tecumseh does not provide test specifications for electric starter motors, so service is limited to replacing components that are known or suspected faulty.

CAUTION: The starter motor field magnets may be made of ceramic material. Do not clamp starter housing in a vise or hit housing as field magnets may be damaged.

Refer to Figs. TP39 and TP40 for exploded views of typical starter motors. Note the following when servicing electric starter motor: Prior to disassembly, mark drive plate, frame and end cap so they can be aligned during assembly. Some motors have alignment notches, in which case alignment marks are not necessary.

Minimum brush length is not specified. If brush wire bottoms against slot in brush holder or brush is less than half its original length, replace brush. Be sure brushes do not bind in holders.

Check strength of brush springs. The spring must force brush against commutator with sufficient pressure to ensure good contact.

Bushings in drive plate and end cap are not available separately, only as a unit assembly with drive plate or end cap.

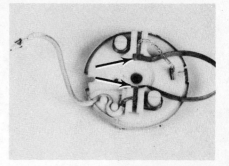

Fig. TP41—A piece of rewind starter spring may be positioned as shown to hold brushes in place during reassembly.

Apply a light coat of grease to helix. All other parts should be assembled dry.

On some motors, a brush holding tool may be helpful to retain brushes and springs during assembly. If end cap has two brushes, a piece of manual rewind starter spring can be modified to hold brushes as shown in Fig. TP41. Use care when installing end cap so commutator and brushes are not damaged.

TECUMSEH
2-STROKE ENGINES

Model	Bore	Stroke	Displacement	Power Rating
HSK840, HXL840	2.44 in. (62 mm)	1.81 in. (46 mm)	8.46 cu.in. (138 cc)	4.0 hp (3.0 kW)
TVS840, TVXL840	2.44 in. (62 mm)	1.81 in. (46 mm)	8.46 cu.in. (138 cc)	4.0 hp (3.0 kW)

ENGINE IDENTIFICATION

All models are two-stroke, single-cylinder, air-cooled engines. Engine model and type numbers are stamped in blower housing as shown in Fig. TP101. Always furnish engine model and type numbers when ordering parts.

MAINTENANCE

LUBRICATION. The engine is lubricated by mixing regular unleaded gasoline with a good quality two-stroke, air-cooled engine oil rated SAE 30 or SAE 40 at a 32:1 ratio. Automotive or multiviscosity type oils are not recommended. Manufacturer states that gasoline containing methanol must not be used, and if gasohol is used, it must not contain more than 10 per cent ethanol. Use a separate container to mix oil and gasoline. Do not mix directly in engine fuel tank.

AIR FILTER. The filter element may be made of foam or paper, or a combination of both foam and paper. The recommended maintenance interval depends on the type of filter element.

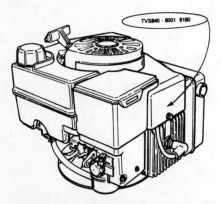

Fig.TP101—Engine model and type numbers are stamped on blower housing in location shown.

Foam type filter elements should be cleaned, inspected and re-oiled after every 25 hours of engine operation, or after three months, whichever occurs first. Clean the filter in soapy water then squeeze the filter until dry (don't twist the filter). Inspect the filter for tears and holes or any other opening. Discard the filter if it cannot be cleaned satisfactorily or if the filter is torn or otherwise damaged. Pour clean engine oil into the filter, then squeeze the filter to remove the excess oil and distribute oil throughout the filter.

Paper type filter elements should be replaced annually or more frequently if the engine operates in a severe environment, such as extremely dusty conditions. A dirty filter element cannot be cleaned and must be discarded.

SPARK PLUG. Recommended spark plug for Models HSK840 and HXL840 is a Champion RCJ8Y. Recommended spark plug for Models TVS840 and TVXL840 is either a Champion CJ8Y or RJ19LM, depending on application. Specified spark plug electrode gap is 0.030 inch (0.76 mm). Tighten spark plug to 16-20 ft.-lbs. (22-27 N•m) torque.

CARBURETOR. All models are equipped with a float type carburetor. Carburetor may be equipped with a fixed high speed jet and an adjustable low speed mixture screw, or fixed high speed and low speed mixture jets. Refer to Fig. TP102.

Initial adjustment of low speed mixture screw on carburetors so equipped is one turn open from a lightly seated position. Make final adjustment with engine running and at operating temperature. Place throttle control in slow speed position and turn low speed mixture screw in a clockwise direction until engine falters. Note screw position, then turn low speed mixture screw in counterclockwise direction until engine starts to sputter or rpm decreases. Turn low speed mixture screw to a position halfway between the two positions to obtain smooth engine operation.

Carburetor With Adjustable Mixture Screw. To disassemble carburetor equipped with low speed mixture screw, remove float bowl retaining bolt, float bowl, float hinge pin, float and fuel inlet needle and spring. Use a wire hook and remove fuel inlet needle seat. Use a small, sharp chisel to remove Welch plug as shown in Fig. TP103. Do not attempt to drill Welch plug. Remove low speed mixture screw. Remove throttle and choke plates. Remove throttle and choke shafts.

When reassembling carburetor, install the new fuel inlet seat into carburetor body with grooved side of seat facing down (Fig. TP104). Apply a light coat of suitable sealer to outer edge of new Welch plug, then install plug by tapping crown of plug with a hammer and punch. A correctly installed Welch plug is flat; a concave plug may leak. When installing dampening spring on float hinge pin, the short leg hooks onto carburetor body and the longer end points toward choke end (Fig. TP105). Use Tecumseh float tool 670253A to set the correct float height (Fig. TP106). The gauge is a go-no go type. Pull the tool in a 90° direction to the float hinge pin. The toe of the float, end opposite the hinge, must be under the first step and can touch the second step without a gap. Carefully bend the float tab holding the

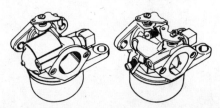

Fig. TP102—View showing fixed jet carburetor (left) and carburetor with adjustable low speed mixture screw (right). Refer to text.

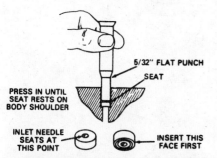

Fig. TP103—Use a sharp chisel to remove carburetor Welch plug.

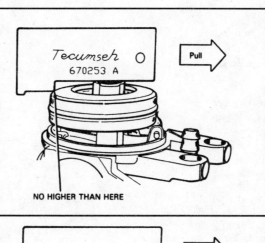

Fig. TP104—Fuel inlet seat must be installed in carburetor body with grooved side facing inward.

fuel inlet needle to obtain correct height.

Carburetor With Fixed Jets. To disassemble carburetor equipped with fixed high speed and low speed jets, note that fuel bowl retaining bolt stamped as shown in Fig. TP107 is manufactured with left-hand threads. Remove float bowl retaining bolt, float bowl, float hinge pin, fuel inlet needle and spring. Use a hooked wire to remove the fuel inlet needle seat. Remove primer bulb retaining ring and primer bulb. Remove throttle plate, throttle shaft, dust seal and spring.

To reassemble carburetor, install the new fuel inlet needle seat with grooved side of seat facing down (Fig. TP104). When installing dampening spring on float, the short leg hooks onto carburetor body and the longer end points toward choke end (Fig. TP105). Use Tecumseh float tool 670253A to set the correct float height (Fig. TP106). The gauge is a go-no go type. Pull the tool in

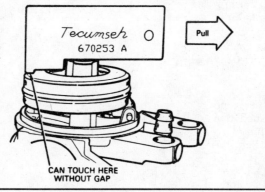

Fig. TP106—Tecumseh go-no go gage 670253 should be used to set correct float height. Refer to text.

ring in a ¾-inch deep socket and install assembly into primer bulb recess in carburetor body until retaining ring is seated.

GOVERNOR. All models are equipped with a mechanical type governor located at pto end of crankshaft. All governor linkage positions should be marked prior to disassembly. To adjust governor, first place the throttle at wide-open position. Then loosen holding

a 90° direction to the hinge pin. The toe of the float, end opposite the hinge, must be under the first step and can touch the second step without a gap. Carefully bend the float tab holding the fuel inlet needle to obtain correct height. Place primer bulb and retaining

screw (Fig. TP108) and turn slotted governor shaft clockwise as far as it will go without forcing it. Tighten holding screw to secure adjustment.

IGNITION SYSTEM. Standard ignition system on all models is a solid-

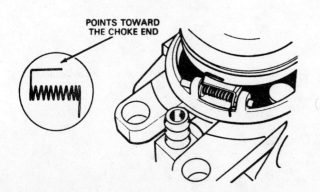

Fig. TP105—Install fuel inlet needle spring as shown. Refer to text.

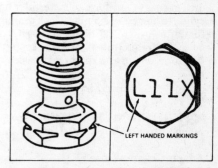

Fig. TP107—Float bowl retaining bolt marked as shown is manufactured with left-hand threads.

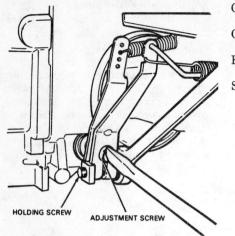

Fig. TP108—View of external governor linkage. Refer to text for adjustment procedure.

state electronic system which does not have breaker points. There is no scheduled maintenance.

To test for spark, remove spark plug cable from spark plug. Insert metal conductor into cable end and hold conductor $\frac{1}{8}$ inch (3 mm) from cylinder shroud. Crank engine and observe spark. A weak spark or no spark indicates a defective ignition coil. Also check for broken, loose or shorted wiring or a faulty spark plug.

Air gap between solid-state module and flywheel should be 0.0125 inch (0.32 mm).

CARBON. Carbon and other combustion deposits should be cleaned from exhaust port and EGR tube after every 75-100 hours of normal operation. Before cleaning the ports, remove the muffler and position the piston at bottom dead center. Clean ports using a pointed $\frac{3}{8}$-inch (9.5 mm) wooden dowel rod. Refer also to EGR tube in REPAIRS section for EGR tube cleaning or removal on Models TVS840 and TVXL840.

GENERAL MAINTENANCE. Periodically check and tighten all loose bolts, nuts or clips. Check for fuel leakage and repair as necessary. Clean dust, dirt, grease or any foreign material from cylinder head and cylinder block cooling fins after every 100 hours of operation or more frequently if needed. Inspect fins for damage and repair if necessary.

REPAIRS

TIGHTENING TORQUES. Recommended tightening torques are as follows:

Adapter-to-crankcase:

(TVS840 & TVXL840). . 13-18 ft.-lbs.
(18-2 4N•m)

Carburetor . . . . 10-12 ft.-lbs.
(13-1 6N•m)
Crankcase. . . . . 10-17 ft.-lbs.
(13-2 3N•m)
Flywheel nut. . . . . 34-36 ft.-lbs.
(46-49N•m)
Spark plug . . . . 16-20 ft.-lbs.
(22-27N•m)

CRANKCASE PRESSURE TEST. An improperly sealed crankcase can cause the engine to be hard to start, run rough, have low power and overheat. Refer to SERVICE SECTION TROUBLE-SHOOTING section of this manual for crankcase pressure test procedure. If crankcase leakage is indi-

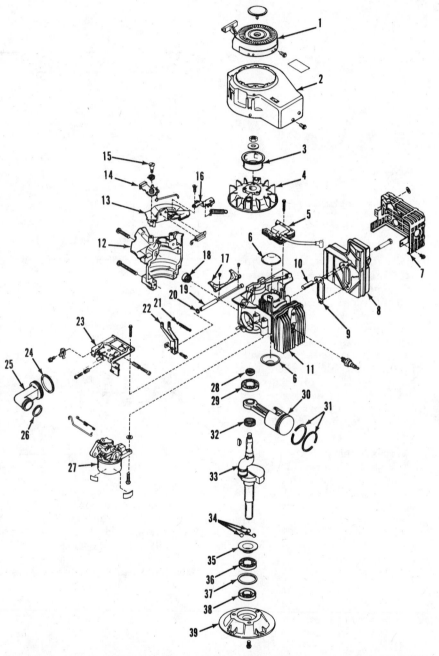

Fig. TP109—Exploded view of vertical crankshaft engine. Engines with horizontal crankshaft are similar, however, adapter (39) and EGR tube (10) are not used.

1. Rewind starter	11. Cylinder & crankcase half	21. Governor spring
2. Blower housing	12. Crankcase half	22. Governor lever
3. Starter cup	13. Brake mounting bracket	23. Governor linkage bracket
4. Flywheel	14. Flywheel brake	24. "O" ring
5. Ignition coil/module	15. Spring	25. Intake pipe
6. Port cover	16. Brake control lever	26. "O" ring
7. Heat shield	17. Governor arm	27. Carburetor
8. Muffler	18. Fuel filter	28. Oil seal
9. Gasket	19. Governor shaft	29. Ball bearing
10. EGR tube	20. Oil seal	30. Piston & rod assy.
		31. Piston rings
		32. Bearing rollers (31)
		33. Crankshaft
		34. Governor balls (3)
		35. Slide ring
		36. Ball bearing
		37. Retainer
		38. Oil seal
		39. Adapter

Illustrations courtesy Tecumseh Products Co.

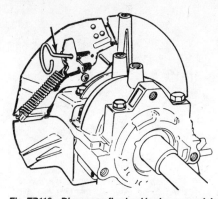

Fig. TP110—Disengage flywheel brake, on models so equipped, by extending brake lever as shown, then install an alignment pin through hole in lever to retain position.

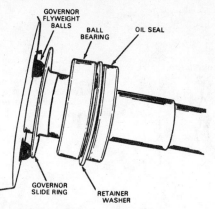

Fig. TP112—Install governor flyweight balls, governor slide ring, ball bearing, retainer washer and oil seal onto flywheel side of crankshaft as shown.

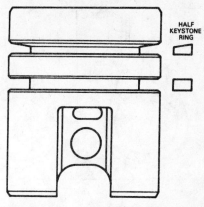

Fig. TP114—View showing piston used on engines equipped with one half-keystone type ring and one rectangular ring. Note correct position of each ring.

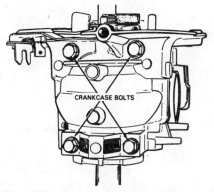

Fig. TP111—View showing location of the four crankcase screws.

cated, pressurize crankcase and use a soap and water solution to check gaskets, seals, carburetor pulse line and casting for leakage.

DISASSEMBLY. Refer to Fig. TP109 for an exploded view of engine. The cylinder is an integral part of one crankcase half. To disassemble engine, drain all fuel and remove fuel tank. Remove air cleaner assembly, muffler and blower housing. Remove solid-state ignition module. Use strap wrench 670305 to hold flywheel, then remove flywheel retaining nut. Pull brake lever away from return spring as far as it will go and place an alignment pin through the lever hole so it catches on the shroud (Fig. TP110). This will hold the brake pad off the flywheel. Use flywheel puller 670306 to remove the flywheel. Remove the three screws retaining the speed control to the cylinder block. Mark all linkage connections and disconnect all governor and carburetor linkage. On Models TVS840 and TVXL840, remove the three screws from the pto mounting adapter. Lightly tap adapter to separate it from crankcase. Remove carburetor, then remove intake elbow and reed valve assembly. Remove brake assem-

bly. Remove the four crankcase bolts (Fig. TP111) and separate crankcase half from cylinder and crankcase assembly. Carefully lift crankshaft and piston assembly from crankcase.

Slide oil seal, retainer washer, ball bearing and governor slide ring off crankshaft (Fig. TP112). Catch the governor flyweight balls as governor slide ring is removed. Remove the oil seal and ball bearing from flywheel end of crankshaft. Slide the piston and connecting rod off toward the flywheel end of crankshaft. The 31 needle bearings in connecting rod are loose and will fall out during connecting rod removal.

Clean and inspect all parts. Refer to following sections for service of engine components. Refer to REASSEMBLY section to reassemble engine.

PISTON AND RINGS. Some engines are equipped with standard rectangular rings (Fig. TP113) and either ring can be installed in either piston ring groove. Most engines are equipped with a piston ring set containing one half-keystone ring and one standard rectangular ring (Fig. TP114). The half-

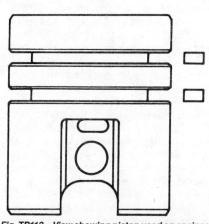

Fig. TP113—View showing piston used on engines equipped with standard rectangular type piston rings.

keystone ring and piston ring groove are beveled on one side and ring must be installed in the upper piston ring groove with beveled side of ring toward top of piston.

Standard piston diameter is 2.4325-2.4330 inches (61.785-61.798 mm). Piston ring groove width should be 0.0645-0.0655 inch (1.638-1.664 mm) for all rectangular type ring grooves. Side clearance of new ring (rectangular rings) in piston ring groove should be 0.002-0.004 inch (0.05-0.10 mm). Renew piston if ring side clearance is excessive. Standard cylinder bore diameter is 2.437-2.438 inches (61.89-61.93 mm).

CRANKSHAFT AND CONNECTING ROD. Crankshaft main bearing journal diameter at pto end should be 0.9833-0.9838 inch (24.976-24.989 mm). Crankshaft main bearing journal diameter at flywheel end should be 0.7864-0.7869 inch (19.975-19.987 mm). Crankpin journal diameter should be 0.9710-0.9715 mm). Crankshaft end play should be 0.0004-0.0244 inch (0.010-0.0244 mm).

Connecting rod and piston assembly must be installed on crankshaft as shown in Fig. TP115.

REED VALVE. The reed plate is located on the intake elbow (Fig. TP116). If clearance between end of reed and intake elbow is 0.010 inch (0.25 mm) or more, intake elbow must be renewed.

EGR TUBE. Models TVS840 and TVXL840 are equipped with an EGR (exhaust gas recirculation) tube located near the cylinder exhaust ports. A controlled amount of exhaust gas is drawn into the crankcase to aid movement of the air-fuel mixture to the combustion chamber. This tube should be cleaned

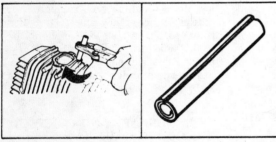

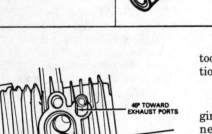

Fig. TP115—Connecting rod must be installed on crankshaft so flanged side of connecting rod is toward magneto end of crankshaft.

INTAKE ELBOW

Fig. TP116—Reed valve is an integral part of the intake elbow. Refer to text.

Fig. TP117—To remove the EGR tube, clamp locking pliers onto tube and rotate in a clockwise direction to collapse tube.

Fig. TP118—EGR tube must be installed with seal toward exhaust ports as shown.

45° TOWARD EXHAUST PORTS

Fig. TP119—Piston and connecting rod assembly are installed by working connecting rod over crankshaft as shown.

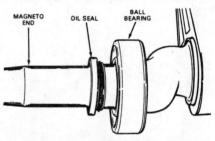

MAGNETO END   OIL SEAL   BALL BEARING

Fig. TP120—Install ball bearing and oil seal onto flywheel side of crankshaft as shown.

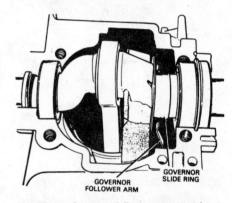

GOVERNOR FOLLOWER ARM   GOVERNOR SLIDE RING

Fig. TP121—Governor follower arm must be on pto side of governor slide ring as shown after crankshaft installation in crankcase.

when cleaning combustion deposits from exhaust ports.

To remove the EGR tube, measure distance EGR tube protrudes from cylinder block, then clamp tube with vise grips and rotate in a clockwise direction. This will collapse the tube and allow removal (Fig. TP117). Install new tube with seam in tube facing 45° to the exhaust port as shown in Fig. TP118 and to the same depth as old tube. Tecumseh

tool 670318 is available to aid installation of EGR tube to correct depth.

REASSEMBLY. To reassemble engine, use heavy grease to retain connecting rod needle bearings (new bearings are retained on a strip) and position piston and connecting rod on crankshaft as shown in Fig. TP119. Make certain flanged side of connecting rod (Fig. TP115) is toward flywheel end of crankshaft, then work connecting rod and piston assembly onto crankshaft. Install the three governor flyweight balls into the crankshaft. Install governor slide ring onto crankshaft so flat portion of ring covers flyweight balls (Fig. TP112). Install ball bearing type main bearing, retainer washer and the large oil seal as shown in Fig. TP112. Install ball bearing type main bearing and smaller oil seal onto flywheel end of crankshaft as shown in Fig. TP120. Stagger ring end gaps and carefully install piston and connecting rod assembly into cylinder and crankcase assembly. Tapered edge at bottom of cyl-

inder will compress piston rings. Position governor follower arm so it rides on the pto side of the governor slide ring as shown in Fig. TP121. Apply a silicon type sealer on mating surfaces of crankcase halves before assembling. Crankcase halves must be aligned at pto end of crankcase as shown in Fig. TP122. Use a straightedge to make certain crankcase halves are aligned, then tighten the four crankcase screws to 10-17 ft.-lbs. (13-23 N·m) torque. It is recommended that the crankcase be pressure tested for leakage before proceeding with remainder of engine reassembly.

Refer to Fig. TP123 and install flywheel brake assembly. Install new "O" rings on intake elbow and reed assembly prior to installation. Position one

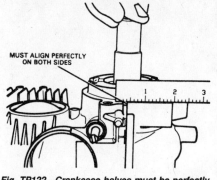

Fig. TP122—Crankcase halves must be perfectly aligned.

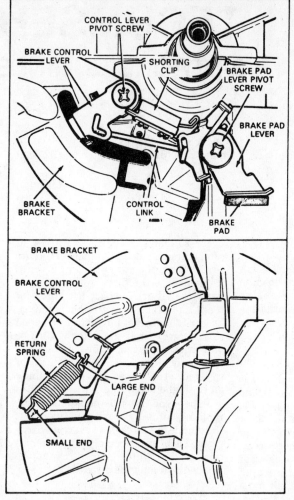

Fig. TP123—View showing assembled position of flywheel brake assembly on models so equipped.

fiber washer on each carburetor post prior to mounting carburetor on cylinder block (Fig. TP124). Float bowl of carburetor must be facing pto side of engine. On Models TVS840 and TVXL840, install the pto adapter (39—Fig. TP109) and tighten retaining screws to 13-18 ft.-lbs. (18-24 N·m) torque. Disengage flywheel brake and install an alignment pin as shown in Fig. TP110. Install flywheel and tighten flywheel retaining nut to 34-36 ft.-lbs. (46-49 N·m) torque. Install solid-state ignition module with a 0.0125 inch (0.32 mm) gap between module and flywheel magnets. Tighten module retaining screws to 30-40 in.-lbs. (3.4-4.5 N·m) torque. Remove brake alignment pin. Install blower housing and rewind starter assembly. Install gasket and muffler, air cleaner assembly and fuel tank.

REWIND STARTER. To disassemble starter, remove rope handle and allow rope to wind slowly into starter. Position a suitable sleeve support (a ¾-inch deep socket) under retainer (8—Fig. TP125), then use a punch to drive pin (11) free of starter. Remove brake spring (9), retainer (8), dogs (6) and springs (5). Wear appropriate safety eyewear and gloves before disengaging pulley from starter as spring may uncoil uncontrolled. Place shop towel around pulley and lift pulley out of housing; spring should remain with pulley.

Inspect components for damage and excessive wear. Standard rope length is 98 inches (249 cm) and rope diameter is 9/64 inch (3.5 mm).

Reverse disassembly procedure to install components. Rewind spring coils wind in clockwise direction from outer end. Wind rope around pulley in counterclockwise direction as viewed from retainer side of pulley. Be sure inner end of rewind spring engages spring retainer adjacent to housing center post. Use two plastic washers (7). Install a new pin (11) so top of pin is ⅛ inch (3.2 mm) below top of starter. Driving pin in too far may damage retainer pawl.

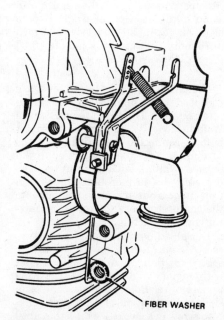

Fig. TP124—One fiber washer must be installed on each carburetor post prior to mounting carburetor on cylinder block.

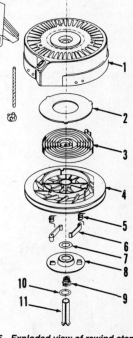

Fig. TP125—Exploded view of rewind starter.

1. Starter housing
2. Cover
3. Rewind spring
4. Pulley
5. Springs (2)
6. Dogs (2)
7. Plastic washers (2)
8. Retainer
9. Brake spring
10. Metal washer
11. Pin

TECUMSEH
2-STROKE ENGINES

Model	Bore	Stroke	Displacement	Rated Power
TC200, TCH200	1.438 in. (36.52 mm)	1.250 in. (31.75 mm)	2.0 cu. in. (32.8 cc)	1.6 hp (1.2 kW)
TC300, TCH300	1.750 in. (44.45 mm)	1.250 in. (31.75 mm)	3.0 cu. in. (49.2 cc)	2.0 hp (1.5 kW)

ENGINE IDENTIFICATION

Engine model and type numbers are stamped in blower housing as shown in Fig. TP201. Always furnish engine model and type numbers when ordering parts.

MAINTENANCE

LUBRICATION. Both models are lubricated by mixing regular unleaded gasoline with a good quality two-stroke, air-cooled engine oil rated SAE 30 or SAE 40 at a 24:1 ratio. Automotive or multiviscosity type oils are not recommended. Manufacturer states that gasoline containing methanol must not be used, and if gasohol is used, it must not contain more than 10 per cent ethanol. Use a separate container to mix oil and gasoline. Do not mix directly in engine fuel tank.

AIR FILTER. The filter element may be made of foam or paper, or a combination of both foam and paper. The recommended maintenance interval depends on the type of filter element.

Foam type filter elements should be cleaned, inspected and re-oiled after every 25 hours of engine operation, or after three months, whichever occurs first. Clean the filter in soapy water then squeeze the filter until dry (don't twist the filter). Inspect the filter for tears and holes or any other opening. Discard the filter if it cannot be cleaned satisfactorily or if the filter is torn or otherwise damaged. Pour clean engine oil into the filter, then squeeze the filter to remove the excess oil and distribute oil throughout the filter.

Paper type filter elements should be replaced annually or more frequently if the engine operates in a severe environment, such as extremely dusty conditions. A dirty filter element cannot be cleaned and must be discarded.

SPARK PLUG. Recommended spark plug for Models TC200 and TCH200 is a Champion CJ8Y. Recommended spark plug for Models TC300 and TCH300 is a Champion CJ6Y. A resistor type plug may be required in some localities. Specified spark plug electrode gap for all models is 0.030 inch (0.76 mm). Tighten spark plug to 16-20 ft.-lbs. (22-27 N•m) torque.

CARBURETOR. All engines are equipped with a diaphragm type carburetor, either a Tillotson Model HU or a Walbro Model WTA. A remote primer bulb is connected to a fitting on the carburetor on some engines equipped with a Walbro WTA carburetor. Depressing the primer bulb forces fuel from the carburetor fuel chamber into the carburetor bore to enrich the mixture during starting.

Initial adjustment of idle mixture screw is one turn open from a lightly seated position. Make final adjustment with engine running and at operating temperature. Place throttle control in slow speed position and turn idle mixture screw in a clockwise direction until engine falters. Note screw position, then turn idle mixture screw in counterclock-wise direction until engine starts to sputter or rpm decreases. Turn idle mixture screw to a position halfway between the two positions to obtain smooth engine operation.

Refer to appropriate following section for carburetor service.

Tillotson HU Carburetor. The carburetor may be disassembled after inspecting unit and referring to exploded view in Fig. TP202. Clean filter screen (6). Carefully remove Welch plug using a sharp chisel or scratch awl while being careful not to damage underlying metal.

Clean all metallic parts with solvent. Blow out all passages with compressed air in the opposite direction of normal fuel flow. Do not soak in solvent over 30 minutes. Do not use wires or drills to clean orifices or passages. Inspect inlet lever spring (24) and renew if stretched or damaged. Inspect diaphragms (5 and 28) for tears, cracks or other damage. Fuel inlet valve (22) has a rubber tip and seats directly on a machined orifice on the carburetor body. Inlet valve and body should be renewed if worn excessively.

When reassembling carburetor be sure that diaphragms and gaskets are arranged as shown in Fig. TP202. Rivet head on metering diaphragm (28) must be toward metering lever (23). Adjust position of fuel inlet control lever (23) so lever is flush with diaphragm chamber floor as shown in Fig. TP203. Bend lever adjacent to spring to obtain correct lever position. Apply a light coat of suitable sealer to outer edge of new Welch plug, then install plug by tapping crown of plug with a hammer and punch that is slightly larger in diameter than the Welch plug. A correctly installed Welch plug is flat; a concave plug may leak.

Connect throttle link from air vane governor in innermost hole on carburetor throttle arm.

Walbro WTA. The carburetor may be disassembled after inspecting unit and referring to exploded view in Fig.

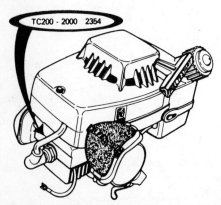

Fig. TP201—Engine model and type number are stamped into blower housing base.

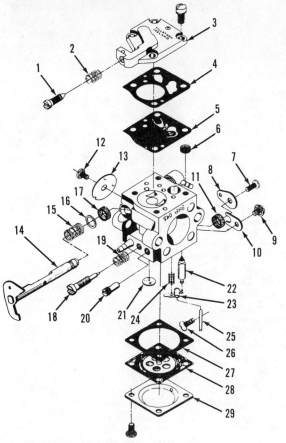

1. Idle speed screw
2. Spring
3. Cover
4. Gasket
5. Fuel pump diaphragm
6. Screen
7. Screw
8. Throttle arm
9. Screw
10. Retainer
11. Felt seal
12. Screw
13. Throttle plate
14. Throttle shaft
15. Spring
16. Washer
17. Felt seal
18. Idle mixture screw
19. Spring
20. Main jet
21. Welch plug
22. Fuel inlet valve
23. Metering lever
24. Spring
25. Pin
26. Screw
27. Gasket
28. Diaphragm
29. Cover

or damaged. Inspect diaphragms (2 and 30) for tears, cracks or other damage.

When reassembling carburetor, install throttle plate so numbers are visible when throttle is in closed position. Apply a small amount of Loctite 262 (red) to fuel fitting (20) and reinstall in original position. Apply a light coat of suitable sealer to outer edge of new Welch plug, then install plug by tapping crown of plug with a hammer and punch that is slightly larger in diameter than the Welch plug. A correctly installed Welch plug is flat; a concave plug may leak. Be sure diaphragms and gaskets are arranged as shown in Fig. TP204. Rivet head on metering diaphragm (2) must be toward metering lever (7). Adjust position of fuel inlet control lever (7) so lever tip is 0.060-0.070 inch (1.52-1.78 mm) below face of carburetor as shown in Fig. TP205. Bend lever adjacent to spring to obtain correct lever position.

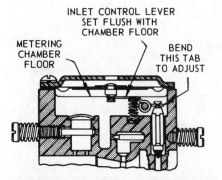

Fig. TP203—The metering lever on Tillotson HU carburetor should be flush with metering chamber floor. Bend metering lever tab to adjust lever height.

TP204. Note that choke detent ball (17) and spring (18) will be loose when the choke shaft is withdrawn. Clean filter screen (23). Carefully remove Welch plug using sharp chisel or scratch awl while being careful not to damage underlying metal. If fuel inlet fitting (20) must be removed, note position before removal.

Clean all metallic parts with solvent. Blow out all passages with compressed air in the opposite direction of normal fuel flow. Do not soak in solvent over 30 minutes. Do not use wires or drills to clean orifices or passages. Inspect inlet lever spring (9) and renew if stretched

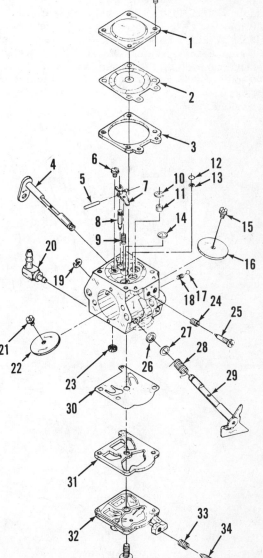

Fig. TP204—Exploded view of Walbro WTA diaphragm carburetor.

1. Cover
2. Diaphragm
3. Gasket
4. Choke shaft
5. Pin
6. Screw
7. Metering lever
8. Fuel inlet valve
9. Spring
10. Welch plug
11. Check valve
12. Clip
13. Screen
14. Welch plug
15. Screw
16. Choke plate
17. Detent ball
18. Spring
19. "E" ring
20. Fuel inlet
21. Screw
22. Throttle plate
23. Screen
24. Spring
25. Idle mixture screw
26. Felt seal
27. Washer
28. Spring
29. Throttle shaft
30. Fuel pump diaphragm
31. Gasket
32. Cover
33. Spring
34. Idle speed screw

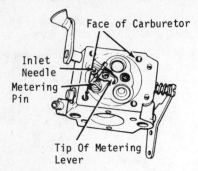

Fig. TP205—Tip of metering lever should be 0.060-0.070 inch (1.52-1.78 mm) from the face of carburetor body.

Connect throttle link from air vane governor in innermost hole on carburetor throttle arm.

IGNITION SYSTEM. All engines are equipped with a solid-state ignition module located adjacent to the outside of the flywheel. Air gap between module laminations and flywheel magnets should be 0.0125 inch (0.32 mm) on TC200 and TCH200 models. Air gap should be 0.0125 inch (0.32 mm) on TC300 engines on rotary mowers. Air gap should be 0.030 inch (0.76 mm) on TC300 and TCH300 engines not used on rotary mowers.

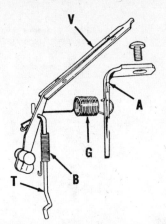

Fig. TP208—Diagram of air vane governor mechanism. Long end of governor spring (G) is attached to air vane (V) arm. Backlash spring (B) is used on later models.

GOVERNOR ADJUSTMENT. The engine is equipped with an air vane type governor (Fig. TP206). Governed speed is adjusted by bending the governor spring anchor (A—Fig. TP207 and Fig. TP208). Bending the anchor toward the spark plug will decrease governed engine speed. Adjust governed engine speed to equipment manufacturer's specification.

Long end of governor spring is attached to the air vane arm as shown in

Fig. TP208. Note that backlash spring (B) surrounding throttle link (T) is not used on early models. Early and late air vane assemblies are not interchangeable due to corresponding differences in blower housing design.

CARBON. The muffler and exhaust ports should be inspected periodically for carbon blockage and a suitable maintenance schedule determined. Use a wooden dowel or other suitable tool to remove carbon.

REPAIRS

TIGHTENING TORQUES. Recommended tightening torques are as follows:

Crankcase cover....... 70-100 in.-lbs.
(8-11 N•m)
Cylinder base.......... 60-75 in.-lbs.
(7-8 N•m)
Carburetor .... 20-32 in.-lbs.
(2.3-3.6 N•m)
Flywheel............. 180-240 in.-lbs.
(20-27 N•m)
Ignition module .... 30-40 in.-lbs.
(3.4-4.5 N•m)
Starter retainer screw... 30-40 in.-lbs.
(3.4-4.5 N•m)

CRANKCASE PRESSURE TEST. An improperly sealed crankcase can cause the engine to be hard to start, run rough, have low power and overheat. Refer to SERVICE SECTION TROUBLE-SHOOTING section of this manual for crankcase pressure test procedure. If crankcase leakage is indicated, pressurize crankcase and use a soap and water solution to check gaskets, seals, carburetor pulse line and casting for leakage.

FLYWHEEL. The flywheel is accessible after removing the blower housing. Tecumseh puller 670299 or other suitable puller should be used to remove the flywheel.

When installing the flywheel, install Belleville washer (2—Fig. TP209) so the concave side is down. Tighten the flywheel nut to 180-240 in.-lbs. (20-27 N•m) torque.

CYLINDER. The cylinder (8—Fig. TP209) can be removed after removing the fuel tank, carburetor, blower shroud, ignition module, flywheel, blower base (6) and muffler. Unscrew cylinder retaining fasteners. Note that later models are equipped with Torx screws while early models use studs and nuts. Lift off cylinder in a straight line so bending force is not exerted against the connecting rod.

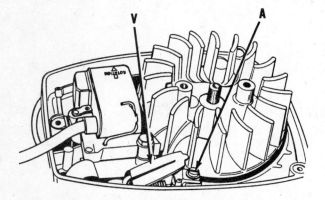

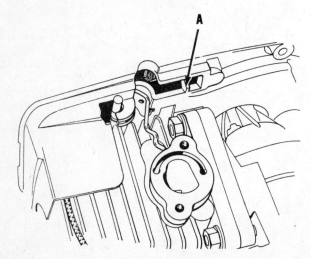

Fig. TP206—View of governor air vane (V) and governor spring anchor (A). Refer also to Fig. TP208.

Fig. TP207—Bend governor spring anchor (A) to adjust governed engine speed. See text.

Fig. TP209—Exploded view of a typical vertical crankshaft engine. Horizontal crankshaft engine is similar. Loose bearing rollers (13) are not used on later models. Cylinder base retaining screws (10) are used on later models in place of studs (14) and nuts (9).

1. Nut
2. Belleville washer
3. Flywheel
4. Ignition module
5. Governor spring anchor
6. Blower base
7. Oil seal
8. Cylinder
9. Nut
10. Screw
11. Piston rings
12. Piston & connecting rod
13. Needle bearing rollers (23)
14. Stud
15. Crankcase
16. Snap ring
17. Needle bearing
18. Crankshaft
19. Needle bearing
20. Snap ring
21. Crankcase cover
22. Oil seal
23. Screw

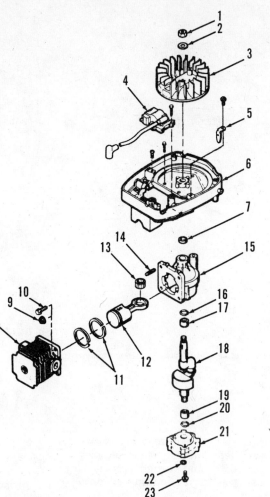

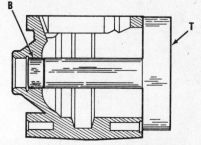

Fig. TP211—Tool 670302 (T) should be used when installing bearing (B) in crankcase. Printed side of bearing must be toward end of tool spindle. Push in until tool bottoms against crankcase.

Inspect cylinder for damage and excessive wear. Cylinder cannot be bored; oversize pistons are not available.

Apply a 1/16-inch bead of Loctite 515 sealant on mating surface of crankcase and around screw holes or studs prior to installing cylinder. When installing cylinder, be aware that twisting or rotating cylinder may bend connecting rod. A slotted 3/8-inch thick board placed so that the rod is positioned in the slot will help support the rod while installing the cylinder. Tighten cylinder base fasteners to 60-75 in.-lbs. (7-8 N•m) torque.

PISTON, CONNECTING ROD AND CRANKSHAFT. The piston and connecting rod are available only as a unit assembly and should not be separated. The crankshaft must be removed before the connecting rod and piston can be removed.

To remove the crankshaft, piston and connecting rod, proceed as follows. Remove the cylinder as previously outlined. Unscrew crankcase cover (21—Fig. TP209) retaining screws and separate crankcase cover from crankcase (15). Rotate crankshaft to top dead

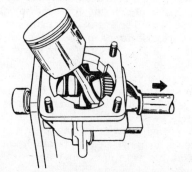

Fig. TP210—Connecting rod must be carefully moved over crankpin when extracting crankshaft from crankcase. On early models, do not lose the 23 loose needle bearing rollers on crankpin.

center position. Note that connecting rod on early models rides on 23 loose bearing rollers on crankpin. Carefully withdraw crankshaft from crankcase and connecting rod (Fig. TP210). On early models, do not lose bearing rollers. Do not attempt to separate piston and connecting rod; they are available only as a unit assembly.

Inspect engine components for excessive wear and damage. If engine is

equipped with loose bearing rollers at crankpin, note that the bearing, piston and connecting rod, and crankshaft are no longer available and must be replaced with a short block assembly. On later models equipped with a caged bearing at the crankpin, the piston and rod assembly and crankshaft are available separately.

Piston skirt diameter on Models TC200 and TCH200 should be 1.4327-1.4340 inches (36.391-36.434 mm). Piston skirt diameter on Models TC300 and TCH300 should be 1.7452-1.7460 inches (44.328-44.348 mm). Piston ring groove width on all models should be 0.0485-0.0495 inch (1.232-1.257 mm) for both ring grooves. Piston ring width should be 0.046-0.047 inch (1.17-1.19 mm).

Crankpin diameter on models with loose bearing rollers should be 0.5985-0.5990 inch (15.202-15.215 mm). Crankpin diameter on models with a caged needle bearing should be 0.6870-0.6875 inch (17.450-17.462 mm).

The crankshaft is supported by roller bearing main bearings in the crankcase and crankcase cover. Crankshaft main bearing diameter on all models at flywheel end should be 0.4998-0.5003 inch (12.695-12.708 mm). Crankshaft main bearing diameter on all models at output end should be 0.6248-0.6253 inch (15.870-15.883 mm).

On early models, the main bearings are secured by snap rings (16 and 20—Fig. TP209). On later models, the bearings abut a shoulder in the crankcase or crankcase cover. To properly locate bearings during installation, tool 670302 should be used when installing bearing in crankcase (Fig. TP211). Tool 670304A should be used when installing bearing in crankcase cover (Fig. TP212). Numbered side of bearing should be toward end of tool.

When assembling engine, note the following. If loose bearing rollers are used on crankpin, hold rollers in place with grease. Install piston and rod on crankshaft so triangle mark (T—Fig.

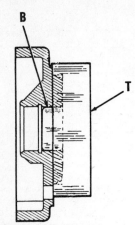

Fig. TP212—Tool 670304A (T) should be used when installing bearing (B) in crankcase cover. Printed side of bearing must be toward end of tool spindle. Push in until tool bottoms against crankcase.

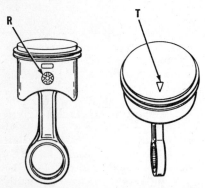

Fig. TP213—Install piston and rod on crankshaft so triangle mark (T) on piston crown is toward exhaust. If there is no mark on piston crown, install piston so piston side with piston pin retainer (R) is toward flywheel end of crankshaft.

TP213) on piston crown is toward exhaust port side of engine. If there is no mark on piston crown, install piston so piston side with piston pin retainer (R) is toward flywheel end of crankshaft.

Apply a $\frac{1}{16}$-inch bead of Loctite 515 sealant on crankcase cover mating surface of the crankcase and around screw holes. Do not allow sealant on inner crankcase surface. Do not rotate crankcase cover after it contacts sealant or a leak may result. Tighten crankcase cover screws to 70-100 in.-lbs. (8-11 N•m) torque. It is recommended that the crankcase be pressure tested for leakage before proceeding with remainder of engine reassembly.

REWIND STARTER. The rewind starter is contained in the blower housing. To disassemble starter, remove staple in rope handle, remove rope handle and allow rope to wind slowly into starter. Unscrew rope pulley retaining screw (9—Fig. TP214). Lift out pawl retainer and pawl (Fig. TP215). Carefully disengage rope pulley from rewind

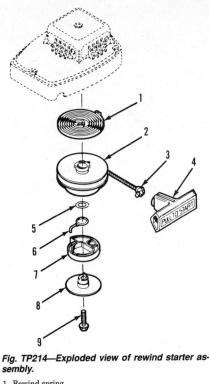

Fig. TP214—Exploded view of rewind starter assembly.

1. Rewind spring
2. Rope pulley
3. Rope
4. Handle
5. Washer
6. Brake spring
7. Pawl
8. Pawl retainer
9. Retaining screw

spring and remove rope pulley. Wear appropriate safety eyewear and gloves before detaching rewind spring from housing as spring may uncoil uncontrolled.

To reassemble starter, apply grease to center post of housing and area of housing that contacts side of rewind spring. Wind rewind spring in housing in counterclockwise direction from outer end as shown in Fig. TP216. Apply grease to side of spring. Wind rope on rope pulley in counterclockwise direction from inner end as shown in Fig. TP217. Install rope pulley in housing and rotate counterclockwise while pushing down so pulley engages rewind spring end. Apply grease to pawl side of retainer, then assemble retainer, pawl and brake spring as shown in Fig. TP215. Brake

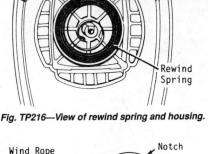

Fig. TP216—View of rewind spring and housing.

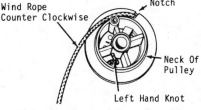

Fig. TP217—View rope and rope pulley.

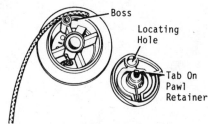

Fig. TP218—Hole in pawl must engage boss on pulley. Tab on retainer must engage notch on center post.

spring end must engage pawl as shown. Install retainer, pawl and spring assembly on starter so hole in pawl engages boss on rope pulley shown in Fig. TP218 and tab on retainer engages notch in starter center post. Install retainer screw (9—Fig. TP214) and tighten to 30-40 in.-lbs. (3.4-4.5 N•m) torque. Apply tension to rewind spring by rotating rope pulley two turns counterclockwise, then route starter rope end through rope outlet in housing. Attach rope handle to rope end.

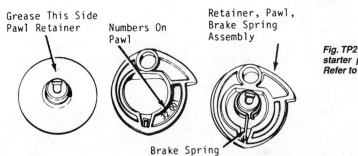

Fig. TP215—View of rewind starter pawl and retainer. Refer to text.

TECUMSEH

4-STROKE ENGINES
(Except Vector & Overhead Valve Engines)

Model	Bore	Stroke	Displacement	Power Rating
ECV100	2.625 in.	1.844 in.	9.98 cu.in.	
	(66.68 mm)	(46.84 mm)	(164 cc)	
H30	2.500 in.	1.844 in.	9.05 cu.in.	3.0 hp
	(63.50 mm)	(46.84 mm)	(148 cc)	(2.2 kW)
H35	2.500 in.	1.938 in.	9.51 cu.in.	3.5 hp
	(63.50 mm)	(49.23 mm)	(156 cc)	(2.6 Kw)
H50, HH50	2.625 in.	2.250 in.	12.18 cu.in.	5.0 hp
	(66.68 mm)	(57.15 mm)	(229 cc)	93.7 Kw)
H60, HH60	2.625 in.	2.500 in.	13.53 cu.in.	6.0 hp
	(66.68 mm)	(63.50 mm)	(222 cc)	(4.5 kW)
HS40	2.625 in.	1.938 in.	10.49 cu.in.	4.0 hp
	(66.68 mm)	(49.23 mm)	(172 cc)	(3.0 kW)
HS50	2.812 in.	1.938 in.	12.04 cu.in.	5.0 hp
	(71.43 mm)	(49.23 mm)	(197 cc)	(3.7 kW)
TNT100	2.625 in.	1.844 in.	9.98 cu.in.	4.0 hp
	(66.68 mm)	(46.84 mm)	(164 cc)	(3 kW)
TNT120	2.812 in.	1.938 in.	12.04 cu.in.	5.0 hp
	(71.43 mm)	(49.23 mm)	(197 cc)	(3.7 kW)
TVM125	2.625 in.	2.250 in.	12.18 cu.in.	5.0 hp
	(66.68 mm)	(57.15 mm)	(229 cc)	(3.7 kW)
TVM140	2.625 in.	2.500 in.	13.53 cu.in.	6.0 hp
	(66.68 mm)	(63.50 mm)	(222 cc)	(4.5 kW)
TVS75	2.313 in.	1.844 in.	7.75 cu.in.	3.0 hp
	(58.74 mm)	(46.84 mm)	(127 cc)	(2.2 kW)
TVS90	2.500 in.	1.844 in.	9.05 cu.in.	3.5 hp
	(63.50 mm)	(46.84 mm)	(148 cc)	(2.6 kW)
TVS100	2.625 in.	1.844 in.	9.98 cu.in.	4.0 hp
	(66.68 mm)	(46.84 mm)	(164 cc)	(3.0 kW)
TVS105	2.625 in.	1.938 in.	10.49 cu.in.	4.0 hp
	(66.68 mm)	(49.23 mm)	(172 cc)	(3.0 kW)
TVS115	2.812 in.	1.844 in.	11.45 cu.in.	4.5 hp
	(71.44 mm)	(46.84 mm)	(188 cc)	(3.3 kW)
TVS120	2.812 in.	1.938 in.	12.04 cu.in.	5.0 hp
	(71.43 mm)	(49.23 mm)	(197 cc)	(3.8 kW)

ENGINE IDENTIFICATION

Engines must be identified by the complete model number, including the specification number in order to obtain correct repair parts. Engine identification numbers, including the model number, specification number and serial number, are located on the blower housing or on a tag attached to the engine. The numbers are stamped in an identification plate or directly in the metal as shown in Fig. T1.

The engine model number identifies the basic engine family. Refer to Fig. T2 for a breakdown and example of a typical Tecumseh engine model number.

The specification number specifies the parts configuration of the engine, as well as cosmetic details such as paint color and decals. The specification number also determines governor speed settings depending on the engine's application, i.e., lawnmower, tractor, pump, etc.

The serial number provides information concerning the manufacturing of

Fig. T1—View of a typical Tecumseh engine identification number.

ECH	Exclusive Craftsman Horizontal Crankshaft
ECV	Exclusive Craftsman Vertical Crankshaft
H	Horizontal Crankshaft
HH	Horizontal Crankshaft Heavy Duty (Cast Iron)
HHM	Horizontal Crankshaft Heavy Duty (Cast Iron) Medium Frame
HM	Horizontal Crankshaft Medium Frame
HS	Horizontal Crankshaft Small Frame
LAV	Lightweight Aluminum Vertical Crankshaft
OHM	Overhead Valve Horizontal Crankshaft Medium Frame
OVM	Overhead Valve Vertical Crankshaft Medium Frame
OVRM	Overhead Valve Vertical Crankshaft Rotary Mower
TNT	Toro N'Tecumseh
TVM	Tecumseh Vertical Crankshft (Medium Frame)
TVS	Tecumseh Styled Vertical Crankshaft
V	Verrtical Crankshaft
VH	Vertical Crankshaft Heavy Duty (Cast Iron)
VLV	Vector Lightweight Vertical Crankshaft
VM	Vertical Crankshaft Medium Frame

EXAMPLE

Engine model and specification numbers: TVS90-43056A

TVS	Tecumseh Styled Vertical Crankshaft
90	Indicates a 9 cubic inch displacement
43056A	Is the specification number used to identify engine parts

Engine serial number: 8310C

8	First digit is the year of manufacture (1978)
310	Indicates calendar day of that year (310th day or November 6, 1978)
C	Represents the line and shift on which the engine was built at the factory

Fig. T2—Table showing Tecumseh engine model number and serial number interpretation.

(EXAMPLE) 8   0   4   1   0   1   A

1st DIGIT	2nd & 3rd DIGIT	4th DIGIT	5th & 6th DIGIT	7th DIGIT
CRANKSHAFT POSITION	HORSEPOWER OR 2 CYCLE	PRIMARY FEATURES	ENGINE VARIATION NUMBER	REVISION LETTER
8 - Vertical 9 - Horizontal	00 = 2 Cycle 02 = 3 H.P. 03 = 3.5 H.P. 04 = 4 H.P. 05 = 5 H.P. 06 = 6 H.P. 07 = 7 H.P. 08 = 8 H.P. 10 = 10 H.P. 12 = 12 H.P. 14 = 14 H.P. 16 = 16 H.P. 18 = 18 H.P.	1 = Rotary Mower 2 = Industrial 3 = Snow King 4 = Mini Bike 5 = Tractor 6 = Tiller 7 = Rider 8 = Rotary Mower	00 thru 99 00 thru 99 00 thru 99 00 thru 99 00 thru 99 00 thru 99 00 thru 99 00 thru 99	

Fig. T3—Reference chart used to select or identify Tecumseh replacement engines.

the engine. Refer to Fig. T2 for a breakdown of a typical serial number.

Note that some engines are classified using the term "frame." Tecumseh classifies the basic engine structure according to the type of metal, either aluminum or cast iron, used to manufacture the engine crankcase and the metal in the cylinder bore. A small frame engine is made of aluminum with an aluminum cylinder bore. A medium frame engine is made of aluminum and has a cast iron liner in the cylinder bore. A heavy frame engine is made of cast iron with a cast iron cylinder bore.

It is important to transfer the blower housing or identification tag from the original engine to a replacement short block assembly so unit can be identified when servicing.

If selecting a replacement engine and model or type number of the old engine is not known, refer to chart in Fig. T3 and proceed as follows:

1. List the corresponding number which indicates the crankshaft position.
2. Determine the horsepower needed.
3. Determine the primary features needed. (Refer to the Tecumseh Engines Specification Book No. 692531 for specific engine variations.)
4. Refer to Fig. T2 for Tecumseh engine model number and serial number interpretation.

Note that new short blocks are identified by a tag marked SBH (Short Block Horizontal) or SBV (Short Block Vertical).

MAINTENANCE

LUBRICATION. Vertical crankshaft engines are equipped with a barrel and plunger type oil pump. Horizontal crankshaft engines are equipped with a dipper type oil slinger attached to the connecting rod.

Oil level should be checked after every five hours of operation. Maintain oil level at lower edge of filler plug or at "FULL" mark on dipstick.

Engine oil should meet or exceed latest API service classification. Use SAE 30 or SAE 10W-30 motor oil for temperatures above 32° F (0° C). Use SAE 5W-30 or SAE 10W for temperatures below 32° F (0° C). Manufacturer explicitly states: DO NOT USE SAE 10W-40 motor oil.

Oil should be changed after the first two hours of engine operation (new or rebuilt engine) and after every 25 hours of operation thereafter.

AIR FILTER. The filter element may be made of foam or paper, or a combination of both foam and paper. The recommended maintenance interval depends on the type of filter element.

Foam type filter elements should be cleaned, inspected and re-oiled after every 25 hours of engine operation, or after three months, whichever occurs first. Clean the filter in soapy water then squeeze the filter until dry (don't twist the filter). Inspect the filter for

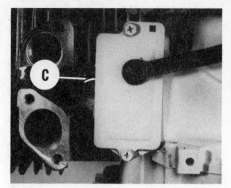

Fig. T4—On some engines, crankcase breather is located in valve tappet cover (C).

Fig. T6—View of top-mounted crankcase breather used on some engines with a vertical crankshaft.

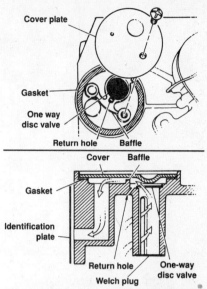

Fig. T8—Drawing of integral type breather used on some ECV series engines.

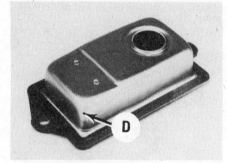

Fig. T5—Drain hole (D) must be down when installing valve tappet cover.

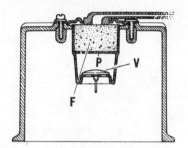

Fig. T7—Cross-section of top-mounted crankcase breather.

ECV100, H30, H35,
  H50, H60, HH50,
  HH60, HS40, HS50,
  TVM125, TVM140 . . . . J8C
TNT100, TNT120, TVS75,
  TVS90, TVS100, TVS105,
  TVS115, TVS120 . . . . J19LM

A resistor type plug may be required in some localities. Specified spark plug electrode gap for all models is 0.030 inch (0.76 mm). Tighten spark plug to 18-20 ft.-lbs. (24.5-27 N•m) torque.

tears and holes or any other opening. Discard the filter if it cannot be cleaned satisfactorily or if the filter is torn or otherwise damaged. Pour clean engine oil into the filter, then squeeze the filter to remove the excess oil and distribute oil throughout the filter.

Paper type filter elements should be replaced annually or more frequently if the engine operates in a severe environment, such as extremely dusty conditions. A dirty filter element cannot be cleaned and must be discarded.

CRANKCASE BREATHER. Three types of crankcase breather have been used: an integral breather, a top-mounted breather and a side-mounted breather.

Side-Mounted Crankcase Breather. This type of breather is located in the valve tappet cover (C—Fig. T4). A disc valve regulates pressure in the crankcase. Unscrew retaining screws to remove breather.

On some engines breather can be separated for access to internal filter element, while on other engines, breather is a unit assembly. On unit type breather, a removable filter element resides inside breather. A barb inside the housing holds element in place. Insert a smooth blade between the barb and filter element to remove

element. Clean filter element in a suitable solvent.

Install either type breather so drain hole (D—Fig. T5) is down. Some units have two gaskets located between breather and engine. Install both gaskets if so equipped.

Top-Mounted Crankcase Breather. Some vertical crankshaft engines are equipped with crankcase breather (B—Fig. T6) mounted on top of crankcase. Remove flywheel for access to the crankcase breather. Unscrew the breather cover and lift out the breather assembly. Remove and clean filter element (F—Fig. T7) in a suitable solvent. Inspect check valve (V) for damage. Removal may damage the valve. Lubricate stem of new check valve to ease installation. Install baffle plate (P) above check valve.

Integral Crankcase Breather. An integral type breather is found on some ECV series engines. This type breather is mounted on top of the crankcase and functions using passages in the crankcase (Fig. T8). Gases are vented out the back of the crankcase, sometimes behind the identification plate.

Remove flywheel for access to crankcase breather. Check for blocked passages and a damaged valve. A replacement parts set is available.

SPARK PLUG. Spark plug recommendations for Champion spark plugs are listed in the following chart.

CARBURETOR. Several different carburetors are used on these engines. Refer to the appropriate paragraph for model being serviced.

Tecumseh Diaphragm Carburetor. Refer to model number stamped on the carburetor mounting flange and to Fig. T9 for exploded view of Tecumseh diaphragm type carburetor.

Initial adjustment of idle mixture and main fuel mixture screws from a lightly seated position is one turn open. Clockwise rotation leans mixture and counterclockwise rotation richens mixture.

Final adjustment is made with engine at operating temperature and running. Operate engine at rated speed and adjust main fuel mixture screw (20—Fig. T9) for smoothest engine operation. Operate engine at idle speed and adjust idle mixture screw (16) for smoothest engine idle. If engine does not accelerate smoothly, slight adjustment of main fuel mixture screw may be required. Engine idle speed should be approximately 1800 rpm.

The fuel strainer in the fuel inlet fitting can be cleaned by reverse flushing with compressed air after the inlet nee-

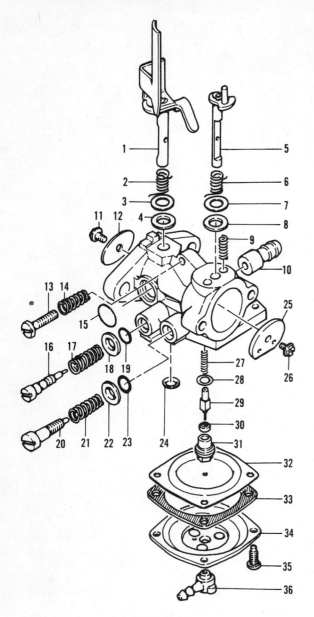

Fig. T11—If carburetor flange is marked with an "F," then diaphragm (32—Fig. T9) and gasket (33) should be installed in order shown in Fig. T9. If carburetor is not marked with an "F," then diaphragm and gasket must be installed in order shown in Fig. T12.

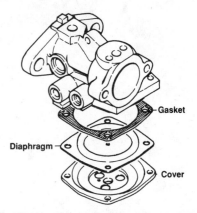

Fig. T12—If carburetor flange is not marked with an "F" (see Fig. T11), then diaphragm and gasket should be installed in order shown.

Fig. T9—Exploded view of Tecumseh diaphragm carburetor. Refer also to Figs. T11 and T12.

1. Throttle shaft
2. Spring
3. Washer
4. Felt washer
5. Choke shaft
6. Spring
7. Washer
8. Felt washer
9. Spring
10. Fuel inlet fitting
11. Screw
12. Throttle plate
13. Idle speed screw
14. Spring
15. Welch plug
16. Idle mixture screw
17. Spring
18. Washer
19. "O" ring
20. High speed mixture screw
21. Spring
22. Washer
23. "O" ring
24. Welch plug
25. Choke plate
26. Screw
27. Spring
28. Washer
29. Fuel inlet valve
30. Fuel inlet valve seat
31. Fuel inlet nut
32. Diaphragm
33. Gasket
34. Cover
35. Screw
36. Fitting

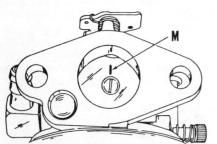

Fig. T10—The mark (M) on throttle plate should be parallel to the throttle shaft and outward as shown. Some models may also have mark at 3 o'clock position.

dle and seat (31—Fig. T9) are removed. The inlet needle seat fitting is metal with a neoprene seat, so the fitting (and enclosed seat) should be removed before carburetor is cleaned with a commercial solvent. Welch plug should be removed to expose drilled passage for cleaning. A small chisel or scratch awl should be used to remove Welch plugs. Do not use wire or drill bit to clean orifices or passages.

Inspect inlet lever spring (27) and renew if stretched or damaged. Inspect diaphragm (32) for tears, cracks or other damage.

When reassembling carburetor, note the following: The stamped line on carburetor throttle plate should be toward top of carburetor, parallel with throttle shaft and facing outward as shown in Fig. T10. Flat side of choke plate should be toward the fuel inlet fitting side of carburetor. Mark on choke plate should be parallel to shaft and should face inward when choke is closed. Apply a light coat of suitable sealer to outer edge of new Welch plug, then install plug by tapping crown of plug with a hammer and punch that is slightly larger in di-

ameter than the Welch plug. A correctly installed Welch plug is flat; a concave plug may leak. Diaphragm (32—Fig. T9) should be installed with rounded head of center rivet up toward the inlet needle (29) regardless of size or placement of washers around the rivet.

On carburetors marked with a "F" as shown in Fig. T11, gasket (33—Fig. T9) must be installed between diaphragm (32) and cover (34). All other models are assembled as shown in Fig. T12, with gasket between diaphragm and carburetor body.

Tecumseh Standard Float Carburetor. Refer to Fig. T13 for exploded view of Tecumseh standard float type carburetor.

Initial adjustment of idle mixture and main fuel mixture screws from a lightly seated position is one turn open. Clockwise rotation leans mixture and counterclockwise rotation richens mixture.

Final adjustment is made with engine at operating temperature and running. Operate engine at rated speed and adjust main fuel mixture screw (35) for smoothest engine operation. Operate

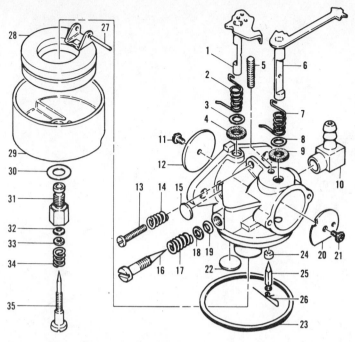

Fig. T13—Exploded view of Tecumseh adjustable float type carburetor.

1. Throttle shaft	10. Fuel inlet fitting	19. "O" ring	28. Float
2. Spring	11. Screw	20. Choke plate	29. Fuel bowl
3. Washer	12. Throttle plate	21. Screw	30. Washer
4. Felt washer	13. Idle speed screw	22. Welch plug	31. Fuel bowl nut
5. Spring	14. Spring	23. Gasket	32. "O" ring
6. Choke shaft	15. Welch plug	24. Fuel inlet valve seat	33. Washer
7. Spring	16. Idle mixture screw	25. Fuel inlet valve	34. Spring
8. Washer	17. Spring	26. Clip	35. High speed
9. Felt washer	18. Washer	27. Float pin	mixture screw

Fig. T16—Float height should be 0.162-0.215 inch (4.11-5.46 mm).

Fig. T17—Float height can be set using Tecumseh float tool 670253A as shown.

Fig. T14—Do not attempt removal of main nozzle (N).

Fig. T15—Fuel inlet valve seat (T) must be installed so grooved side is toward body and flat side is out as shown.

engine at idle speed and adjust idle mixture screw (16) for smoothest engine idle. If engine does not accelerate smoothly, slight adjustment of main fuel mixture screw may be required.

Carburetor must be disassembled and all neoprene or Viton rubber parts removed before carburetor is immersed in cleaning solvent. Do not attempt to reuse any expansion plugs. Install new plugs if any are removed for cleaning. Do not attempt to remove nozzle (N—Fig. T14)) as it is pressed into position and movement will affect carburetor operation.

The fuel inlet valve needle (25—Fig. T13) seats against a Viton seat which must be removed before cleaning. The seat can be removed by blowing compressed air in from the fuel inlet fitting or by using a hooked wire. The grooved face of valve seat should be in toward bottom of bore and the valve needle should seat on smooth side of the Viton seat (Fig. T15).

Install the throttle plate (12—Fig. T13) with the two stamped marks out and at 12 and 3 o'clock positions. The 12 o'clock line should be parallel with the throttle shaft and toward top of carburetor. Install choke plate (20) with flat side down toward bottom of carburetor. Fuel inlet fitting (10) is pressed into body on some models. Start fitting into

body, then apply a light coat of Loctite sealant to shank and press fitting into position. Install the fuel valve retaining clip (26) on the float tab so the long end is toward the choke end of the carburetor.

With carburetor inverted, measure float height (H—Fig. T16) at a point opposite the fuel inlet valve. The float height should be 0.162-0.215 inch (4.11-5.46 mm). Bend the float tab to adjust the float height. Tecumseh has a gauge (Tecumseh part 670253A) that can be used to determine the correct float height. With carburetor inverted, position gauge 670253A at a 90° angle to the carburetor bore and resting on the nozzle stanchion as shown in Fig. T17. The toe of the float should not be higher than the first step on the gauge or lower than the second step. If equipped with a fiber washer between the fuel bowl and nozzle stanchion, place the washer on the stanchion then place the gauge on the washer to measure float height.

Install fuel bowl so indent (N—Fig. T18) on fuel bowl is under the fuel inlet fitting.

Be sure to use correct parts when servicing the carburetor. Some fuel bowl gaskets are square section, while others are round. The bowl retainer nut (Fig. T19) contains a drilled passage for fuel to the high speed metering needle. The fuel bowl retaining nut may have one or two holes adjacent to the hex. If a replacement nut is required, the new nut

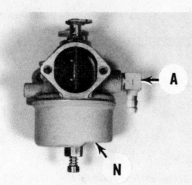

Fig. T18—Indented part of float bowl (N) should be located under the fuel inlet fitting (A).

Fig. T19—Fuel bowl retaining nut has drilled fuel passages. Be sure replacement nut is same as defective nut.

must have the same number of holes as the original nut.

Tecumseh Automagic And Dual System Float Carburetors. Refer to Fig. T20 or Fig. T21 for an exploded view of Automagic or Dual System carburetor. Neither type carburetor is equipped with a choke plate. Dual System carburetors are equipped with a primer bulb (10—Fig. T21) on the side while some Automagic carburetors are equipped with a primer bulb (26—Fig. T20) on the fuel bowl retaining nut. Automagic carburetors are not equipped with mixture screws while some Dual System carburetors are equipped with a high speed mixture screw (27—Fig. T21) in the fuel bowl retaining nut. Refer to Fig. T22 for operating principles of Automagic carburetor. Dual System carburetor operation is similar. On carburetors equipped with a primer bulb, air pressure from the primer bulb forces fuel through the main jet and up the nozzle, thereby enriching the air:fuel mixture for starting.

Follow service procedures outlined in previous section on Tecumseh float carburetors when servicing Automagic or Dual System carburetor. On Dual System carburetors, install the primer bulb and retainer by pushing the bulb and retainer into the carburetor using a ¾-inch deep well socket.

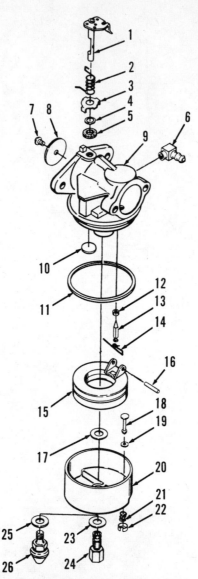

Fig. T20—Exploded view of Automagic carburetor.

1. Throttle shaft		14. Clip	
2. Spring		15. Float	
3. Plate		16. Pin	
4. Washer		17. Washer	
5. Felt washer		18. Drain valve	
6. Fuel inlet fitting		19. Washer	
7. Screw		20. Fuel bowl	
8. Throttle plate		21. Spring	
9. Body		22. Clip	
10. Welch plug		23. Washer	
11. Gasket		24. Fuel bowl nut	
12. Fuel inlet seat		25. Washer	
13. Fuel inlet valve		26. Primer	

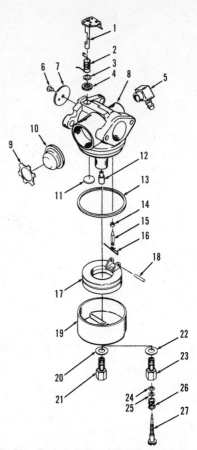

Fig. T21—Exploded view of Dual System carburetor.

1. Throttle shaft		15. Fuel inlet valve	
2. Spring		16. Clip	
3. Washer		17. Float	
4. Felt washer		18. Pin	
5. Fuel inlet fitting		19. Fuel bowl	
6. Screw		20. Washer	
7. Throttle plate		21. Fuel bowl nut	
8. Body		22. Washer	
9. Retainer		23. Fuel bowl nut	
10. Primer bulb		24. "O" ring	
11. Welch plug		25. Washer	
12. Spacer		26. Spring	
13. Gasket		27. High speed	
14. Fuel inlet seat		mixture screw	

SPEED CONTROL PANEL. Engines with a vertical crankshaft may be equipped with a speed control panel that is a separate component adjacent to the carburetor (Fig. T23). The idle speed screw (16—Fig. T24) and high

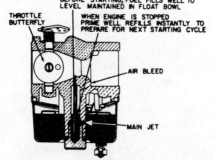

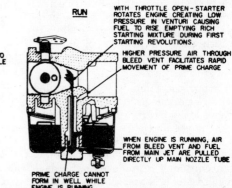

Fig. T22—The "Automagic" carburetor provides a rich starting mixture without using a choke plate. Mixture will be changed by operating with a dirty air filter or by incorrect float setting.

Illustrations courtesy Tecumseh Products Co.

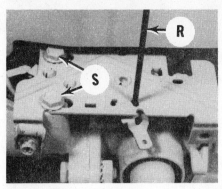

Fig. T23—View of speed control panel. Refer to text for adjustment.

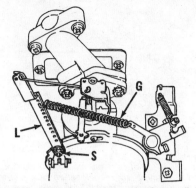

Fig. T26—View of governor mechanism on light frame, horizontal crankshaft engines for recreational vehicle application showing location of governor spring (G), governor lever (L) and adjusting screw (S).

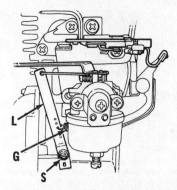

Fig. T29—View of governor mechanism on horizontal crankshaft Snow King engines showing location of governor spring (G), governor lever (L) and adjusting screw (S).

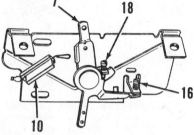

Fig. T24—Drawing of underside of speed control panel.

7. Control lever	16. Idle speed stop
10. Ignition stop switch	18. High speed stop screw

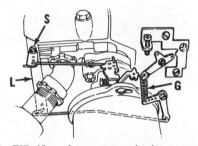

Fig. T27—View of governor mechanism on medium frame, horizontal crankshaft engines showing location of governor spring (G), governor lever (L) and adjusting screw (S).

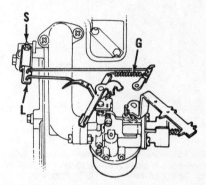

Fig. T30—View of governor mechanism on medium frame, horizontal crankshaft Snow King engines showing location of governor spring (G), governor lever (L) and adjusting screw (S).

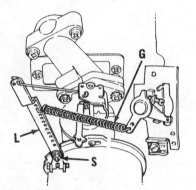

Fig. T25—View of governor mechanism on light frame, horizontal crankshaft engines showing location of governor spring (G), governor lever (L) and adjusting screw (S).

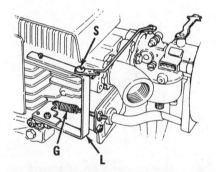

Fig. T28—View of governor mechanism on horizontal crankshaft engines for constant speed application showing location of governor spring (G), governor lever (L) and adjusting screw (S).

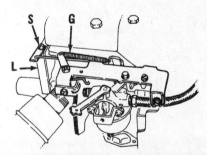

Fig. T31—View of governor mechanism on horizontal crankshaft Snow King engines showing location of governor spring (G), governor lever (L) and adjusting screw (S).

speed stop screw (18) are located on the panel.

The position of the panel must be adjusted so the linkage is synchronized. To adjust panel position, loosen mounting screws (S—Fig. T23). Insert rod (R) through holes in panel, choke actuating lever and choke as shown, then retighten mounting screws. Check operation.

GOVERNOR. All engines are equipped with a mechanical (flyweight) type governor. To adjust the governor linkage, refer to Figs. T25 through T39 and loosen governor lever screw (S). Ro-

tate governor shaft clamp counterclockwise as far as possible on vertical crankshaft engines; clockwise on horizontal crankshaft engines. On all models, move the governor lever (L) until carburetor throttle shaft is in wide open position, then tighten governor lever clamp screw.

Binding or worn governor linkage will result in hunting or unsteady engine operation. An improperly adjusted carburetor will also cause a surging or hunting condition.

Refer to Figs. T25 through T39 for views of typical mechanical governor speed control linkage installations.

IGNITION SYSTEM. A magneto ignition system with breaker points or capacitor-discharge ignition (CDI) may be used according to model and application. Refer to appropriate paragraph for model being serviced.

Breaker-Point Ignition System. Breaker-point gap at maximum opening should be 0.020 inch (0.51 mm) for all models. Marks are usually located on stator and mounting post to facilitate timing (Fig. T40).

Ignition timing can be checked and adjusted to occur when piston is at specific location (BTDC) if marks are miss-

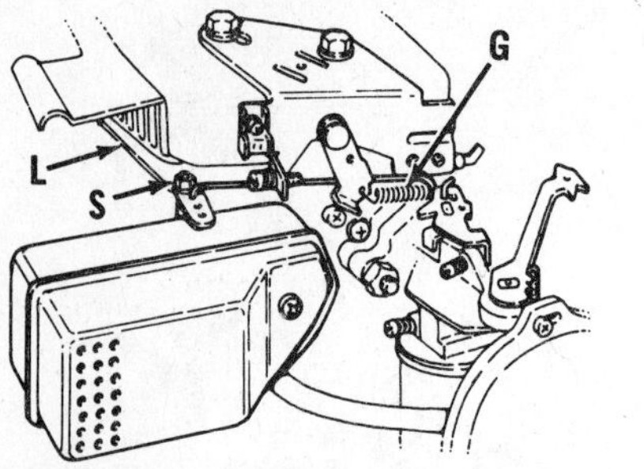

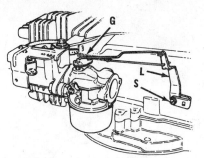

Fig. T32—View of governor mechanism on medium frame, horizontal crankshaft engines showing location of governor spring (G), governor lever (L) and adjusting screw (S).

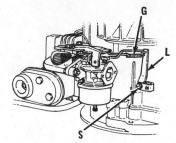

Fig. T35—View of governor mechanism on Model TNT100 engine showing location of governor spring (G), governor lever (L) and adjusting screw (S).

Fig. T38—View of governor mechanism on Model TVS engines with a fully adjustable carburetor showing location of governor spring (G), governor lever (L) and adjusting screw (S).

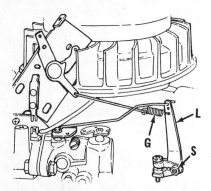

Fig. T33—View of governor mechanism on vertical crankshaft engines showing location of governor spring (G), governor lever (L) and adjusting screw (S).

Fig. T36—View of governor mechanism on Model TNT100 engine showing location of governor spring (G), governor lever (L) and adjusting screw (S).

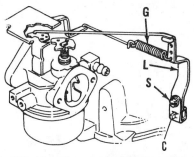

Fig. T39—View of governor mechanism on Model TVM engines showing location of governor spring (G), governor lever (L) and adjusting screw (S).

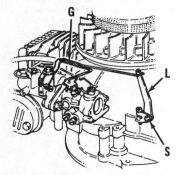

Fig. T34—View of governor mechanism on vertical crankshaft engines showing location of governor spring (G), governor lever (L) and adjusting screw (S).

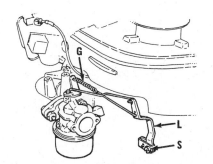

Fig. T37—View of governor mechanism on Model TNT120 engine showing location of governor spring (G), governor lever (L) and adjusting screw (S).

Fig. T40—View of ignition timing marks on engines with a breaker-point ignition system.

ing. Refer to the following specifications for recommended timing.

Model	Piston Position BTDC
ECV100, HS40, TNT100, TNT120, TVS105	0.035 in. (0.89 mm)
HS50, TVS120	0.050 in. (1.27 mm)
H30, H35, TVS75, TVS90	0.065 in. (1.65 mm)
H50, HH50, H60, HH60, TVM125, TVM140	0.080 in. (2.03 mm)

Some models may be equipped with the coil and laminations mounted outside the flywheel. Engines equipped with breaker-point ignition have the ignition points and condenser mounted under the flywheel, and the coil and laminations mounted outside the flywheel. This system is identified by the round shape of the coil (Fig. T41) and a stamping "Grey Key" in the coil to identify the correct flywheel key.

The correct air gap setting between the flywheel magnets and the coil laminations is 0.0125 inch (0.32 mm). Use Tecumseh gauge 670297 or equivalent thickness plastic strip to set gap.

Solid-State Ignition System. The Tecumseh solid-state ignition system does not use ignition breaker points. The only moving part of the system is the rotating flywheel with the charging magnets. Engines with solid-state (CDI) ignition have all the ignition components sealed in a module and located outside the flywheel. There are no components under the flywheel except a spring clip to hold the flywheel key in position. This system is identified by the square shape module and a stamping "Gold Key" to identify the correct flywheel key. See Fig. T42.

The correct air gap setting between the flywheel magnets and the laminations on ignition module is 0.0125 inch (0.32 mm). Insert 0.0125 inch (0.32 mm) feeler gauge (such as Tecumseh gauge 670297) between the flywheel magnet and ignition module armature legs. Loosen armature retaining screws and push armature legs against feeler

Illustrations courtesy Tecumseh Products Co.

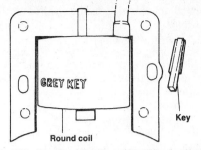

Fig. T41—Drawing of ignition coil and flywheel key used on engines equipped with a breaker-point ignition system and an external ignition coil. Coil is marked "GREY KEY" and key is colored grey.

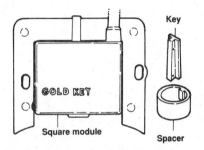

Fig. T42—Drawing of ignition coil, flywheel key and spacer used on some engines with a solid-state ignition system. Coil is marked "GOLD KEY" and key is colored gold.

gauge, then tighten armature retaining screws.

FLYWHEEL BRAKE. A flywheel brake is used on some engines that will stop the engine within three seconds when the mower safety handle is released. The ignition circuit is grounded also when the brake is actuated. On electric start models, an interlock switch prevents energizing the starter motor if the brake is engaged.

Refer to FLYWHEEL BRAKE in REPAIRS section for adjustment and service.

VALVE ADJUSTMENT. Clearance between valve tappet and valve stem (engine cold) is 0.010 inch (0.25 mm) for intake and exhaust valves for Models H50, HH50, H60, HH60, TVM125 and TVM140. Valve tappet clearance for all other models is 0.008 inch (0.03 mm) for intake and exhaust valves. Check clearance with piston at TDC on compression stroke. Grind valve stem end as necessary to obtain specified clearance.

REPAIRS

TIGHTENING TORQUES. Recommended tightening torque specifications are as follows:

Carburetor to
  intake pipe . . . . 48-72 in.-lbs.
                         (5-8 N•m)

Connecting rod:
  5 hp medium
    frame . . . . 160-180 in.-lb.
               (18-20.3 N•m)
  All other models . . . . . . 95-110 in.-lb.
               (11-12 N•m)
Cylinder head . . . . 160-200 in.-lbs.
               (18-23 N•m)
Flywheel:
  Light frame . . . . 30-33 ft.-lbs.
               (41-45 N•m)
  Medium frame . . . . 36-40 ft.-lbs.
               (49-54 N•m)
External ignition . . . . . . 33-36 ft.-lbs.
               (45-50 N•m)
Gear reduction cover . . 75-110 in.-lbs.
               (9-12 N•m)
Gear reduction
  housing . . . . 100-144 in.-lbs.
               (11-16 N•m)
Intake pipe to cylinder . . 72-96 in.-lbs.
               (8-11 N•m)
Magneto stator . . . . 40-90 in.-lbs.
               (5-10 N•m)
Crankcase cover/
  oil pan . . . . 75-110 in.-lbs.
               (9-12 N•m)
Spark plug . . . . 18-20 ft.-lbs.
               (24.5-27 N•m)

FLYWHEEL. On models so equipped, disengage flywheel brake as outlined in FLYWHEEL BRAKE section. If flywheel has tapped holes, use a suitable puller to remove flywheel. If no holes are present, screw a knock-off nut onto crankshaft so there is a small gap between nut and flywheel. Gently pry against bottom of flywheel while tapping sharply on nut.

On engines originally equipped with a breaker-point ignition system and an external ignition coil, the ignition coil is round and stamped "GREY KEY" as shown in Fig. T41. The flywheel key is colored grey.

The ignition coil on some engines equipped with an electronic ignition (often used as a replacement for the external coil on the breaker-point ignition system) has a square shape and is stamped "GOLD KEY" as shown in Fig. T42. A gold colored flywheel key must be used with this ignition coil for proper ignition timing. A spacer must also be installed on the crankshaft.

Install a stepped flywheel key so the stepped end is toward the engine. Install a tapered flywheel key so the big end is toward the engine. If a spacer is used, install the spacer so the protrusion fits in the keyway and is toward the end of the crankshaft.

After installing flywheel, tighten flywheel nut to torque listed in TIGHTENING TORQUES table.

CYLINDER HEAD. To remove cylinder head (16—Fig. T43 or 22—Fig. T44), remove all interfering shrouds and brackets. Clean area around cylinder head to prevent entrance of foreign material. Unscrew cylinder head screws and remove cylinder head.

Clean carbon from cylinder head being careful not to damage gasket mating surface. Use a straightedge to check cylinder head for distortion. If warped excessively, renew the cylinder head.

A new head gasket should be installed when installing cylinder head. Do not apply any type of sealer to the head gasket. Some cylinder head screws are equipped with flat washers, while some screws are equipped with a Belleville washer (B—Fig. T45) as well as a flat washer (W). The Belleville washer must be installed so the concave side is toward the screw threads. On H50 and H60 series engines, install cylinder head screws equipped with a Belleville washer in the locations shown in Fig. T46.

Tighten the cylinder head screws on engines with eight cylinder head screws in the sequence shown in Fig. T47. Tighten the cylinder head screws on engines with nine cylinder head screws in the sequence shown in Fig. T48. Tighten cylinder head retaining screws evenly to a torque of 160-200 in.-lbs. (18-23 N•m).

CRANKCASE COVER. Some horizontal crankshaft models are equipped with a ball bearing at the pto end of the crankshaft. On Models H30 through HS50 the ball bearing is retained in the crankshaft cover, while on Models H50, H60, HH50 and HH60, the bearing is pressed on the crankshaft.

Before removing the crankcase cover on Models H30 through HS50, measure oil seal depth then remove the oil seal. Detach snap ring (Fig. T49) from the crankshaft and remove crankcase cover. The bearing is secured by a screw on the inside of the cover. When installing cover, press new seal in to same depth as old seal before removal.

To remove the crankcase cover on Models H50, H60, HH50 and HH60, refer to Fig. T50 and note location of bearing lock screws. Loosen locknuts and rotate protruding ends of lock screws counterclockwise. The inner ends of the lock screws will move away from the bearing (Fig. T51) thereby allowing withdrawal of the cover from the crankshaft and bearing.

VALVE SYSTEM. Valve face angle is 45° and valve seat angle is 46° for intake and exhaust.

Valve seat width should be 0.042-0.052 inch (1.07-1.32 mm) for Models

mechanism to clear the exhaust valve tappet.

Renew camshaft if lobes or journals are worn or scored. Spring on compression release mechanism should snap weight against camshaft. Compression release mechanism and camshaft are serviced as an assembly only.

Standard camshaft journal diameter is 0.6230-0.6235 inch (15.824-15.837 mm) for Models H50, HH50, H60, HH60, TVM125 and TVM140, and 0.4975-0.4980 inch (12.637-12.649 mm) for all other models.

On models equipped with the barrel and plunger type oil pump, the pump is operated by an eccentric on the camshaft. Refer to OIL PUMP paragraph.

When installing camshaft, align crankshaft and camshaft timing marks as shown in Fig. T53. If there is no timing mark on the crankshaft gear, align the keyway in the crankshaft with the timing mark on the camshaft gear. If the crankshaft is equipped with a ball bearing or the crankshaft gear is pressed on (no keyway), align the crankshaft gear tooth that is beveled or has a punch mark (Fig. T54) with the mark on the camshaft.

PISTON, PIN AND RINGS. Piston and connecting rod assembly is removed from cylinder head end of engine. Before removing piston, remove any carbon or ring ridge from top of cylinder to prevent ring breakage. Aluminum alloy pistons are equipped with two compression rings and one oil control ring.

Piston ring end gap is 0.007-0.017 inch (0.18-0.43 mm) on models without external ignition and displacement of 12.04 cu. in. (197 cc) or less. If engine is equipped with an external ignition, or displacement is 12.18 cu.in. (229 cc) or greater, piston ring end gap should be 0.010-0.020 inch (0.25-0.50 mm).

Piston skirt-to-cylinder clearances are listed in the following table (if engine has external ignition, check list for a different specification than engines not so equipped):

Model	Piston Clearance
ECV100, H30, HS40, HS50, TNT100, TNT120, TVS90 TVS100, TVS105, TVS115, TVS120	0.0040-0.0058 in. (0.102-0.147 mm)
H50, H60, HH50, HH60, TVM125, TVM140	0.0035-0.0050 in. (0.089-0.127 mm)
H50(external ignition), H60(external ignition), TVM125(external ignition), TVM140(external	

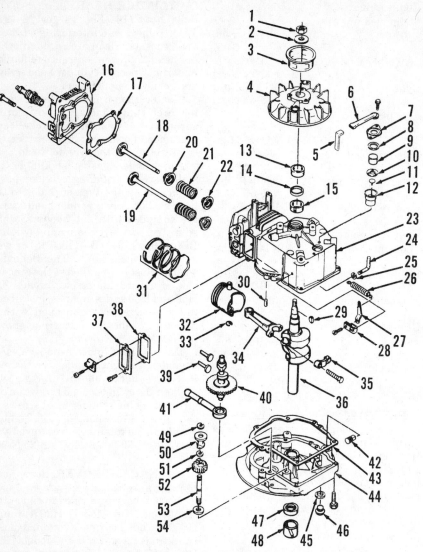

Fig. T43—Exploded view of typical engine with a vertical crankshaft.

1. Nut	15. Bushing		42. Drain plug
2. Belleville washer	16. Cylinder head	29. Key	43. Gasket
3. Starter cup	17. Gasket	30. Dowel pin	44. Oil pan
4. Flywheel	18. Intake valve	31. Piston rings	45. Gasket
5. Elbow	19. Exhaust valve	32. Piston	46. Drain plug
6. Breather tube	20. Valve seal	33. Retaining ring	47. Oil seal
7. Cover	21. Spring	34. Connecting rod	48. Bushing
8. Gasket	22. Retainer	35. Rod cap	49. Snap ring
9. Filter	23. Crankcase	36. Crankshaft	50. Governor spool
10. Baffle	24. Governor shaft	37. Tappet cover	51. Snap ring
11. Valve	25. Washer	38. Gasket	52. Flyweight assy.
12. Housing	26. Governor spring	39. Tappets	53. Governor shaft
13. Spacer	27. Governor lever	40. Camshaft	54. Washer
14. Oil seal	28. Clamp	41. Oil pump	

H50, HH50, H60, HH60, TVM125 and TVM140; and 0.035-0.045 inch (0.89-1.14 mm) for all other models.

Valve stem guides are cast into cylinder block and are not renewable. If excessive clearance exists between valve stem and guide, guide should be reamed and a new valve with an oversize stem installed.

CAMSHAFT. The camshaft and camshaft gear are an integral part which rides on journals at each end of camshaft. Camshaft on some models also has a compression release mechanism mounted on camshaft gear which lifts exhaust valve at low cranking rpm to reduce compression and aid starting (Fig. T52).

When removing camshaft, align timing marks (Fig. T53 or T54) on camshaft gear and crankshaft gear to relieve valve spring pressure on camshaft lobes. On models with compression release, it is necessary to rotate crankshaft three teeth past the aligned position to allow compression release

Illustrations courtesy Tecumseh Products Co.

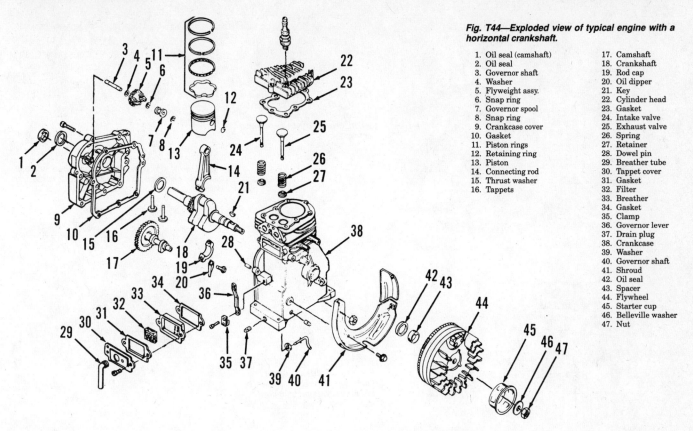

Fig. T44—Exploded view of typical engine with a horizontal crankshaft.

1. Oil seal (camshaft)	17. Camshaft
2. Oil seal	18. Crankshaft
3. Governor shaft	19. Rod cap
4. Washer	20. Oil dipper
5. Flyweight assy.	21. Key
6. Snap ring	22. Cylinder head
7. Governor spool	23. Gasket
8. Snap ring	24. Intake valve
9. Crankcase cover	25. Exhaust valve
10. Gasket	26. Spring
11. Piston rings	27. Retainer
12. Retaining ring	28. Dowel pin
13. Piston	29. Breather tube
14. Connecting rod	30. Tappet cover
15. Thrust washer	31. Gasket
16. Tappets	32. Filter
	33. Breather
	34. Gasket
	35. Clamp
	36. Governor lever
	37. Drain plug
	38. Crankcase
	39. Washer
	40. Governor shaft
	41. Shroud
	42. Oil seal
	43. Spacer
	44. Flywheel
	45. Starter cup
	46. Belleville washer
	47. Nut

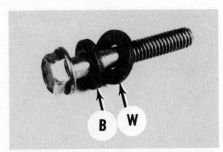

Fig. T45—Install Belleville washer (B) and flat washer (W) on cylinder head screw as shown.

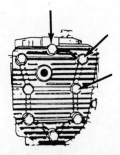

Fig. T46—On H50 and H60 series engines, install cylinder head screws equipped with a Belleville washer in the locations shown.

ignition) . . . . 0.0030-0.0048 in.
(0.076-0.123 mm)
HH50, HH60 . . . . . . 0.0015-0.0055 in.
(0.038-0.140 mm)
TVS75 . . . . 0.0025-0.0043 in.
(0.064-0.110 mm)

Standard piston diameters measured at piston skirt 90° from piston pin bore are listed in the following table (if engine has external ignition, check list for a different specification than engines not so equipped):

Model	Piston Diameter
ECV100, HS40,	
TNT100, TVS100,	
TVS105	2.6202-2.6210 in.
	(66.553-66.573 mm)
H30, H35, TVS90 . .	2.4952-2.4960 in.
	(63.378-63.398 mm)
H50, HH50, H60,	
HH60, TVM125,	
TVM140	2.6210-2.6215 in.
	(66.573-66.586 mm)
HH50(external ignition),	
HH60(external	
ignition)	2.6205-2.6235 in.
	(66.561-66.637 mm)
HS50, TNT120,	
TVS115, TVS120 . .	2.8072-2.8080 in.
	(71.303-71.323 mm)
TVM125(external ignition),	
TVM140(external	
ignition)	2.6212-2.6220 in.
	(66.578-66.599 mm)
TVS75	2.3092-2.3100 in.
	(58.654-58.674 mm)

Standard ring side clearance in ring grooves is listed in the following table (if engine has external ignition, check list for a different specification than engines not so equipped):

Model	Ring Side Clearance
H30, H35, TVS75, TVS90:	
Compression rings. . .	0.002-0.005 in.
	(0.05-0.13 mm)
Oil control ring . . .	0.0005-0.0035 in.
	(0.013-0.089 mm)
ECV100, H50(external ignition),	
H60(external ignition),	
HH50(external ignition),	
HH60(external ignition),	
HS40, HS50, HS50	
(external ignition), TNT100,	
TNT120, TVM125	
(external ignition),	
TVM140(external ignition),	
TVS100, TVS105,	
TVS115, TVS120:	
Compression rings .	0.002-0.005 in.
	(0.05-0.13 mm)
Oil control ring	0.001-0.004 in.
	(0.03-0.10 mm)
H50, H60, HH50,	
HH60, TVM125, TVM140:	
Compression rings .	0.002-0.004 in.
	(0.05-0.10 mm)
Oil control ring	0.002-0.004 in.
	(0.05-0.10 mm)

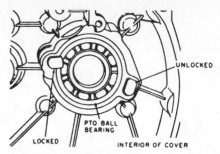

Fig. T47—Tighten cylinder head screws in sequence shown on engines equipped with eight cylinder head screws.

Fig. T48—Tighten cylinder head screws in sequence shown on engines equipped with nine cylinder head screws.

Fig. T51—Interior view of crankcase cover and ball bearing locks used on Models H50, H60, HH50 and HH60.

Fig. T52—View of compression release actuating pin (P) and weight (W).

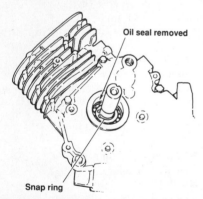

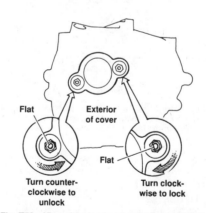

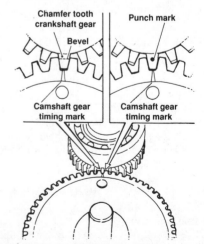

Fig. T49—On Models H30 through HS50 equipped with a ball bearing at pto end, a snap ring retains bearing on crankshaft.

Fig. T50—View showing bearing locks on Models H50, H60, HH50 and HH60 equipped with ball bearing main bearings. Locks must be released before removing crankcase cover. Refer to Fig. T51 for interior view of cover and locks.

Fig. T53—View of timing marks (M) on crankshaft and camshaft gears.

Fig. T54—Drawing of timing marks used on crankshaft equipped with a ball bearing.

Piston pin should be a tight push fit in piston pin bore and connecting rod pin bore. Pin is retained by retainer clips at each end of piston pin bore. If piston crown is marked with an arrow, note correct assembly of rod and piston in Fig. T55. Install top and second piston rings with beveled edge toward piston crown. When installing connecting rod and piston assembly, align the match marks (casting projections) on connecting rod and cap as shown in Fig.

T56 or T57. Install piston so match marks on connecting rod are toward power take-off (pto) end of crankshaft. Stagger ring end gaps equally around circumference of piston before installation. If the area adjacent to the valves has been machined (trenched) as shown in Fig. T58, position the piston rings on the piston so the end gaps are staggered

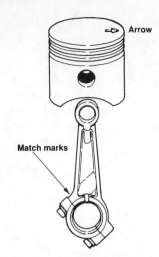

Fig. T55—If piston crown is marked with an arrow, assemble piston and rod as shown.

Fig. T56—View of rod match marks used on engines.

Fig. T57—View of rod match marks used on engines.

Fig. T58—On some engines the area adjacent to the valves has been machined (trenched) as shown.

Fig. T59—View of governor used on engines with vertical crankshaft.

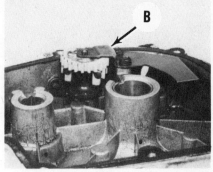

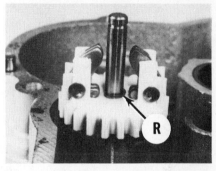

Fig. T61—Detach snap ring (R) to remove flyweight assembly.

Fig. T62—A thrust washer is located under the governor gear.

and none of the end gaps will coincide with the machined area. This will prevent a piston ring end from catching the machined surface during installation.

CONNECTING ROD. The aluminum alloy connecting rod rides directly on crankshaft crankpin.

Refer to the following table for standard crankpin journal diameter:

Model	Crankpin Diameter
ECV100, H30, TNT100, TVS75, TVS90, TVS100, TVS115	0.8610-0.8615 in. (21.869-21.882 mm)
H35, HS40, HS50, TNT120, TVS105, TVS120	0.9995-1.0000 in. (25.390-25.400 mm)
H50, HH50, H60, HH60, TVM125, TVM140	1.0615-1.0620 in. (26.962-26.975 mm)

Standard inside diameter for connecting rod big end is listed in the following table (if engine has external ignition, check list for a different specification than engines not so equipped):

Model	Rod Big End Diameter
ECV100, H30, TNT100, TVS75, TVS90, TVS100, TVS115	0.8620-0.8625 in. (21.895-21.908 mm)
H35, HS40, HS50, TNT120, TVS105, TVS120	1.0005-1.0010 in. (25.413-25.425 mm)
H50, HH50, H60, HH60, TVM125, TVM140	1.0630-1.0636 in. (27.000-27.013 mm)

Connecting rod bearing-to-crankpin journal clearance should be 0.0005-0.0015 inch (0.013-0.038 mm) for all models.

Assemble connecting rod and piston as outlined in previous section. Tighten connecting rod screws to 160-180 in.-lb. (18-20.3 N•m) torque on 5 hp medium frame engines or to 95-110 in.-lb. (11-12 N•m) torque on all other models.

GOVERNOR. On most engines with a vertical crankshaft and low horsepower horizontal crankshaft engines, the governor is retained on the shaft by snap ring (R—Fig. T59). On larger horizontal crankshaft engines, the governor is retained by bracket (B—Fig. T60). To remove spool, detach either upper snap ring or bracket. Detach lower snap ring (R—Fig. T61) to remove flyweight assembly. A washer (W—Fig. T62) is located under the flyweight assembly. On some engines, a spacer (S—Fig. T63) is located under washer (W).

On later small frame engines and replacement shafts, no snap rings are

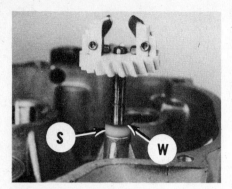

Fig. T63—Some engines are equipped with a spacer (S) under thrust washer (W).

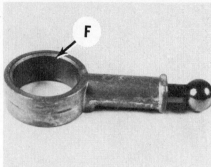

Fig. T65—Install oil pump so chamfer (F) is toward camshaft.

used on the governor shaft. The governor is held in place by a boss on the governor shaft. The flyweights and gear are available only as a unit assembly.

The governor shaft is pressed into the crankcase cover or oil pan and may be replaced if the mounting boss is not damaged or the hole is not enlarged. To remove the governor shaft, clamp the shaft in a vise and using a soft mallet, drive the crankcase cover or oil pan off the shaft. Do not attempt to twist governor shaft out of boss. Twisting will enlarge hole. Apply Loctite 271 (red) to shaft end. If shaft is retainerless design, position washer and governor gear assembly on shaft before installing shaft, then press in shaft until the governor gear has 0.010-0.020 inch (0.25-0.50 mm) axial play. If snap rings are used on the governor shaft, press shaft into crankcase cover or oil pan until the height above boss (Fig. T64) is as specified in the following table:

Model	Governor Shaft Height
H50, HH50, H60, HH60	$1\frac{7}{16}$ in. (36.51 mm)
TVM125, TVM140	$1\frac{19}{32}$ in. (40.48 mm)
All other models	$1\frac{21}{64}$ in. (34.92 mm)

OIL PUMP. Vertical crankshaft engines are equipped with a barrel and

Fig. T64—Install governor shaft so height is as specified in text.

plunger type oil pump. The barrel and plunger type oil pump is driven by an eccentric on the camshaft. Chamfered side of drive collar (Fig. T65) should be toward camshaft gear. Oil pumps may be equipped with two chamfered sides, one chamfered side or with flat boss as shown. Be sure installation is correct.

CRANKSHAFT, MAIN BEARINGS AND SEALS. Always note oil seal depth and seal lip direction before removing oil seal from crankcase or cover. New seals must be pressed into seal bores to the same depth as old seal before removal on all models.

Refer to CONNECTING ROD section for standard crankshaft crankpin journal diameters.

Crankshaft main bearing journals on some models ride directly in the aluminum alloy bores in the cylinder block and crankcase cover or oil pan. Other engines were originally equipped with renewable steel backed bronze bushings and some were originally equipped with a ball type main bearing at the pto end of crankshaft. On Models H50, H60, HH50 and HH60, the bearing is pressed on the crankshaft.

Standard main bearing bore diameters for main bearings are listed in the following table (if engine has external ignition, check list for a different specification than engines not so equipped):

Model	Main Bearing Inside Diameter
ECV100, TVS75, TVS90	0.8755-0.8760 in. (22.238-22.250 mm)
ECV100(external ignition), H30, TVS75(external ignition), TVS90(external ignition), TNT100, TVS100:	
Crankcase side	1.0005-1.0010 in. (25.413-25.425 mm)
Cover (flange) side	0.8755-0.8760 in. (22.238-22.250 mm)
H35, H50, H60, HH50, HH60, HS40, HS50, TNT120, TVM125, TVM140, TVS105, TVS115, TVS120	1.0005-1.0010 in. (25.413-25.425 mm)

Standard diameters for crankshaft main bearing journals are shown in the following table (if engine has external ignition, check list for a different specification than engines not so equipped):

Model	Main Bearing Journal Diameter
ECV100, TVS75, TVS90	0.8735-0.8740 in. (22.187-22.200 mm)
ECV100(external ignition), H30, TVS75(external ignition), TVS90(external ignition), TNT100, TVS100:	
Crankcase side	0.9985-0.9990 in. (25.362-25.375 mm)
Cover (flange) side	0.8735-0.8740 in. (22.187-22.200 mm)
H35, H50, H60, HH50, HH60, HS40, HS50, TNT120, TVM125, TVM140, TVS105, TVS115, TVS120	0.9985-0.9990 in. (25.362-25.375 mm)

Main bearing clearance should be 0.0015-0.0025 inch (0.038-0.064 mm). Crankshaft end play should be 0.005-0.027 inch (0.13-0.069 mm) for all models.

On models equipped with a ball bearing, inspect the bearing and renew if rough, loose or damaged. On Models H50, H60, HH50 and HH60, the bearing must be pressed on or off of crankshaft journal using a suitable press or puller.

When installing crankshaft, align crankshaft and camshaft gear timing marks as shown in Fig. T53. If there is no timing mark on the crankshaft gear, align the keyway in the crankshaft with the timing mark on the camshaft gear. If the crankshaft is equipped with a ball bearing or the crankshaft gear is pressed on (no keyway), align the crankshaft gear tooth that is beveled or has a punch mark (Fig. T54) with the mark on the camshaft.

CYLINDER AND CRANKCASE. Cylinder and crankcase are an integral casting on all models. Cylinder should be honed and fitted to nearest oversize for which piston and ring set are available if cylinder is scored, tapered or out-of-round more than 0.005 inch (0.13 mm).

Standard cylinder bore diameters are listed in the following table (if engine has external ignition, check list for a different specification than engines not so equipped):

Model	Cylinder Bore Diameter
H30, TVS75	2.3125-2.3135 in. (58.738-58.763 mm)
H30, H35, TVS90 . .	2.5000-2.5010 in. (63.500-63.525 mm)
ECV100, H50, H60, HH50, HH60, HS40, TNT100, TVM125, TVS100, TVS105. .	2.6250-2.6260 in. (66.675-66.700 mm)
HS50, TNT120, TVS115, TVS120. . .	2.8120-2.8130 in. (71.425-71.450 mm)

Refer to PISTON, PIN AND RINGS section for correct piston-to-cylinder block clearance. Note also that cylinder block used on Models H50, HH50, TVM125 and TVM140 has been "trenched" to improve fuel flow and power (Fig. T58).

REDUCED SPEED PTO SHAFTS. A pto (power take-off) shaft which rotates at half the speed of the crankshaft is available by extending the camshaft through the oil pan. Refer to Fig. T66. Except for the seal around the extended

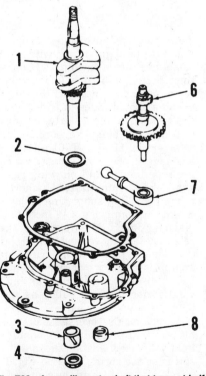

Fig. T66—An auxiliary pto shaft that turns at half the speed of the crankshaft is available using a special extended camshaft. The camshaft is sealed using lip type seal (8). Oil pump (7) is not used on engines with horizontal crankshaft.

1. Crankshaft
2. Thrust washer
3. Bushing
4. Oil seal
6. Camshaft
7. Oil pump
8. Oil seal

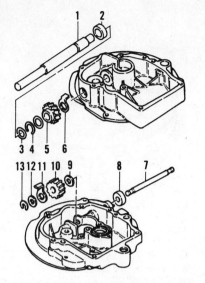

Fig. T67—Exploded views of 8.5:1 auxiliary pto drive systems that may be used.

1. Shaft
2. Seal
3. Washers (2)
4. Snap ring
5. Gear
6. Tang washer
7. Shaft
8. Seal
9. Thick washer
10. Gear
11. Tang washer
12. Washer
13. Snap ring

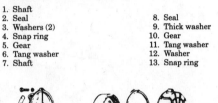
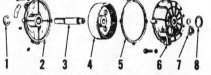

Fig. T68—Exploded view of the 6:1 gear reduction assembly. Housing (6) is attached to crankcase cover and the pinion gear is machined on end of crankshaft.

1. Seal
2. Cover
3. Output shaft
4. Gear
5. Gasket
6. Housing
7. Seal
8. Cork gasket

camshaft, service is similar to standard models.

A slow speed (8.5:1) auxiliary pto shaft is used on some vertical shaft engines. Two designs have been used as shown in Fig. T67. A worm gear on the crankshaft turns the pto gear and pto shaft. Several different versions of this unit have been used.

Disassembly and repair procedure for the 6:1 reduction used on some horizontal shaft engines will be evident after examination of the unit and reference to Fig. T68.

FLYWHEEL BRAKE. A flywheel brake is used on some engines. Two configurations have been used. The brake shown in Fig. T69 contacts the bottom surface of the flywheel while the brake shown in Fig. T70 contacts the inside of the flywheel. Before the flywheel can be removed for either type, the brake must be disengaged from the flywheel. On bottom surface type, unhook brake spring (B—Fig. T69). On inside surface

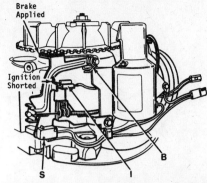

Fig. T69—View of bottom-surface type flywheel brake showing location of brake spring (B), ignition cut-out switch (I) and interlock switch (S) for electric starter.

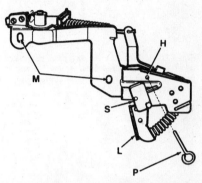

Fig. T70—Push against lever (L) and insert pin (P) through holes (H) to hold brake pad away from flywheel. Interlock switch (S) is used on engines with an electric starter.

type, push lever (L—Fig. T70) toward spark plug so brake pad moves away from flywheel, then insert Tecumseh tool 670298 or a suitable pin in hole (H) to hold lever.

Inspect mechanism for excessive wear and damage. Tighten mounting screws on bottom surface type brake to 60-70 in.-lbs. (6.8-7.9 N•m) torque. On inside surface type, minimum allowable thickness of brake pad at narrowest point is 0.060 inch (1.52 mm). Install brake mechanism and push up on bracket so bracket mounting screws are at bottom of slotted holes (M) in bracket. Tighten mounting screws to 90 in.-lbs. (10.2 N•m) torque.

REWIND STARTER. The rewind starter may be mounted on the blower housing or attached to the side of the engine. Refer to appropriate following paragraphs for service.

Starters Mounted On Blower Housing. TEARDROP HOUSING. Note shape of starter housing in Fig. T71. The pulley may be secured with either a retainer screw or retainer pin. Most starters use a rope that is 54 inches (137 cm) long. Refer to following paragraphs for service.

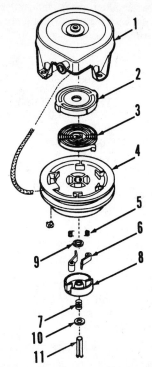

Fig. T71—Exploded view of typical pawl type rewind starter with teardrop shaped housing (1) using retainer screw (9). Some starters may have three starter pawls (6).

1. Housing	8. Retainer
2. Spring keeper	9. Screw
3. Rewind spring	10. Centering pin
4. Pulley	11. Nylon bushing
5. Spring	12. Rope coupler
6. Pawl	13. Handle
7. Brake spring	14. Insert

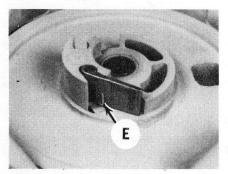

Fig. T72—Spring end (E) should force pawl toward center of pulley.

To disassemble starter equipped with retainer screw (Fig. T71), release preload tension of rewind spring by removing rope handle and allowing rope to wind slowly into starter. Remove retainer screw (9), retainer (8) and spring (7). Remove pawl (6) and spring (5). Remove pulley with spring. Wear appropriate safety eyewear and gloves before disengaging keeper (2) and rewind spring (3) from pulley as spring may uncoil uncontrolled.

To reassemble, reverse the disassembly procedure. Spring (3) should be lightly greased. Install the pawl and spring so the spring end (E—Fig. T72) forces the pawl toward the center of the pulley.

Fig. T73—Exploded view of starter drive used on Snow King engines.

1. Rope pulley	5. Screw
2. Pawl spring	6. Brake spring
3. Pawl	7. Cam
4. Pawl retainer	8. Screw

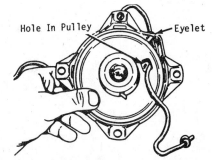

Fig. T74—Insert rope through starter housing eyelet and hole in pulley, then tie knot in rope end.

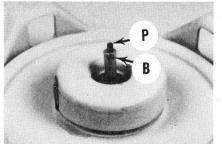

Fig. T75—View of centering pin (P) and bushing (B).

NOTE: On Snow King engines, refer to Fig. T73 for installation of the drive mechanism. Note that the pawl (3) is secured by a retainer (4) and screw (5).

With starter assembled, except for rope, install rope as follows: Turn pulley counterclockwise until tight, then allow to unwind so hole in pulley aligns with rope outlet as shown in Fig. T74. Insert rope through starter housing and pulley hole, tie a knot in rope end, allow rope to wind onto pulley and install rope han-

Fig. T76—Exploded view of typical pawl type rewind starter with teardrop shaped housing (1) using retainer pin (11).

1. Housing	
2. Spring keeper	7. Brake spring
3. Rewind spring	8. Retainer
4. Pulley	9. Washer
5. Spring	10. Washer
6. Pawl	11. Pin

dle. Some models use centering pin (P—Fig. T75) to align starter with starter cup. Place nylon bushing (B) on pin, then bottom pin in hole in retainer screw as shown in Fig. T75. Pin and bushing should index in end of crankshaft when installing starter on engine.

To disassemble starter equipped with retainer pin (11—Fig. T76), release preload tension of rewind spring by removing rope handle and allowing rope to wind slowly into starter. Remove retainer pin (11) by supporting pulley and driving out pin using a ¼ inch (6 mm) diameter punch. Remove spring (7) and retainer (8). Remove pawl (6) and spring (5). Remove pulley with spring. Wear appropriate safety eyewear and gloves before disengaging keeper (2) and rewind spring (3) from pulley as spring may uncoil uncontrolled.

To reassemble, reverse the disassembly procedure. Spring (3) should be lightly greased. Install the pawl and spring so the spring end (E—Fig. T72) forces the pawl toward the center of the pulley. Drive in retainer pin (11—Fig. T76) until seated against shoulder of housing.

NOTE: On engines with an alternator mounted under the rewind starter,

Illustrations courtesy Tecumseh Products Co.

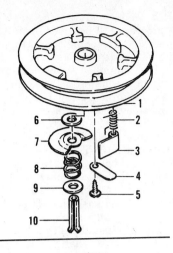

Fig. T77—Exploded view of starter drive used on Snow King engines.

1. Rope pulley	6. Washer
2. Pawl spring	7. Cam
3. Pawl	8. Brake spring
4. Pawl retainer	9. Washer
5. Screw	10. Pin

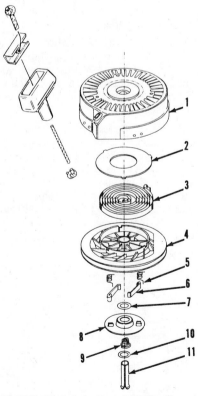

Fig. T78—Exploded view of "stylized" rewind starter.

1. Starter housing	
2. Cover	7. Plastic washers (2)
3. Rewind spring	8. Retainer pawl
4. Pulley	9. Brake spring
5. Springs (2)	10. Metal washer
6. Pawls (2)	11. Pin

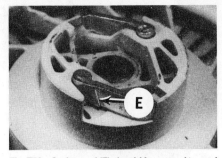

Fig. T79—Spring end (E) should force pawl toward center of pulley.

Fig. T80—Tabs (T) on retainer must fit inside pawls during assembly.

refer to Fig. T77 for installation of the drive mechanism. Note that the pawl (3) is secured by a retainer (4) and screw (5). The drawing in the inset shows the components assembled.

With starter assembled, except for rope, install rope as follows: Rotate pulley counterclockwise until tight, then allow to unwind so hole in pulley aligns with rope outlet as shown in Fig. T74. Insert rope through starter housing and pulley hole, tie a knot in rope end, allow rope to wind onto pulley and install rope handle.

STYLIZED STARTER. The "stylized" starter is shown in Fig. T78. Typical rope lengths are 69 inches (175 cm), 98 inches (249 cm) and 114 inches (290 cm).

To disassemble starter, remove rope handle and allow rope to wind slowly into starter. Position a suitable sleeve support under retainer pawl (8) and using a $5/16$ inch (8 mm) punch, drive pin (11) free of starter. Remove brake spring (9), retainer (8), pawls (6) and springs (5). Wear appropriate safety eyewear and gloves before disengaging pulley from starter as spring may uncoil uncontrolled. Place shop towel around pulley and lift pulley out of housing; spring should remain with pulley.

Inspect components for damage and excessive wear. Reverse disassembly procedure to install components. Rewind spring coils wind in counterclockwise direction from outer end. When

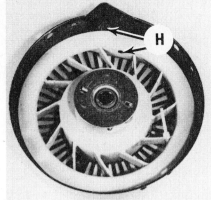

Fig. T81—Align holes (H) in pulley and starter housing then insert rope through holes.

installing pulley, be sure inner end of rewind spring engages spring retainer adjacent to housing center post. Install the pawl and spring so the spring end (E—Fig. T79) forces the pawl toward the center of the pulley. Place the retainer on the pulley hub so tabs (T—Fig. T80) are inside ends of pawls. Install pin (11—Fig. T78) so top of pin is $1/8$ inch (3.2 mm) below top of starter. Driving pin in too far may damage retainer pawl.

With starter assembled, except for rope, install rope as follows: Rotate pulley counterclockwise until tight, then allow to unwind so hole (H—Fig. T81) in pulley aligns with rope outlet. Insert rope through starter housing and pulley hole, tie a knot in rope end, allow rope to wind onto pulley and install rope handle.

Vertical-Pull Rewind Starters. Two types of vertical-pull starters have been used. The types are identified by the manner of gear engagement, either horizontal or vertical.

HORIZONTAL ENGAGEMENT GEAR STARTER. When the rope handle is pulled, the starter gear moves horizontally to engage the gear teeth on the flywheel. An exploded view of the starter is shown in Fig. T82. Most starters use a rope that is 61 inches (155 cm) long.

After installing starter assembly, clearance between teeth on gear (6) and teeth on flywheel should be checked when starter is operated. When teeth are fully engaged, there should be at least $1/16$ inch (1.6 mm) clearance from top of gear tooth to base of opposite gear teeth. Remove spark plug wire, operate starter several times and check gear engagement. Insufficient gear tooth clearance could cause starter gear to hang up on flywheel gear when engine starts, which could damage starter.

To disassemble starter, proceed as follows: Detach rope from insert (8—Fig. T82) and handle (9), then allow the rope to wind slowly onto the pulley to relieve spring tension. Unscrew screws securing spring cover (13) and carefully remove cover without disturbing rewind spring. Remove rewind spring. Safety eyewear and gloves should be worn when working on or around the rewind spring. Remove hub screw (15) and spring hub (14), then withdraw pulley and gear assembly. Remove snap ring (3), washer (4), brake spring (5) and gear (6) from pulley.

Reassemble starter using the following procedure: If removed, install rope guide (2) on starter bracket so dimple on guide fits in depression on bracket. Install rope on pulley. Wrap rope around pulley in a counterclockwise direction when viewing pulley from rewind spring side of pulley. Place gear on pulley. Do not lubricate helix in gear or on pulley shaft. Install brake spring (5) in groove of gear (6). The bent end of brake spring should point away from gear. The brake spring should fit snugly in groove. Do not lubricate brake spring. Install washer (4) and snap ring (3) on pulley shaft. Lightly lubricate shaft on starter bracket then install pulley and gear assembly on bracket shaft. The closed end of brake spring (5) must fit around tab (A) on bracket. Install spring anchor (14) and screw (15). Tighten screw to 44-55 in.-lb. (5.08-6.21 N•m) torque. Install rewind spring. The spring coils should wind in a clockwise direction from outer end. A new spring is contained in a holder that allows spring installation by pushing spring from holder into spring cavity on pulley. Pass outer end of rope through rope bracket (10) and install rope handle and insert. Pull a portion of rope out past rope guide and wrap any excess rope around pulley, then turn pulley 2-2½ turns against spring tension to preload rewind spring. Check starter operation.

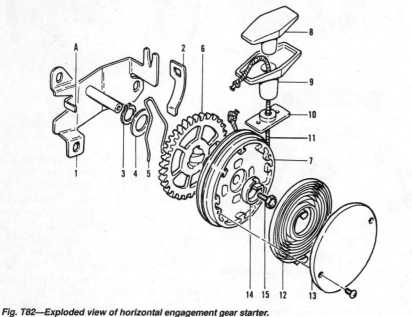

Fig. T82—Exploded view of horizontal engagement gear starter.

1. Mounting bracket
2. Rope guide
3. Snap ring
4. Washer
5. Brake spring
6. Gear
7. Pulley
8. Insert
9. Handle
10. Bracket
11. Rope
12. Rewind spring
13. Cover
14. Spring anchor
15. Hub screw

VERTICAL ENGAGEMENT GEAR STARTER. When the rope handle is pulled, the starter gear moves vertically to engage the gear teeth on the flywheel. An exploded view of the starter is shown in Fig. T83.

To replace the starter rope, proceed as follows: If the starter bracket has a "V" notch (V—Fig. T84), then the inner end of the rope is accessible and the rope can be replaced without disassembling the starter. Typical rope lengths are 65 inches (165 cm) and 98 inches (249 cm). If rope is unbroken, remove rope handle and let rope wind onto rope pulley. Note

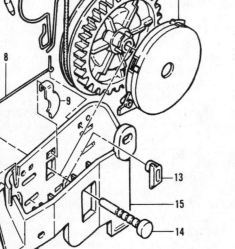

Fig. T83—Exploded view of vertical engagement gear starter.

1. Handle
2. Brake spring
3. Staple
4. Rope
5. Rope pulley
6. Rewind spring & housing
7. Rope clip
8. Pawl spring
9. Pawl
10. Key (not all models)
11. Rope clip
12. Clip (not all models)
13. Lock pawl (not all models)
14. Pin
15. Mounting bracket

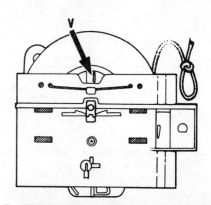

Fig. T84—Drawing of vertical engagement gear starter. Note location of "V" notch.

Illustrations courtesy Tecumseh Products Co.

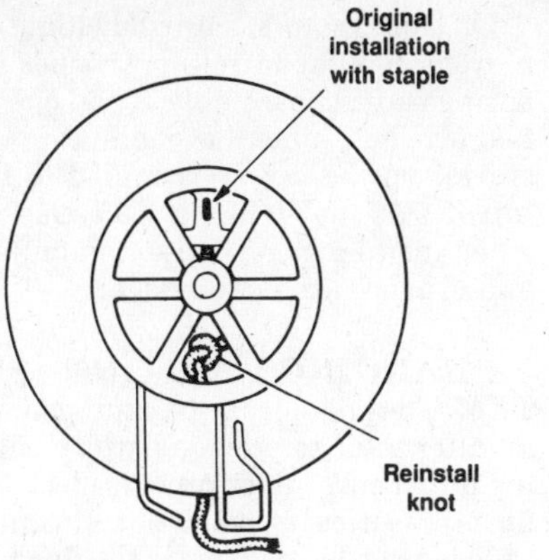

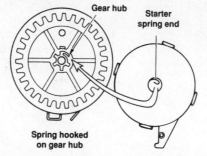

Fig. T87—If removed, place spring housing on rope pulley while being sure inner spring end engages anchor on pulley.

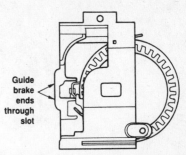

Fig. T89—Install rope pulley assembly in starter bracket while inserting brake spring legs into slots in starter bracket.

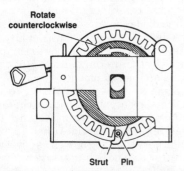

Fig. T85—Original rope was attached to pulley with a staple. Replacement rope must be secured with a knot.

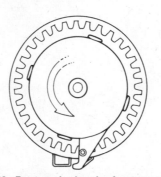

Fig. T86—Housing rotation may be prevented by inserting a pin or rod through strut hole and into gear teeth.

Fig. T88—Rotate spring housing four turns counterclockwise then align legs on brake spring with hole in strut.

in Fig. T85 that inner end of rope was originally retained by a staple, while inner end on replacement ropes is inserted through a hole in pulley and knotted. Rotate pulley so that either the stapled rope end or knotted rope end is visible (Fig. T84) in the "V" notch. Pry out staple or untie knot and pull out rope. Rotate pulley counterclockwise until rewind spring is tight, then allow pulley to turn clockwise until rope hole in pulley is visible in "V" notch. Route new rope through pulley hole and tie a knot in rope end. Pull the knot into pulley cavity so rope end does not protrude. Allow pulley to wind rope onto pulley. Attach rope handle to rope end.

To disassemble starter, proceed as follows: If rope is unbroken, remove rope handle and let rope wind onto rope pulley. Drive or press out pulley spindle (14—Fig. T83) by placing starter over a deep-well socket and forcing spindle into socket. Turn spring housing (6) so strut aligns with legs on brake spring (2). Prevent housing rotation by inserting a pin or rod through strut hole and into gear teeth as shown in Fig. T86. Remove pulley assembly from starter bracket.

CAUTION: Do not allow spring housing to separate from rope pulley until spring tension has been relieved.

Hold spring housing (6—Fig. T83) against rope pulley so spring housing cannot rotate. Withdraw pin or rod in spring housing strut (Fig. T86) and allow spring housing to rotate thereby relieving rewind spring tension. If necessary, separate spring housing from rope pulley. Do not attempt to remove rewind spring from spring housing. The spring and housing are available only as a unit assembly.

If rope replacement is necessary, note in Fig. T85 that inner end of rope was originally retained by a staple, while inner end on replacement ropes is inserted through a hole in pulley and knotted. Pry out staple, if so equipped, to release rope. Route new rope through pulley hole and tie a knot in rope end. The knot must be positioned in pulley cavity so rope end does not protrude.

To assemble starter, proceed as follows: Do not lubricate any starter components. When viewed from gear side of rope pulley, wind rope onto pulley in a clockwise direction. Install brake spring (2—Fig. T83) on rope pulley while being careful not to distort spring legs. If removed, place spring housing on rope pulley while being sure inner spring end engages anchor on pulley. See Fig. T87. Rotate spring housing four turns counterclockwise (Fig. T88) then align legs

on brake spring with hole in strut. Prevent housing rotation by inserting a pin or rod through strut hole and into gear teeth. If so equipped, install clip (7—Fig. T83), key (10) and pawl (9). Install rope pulley assembly in starter bracket while inserting brake spring legs into slots in starter bracket (Fig. T89). Route outer rope end past rope guide and install rope handle. Withdraw pin or rod in strut hole. The strut will rotate until it contacts starter bracket. Press or drive in a new pulley spindle (14—Fig. T83).

ELECTRIC STARTER. While several electric starter motors have been used on Tecumseh engines, the motors are basically divided into those with a round frame or a square frame. Several variations of each type have been produced, but service is basically similar except as noted in following paragraphs.

Tecumseh does not provide test specifications for electric starter motors, so service is limited to replacing components that are known or suspected faulty.

CAUTION: The starter motor field magnets may be made of ceramic material. Do not clamp starter housing in a vise or hit housing as field magnets may be damaged.

Refer to Figs. T90, T91 and T92 for exploded views of typical starter motors. Note that starter drive assemblies shown may be used on other types of motors. Some motors may be equipped with a covered drive (Fig. T93).

Note the following when servicing electric starter motor: Prior to disassembly, mark drive plate, frame and end cap so they can be aligned during assembly. Some motors have alignment notches and marks are not necessary.

Minimum brush length is not specified. If brush wire bottoms against slot in brush holder or brush is less than half its original length, replace brush. Be sure brushes do not bind in holders. On

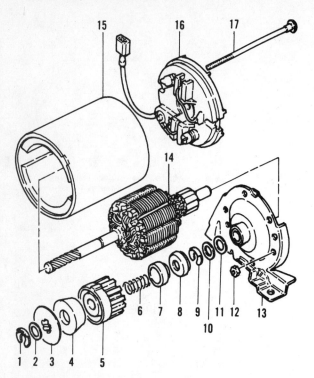

Fig. T90—Exploded view of round frame starter motor with inward-acting starter drive.

1. "E" ring
2. Plastic washer
3. Drive hub
4. Rubber driver
5. Pinion gear
6. Spring
7. Plastic spring cup
8. Metal cup
9. "E" ring
10. Metal washer
11. Plastic washer
12. Nut
13. Drive plate
14. Armature
15. Frame
16. End cap assy.
17. Through-bolt

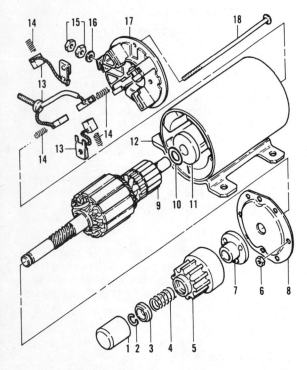

Fig. T91—Exploded view of round frame starter motor with outward-acting starter drive.

1. Boot
2. Retaining ring
3. Spring cup
4. Spring
5. Pinion gear
6. Nut
7. Drive hub
8. Drive plate
9. Armature
10. Washer
11. Washer
12. Frame
13. Brushes
14. Brush springs
15. Nuts
16. Washer
17. End cap
18. Through-bolt

On some motors, a brush holding tool may be helpful to retain brushes and springs during assembly. If end cap has two brushes, a piece of manual rewind starter spring can be modified to hold brushes as shown in Fig. T95. Use care when installing end cap so commutator and brushes are not damaged.

CHARGING/LIGHTING SYSTEM. The engine may be equipped with an alternator to provide battery charging direct current, or provide electricity for accessories, or both. The alternator coils may be located under the flywheel, or on some models with an external ignition module, the alternator coils are attached to the legs of the ignition coil. Rectification is accomplished either with a rectifier panel, regulator-rectifier unit, external or internal, or by an inline diode contained in the harness.

A lighting system consists of just the alternator. The alternating current produced by the alternator is used to power lights on the equipment.

Some systems are designed to use alternator output for both a charging system and a lighting system. The charging system provides direct current for battery charging as well as powering accessories. The lighting system powers the lights.

Refer to the following sections.

External Ignition Module Alternator. Engines with an external ignition may have an alternator coil attached to the ignition module (Fig. T96). The alternator produces approximately 350 milliamperes for battery charging. To check alternator output, connect a DC voltmeter to the battery (battery must be in normal circuit) as shown in Fig. T97. Run the engine. Voltage should be higher when the engine is running or the alternator is defective.

1 Amp Add-On Lighting System. The alternator is attached to rewind starter and is used to power an AC lighting system. The alternator is driven by a shaft that attaches to flywheel nut and extends through rewind starter. Output is determined by engine speed. An exploded view of system is shown in Fig. T98.

To check alternator output, disconnect alternator lead and connect a 4414, 18-watt bulb to connector terminals as shown in Fig. T99. Connect test leads of an AC voltmeter to bulb leads or connector terminals as shown in Fig. T99. With the engine running at 3600 rpm, voltmeter should indicate at least 12 volts. If voltage reading is insufficient, then alternator is faulty.

3 Amp Lighting System. This AC alternator is used to power 12-volt lights. Output is determined by engine

some round frame motors, the brushes are available only as a unit assembly with end cap.

On square frame motors, brush leads connected to frame field coils must be cut so new brushes can be installed. The new brushes must be connected and soldered to field coil leads using rosin core solder. See Fig. T94.

Check strength of brush springs. The spring must force brush against com-mutator with sufficient pressure to ensure good contact.

On square frame motors, replace brush holder card (14—Fig. T92) if card is warped or otherwise damaged.

Bushings in drive plate and end cap are not available separately, only as a unit assembly with drive plate or end cap.

Apply a light coat of grease to helix. All other parts should be assembled dry.

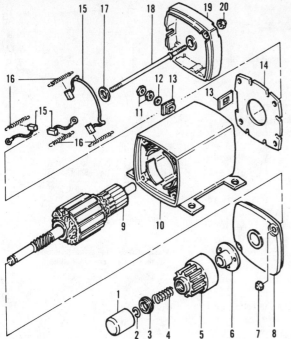

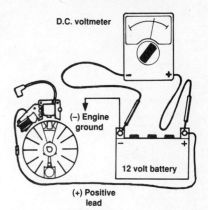

Fig. T92—Exploded view of square frame starter motor.

1. Boot
2. Snap ring
3. Spring cup
4. Spring
5. Pinion gear
6. Drive hub
7. Nut
8. Drive plate
9. Armature
10. Frame
11. Nuts
12. Washer
13. Grommet
14. Brush card
15. Brushes
16. Brush springs
17. Washer
18. Through stud
19. End cap
20. Nut

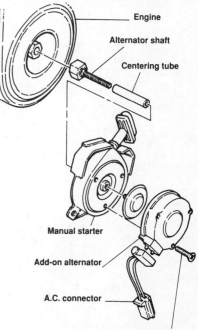

Fig. T97—To test external ignition module alternator, refer to drawing and text.

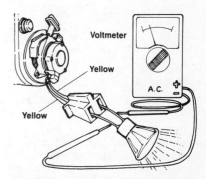

Fig. T98—Exploded view of 1 amp add-on lighting system.

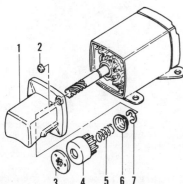

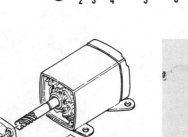

Fig. T93—Exploded view of covered starter drive used on square frame starter motor.

1. Drive cap
2. Nut
3. Drive hub
4. Pinion gear
5. Spring
6. Spring cup
7. "E" ring

Fig. T95—A piece of rewind starter spring may be positioned as shown to hold brushes in place.

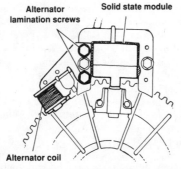

Fig. T96—Drawing of external ignition module alternator.

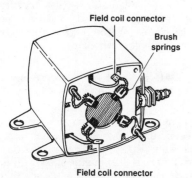

Fig. T94—Drawing of brush holder card and brush leads on square frame starter motor.

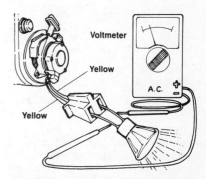

Fig. T99—To test 1 amp add-on lighting system, refer to drawing and text.

speed. A typical wiring diagram is shown in Fig. T100. Recommended bulbs are 4416 or 4420 for headlights and 1157 for tail light/stop light.

To check output, disconnect wiring connector and connect red lead of a voltmeter to connector terminal for stop light circuit (red wire). Run engine at 3600 rpm. Voltmeter should indicate at least 11.5 volts, otherwise, alternator stator must be replaced.

Inline Diode System. The inline diode system has a diode connected into alternator wire leading from engine.

Two systems using an inline diode have been used. The system diagrammed in Fig. T101 provides direct current only, while system in Fig. T102 provides both alternating and direct current.

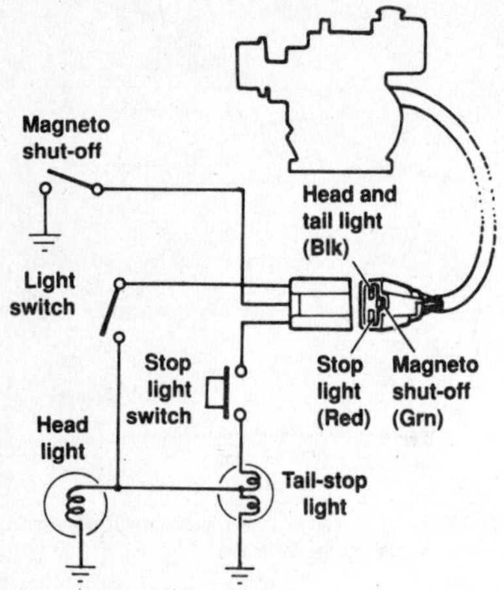

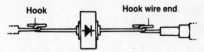

Fig. T103—Install diode so arrowhead on diode points toward output end of wire.

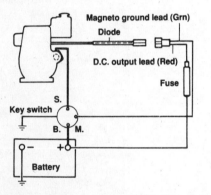

Fig. T100—Wiring diagram of 3 amp lighting system.

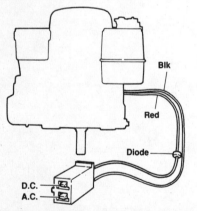

Fig. T101—Wiring diagram of 3 amp DC system with inline diode.

Fig. T102—Wiring diagram of 3 amp DC and 5 amp AC system with inline diode.

The system in Fig. T101 produces approximately 3 amps direct current. The diode rectifies alternator alternating current into direct current and a 6 amp fuse provides overload protection. To check system, disconnect harness connector and using a DC voltmeter connect tester positive lead to red wire connector terminal and ground negative tester lead to engine. At 3600 rpm

engine speed, voltmeter reading should be at least 11.5 volts. If engine speed is less, voltmeter reading will be less. If voltage reading is unsatisfactory, check alternator coils by taking an AC voltage reading. Connect one tester lead between diode and engine and ground other tester lead to engine. At an engine speed of 3600 rpm, voltage reading should be 26 volts, otherwise alternator is defective. If engine cannot attain 3600 rpm, voltage reading will be less.

The inline diode type system in Fig. T102 provides 3 amps direct current and 5 amps alternating current. This system has a two-wire pigtail consisting of a red wire and a black wire; diode is inline with red wire and covered by sheathing. To test system, check voltage at pigtail connector. At an engine speed of 3600 rpm (less engine speed will produce less voltage), voltage at red wire terminal should be 13 volts DC, while voltage at black wire terminal should be 13 volts AC. If voltage reading is unsatisfactory, check alternator coils by taking an AC voltage reading. Pull back wire sheathing and connect one tester lead to red wire between diode and engine and ground other tester lead to engine. At an engine speed of 3600 rpm, voltage reading should be at least 29 volts, otherwise alternator is defective. If engine cannot attain 3600 rpm, voltage reading will be less. If at least 29 volts is obtained, then the diode is defective.

To replace an inline diode proceed as follows: Pull back wire sheathing so diode is accessible and cut diode wires. If using heat-shrink tubing, slide tubing over wires. Bend wire ends into hooks and install new diode as shown in Fig. T103. The diode must be installed so arrowhead on diode points toward output end of wire. Tightly squeeze wire ends together. Solder wire ends together using rosin core solder. If not using heat-shrink tubing, apply insulating tape.

7 Amp Charging System. The system shown in Fig. T104 produces 7 amperes at full throttle and uses a solid-state regulator-rectifier outside flywheel that converts generated alternating current to direct current for charging battery. The regulator-rectifier also allows only required amount of current flow for existing battery conditions. When battery is fully charged, current output is decreased to prevent

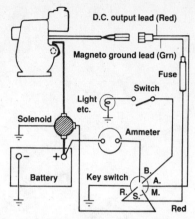

Fig. T104—Wiring diagram of 7 amp DC system.

Insert probes into connector slots. Do not remove connector wires.

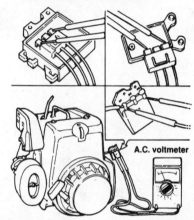

Fig. T105—Compare shape of regulator being tested with types shown and refer to text for testing procedure.

overcharging battery. The regulator-rectifier unit is epoxy covered or epoxied in an aluminum box and mounted under the blower housing. Units are not interchangeable.

It is not possible to perform an open-circuit DC test. To check alternator stator, remove regulator-rectifier unit from blower housing, but do not disconnect wire connector from regulator-rectifier. Reinstall blower housing.

CAUTION: Do not run engine without blower housing installed.

Connect AC voltmeter leads to "AC" terminals of regulator-rectifier unit as shown in Fig. T105. At an engine speed of 3600 rpm (less engine speed will produce less voltage), voltmeter should indicate at least 23 volts. If voltage reading is insufficient, then alternator is defective. If voltage reading is satisfactory and a known to be good battery is not charged by system, then regulator-rectifier unit is defective.

TECUMSEH
4-STROKE VECTOR ENGINES

Model	Bore	Stroke	Displacement	Rated Power
VLV50, VLXL50	2.797 in. (71.04 mm)	2.047 in. (51.99 mm)	12.58 cu.in. (206 cc)	5.0 hp (3.8 kW)
VLV55, VLXL55	2.797 in. (71.04 mm)	2.047 in. (51.99 mm)	12.58 cu.in. (206 cc)	5.5 hp (4.1 kW)

ENGINE IDENTIFICATION

Engine is identified by complete model number (M—Fig. T201) and specification number (S) stamped in the blower housing. These numbers are necessary in order to obtain correct repair parts.

The model number specifies the engine design and horsepower rating. For instance, model number VLV55 indicates the engine is a Vector Lightweight Vertical with 5.5 horsepower. Model numbers with "XL" indicate the engine has a cast iron cylinder sleeve.

It is important to transfer the blower housing from an original engine to a replacement short block assembly so the engine can be identified when servicing.

MAINTENANCE

LUBRICATION. All engines are equipped with a barrel and plunger type oil pump. Oil is forced through the camshaft into a passage on top of the crankcase and through holes to the upper main bearing and an oil spray hole.

Oil level should be checked after every five hours of operation. Maintain oil level at lower edge of filler plug or at "FULL" mark on dipstick.

Engine oil should meet or exceed latest API service classification. Use SAE 30 or SAE 10W-30 motor oil for tem-

peratures above 32° F (0° C). Use SAE 5W-30 or SAE 10W for temperatures below 32° F (0° C). SAE 10W-40 motor oil should not be used.

Oil should be changed after the first two hours of engine operation (new or rebuilt engine) and after every 25 hours of operation thereafter.

AIR FILTER. The engine is equipped with a paper filter and a foam precleaner filter.

The foam filter should be cleaned, inspected and re-oiled after every 25 hours of engine operation, or after three months, whichever occurs first. Clean the filter in soapy water then squeeze the filter until dry (don't twist the filter). Inspect the filter for tears and holes or any other opening. Discard the filter if it cannot be cleaned satisfactorily or if the filter is torn or otherwise damaged. Pour clean engine oil into the filter, then squeeze the filter to remove the excess oil and distribute oil throughout the filter.

The paper filter should be replaced annually or more frequently if the en-

gine operates in a severe environment, such as extremely dusty conditions. A dirty filter cannot be cleaned and must be discarded.

SPARK PLUG. Recommended spark plug is Champion RJ19LM or equivalent. Tighten spark plug to 180 in.-lbs. (20.3 N•m).

CARBURETOR. Refer to Fig. T202 for an exploded view of the carburetor. The body is extruded aluminum while the plastic fuel bowl is secured by a bail wire (22). Some carburetors may be equipped with a removable main jet (18), however, only one size is available.

A primer bulb (8—Fig. T203) is located on the side of the air cleaner body. A tube connects the primer bulb to the carburetor. Air pressurized by the primer bulb forces fuel up the carburetor nozzle into the carburetor bore for starting.

Fuel mixture is not adjustable. Engine idle speed is adjusted by turning idle speed screw (S—Fig. T204) which is accessible through a slot in the control

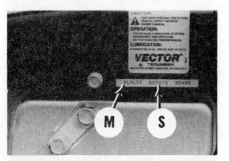

Fig. T201—Model number (M) and specification number (S) are stamped in blower housing.

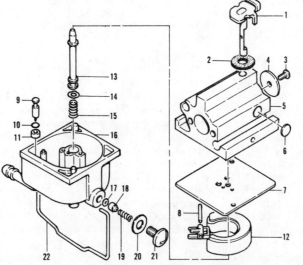

Fig. T202—Exploded view of carburetor. Main jet (18) is not used on all models.

1. Throttle shaft
2. Felt washer
3. Screw
4. Throttle plate
5. Body
6. Welch plug
7. Gasket
8. Float pin
9. Fuel inlet valve
10. Retaining ring
11. Fuel inlet valve seat
12. Float
13. Nozzle
14. "O" ring
15. Spring
16. Fuel bowl
17. "O" ring
18. Main jet
19. Spring
20. Washer
21. Drain screw
22. Bail

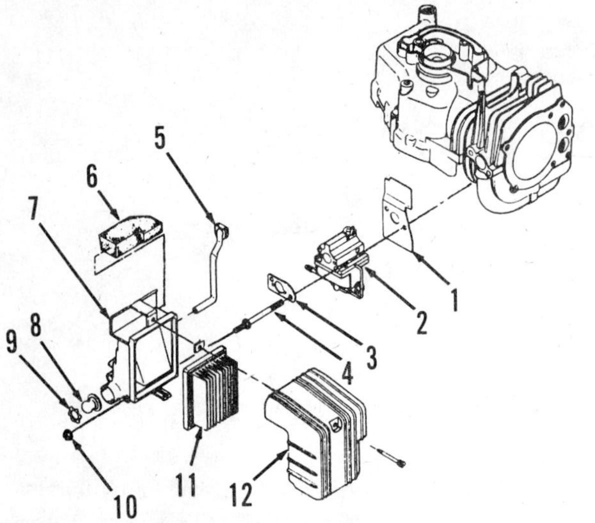

Fig. T203—Exploded view of intake system.

1. Spacer
2. Carburetor
3. Gasket
4. Stud
5. Breather tube
6. Precleaner element
7. Air cleaner box
8. Primer bulb
9. Retainer
10. Nut
11. Air filter element
12. Cover

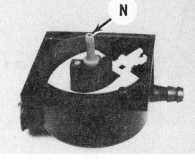

Fig. T207—Install nozzle (N) as shown. Apply a small amount of oil to "O" ring to ease insertion.

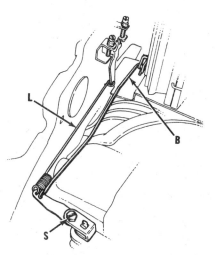

Fig. T204—View showing location of idle speed screw (S) and maximum governed speed screw (M).

Fig. T205—Install throttle plate so line is as shown.

Fig. T208—Note installation of governor spring and throttle link. Short bend (B) of throttle link attaches to governor lever. Long end (L) of governor spring attaches to speed control lever. If equipped with an electric starter motor, see also Fig. T225.

Fig.T206—Install fuel inlet valve so groove side is down and ridged side is up as shown.

panel. Engine idle speed should be adjusted to the idle speed specified by the equipment manufacturer, or if unavailable, adjust the idle speed so the engine idles smoothly (approximately 1500 rpm).

Because most major carburetor components are contained in the fuel bowl, most service can be performed by removing the fuel bowl without removing the carburetor body. Push the bail that secures the fuel bowl towards the engine to release the fuel bowl.

When disassembling the carburetor, refer to Fig. T202 and note the following: Detach fuel bowl retaining bail by pushing towards throttle end of carburetor. Don't lose spring (15) under nozzle (13). Insert a screwdriver under float arm and carefully pry float pin free from posts in fuel bowl. A retaining ring (10) is located above the fuel inlet seat which should be extracted before removing the fuel inlet valve seat. Compressed air can be used to dislodge the inlet valve seat, but cover the fuel bowl so the seat will not be ejected uncontrolled. Some engines are equipped with a removable main jet (18). When removing Welch

plug (6) be sure underlying metal is not damaged.

Inspect components for excessive wear and damage. Float height is not adjustable.

When assembling carburetor note the following: Install Welch plug (6) with concave side toward carburetor. Do not indent plug; plug should be flat after installation. Install throttle plate so line is visible and toward top of carburetor as shown in Fig. T205. Install fuel valve inlet seat with grooved side down and ridge up as shown in Fig. T206. Apply a small amount of oil to the outside diameter of the seat to ease insertion. Push the seat in using a tool with a $5/32$-inch (4 mm) diameter so seat bot-

toms in bore. Be careful not to scratch bore. Install and push down retaining ring against valve seat. Apply a small amount of oil to "O" ring on nozzle to ease insertion of the nozzle, which must be installed as shown in Fig. T207. The float pin snaps into the fuel bowl posts.

To install the primer bulb, place the retainer ring around the bulb then push the bulb and retainer into the carburetor using a deep well socket.

GOVERNOR. All engines are equipped with a mechanical (flyweight) type governor. Maximum governed speed is adjusted by turning adjusting screw (M—Fig. T204) which is accessible through a slot in the control panel.

To adjust the governor linkage, refer to Fig. T208 and loosen governor lever screw (S). Rotate governor shaft clamp clockwise as far as possible. Move the governor lever until carburetor throttle shaft is in wide open position, then tighten governor lever clamp screw.

Binding or worn governor linkage will result in hunting or unsteady engine operation. An improperly adjusted

Illustrations courtesy Tecumseh Products Co.

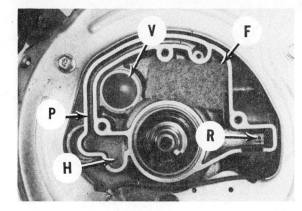

Fig. T209—View of crankcase breather showing breather valve (V), filter element (F) and return passage (R). Oil from the camshaft is routed through passage (P) to an oil spray hole and an oil hole for the upper main bearing.

carburetor will also cause a surging or hunting condition.

Note attachment of governor spring and throttle link to governor lever in Fig. T208. The throttle link end with the short bend should be attached to the governor lever.

CRANKCASE BREATHER. The crankcase breather is located under the flywheel beneath a cover on the top of the crankcase. Disc type valve (V—Fig. T209) maintains a vacuum in the crankcase. Crankcase gases are routed via a breather tube to the air cleaner. Filter (F) traps contaminants, while any oil is returned to the crankcase through return passage (R) that exits through a hole in the cylinder bore.

Clean the filter element (F) in a suitable solvent.

A replacement parts set is available. Do not remove the check valve (V) unless faulty as removal may break off the valve stem, which will fall into the crankcase. The edge of the valve should fit snugly against the valve seat, which is pressed into the crankcase. Inspect the valve for cracks and other damage. Check for a blocked return passage. Lubricate stem of new check valve to ease installation.

Oil leaking around the cover may be due to the filter element trapped under outer edge of cover or a plugged lubrication hole (H).

IGNITION SYSTEM. Standard ignition system on all models is a solid-state electronic system which does not have breaker points. The only moving part of the system is the rotating flywheel with the charging magnets. There is no scheduled maintenance.

Air gap between solid-state module and flywheel should be 0.0125 inch (0.32 mm). Use Tecumseh gage 670297 or equivalent thickness plastic strip to set gap.

VALVE ADJUSTMENT. Clearance between valve tappet and valve stem (engine cold) is 0.008 inch (0.03 mm) for intake and exhaust valves. To check clearance, remove valve tappet chamber cover and rotate crankshaft so piston is at TDC on compression stroke. Use a feeler gauge to measure clearance between each tappet and valve stem end. Grind valve stem end as necessary to increase clearance. Cut valve seat deeper or renew valve and/or tappet to reduce clearance.

REPAIRS

TIGHTENING TORQUES. Recommended tightening torque specifications are as follows:

Carburetor studs ...... 50-75 in.-lbs.
(5.6-8.5 N•m)
Connecting rod .... 95-110 in.-lb.
(11-12 N•m)
Cylinder head .... 180-220 in.-lbs.
(20.3-24.8 N•m)
Flywheel .... 33-36 ft.-lbs.
(45-50 N•m)
Oil pan............. 100-130 in.-lbs.
(11.3-14.7 N•m)
Spark plug .... 180 in.-lbs.
(20.3 N•m)

FLYWHEEL. Before removing flywheel, disengage flywheel brake as outlined in FLYWHEEL BRAKE section. Use a suitable puller or knockoff nut to remove flywheel. Gently pry against bottom of flywheel while tapping sharply on nut.

After installing flywheel, tighten flywheel nut to 33-36 ft.-lbs. (45-50 N•m). Note that removable plastic fan has a boss on the bottom that must fit in a recess on top of the flywheel.

CYLINDER HEAD. To remove cylinder head (17—Fig. T210), remove all interfering shrouds and brackets. Clean area around cylinder head to prevent entrance of foreign material. Unscrew

cylinder head screws and remove cylinder head.

A new head gasket should be installed when installing cylinder head. Tighten cylinder head retaining screws evenly in the sequence shown in Fig. T211 to a torque of 180-220 in.-lbs. (20.3-24.8 N•m).

VALVE SYSTEM. Exhaust valve is marked "EX" or "X" while intake valve is marked "I". Valve face angle is 45° and valve seat angle is 46°. Valve seat width should be 0.035-0.045 inch (0.89-1.14 mm).

Valve stem guides are cast into cylinder block and are not renewable. If excessive clearance exists between valve stem and guide, guide should be reamed and a new valve with an oversize stem installed. Oversize valve guide diameter should be 0.2807-0.2817 inch (7.130-7.155 mm).

CAMSHAFT. The camshaft and camshaft gear are an integral part which rides on journals at each end of camshaft. Camshaft is equipped with a compression release mechanism which lifts exhaust valve at low cranking rpm to reduce compression and aid starting.

When removing camshaft, align timing marks on camshaft gear and crankshaft gear as shown in Fig. T212 to relieve valve spring pressure on camshaft lobes.

Renew camshaft if lobes or journals are worn or scored. Spring on compression release mechanism should hold weight against camshaft. Compression release mechanism and camshaft are serviced as an assembly only. Standard camshaft journal diameter is 0.4975-0.4980 inch (12.637-12.649 mm).

The oil pump is operated by an eccentric on the camshaft. Refer to OIL PUMP paragraph.

When installing camshaft, align crankshaft and camshaft timing marks as shown in Fig. T212.

PISTON, PIN AND RINGS. Piston and connecting rod assembly is removed from cylinder head end of engine. Before removing piston, remove any carbon or ring ridge from top of cylinder to prevent ring breakage. Aluminum alloy pistons are equipped with two compression rings and one oil control ring. Oversize pistons and rings are available for all engines.

Piston ring end gap is 0.007-0.020 inch (0.18-0.50 mm).

Piston skirt-to-cylinder clearance is 0.004-0.006 inch (0.10-0.15 mm). Standard piston diameter measured at piston

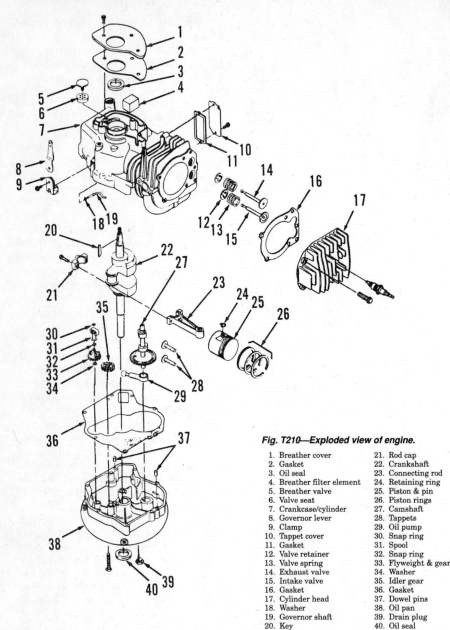

Fig. T210—Exploded view of engine.

1. Breather cover	21. Rod cap
2. Gasket	22. Crankshaft
3. Oil seal	23. Connecting rod
4. Breather filter element	24. Retaining ring
5. Breather valve	25. Piston & pin
6. Valve seat	26. Piston rings
7. Crankcase/cylinder	27. Camshaft
8. Governor lever	28. Tappets
9. Clamp	29. Oil pump
10. Tappet cover	30. Snap ring
11. Gasket	31. Spool
12. Valve retainer	32. Snap ring
13. Valve spring	33. Flyweight & gear
14. Exhaust valve	34. Washer
15. Intake valve	35. Idler gear
16. Gasket	36. Gasket
17. Cylinder head	37. Dowel pins
18. Washer	38. Oil pan
19. Governor shaft	39. Drain plug
20. Key	40. Oil seal

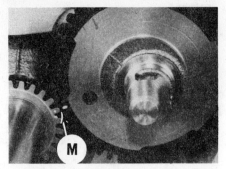

Fig. T212—View of aligned timing marks (M) on crankshaft and camshaft gears.

Fig. T213—Match marks on connecting rod and rod cap must be aligned and toward pto end of crankshaft after installation.

skirt 90° from piston pin bore is 2.7900-2.7910 inches (70.866-70.891 mm).

Standard ring side clearance in ring grooves is 0.002-0.005 inch (0.05-0.13 mm) for compression rings and 0.0005-0.0035 inch (0.013-0.089 mm) for oil ring.

Piston pin should be a tight push fit in piston pin bore and connecting rod pin bore. Piston pin is retained by retaining rings at each end of piston pin bore. Install top and second piston rings with beveled edge toward piston crown. When installing connecting rod and piston assembly, align the match marks on connecting rod and cap as shown in Fig. T213. Install piston so match marks on connecting rod are toward power take-off (pto) end of crankshaft. Stagger ring end gaps equally around circumference of piston before installation.

CONNECTING ROD. The aluminum alloy connecting rod rides directly on crankshaft crankpin.

Crankpin diameter should be 1.0230-1.0235 inch (25.088-25.103 mm). Standard inside diameter for connecting rod big end should be 1.0240-1.0246 inch (25.088-25.103 mm). Connecting rod bearing-to-crankpin journal clearance should be 0.0005-0.0016 inch (0.013-0.041 mm) for all models.

Assemble connecting rod and piston as outlined in previous section. Tighten connecting rod screws to 95-110 in.-lb. (11-12 N•m).

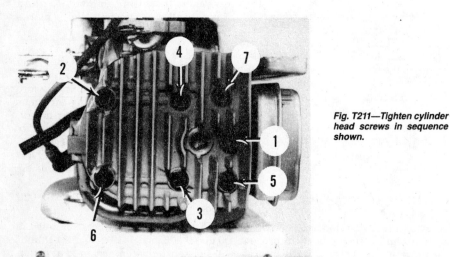

Fig. T211—Tighten cylinder head screws in sequence shown.

Illustrations courtesy Tecumseh Products Co.

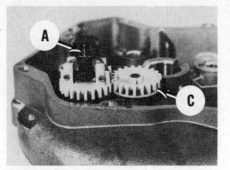

Fig. T214—Some governor spools are retained by a snap ring (A). See exploded view in Fig. T210.

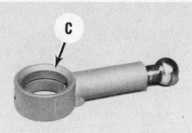

Fig. T216—Install oil pump so chamfered (C) side is toward camshaft gear.

Fig. T215—A governor shaft using snap rings to retain governor assembly is shown in the picture above. Governor shaft can be renewed if damaged or worn excessively.

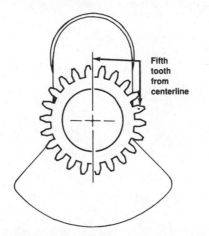

Fifth tooth from centerline

Fig. T217—Timing mark on crankshaft gear must be in 2:30 position as shown.

GOVERNOR. On early models, the governor is retained on the shaft by snap ring (A—Fig. T214). To remove spool, detach upper snap ring. Detach lower snap ring to remove flyweight and gear assembly. Washer (34—Fig. T210) is located under the gear.

On later engines and replacement governor shafts, no snap rings are used on the governor shaft. The governor is held in place by a boss on the governor shaft. The flyweights and gear are available only as a unit assembly.

The governor gear is driven by idler gear (C—Fig. T214) that meshes with the camshaft gear. The idler gear shaft is pressed into the oil pan and is not renewable.

The governor shaft is pressed into the oil pan and may be replaced if the mounting boss is not damaged or the hole is not enlarged. To remove the governor shaft, clamp the shaft in a vise and using a soft mallet, drive the oil pan off the shaft. Do not attempt to twist governor shaft out of boss. Twisting will enlarge hole. Apply Loctite 271 (red) to shaft end. When installing the retainerless design shaft, position washer and governor gear assembly on shaft before installing shaft, then press shaft into oil pan until the governor gear has 0.010-0.020 inch (0.25-0.50 mm) axial play.

OIL PUMP. The engine is equipped with a barrel and plunger type oil pump that is driven by an eccentric on the camshaft. Pump is available only as a unit assembly. Chamfered side of drive collar (C—Fig. T216) should be toward camshaft gear.

CRANKSHAFT, MAIN BEARINGS AND SEALS. Always note oil seal depth and direction before removing oil seal from crankcase or cover. New seals must be pressed into seal bores to the same depth as old seal before removal on all models.

The crankshaft gear is pressed on the crankshaft. The crankshaft gear timing mark should be in the 2:30 position relative to the crankpin (see Fig. T217), otherwise the gear has moved. In some instances the gear can be relocated satisfactorily, but if not, the crankshaft must be replaced.

Refer to CONNECTING ROD section for standard crankshaft crankpin journal diameters.

Crankshaft main bearing journals ride directly in the aluminum alloy bores in the cylinder block and crankcase cover or oil pan. Standard main bearing bore diameter for main bearings is 1.0257-1.0262 inch (26.053-26.066 mm).

Standard diameter for crankshaft both main bearing journals is 1.0237-1.0242 inch (26.002-26.015 mm).

When installing crankshaft, align crankshaft and camshaft gear timing marks as shown in Fig. T212.

CYLINDER AND CRANKCASE. Cylinder and crankcase are an integral casting on all models. Models VLXL50 and VLXL55 are equipped with a cast iron sleeve in the cylinder bore. Oversize pistons and rings are available for all engines. Cylinder should be honed and fitted to nearest oversize for which piston and ring set are available if cylinder is scored, tapered or out-of-round more than 0.005 inch (0.13 mm).

Standard cylinder bore diameter is 2.7950-2.7960 inch (70.993-71.018 mm). Note that some engines were manufactured with cylinders that are oversize. The piston is marked on the piston crown with the oversize on the piston crown.

FLYWHEEL BRAKE. The engine is equipped with a flywheel brake that simultaneously stops the flywheel and grounds the ignition. The brake should stop the engine within three seconds when the operator releases the mower safety control and the speed control is in high speed position. Engine rotation is stopped by a pad type brake that contacts the inside of the flywheel when the operator's handle is released.

On engines with an electric starter, a switch mounted on the flywheel brake lever prevents starter engagement unless the flywheel brake is disengaged.

To hold the flywheel brake in the disengaged position, bend a piece of 3/32-inch (25 mm) diameter metal rod so there is approximately one inch (25 mm) between the bends. Remove the left screw securing the speed control cover. Move the flywheel brake lever to the disengaged position and insert the holder rod into the outer hole of the brake lever and the speed control cover screw hole as shown in Fig. T218. Tecumseh offers tool 36114 (T—Fig. T219) to hold the brake lever in the disengaged position.

To disassemble flywheel brake assembly, hold flywheel brake in disen-

Fig. T218—Fabricate a piece of stiff rod and install as shown to hold flywheel brake in disengaged position.

Fig. T219—Flywheel brake may be held in disengaged position by installing Tecumseh tool 36114 as shown.

Fig. T220—Flywheel brake lever spring end should be installed in original hole (H) in baffle plate.

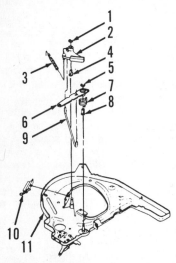

Fig. T221—Exploded view of flywheel brake.

1. "E" ring
2. Brake arm
3. Brake spring
4. Bushing
5. "E" ring
6. Brake lever
7. Spring
8. Bushing
9. Link
10. Switch
11. Baffle plate

gaged position as previously outlined. Remove flywheel then disassemble brake components. For reference during assembly, mark hole (H—Fig. T220) on bottom of baffle plate that holds brake lever spring end. If equipped with electric starter, mark and disconnect wires to starter switch on brake lever.

Replace the brake pad and arm if the pad is damaged, contaminated by oil, or worn to a thickness less than 0.060 inch (1.52 mm).

Fig. T222—Note bosses on brake link and connect end with four bosses to brake pad arm and end with three bosses to brake lever.

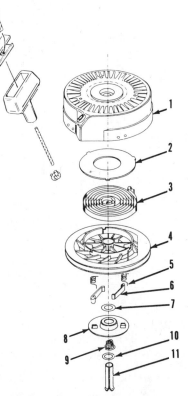

Fig. T223—Exploded view of rewind starter.

1. Starter housing
2. Cover
3. Rewind spring
4. Pulley
5. Springs (2)
6. Pawls (2)
7. Plastic washers (2)
8. Retainer
9. Brake spring
10. Metal washer
11. Pin

To reassemble flywheel brake assembly, reverse removal procedure while noting the following: The brake pad arm and brake lever ride on renewable plastic bushings. Install bushings (4 and 8—Fig. T221) so flange is toward bottom of brake arm (2) or lever (6). Correct installation of brake link (9) is determined by number of bosses on link end. The end that attaches to brake pad arm has four bosses (Fig. T222), while brake lever end has three bosses. The brake lever spring end must be inserted in original hole (H—Fig. T220) in baffle plate. Note that brake spring (3—Fig.

T221) end with the short hook must be attached to brake arm (2).

REWIND STARTER. An exploded view of starter is shown in Fig. T223. To disassemble starter, remove rope handle and allow rope to wind into starter. Position a suitable sleeve support under pawl retainer (8), then use a $\frac{5}{16}$ inch (8 mm) punch to drive pin (11) out of starter. Remove brake spring (9), retainer (8), pawls (6) and springs (5). Wear appropriate safety eyewear and gloves before disengaging pulley from starter as spring may uncoil uncontrolled. Place shop towel around pulley and lift pulley out of housing; spring should remain with pulley.

Inspect components for damage and excessive wear. Rope length is 98 inches (249 cm) and rope diameter is $\frac{9}{64}$ inch. Reverse disassembly procedure to install components. Rewind spring coils wind in clockwise direction from outer end. Wind rope around pulley in counterclockwise direction as viewed from retainer side of pulley. Be sure inner end of rewind spring engages spring retainer adjacent to housing center post. Use two plastic washers (7). Install a new pin (11) so top of pin is $\frac{1}{8}$ inch (3.2 mm) below top of starter. Driving pin in too far may damage retainer pawl.

ELECTRIC STARTER. Tecumseh does not provide test specifications for electric starter motors, so service is limited to replacing components that are known or suspected faulty.

Refer to Fig. T224 for an exploded view of the starter motor. Disassembly and reassembly is evident after inspection of motor and referral to Fig. T224 Minimum brush length is not specified. Renew brushes if the brush wire bottoms against the slot in the brush holder. Brushes must not bind in the holders. Brushes are available only as a unit assembly with end cap. Bushings are not available separately; only as a unit assembly with drive plate or end cap.

When installing starter motor, place ground wire on starter motor stud near governor lever (B—Fig. T225) before installing motor on baffle plate (C). The ground wire (D) fits between two adjacent index tabs on drive end plate of starter motor. The index tabs must fit into slots (E) on baffle plate. Note that governor spring (F) passes between tabs on starter motor.

A switch mounted on the flywheel brake lever prevents starter engagement unless the flywheel brake is disengaged.

Illustrations courtesy Tecumseh Products Co.

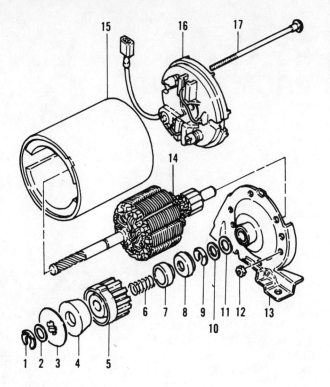

Fig. T224—Exploded view of electric starter motor.

1. "E" ring
2. Plastic washer
3. Drive hub
4. Rubber driver
5. Pinion gear
6. Spring
7. Plastic spring cup
8. Metal cup
9. "E" ring
10. Metal washer
11. Plastic washer
12. Nut
13. Drive plate
14. Armature
15. Frame
16. End cap & brush assy.
17. Through-bolt

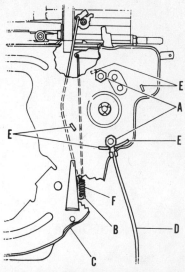

Fig. T225—Drawing showing installation of starter motor. Index tabs on motor fit in slots (E) in baffle plate. Governor spring (F) passes between index tabs on motor.

A. Starter nuts
B. Governor lever
C. Baffle plate
D. Ground wire
E. Slots
F. Governor spring

TECUMSEH
4-STROKE OHV ENGINES

Model	Bore	Stroke	Displacement	Rated Power
OVRM40	2.5 in.	1.844 in.	9.05 cu.in.	4.0 hp
	(63.5 mm)	(46.8 mm)	(148.3 cc)	(3.0 kW)
OVRM50	2.625 in.	1.938 in.	10.49 cu.in.	5.0 hp
	(66.68 mm)	(49.23 mm)	(172 cc)	(3.7 kW)
OVRM55	2.625 in.	1.938 in.	10.49 cu.in.	5.5 hp
	(66.68 mm)	(49.23 mm)	(172 cc)	(4.1 kW)

ENGINE INFORMATION

All engines are air-cooled, four-stroke, single-cylinder engines. The engine has a vertical crankshaft and utilizes an overhead valve system.

Engine is identified by model number and specification number stamped in the blower housing (Fig. T301). These numbers are necessary in order to obtain correct repair parts.

The model number specifies the engine design and horsepower rating. For instance, model number OVRM55 indicates the engine is an overhead valve rotary mower model with 5.5 horsepower.

It is important to transfer the blower housing from an original engine to a replacement short block assembly so the engine can be identified when servicing.

MAINTENANCE

LUBRICATION. All engines are equipped with a barrel and plunger type oil pump.

Oil level should be checked after every five hours of operation. Maintain oil level at lower edge of filler plug or at "FULL" mark on dipstick.

Engine oil should meet or exceed latest API service classification. Use SAE 30 or SAE 10W-30 motor oil for temperatures above 32° F (0° C). Use SAE 5W-30 or SAE 10W for temperatures below 32° F (0° C). SAE 10W-40 motor oil should not be used.

Oil should be changed after the first two hours of engine operation (new or rebuilt engine) and after every 25 hours of operation thereafter.

AIR FILTER. The filter element may be made of foam or paper, or a combination of both foam and paper. The recommended maintenance interval depends on the type of filter element.

Foam type filter elements should be cleaned, inspected and re-oiled after every 25 hours of engine operation, or after three months, whichever occurs first. Clean the filter in soapy water then squeeze the filter until dry (don't twist the filter). Inspect the filter for tears and holes or any other opening. Discard the filter if it cannot be cleaned satisfactorily or if the filter is torn or otherwise damaged. Pour clean engine oil into the filter, then squeeze the filter to remove the excess oil and distribute oil throughout the filter.

Paper type filter elements should be replaced annually or more frequently if the engine operates in a severe environment, such as extremely dusty conditions. A dirty filter element cannot be cleaned and must be discarded.

SPARK PLUG. Recommended spark plug is a Champion RN4C or equivalent. Specified spark plug electrode gap is 0.030 inch (0.76 mm). Tighten spark plug to 18-23 ft.-lbs. (25-31 N•m).

CARBURETOR. A float type carburetor is used on all engines. No adjust-

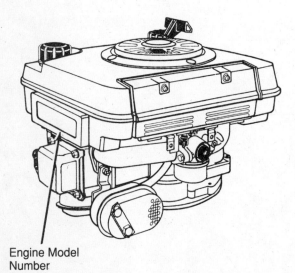

Engine Model Number

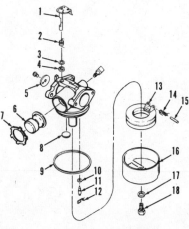

Fig. T302—Exploded view of carburetor.

1.	Throttle shaft	10.	Inlet valve seat
2.	Spring	11.	Fuel inlet valve
3.	Washer	12.	Clip
4.	Seal	13.	Float
5.	Throttle plate	14.	Dampener spring
6.	Primer bulb	15.	Pin
7.	Retainer	16.	Fuel bowl
8.	Welch plug	17.	Washer
9.	"O" ring	18.	Main jet/nut

Fig. T301—Model and specification number are stamped in blower housing.

Illustrations courtesy Tecumseh Products Co.

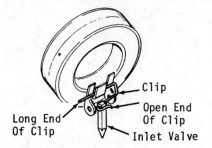

Fig. T303—Install inlet valve clip so long end points toward intake end of carburetor.

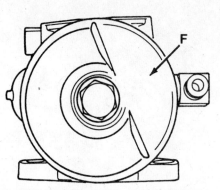

Fig. T306—Install fuel bowl so flat area (F) is positioned over fuel inlet valve and parallel with float hinge pin.

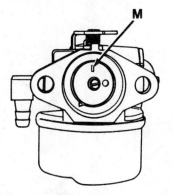

Fig. T304—Mark (M) on throttle plate must point up.

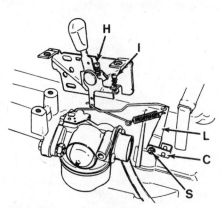

Fig. T307—Adjust idle speed by turning screw (I) and high speed by turning screw (H).

ments are possible. A primer bulb is located on the side of the carburetor for cold starting enrichment.

Refer to Fig. T302 for an exploded view of carburetor. Disassembly and reassembly is evident after inspection of carburetor while noting the following. To remove primer bulb (6), grasp bulb with pliers and roll bulb out of carburetor body. Wear safety eyewear when prying primer bulb retainer (7) away from carburetor body. The inlet valve seat (10) can be removed by blowing compressed air through the fuel inlet fitting into carburetor body or by using a hooked wire.

Install seat (10) with grooved side down so it bottoms in bore. Install inlet valve clip as shown in Fig. T303 so long end of clip will point toward intake end of carburetor (away from throttle plate). Install throttle plate so scribe mark points up as shown in Fig. T304; no light should be visible around throttle plate when closed. A new throttle plate retaining screw should be used. To measure

float level, position Tecumseh tool 670253A at a 90° angle to carburetor bore and resting on nozzle boss as shown in Fig. T305. Toe of float should not be higher than first step on tool, but not lower than second step. Bend tab on float arm to adjust float height. Install fuel bowl so flat area is located over fuel inlet valve and crease runs parallel to float pin. See Fig. T306.

CARBURETOR CONTROL. The engine may be equipped with the control unit similar to that shown in Fig. T307 that controls the throttle. Turn screw (I) to adjust idle speed and screw (H) to set high speed.

GOVERNOR. The engine is equipped with a mechanical, flyweight type governor. To adjust governor linkage, proceed as follows: Loosen governor

lever clamp screw (S—Fig. T307), rotate governor lever (L) so throttle plate is fully open and hold lever in place. Turn governor shaft clamp (C) counterclockwise as far as possible, then tighten lever clamp screw.

If internal governor assembly must be serviced, refer to REPAIRS section.

CRANKCASE BREATHER. The engine is equipped with a crankcase breather that provides a vacuum for the crankcase. Vapor from the crankcase is evacuated to the intake manifold. A check valve in the breather vents positive pressure pulsations into the breather element. A tube directs crankcase gases to the air cleaner. The breather system must operate properly or excessive oil consumption may result.

The breather assembly is mounted in the top rear side of cylinder block. Periodically remove and clean the breather element.

IGNITION SYSTEM. A one-piece ignition coil/module is located adjacent to the outer periphery of the flywheel. Ignition timing is not adjustable. Set air gap between module leg laminations and flywheel by loosening mounting screws and positioning module to obtain an air gap of 0.0125 inch (0.32 mm). See Fig. T308. Tighten mounting screws to 30-40 in.-lbs. (3.4-4.5 N•m).

VALVE ADJUSTMENT. Clearance between rocker arm and valve stem is checked and adjusted with engine cold. Remove rocker arm cover. Rotate crankshaft so piston is at top dead center on compression stroke. Valve clearance should be 0.004 inch (0.10 mm) for intake and exhaust. To adjust clearance, loosen locknut (N—Fig. T309) and using an Allen wrench (W), turn rocker pivot stud to obtain desired clearance. Tighten lock nut to 65-85 in.-lbs. (7.3-9.6 N•m).

REPAIRS

TIGHTENING TORQUES. Recommended tightening torque specifications are as follows:

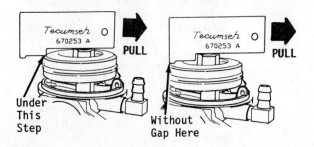

PULL   PULL

Under This Step   Without Gap Here

Fig. T305—Use procedure described in text to measure float height using tool 670253A.

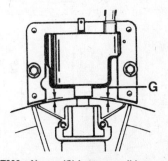

Fig. T308—Air gap (G) between coil legs and flywheel should be 0.0125 inch (0.32 mm).

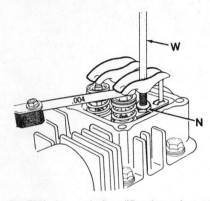

Fig. T309—Loosen locknut (N) and turn pivot stud using Allen wrench (W) to adjust valve clearance which should be 0.004 inch (0.10 mm).

Connecting rod . . . . 95-110 in.-lbs.
(11-12 N•m)

Crankcase cover . . . . . 100-130 in.-lbs.
(11.3-14.7 N•m)

Cylinder head . . . . 220-240 in.-lbs.
(25-27 N•m)

Flywheel nut . . . . 33-36 ft.-lbs.
(45-49 N•m)

Rocker arm cover . . . . . . 30-50 in.-lbs.
(3.4-5.6 N•m)

Rocker stud lock nut . . . . 65-85 in.-lbs.
(7.3-9.6 N•m)

Spark plug . . . . 18-23 ft.-lbs.
(25-31 N•m)

FLYWHEEL. Before flywheel can be removed, first remove fuel tank assembly, rewind starter and blower shroud. Disengage flywheel brake as outlined in FLYWHEEL BRAKE section. If flywheel has tapped holes, use a suitable puller to remove flywheel. If no holes are present, screw a knockoff nut onto crankshaft as shown in Fig. T310 so there is a small gap between nut and flywheel. Gently pry against bottom of flywheel while tapping sharply on nut. After installing flywheel, tighten flywheel nut to 33-36 ft.-lbs. (45-49 N•m).

CYLINDER HEAD AND VALVE SYSTEM. Refer to Fig. T311 for an ex-

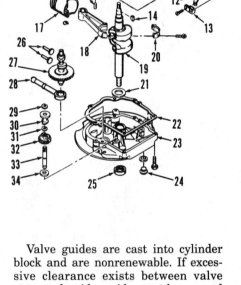

Fig. T311—Exploded of Model OVRM50 engine. Model OVRM40 is similar.

1. Flywheel nut
2. Belleville washer
3. Starter cup
4. Flywheel
5. Spacer
6. Oil seal
7. Breather tube
8. Crankcase breather assy.
9. Washer
10. Governor shaft
11. Governor spring
12. Governor lever
13. Clamp
14. Key
15. Dowel pin
16. Piston rings
17. Piston
18. Connecting rod
19. Crankshaft
20. Connecting rod cap
21. Thrust washer
22. Gasket
23. Crankcase cover (lower mounting flange)
24. Drain plug
25. Oil seal
26. Tappets
27. Camshaft
28. Oil pump
29. Snap ring
30. Governor spool
31. Snap ring
32. Governor gear
33. Shaft
34. Washer
35. Rocker arm cover
36. Pivot stud
37. Rocker arm
38. Locknut
39. Push rod guide plate
40. Valve spring retainer
41. Valve spring
42. Gasket
43. Cylinder head
44. Head gasket
45. Intake valve
46. Exhaust valve
47. Push rod

ploded view of cylinder head and valve system. To remove the cylinder head, first remove intake manifold, exhaust manifold and rocker arm cover. Remove rocker arms and push rods; mark them so they can be returned to original location. Unscrew cylinder head bolts and remove cylinder head.

Valve face angle is 45° while valve seat angle is 46°. Specified seat width is 0.045 inch (1.15 mm). Minimum allowable valve margin is 0.030 inch (0.76 mm).

Valve guides are cast into cylinder block and are nonrenewable. If excessive clearance exists between valve stem and guide, guide must be reamed and a new valve with an oversize stem installed. Oversize valve guide diameter should be 0.2807-0.2817 inch (7.130-7.155 mm) for intake and 0.2787-0.2797 in. (7.079-7.104 mm) for exhaust to fit valve with 1/32 inch (0.8 mm) oversize stem.

Note the following when reinstalling cylinder head. Make certain that mating surfaces of cylinder block and head are clean. The metal head gasket must be renewed whenever cylinder head is removed. Do not apply sealer to cylinder head gasket; new gasket is precoated with a sealing substance. If coating is scratched or if gasket is bent, gasket should be discarded and another gasket installed. Leakage may occur if a damaged gasket is installed. Tighten cylinder head bolts in steps of 60 in.-lbs. (6.8 N•m) using sequence shown in Fig. T312 until final torque reading of 220-240 in.-lbs. (25-27 N•m) is obtained.

When installing push rod guide plate (39—Fig. T311), note that tabs around push rod guide slots must be out toward rocker arms.

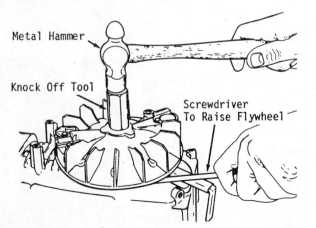

Metal Hammer

Knock Off Tool

Screwdriver To Raise Flywheel

Fig. T310—View showing use of a knock-off tool (nut) to separate flywheel from crankshaft end.

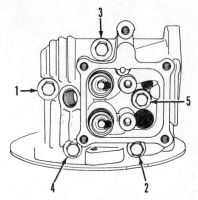

Fig. T312—Tighten cylinder head bolts in sequence shown.

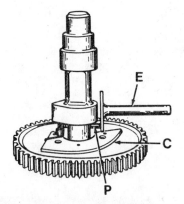

Fig. T313—Compression release pin (P) extends during starting to hold exhaust valve off its seat.

CAMSHAFT. Camshaft and camshaft gear are an integral casting which is equipped with a compression release mechanism. The compression release pin (P—Fig. T313) extends at cranking speed to hold the exhaust valve open slightly thereby reducing compression pressure.

To remove camshaft, remove engine from equipment. Remove rocker arm cover (35—Fig. T311) and disengage push rods (47) from rocker arms (37). Drain engine oil, then remove crankcase cover (23) and withdraw camshaft.

Specified camshaft bearing journal diameter is 0.4975-0.4980 inch (12.636-12.649 mm). Renew camshaft if lobes or journals are excessively worn or scored. Inspect compression release mechanism and check for proper operation. Compression release components and camshaft are available only as a unit assembly.

When installing camshaft, align crankshaft and camshaft timing marks shown in Fig. T314.

PISTON, PIN, RINGS AND CONNECTING ROD. To remove piston and rod assembly, drain engine oil and remove engine from equipment. Remove cylinder head as previously outlined.

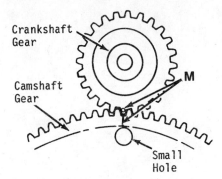

Fig. T314—View of crankshaft and camshaft timing marks (M).

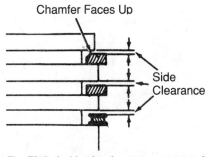

Fig. T315—Inside chamfer on top compression ring must face top of piston.

Clean pto end of crankshaft and remove any burrs or rust. Unscrew fasteners and remove crankcase cover. Remove camshaft. Remove carbon or ring ridge (if present) from top of cylinder before removing piston. Unscrew connecting rod screws and push piston and rod assembly out top of cylinder.

Position piston rings squarely in cylinder bore and measure ring end gap using a feeler gauge. Specified piston ring end gap is 0.010-0.020 inch (0.25-0.50 mm). If ring end gap exceeds specifications, cylinder should be rebored for installation of oversize piston and rings.

Insert new rings in piston ring grooves and measure side clearance between ring and ring land (Fig. T315) using a feeler gauge. Piston ring side clearance should be .002-0.005 inch (0.05-0.13 mm) for compression rings and 0.0005-0.0035 inch (0.013-0.089 mm) for oil ring. Renew piston if side clearance is excessive.

Standard piston diameter measured at bottom of skirt 90° from piston pin hole is 2.4950-2.4952 inches (63.373-63.378 mm) for Model OVRM40 and 2.6202-2.6210 inches (66.553-66.573 mm) for Models OVRM50 and OVRM55. Piston-to-cylinder clearance should be 0.0040-0.0058 inch (0.10-0.15 mm). Oversize pistons are available in sizes of 0.010 and 0.020 inch.

The connecting rod rides directly on crankshaft. Connecting rod big end di-

ameter should be 0.8620-0.8625 inch (21.895-21.907 mm) for OVRM40 and 1.0005-1.0010 inch (25.413-25.425 mm) for OVRM50 and OVRM55.

When installing piston rings, place top compression ring on piston so inside chamfer is toward piston crown (Fig. T315). Install connecting rod so match marks on rod and cap are aligned and toward pto end of crankshaft. Tighten connecting rod screws to 95-110 in.-lbs. (11-12 N•m).

GOVERNOR. The governor gear and flyweight assembly is located on the inside of the crankcase cover. The flyweights on the governor gear (32—Fig. T311) actuate spool (30) which contacts the governor arm and shaft (10) in the crankcase. The governor shaft and arm transfer governor action to the external governor linkage.

On early models, the governor is retained on the shaft by snap ring (29—Fig. T311). To remove spool, detach upper snap ring. Detach lower snap ring (31) to remove flyweight and gear assembly. Washer (34) is located under the gear.

On later engines and replacement governor shafts, no snap rings are used on the governor shaft. The governor is held in place by a boss on the governor shaft. The flyweights and gear are available only as a unit assembly.

The governor shaft is pressed into the oil pan and may be replaced if the mounting boss is not damaged or the hole is not enlarged. To remove the governor shaft, clamp the shaft in a vise and using a soft mallet, drive the oil pan off the shaft. Do not attempt to twist governor shaft out of boss. Twisting will enlarge hole. Apply Loctite 271 (red) to shaft end. When installing the retainerless design governor shaft, position washer and governor gear assembly on shaft before installing shaft, then press in shaft until the governor gear has 0.010-0.020 inch (0.25-0.50 mm) axial play.

OIL PUMP. The engine is equipped with a barrel and plunger type oil pump. The pump is driven by an eccentric on the camshaft. Chamfered side of drive collar (C—Fig. T316) must be toward camshaft. Pump is available only as a unit assembly.

CRANKSHAFT, MAIN BEARINGS AND OIL SEALS. The crankshaft rides directly in aluminum alloy bores of crankcase.

Note oil seal depth and seal lip direction before removing old seals (6 and 25—Fig. T311) from crankcase or cover. Install new seals to depth of old seals.

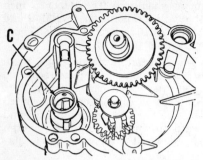

Fig. T316—Oil pump must be installed so chamfer (C) is toward camshaft.

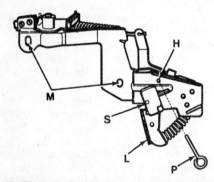

Fig. T317—View of flywheel brake mechanism.

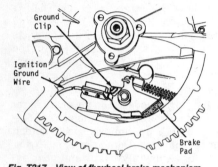

Fig. T318—Push against lever (L) and insert pin (P) through holes (H) to hold brake pad away from flywheel. Interlock switch (S) is used on engines with an electric starter.

Standard crankshaft main bearing journal diameter for both ends is 0.9985-0.9990 inch (25.362-25.375 mm). Crankpin diameter is 0.8610-0.8615 inch (21.869-21.882 mm) for OVRM40 and 0.9995-1.000 inch (25.387-25.40 mm) for OVRM50 and OVRM55. Crankshaft end play should be 0.005-0.027 inch (0.13-0.69 mm).

Crankcase main bearing inner diameter is 1.0005-1.0010 inch (25.413-25.425 mm) for both bearing bores.

CYLINDER. Standard cylinder bore diameter is 2.500-2.501 inches (63.500-63.525 mm) for Model OVRM40 and 2.625-2.626 inches (66.675-66.700 mm) for Models OVRM50 and OVRM55. If cylinder bore wear or out-of-round exceeds 0.005 inch (0.13 mm), cylinder should be bored to the next oversize.

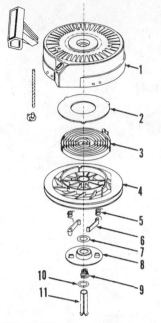

Fig. T319—Exploded view of rewind starter.

1. Starter housing
2. Cover
3. Rewind spring
4. Pulley
5. Springs (2)
6. Pawls (2)
7. Plastic washers (2)
8. Pawl retainer
9. Brake spring
10. Metal washer
11. Pin

FLYWHEEL BRAKE. The engine is equipped with a flywheel brake that utilizes a pad which is forced against the inner surface of the flywheel. The brake should stop the engine within three seconds when the operator releases mower safety control and the speed control is in high speed position. A ground clip kills the ignition when the brake is actuated. See Fig. T317. On models with electric start, a switch prevents starting when the brake pad contacts the flywheel.

The flywheel must be removed to service the brake pad. Before removing flywheel, push lever (L—Fig. T318) towards spark plug so brake pad moves away from flywheel. Insert Tecumseh tool 670298 or a suitable pin in hole (H) to hold lever. Inspect mechanism for damage and excessive wear. Minimum allowable thickness of brake pad at nar-

rowest point is 0.060 inch (1.52 mm). Install brake mechanism and push up on bracket so bracket mounting screws are at bottom of slotted holes (M) in bracket. Tighten mounting screws to 90 in.-lbs. (10.2 N•m).

REWIND STARTER. To disassemble starter, remove rope handle and allow rope to wind into starter. Position a suitable sleeve support (a ¾ inch deep socket) under pawl retainer (8—Fig. T319) and using a 5/16 inch diameter punch, drive pin (11) free of starter. Remove brake spring (9), retainer (8), pawls (6) and springs (5). Wear appropriate safety eyewear and gloves before disengaging pulley from starter as spring may uncoil uncontrolled. Place shop towel around pulley and lift pulley out of housing; spring should remain with pulley.

Inspect components for damage and excessive wear. Rope length is 98 inches (249 cm) and rope diameter is 9/64 inch (3.5 mm). Reverse disassembly procedure to install components. Rewind spring coils wind in clockwise direction from outer end. Wind rope around pulley in counterclockwise direction as viewed from retainer side of pulley. Be sure inner end of rewind spring engages spring retainer adjacent to housing center post. Use two plastic washers (7). Install a new pin (11) so top of pin is ⅛ inch (3.2 mm) below top of starter. Driving pin in too far may damage retainer pawl.

ELECTRIC STARTER AND ALTERNATOR. Some engines may be equipped with an electric starter and an alternator for battery charging.

Use standard electric motor testing procedures to check starter. If brushes require replacement, entire end cap assembly (14—Fig. T320) must be renewed.

Alternator coil is attached to side of ignition module. Alternator is considered satisfactory if output exceeds battery voltage.

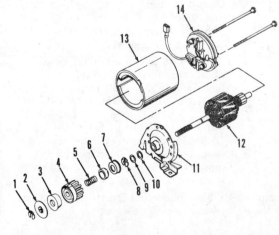

Fig. T320—Exploded view of optional electric starter motor.

1. Retainer ring
2. Drive nut
3. Pinion driver
4. Pinion gear
5. Anti-drift pin
6. Spring retainer
7. Cup
8. Retainer ring
9. Thrust washer (metal)
10. Washer (plastic)
11. End cover
12. Armature
13. Field coil housing
14. End cap & brushes

Illustrations courtesy Tecumseh Products Co.

TECUMSEH CENTRAL PARTS DISTRIBUTORS

(Arranged Alphabetically by States)
**These franchised firms carry extensive stocks of repair parts.
Contact them for name of the nearest service distributor.**

Charlie C. Jones, Inc.
Phone (602) 272-5621
2440 W. McDowell Rd.
P.O. Box 6654
Phoenix, Arizona 85005

Billou's, Inc.
Phone (209) 784-4102
1343 S. Main
Porterville, California 93257

Pacific Power Equipment Company
Phone (303) 744-7891
1441 Bayaud Ave., Unit 4
Denver, Colorado 80223

Radco Distributors, Inc.
Phone (904) 731-7957
6261 Powers Ave., P.O. Box 5459
Jacksonville, Florida 32207

Small Engine Clinic, Inc.
Phone (808) 488-0711
98019 Kam Highway, P.O. Box 427
Pearl City, Hawaii 96782

Industrial Engine & Parts
Phone (708) 263-0500
50 Noll Street
Waukegan, Illinois 60085

Medart Engines of Kansas
Phone: (913) 888-8828
15500 West 109th Street
Lenexa, Kansas 66219

Engines Southwest
Phone (318) 222-3871
215 Spring St., P.O. Box 67
Shreveport, Louisiana 71161

W.J. Connell Company
Phone (508) 543-3600
65 Green St., Rt. 106
Foxboro, Massachusetts 02035

Central Power Distributors, Inc.
Phone (612) 633-5179
2976 N. Cleveland
St. Paul, Minnesota 55113

Medart Engines of St. Louis
Phone: (314) 343-0505
100 Larkin Williams Industrial Ct.
Fenton, Missouri 63026

Original Equipment, Inc.
Phone: (406) 245-3081
905 Second Avenue, North
Billings, Montana 59103

Gardner, Inc.
Phone: (609) 860-8060
12 Melrich Rd., Rd. 3
Cranbury, New Jersey 08512

Smith Engines & Irrigation, Inc.
Phone (704) 392-3100
4250 Golf Acres Dr., P.O. Box 668985
Charlotte, North Carolina 28266

Gardner, Inc.
Phone (614) 488-7951
1150 Chesapeake Ave.
Columbus, Ohio 43212

Medart Engines of Tulsa
Phone: (918) 627-1448
7450 East 46th Place
Tulsa, Oklahoma 74115

Power Equipment Systems
Phone (503) 585-6120
2257 McGilchrist S.E., P.O. Box 629
Salem, Oregon 97308

Pitt Auto Electric Company
Phone: (412) 766-9112
2900 Stayton Street
Pittsburgh, Pennsylvania 15212

Medart Engines of Memphis
Phone: (901) 795-4365
4365 Old Lamar
Memphis, Tennessee 38118

Frank Edwards Company
Phone (801) 972-0128
1284 S. 500 West
Salt Lake City, Utah 84101

Power Equipment Systems
Phone (206) 763-8902
88 South Hudson, P.O. Box 3901
Seattle, Washington 98124

CANADIAN DISTRIBUTORS

CPT Canada Power Technology, Ltd.
Phone (403) 453-5791
13315 146th Street
Edmonton, Alberta T5L 4S8

CPT Canada Power Technology, Ltd.
Phone (416) 890-6900
161 Watline Ave.
Mississauga, Ontario L4W 2T7

TORO

THE TORO COMPANY
8111 Lyndale Ave. South
Bloomington, MN 55420

Model Series	Bore	Stroke	Displacement	Rated Power
47P	58 mm	46 mm	121 cc	2.6 kW
	(2.3 in.)	(1.8 in.)	(7.4 cu.in.)	(3.5 hp)

NOTE: Metric fasteners are used throughout engine.

ENGINE INFORMATION

The Toro (Suzuki) 47P engine model series is used on Toro walk-behind lawn mowers. The engine is an air-cooled, two-stroke, single-cylinder with a vertical crankshaft. The engine model number is stamped in the blower housing above the spark plug.

MAINTENANCE

LUBRICATION. The engine is lubricated by mixing oil with unleaded gasoline (gasoline blended with alcohol is not recommended). A good quality oil designed for two-stroke engines must be used. Fuel:oil ratio should be 50:1.

AIR CLEANER. Engine is equipped with a foam type air cleaner which should be removed and cleaned after every 50 hours of operation. Wash element in a mild detergent and water solution and squeeze out excess water. Allow element to air dry. Apply five teaspoons of SAE 30 oil to element and

squeeze element to distribute oil. Reinstall filter element.

FUEL FILTER. A fuel filter is located in the fuel tank. The filter is molded in the tank. If filter is damaged or cleaning will not remove dirt or debris, the fuel tank must be renewed. Some engines may be equipped with an inline fuel filter in the fuel hose.

SPARK PLUG. Recommended spark plug is a NGK BPMR4A spark plug or equivalent. Electrode gap should be 0.8 mm (0.032 in.). Tighten spark plug to 9-14 N·m (80-124 in.-lbs.).

CARBURETOR. The engine is equipped with a Mikuni BV 18-15 float type carburetor shown in Fig. TO11. Fuel mixture is not adjustable other than replacement of fixed jets.

To disassemble carburetor, refer to Fig. TO11 and remove retaining nut (11) and float bowl (9). Remove float pin (3) by pushing against round end of pin towards the square end of pin. Remove float (4) and fuel inlet needle (5). Unscrew pilot jet (1) and main jet (6). Note that no other parts of carburetor are

serviceable separately. Disassembly of throttle and choke plates is not recommended as the retaining screws are locked in place and the threads may be damaged during removal.

Metal parts should be cleaned in carburetor cleaner. Use compressed air to blow out all orifices and passageways. Do not use wires or drill bits to clean orifices as enlargement of the orifice could affect the calibration of the carburetor.

To check float height, invert carburetor and measure float height as shown in Fig. TO12. Float height should be 17.5 mm (11/16 in.). Bend float tab to adjust float height. Reassembly of the carburetor is the reverse of disassembly procedure.

IGNITION. A solid-state ignition system is used. Ignition timing is fixed and not adjustable. Air gap between ignition module and flywheel is adjustable. Air gap between flywheel and module should be 0.38-0.50 mm (0.015-0.020 in.).

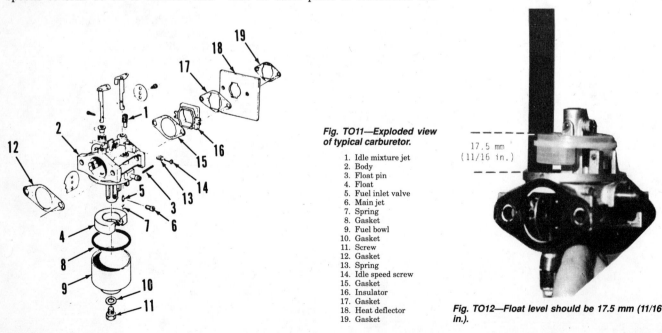

Fig. TO11—Exploded view of typical carburetor.

1. Idle mixture jet
2. Body
3. Float pin
4. Float
5. Fuel inlet valve
6. Main jet
7. Spring
8. Gasket
9. Fuel bowl
10. Gasket
11. Screw
12. Gasket
13. Spring
14. Idle speed screw
15. Gasket
16. Insulator
17. Gasket
18. Heat deflector
19. Gasket

17.5 mm
(11/16 in.)

Fig. TO12—Float level should be 17.5 mm (11/16 in.).

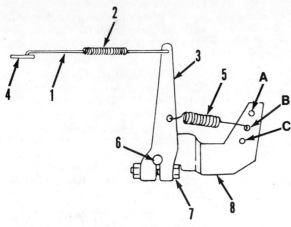

Fig. TO13—Drawing of governor linkage.

1. Rod
2. Spring
3. Governor lever
4. Throttle lever
5. Governor spring
6. Governor shaft
7. Clamp bolt
8. Bracket
A. Upper hole
B. Middle hole (3000 rpm)
C. Lower hole

GOVERNOR. To adjust governor linkage, first make certain all linkage is in good condition and tension spring (5—Fig. TO13) is not stretched or damaged. Spring (2) must pull governor lever (3) and throttle pivot (4) toward each other. Loosen clamp bolt (7) and move governor lever (3) to the right. Hold governor lever in this position and rotate governor shaft (6) in clockwise direction until it stops. Tighten clamp screw.

Correct engine speed is 3000 rpm and is obtained with tension spring (5) connected in center hole (B) of bracket (8). Connecting spring in hole (A) increases engine speed by 150 rpm and connecting spring in hole (C) decreases engine speed by 150 rpm.

REPAIRS

TIGHTENING TORQUES. Recommended tightening torque specifications are as follows:

Crankcase . . . . 9-12 N•m
(80-106 in.-lbs.)
Flywheel nut. . . . . 39-49 N•m
(29-36 ft.-lbs.)
Spark plug . . . . 9-14 N•m
(80-124 in.-lbs.)

CRANKCASE. The engine is equipped with a split crankcase design. The cylinder (2—Fig. TO14) is integral with a crankcase half. It is necessary to separate crankcase halves for access to internal engine components.

To disassemble crankcase, remove rewind starter, disconnect fuel line at carburetor and lift fuel tank from blower shroud. Remove muffler and heat shield. Remove air cleaner and housing, then remove blower shroud.

Compress mower control bar against handlebar to compress flywheel brake spring, remove cap screw retaining ignition switch and brake assembly, disconnect switch and slowly release control bar to release flywheel brake spring. Disconnect brake control cable. Remove starter cup and flywheel (use Toro puller 41-7650 or equivalent). Remove governor lever, control rod and springs, carburetor, spacer and gaskets. Remove blade, blade adapter and self-propelled components, if so equipped. Remove engine from mower.

Remove engine adapter plate (1—Fig. TO14). Unscrew crankcase retaining screws (note location of different length screws). Carefully separate crankcase halves. Remove crankshaft and piston assembly. Remove seals (9 and 18), bearing locating rings (10 and 17) and bearings (11 and 16) from crank-

Fig. TO14—Exploded view of engine.

1. Plate
2. Cylinder
3. Dowel pin
4. Piston rings
5. Retaining ring
6. Piston pin
7. Piston
8. Bearing
9. Oil seal
10. Locating ring
11. Bearing
12. Crankshaft assy.
13. Locating pin
14. Governor collar
15. Flyweight
16. Bearing
17. Locating ring
18. Oil seal
19. Crankcase half

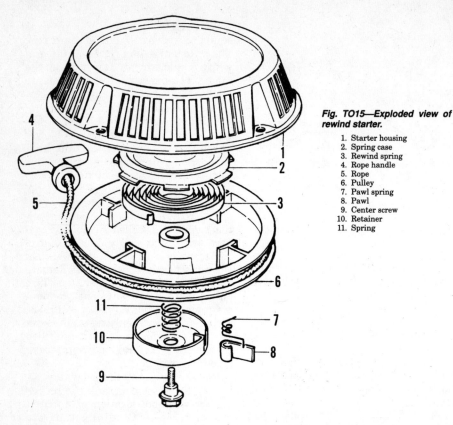

Fig. TO15—Exploded view of rewind starter.

1. Starter housing
2. Spring case
3. Rewind spring
4. Rope handle
5. Rope
6. Pulley
7. Pawl spring
8. Pawl
9. Center screw
10. Retainer
11. Spring

shaft ends. Slide governor assembly from crankshaft; use care not to lose locating pin (13) for governor flyweight collar (14). Remove retaining rings (5) from piston pin bores and remove piston pin (6) and piston (7) from connecting rod. DO NOT attempt to disassemble the three-piece crankshaft assembly (12) as crankshaft damage will occur. The crankshaft and connecting rod are serviced as an assembly.

When reassembling engine, note the following special instructions: Lubricate piston, connecting rod needle bearings and crankshaft main bearings with two-cycle engine oil. Arrow on piston crown must face toward exhaust port side of cylinder. Make certain that the piston ring end gaps are aligned with the locating studs in piston ring grooves before pushing the piston into the cylinder. Be sure that the locating pin (13) is in place in the crankshaft and that the governor collar (14) engages the pin. Apply small amount of #2 grease to lip of seals (9 and 18). Remove any burrs from either end of crankshaft, then install seals with lip facing into the crankcase. Apply a thin coat of Loctite 515 sealant to crankcase mating surfaces before mating crankcase halves. Tighten crankcase screws to 9-12 N•m (80-106 in.-lbs.).

PISTON, PIN AND RING. The engine is equipped with a cam-ground aluminum piston and two piston rings. Disassemble engine as outlined in CRANKCASE section for access to piston. When removing and installing piston rings, be careful not to overstretch and distort the rings.

Standard piston ring end gap is 0.15-0.35 mm (0.006-0.014 in.) and maximum allowable end gap is 0.7 mm (0.027 in.). Specified piston ring side clearance in piston ring grooves is 0.03-0.15 mm (0.001-0.006 in.) measured at bottom side of ring. Renew piston if side clearance is excessive.

Standard piston-to-cylinder clearance is 0.025-0.055 mm (0.0010-0.0022 in.). Piston-to-cylinder bore clearance should not exceed 0.1 mm (0.004 in.). Measure piston diameter at points perpendicular to piston pin and 2.5 mm (1 in.) from bottom of piston skirt. Measure cylinder diameter 110 mm (4.3 in.) from bottom of cylinder. Standard cylinde bore diameter is 58.000-58.115 mm (2.284-2.288 in.).

Specified piston pin diameter is 11.955-12.005 mm (0.4707-0.4726 in.).

The top piston ring has chrome plating on the outer diameter. Install piston rings with side marked "R" toward top of piston. Align the ring end gaps with the locating studs in piston ring grooves

before installing piston in cylinder. Lubricate piston pin, rings and cylinder with two-cycle engine oil when reassembling.

CYLINDER. Cylinder (2—Fig. TO14) is integral with upper crankcase half. Cylinder should be renewed if scored, excessively worn, out-of-round or otherwise damaged. Measure cylinder diameter at top of cylinder, just above intake port and just below third port at bottom. Specified cylinder diameter is 58.000-58.115 mm (2.284-2.288 in.).

CRANKSHAFT, MAIN BEARINGS AND SEALS. Crankshaft and connecting rod are serviced as a unit assembly. Crankshaft and connecting rod are not available separately. Refer to CRANKCASE section for disassembly procedure.

Crankshaft runout should be checked at a point 38.1 mm (1.5 in.) from each end of crankshaft. Standard runout is 0-0.05 mm (0-0.002 in.). Renew crankshaft if total indicated runout exceeds 0.20 mm (0.008 in.). Standard crankshaft end play is 0-0.94 mm (0-0.037 in.) and maximum allowable end play is 1.5 mm (0.060 in.). Standard crankshaft diameter at flywheel end is 19.993-20.00 mm (0.7871-0.7874 in.). Crankshaft diameter at pto end is 24.972-24.993 mm (0.9831-0.9840 in.).

REWIND STARTER. To remove rewind starter, unscrew four retaining screws and lift starter assembly off of blower housing.

To disassemble starter, remove rope handle and allow rope to wind slowly into starter housing. Remove center screw (9—Fig. TO15), then remove pawl retainer (10), spring (11), pawl (8) and spring (7). Wear appropriate safety eyewear and gloves when removing pulley from starter or when separating rewind spring (3) and spring case (2) from pulley.

When reassembling starter, install spring (7) and pawl (8) so the spring end forces the pawl toward the center of the pulley. Apply Loctite 242 to threads of center screw (9). Tighten screw to 8-11 N•m (71-97 in.-lbs.).

FLYWHEEL BRAKE. Some engines are equipped with a pad type flywheel brake. The brake should stop the engine within three seconds when the operator releases mower safety control and the speed control is in high speed position. Adjust cable as needed for proper operation.

TORO

Model Series	Bore	Stroke	Displacement	Rated Power
VM	64.0 mm	44.0 mm	141 cc	3.0 kW
	(2.52 in.)	(1.73 in.)	(8.6 cu.in.)	(4 hp)

NOTE: Metric fasteners are used throughout engine.

ENGINE INFORMATION

The Toro (Suzuki) VM engine model series is used on Toro walk-behind lawn mowers. The engine is an air-cooled, four-stroke, single-cylinder engine with a vertical crankshaft and an overhead valve system. Engine identification numbers are stamped in the blower housing.

MAINTENANCE

LUBRICATION. The engine is lubricated by oil supplied by a rotor type oil pump located in the bottom of the crankcase, as well as by a slinger driven by the camshaft gear.

Change oil after first two hours of operation and after every 25 hours of operation or at least once each operating season.

Engine oil level should be maintained at full mark on dipstick. Engine oil should meet or exceed latest API service classification. Manufacturer recommends SAE 30 or SAE 10W-30 oil.

Crankcase capacity is 0.55 L (18.6 fl. oz.). Fill engine with oil so oil level reaches, but does not exceed, full mark on dipstick.

AIR CLEANER. Engine may be equipped with a foam filter element or with a paper filter element and a foam precleaner element.

Foam filter elements should be cleaned after every 25 hours of operation. Paper filter elements should be cleaned or replaced after every 50 hours of operation. Service either type filter element more frequently if severe operating conditions are encountered.

Clean a foam filter in soapy water and squeeze until dry. Inspect filter for tears and holes or any other opening. Discard filter if it cannot be cleaned satisfactorily or if it is torn or otherwise damaged. Pour clean engine oil into the filter, then squeeze to remove excess oil and distribute oil throughout.

Clean a paper filter by tapping gently to dislodge accumulated dirt. Renew filter if dirty or damaged. Do not apply oil to the foam precleaner element or the paper element.

FUEL FILTER. The engine is equipped with an inline fuel filter in the fuel hose. Renew filter if dirty or damaged.

CRANKCASE BREATHER. The engine is equipped with a crankcase breather that provides a vacuum for the crankcase. A reed valve located in the top of the crankcase acts as a one-way valve to maintain crankcase vacuum. The breather system must operate properly or excessive oil consumption may result. Remove the flywheel for access to breather cover.

SPARK PLUG. Recommended spark plug is NGK BPR6ES. Specified spark plug electrode gap is 0.8 mm (0.032 in.).

CARBURETOR. The engine is equipped with a Mikuni BV 18-13 float type carburetor. Refer to Fig. TO51 for exploded view of carburetor. Carburetor is equipped with a fixed idle mixture jet (7) and a fixed high speed jet (22).

Initial adjustment of idle pilot screw (10) should be one turn open from a lightly seated position. Make final carburetor adjustment with engine at normal operating temperature and running. Adjust pilot screw to obtain smoothest idle operation and acceleration.

To remove carburetor, drain fuel and remove fuel tank. Remove air cleaner assembly. Note the connecting points of carburetor control linkage and springs to insure correct reassembly, then disconnect choke rod and throttle rod and remove carburetor. To install carburetor, reverse the removal procedure. Note that a gasket should be installed on both sides of insulator block (9—Fig. TO51) and raised rib on air cleaner gasket (24) should be toward the air cleaner.

Remove float pin by pushing against round end of pin towards the square end of pin. Float is plastic and float level is not adjustable. Fuel inlet valve (21) is renewable but valve seat is not remov-

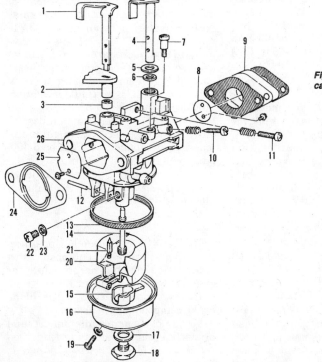

Fig. TO51—Exploded view of carburetor.

1. Choke shaft
2. Choke detent
3. Spacer
4. Throttle shaft
5. Washer
6. Dust seal
7. Idle pilot jet
8. Choke plate
9. Insulator block
10. Pilot screw
11. Idle speed screw
12. Float pin
13. Gasket
14. Nozzle
15. Spacer
16. Float bowl
17. Gasket
18. Screw
19. Drain screw
20. Float
21. Fuel inlet valve
22. Main jet
23. Spacer
24. Gasket

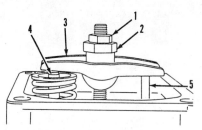

Fig. TO52—Drawing of governor and speed control linkage.

1. Screw	6. Speed control lever
2. Governor lever	7. Maximum speed screw
3. Governor shaft	8. Stop screw
4. Nut	9. Tang
5. Alignment hole	10. Governor spring

Fig. TO53—View of rocker arm and related parts.

1. Jam nut	
2. Adjustment nut	4. Valve stem clearance
3. Rocker arm	5. Push rod

able and body must be renewed if seat is damaged.

Metal parts should be cleaned in carburetor cleaner. Use compressed air to blow out all orifices and passageways. Do not use wires or drill bits to clean orifices as enlargement of the orifice could affect the calibration of the carburetor.

Apply Loctite 271 thread locking compound to threads of choke and throttle plate screws when assembling carburetor.

SPEED CONTROL CABLE. When the speed control is in the "FAST" position, holes (5—Fig. TO52) in speed control lever (6) and bracket should be aligned. Unscrew throttle cable clamp screw (1) and relocate cable to align holes, then retighten clamp screw.

GOVERNOR. All engines are equipped with a mechanical (flyweight) type governor. Maximum governed speed is adjusted by turning adjusting screw (7—Fig. TO52). With speed control in the "FAST" position so holes (5) in speed control lever (6) and bracket are aligned, rotate adjusting screw so engine runs at 3000 rpm. Rotate stop screw (8) so screw end is 0.00-0.05 mm (0.000-0.020 in.) from tang (9).

To adjust the governor linkage, refer to Fig. TO52 and loosen governor lever clamp nut (4). Rotate governor shaft (3) clockwise as far as possible. Move the governor lever until carburetor throttle shaft is in wide open position, then tighten governor lever clamp nut.

IGNITION. The engine is equipped with a breakerless ignition system. All components are located outside the flywheel. Armature air gap should be 0.38-0.50 mm (0.015-0.020 in.).

VALVE ADJUSTMENT. Specified clearance between rocker arm (3—Fig. TO53) and end of valve stem (4) is 0.025-0.13 mm (0.001-0.005 in.) for both valves. To adjust valve clearance, remove rocker arm cover. Rotate crankshaft so piston is at top dead center (TDC) on compression stroke. Insert a suitable thickness feeler gauge between rocker arm and end of valve stem. Loosen rocker arm jam nut (1) and turn adjusting nut (2) to obtain desired clearance (a slight drag should be felt when withdrawing the feeler gauge). Tighten jam nut and recheck clearance. Install rocker arm cover. Note that cutaway portion of cover mounting flange must be adjacent to spark plug.

REPAIRS

TIGHTENING TORQUES. Recommended tightening torque specifications are as follows:

Connecting rod . . . . 5.9-9.8 N•m
　　　　　　　　　　　　　　(54-118 in.-lbs.)
Cylinder head . . . . 17.6-27.4 N•m
　　　　　　　　　　　　　　(156-242 in.-lbs.)
Flywheel . . . . 55-63 N•m
　　　　　　　　　　　　　　(41-46 ft.-lbs.)
Oil pan . . . . 3.9-6.8 N•m
　　　　　　　　　　　　　　(35-60 in.-lbs.)

CYLINDER HEAD. To remove the cylinder head, first remove fuel tank, air cleaner, blower housing, carburetor, speed control bracket, muffler, and rocker arm cover. Loosen jam nuts, and remove rocker arm pivots (4—Fig. TO54), rocker arms (5) and push rods (6); mark all parts so they can be returned to original location. Unscrew cylinder head screws and remove cylinder head (8).

Clean cylinder head thoroughly, then check for cracks, distortion or other damage. Cylinder head warpage should not exceed 0.030 mm (0.0012 in.).

Install cylinder head using a new head gasket. Tighten cylinder head screws adjacent to spark plug hole first, then tighten screws adjacent to push rod opening in a crossing pattern. Final

torque reading should be 17.6-27.4 N•m (156-242 in.-lbs.). Adjust valve clearance as outlined in VALVE ADJUSTMENT section. Install rocker arm cover with the cut out side of the cover toward the spark plug. Be sure that a flat washer is installed on each rocker cover retaining screw, and tighten screws to 4.1-6.7 N•m (36-60 in.-lbs.).

VALVE SYSTEM. To remove valves (11 and 12—Fig. TO54) from cylinder head, compress valve springs (3) by hand or with a suitable spring compressor tool and remove slotted retainers (2). The intake and exhaust valve springs and retainers are identical and interchangeable.

Valve face and seat angles are 45°. Standard valve seat width is 0.80-1.00 mm (0.032-0.039 in.).

Standard valve stem diameter is 5.460-5.475 mm (0.2150-0.2156 in.) for intake valve and 5.440-5.455 mm (0.2142-0.2148 in.) for exhaust valve.

Standard valve guide inside diameter is 5.500-5.512 mm (0.2165-0.2170 in.) for both valves. Standard valve stem-to-guide clearance is 0.025-0.052 mm (0.0010-0.0020 in.) for intake and 0.045-0.072 mm (0.0018-0.0028 in.) for exhaust. Maximum valve stem-to-guide clearance is 0.080 mm (0.0032 in.) for intake valve and 0.100 mm (0.0039 in.) for exhaust valve. Oversize valve guides may be installed.

Valve guides (9—Fig. TO54) can be renewed using Toro valve guide driver 81-4880 or other suitable tool. Press or drive guide out towards rocker arm side of head. Use Toro reamer 81-4850 or other suitable reamer so valve guide bore diameter in head is 9.300-9.315 mm (0.3661-0.3667 in.). Press or drive guide in from rocker arm side of cylinder head so that distance from head gasket surface of cylinder head to lower end of valve guide is 27.5 mm (1.08 in.). Ream guide with Toro reamer 81-4840 or other suitably sized reamer to obtain desired valve stem clearance. Valve guide finished inside diameter should be 5.500-5.512 mm (0.2165-0.2170 in.).

Valve spring free length should be 32.0-34.0 mm (1.26-1.34 in.). Minimum allowable valve spring length is 31.0 mm (1.22 in.).

R&R ENGINE. The following procedure applies to engines equipped with a blade brake clutch. The procedure for engines with a flywheel brake (zone start) is similar, although the blade brake components are absent.

To remove engine, disconnect and properly ground spark plug lead. Disconnect any electrical wires to engine. Remove blade mounting assembly and

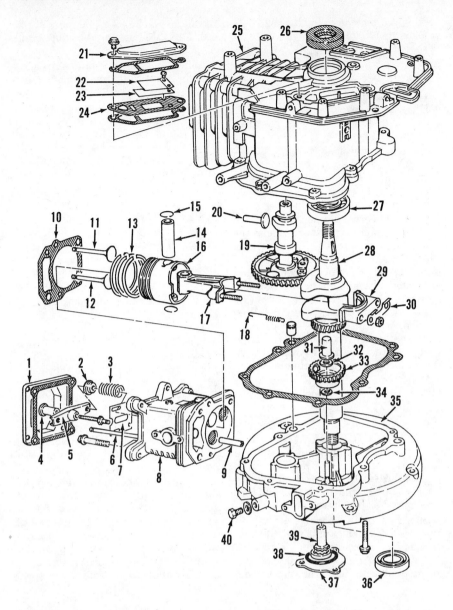

Fig. TO54—Exploded view of engine assembly.

1. Rocker cover	21. Breather cover
2. Valve retainer	22. Reed stop
3. Valve spring	23. Reed valve
4. Rocker arm pivot	24. Breather plate
5. Rocker arm	25. Engine block
6. Push rod	26. Oil seal
7. Push rod guide	27. Ball bearing
8. Cylinder head	28. Crankshaft
9. Valve guide	29. Rod cap
10. Head gasket	30. Lock plate
11. Exhaust valve	31. Thrust sleeve
12. Intake valve	32. Washer
13. Piston rings	33. Governor
14. Piston pin	34. Washer
15. Retaining ring	35. Oil pan
16. Piston	36. Oil seal
17. Connecting rod	37. Oil pump cover
18. Compression release spring	38. "O" ring
19. Camshaft	39. Oil pump rotors
20. Tappet	40. Oil drain plug

bottom cover. Unscrew crankshaft nut, push back belt idler, and remove bearing block, brake drum and flywheel plate. Separate transmission drive belt from drive pulley and remove drive pulley. If desired, drain engine oil. Unscrew engine mounting screws and remove engine while disconnecting throttle cable.

Reverse removal procedure to reinstall engine while noting the following: Pegs on drive pulley must be toward bottom of crankshaft. Note that center of flywheel is tapered to fit crankshaft. Install flywheel so pegs on drive pulley fit in slots in flywheel plate.

CAUTION: Drive pulley will be damaged if pegs on pulley do not fit properly in flywheel slots during assembly. Before tightening crankshaft nut, rotate flywheel slightly. If pegs are in slots, flywheel rotation will be stopped by pegs.

Stepped side of bearing block fits in center of brake drum. Note that flat side on head of blade mounting bolts must be toward center of brake drum. Tighten bearing block retaining nut to 79 N•m (58 ft.-lbs.). When tightening nut, be sure blade brake drive belt is not trapped between drum and flywheel. Tighten blade retaining nuts to 37 N•m (27 ft.-lbs.).

CAMSHAFT. To remove camshaft (19—Fig. TO54) proceed as follows: Remove engine as previously outlined. Drain oil from crankcase. Clean pto end of crankshaft and remove any burrs or rust. Remove rocker arm cover (1), rocker arms (5) and push rods (6); mark all parts so they can be returned to original position. Unscrew fasteners and remove oil pan (35). Rotate crankshaft so timing marks on crankshaft and camshaft gears are aligned (this

will position valve tappets out of way). Withdraw camshaft and remove tappets (20). Mark the tappets so they can installed in their original positions if reused.

Specified camshaft lobe height is 30.517-30.577 mm (1.2014-1.2038 in.). Renew camshaft and tappets if lobes are excessively worn or scored.

Reverse removal procedure to reassemble components. Lubricate tappets and camshaft with engine oil during assembly. Install camshaft while aligning timing marks on crankshaft and camshaft gears. Note that roll pin in end of camshaft must engage oil pump drive shaft in oil pan during assembly. Tighten oil pan screws to 3.9-6.8 N•m (35-60 in.-lbs.) in a crossing pattern. Do not force mating of oil pan with crankcase. Be sure thrust sleeve (31) on governor does not fall off during assembly. Reassemble remainder of components.

PISTON, PIN AND RINGS. To remove piston and rod assembly, remove engine as previously outlined. Remove cylinder head and camshaft as previously outlined. Unscrew connecting rod cap nuts and remove piston and rod.

Measure piston diameter at points perpendicular to piston pin and 14 mm (0.55 in.) from bottom of piston skirt. Specified piston diameter is 63.960-63.975 mm (2.5181-2.5187 in.) with a wear limit of 63.915 mm (2.5163 in.). Specified piston-to-cylinder bore clearance is 0.025-0.055 mm (0.0010-0.0022 in.) with a wear limit of 0.120 mm (0.0047 in.). Oversize pistons are not available.

Specified piston ring end gap for compression rings is 0.2-0.4 mm (0.008-0.016 in.) with a limit of 0.70 mm (0.028 in.). Piston ring groove width for compression rings should be 1.52-1.54 mm (0.060-0.061 in.) and ring thickness should be 1.47-1.49 mm (0.058-0.059 in.). Specified piston ring side clearance in groove is 0.03-0.07 mm (0.001-0.003 in.) with a limit of 0.10 mm (0.004 in.).

Specified piston pin diameter is 14.995-15.000 mm (0.5904-0.5906 in.). Specified piston pin bore diameter is 15.006-15.014 mm (0.5908-0.5911 in.).

Top compression ring is chrome plated. One side of piston ring is marked with the letter "N" to indicate correct installation position. Install the piston rings on piston so that "N" on side of ring is toward piston crown.

When assembling piston and connecting rod, note that arrow on piston crown and arrow on side of connecting rod must point in same direction. Lubricate piston rings and pin with engine oil, then install piston and rod assembly in engine with arrow on piston crown toward push rod side of engine. Install rod cap on connecting rod so match marks on rod and cap are aligned. Tighten connecting rod nuts to 5.9-9.8 N·m (54-118 in.-lbs.).

Install camshaft and cylinder head as previously outlined.

CONNECTING ROD. Connecting rod rides directly on crankshaft journal. Connecting rod and piston are removed as an assembly as outlined in previous section.

Specified clearance between piston pin and connecting rod small end is 0.006-0.019 mm (0.0002-0.0007 in.) with a maximum allowable clearance of 0.050 mm (0.0020 in.).

Specified clearance between crankpin journal and connecting rod bore is 0.015-0.035 mm (0.0006-0.0014 in.) with a maximum allowable clearance of 0.080 mm (0.0031 in.). Specified connecting rod big end diameter is 26.015-26.025 mm (1.0242-1.0246 in.).

Install connecting rod as outlined in previous section.

GOVERNOR. The internal centrifugal flyweight governor is mounted on the oil pan. The governor is driven by the camshaft gear.

To remove governor assembly (33—Fig. FO54), first separate oil pan (35) from crankcase. Use two screwdrivers to snap governor gear and flyweight assembly off governor stub shaft.

Install governor by pushing it down on the stub shaft until it snaps onto the shaft locating groove. Be sure that thrust washers (32 and 34) are positioned on either side of governor. A small amount of grease may be used to hold the thrust washer (32) and sleeve (31) in place on stub shaft. When installing oil pan, it may be necessary to turn crankshaft slightly to mesh the camshaft gear teeth with the governor gear. Do not force the oil pan into place.

Refer to MAINTENANCE section for external governor linkage adjustment.

CRANKSHAFT, MAIN BEARINGS AND OIL SEALS. The crankshaft is supported at flywheel end by a ball bearing (27—Fig. TO54) located in the crankcase. The crankshaft may be removed after removing piston and connecting rod as previously outlined.

Specified main bearing journal diameter is 21.960-21.980 mm (0.8646-0.8654 in.) for flywheel end and 24.959-24.980 mm (0.9826-0.9835 in.) for pto end. Specified crankpin journal diameter is 25.99-26.00 mm (1.023-1.024 in.). Maximum allowable crankshaft runout is 0.05 mm (0.002 in.).

When installing crankshaft, make certain crankshaft and camshaft gear timing marks are aligned.

Inspect ball bearing and renew if rough, loose or damaged. Install oil seals (26 and 36) so lip is toward main bearing. Lubricate oil seals with engine oil prior to installing crankshaft. Be sure to align timing marks on crankshaft gear and camshaft gear.

CYLINDER. Renew cylinder if wear in bore exceeds 0.100 mm (0.0039 in.) or if out-of-round exceeds 0.030 mm (0.0012 in.).

OIL PUMP. A rotor type oil pump driven by the camshaft is located in the bottom of the oil pan.

Remove engine from equipment for access to oil pump cover (37—Fig. TO54). Remove cover and extract pump rotors (39). Mark rotors so they can be reinstalled in their original position. Renew any components which are damaged or excessively worn.

REWIND STARTER. Refer to Fig. TO55 for exploded view of rewind starter.

To disassemble starter, remove rope handle and allow rope to wind into starter. Unscrew retainer nut (14). Remove retainer (13), pawl guide (12), spring (11), spacer (10), washer (9), pawl (8) and pivot pin (7). Wear appropriate safety eyewear and gloves before disengaging pulley (6) from starter as spring

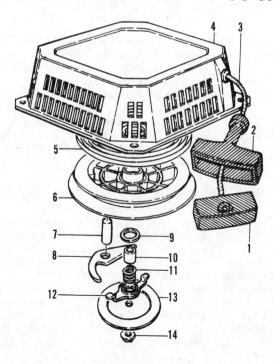

Fig. TO55—Exploded view of rewind starter.

1. Rope retainer
2. Rope handle
3. Rope
4. Starter housing
5. Rewind spring
6. Pulley
7. Pivot pin
8. Pawl
9. Washer
10. Spacer
11. Spring
12. Pawl guide
13. Retainer
14. Locknut

(5) may uncoil uncontrolled. Place shop towel around pulley and lift pulley out of housing; spring should remain with pulley.

Inspect components for damage and excessive wear. A new starter rope is $5/32$ inch (4 mm) in diameter and 70 inches (175 cm) long. A new rewind spring is provided with a retainer to hold it in the coiled position. When replacing the spring, first position it in the pulley, then remove the retainer. Note that rewind spring coils wind in counterclockwise direction from outer end. Install rope on pulley in a counterclockwise direction as viewed from engagement side of pulley. Be sure that pawl guide (12) is positioned as shown in Fig. TO55 when installing in pulley (6). Apply Loctite 242 to threads of retaining nut (14).

With starter assembled, pass rope through rope outlet and install rope handle. To apply spring tension to pulley, pull a loop of rope into notch in pulley and rotate pulley counterclockwise a couple of turns, then pull rope out of notch and allow rope to wind into starter housing. Rope handle should be snug against housing, if not, increase spring tension by rotating pulley another turn.

ELECTRIC STARTER. Some models may be equipped with an electric starter motor. Individual parts of starter drive assembly are available, however, motor unit is available only as a unit assembly. If starter malfunctions, inspect and test components of starter circuit before replacing starter motor.

ALTERNATOR. Some engines may be equipped with an alternator. To test alternator output, disconnect alternator lead and connect a voltmeter to alternator lead and engine ground. With engine running at 3000 rpm, alternator output should be 13.2 volts DC. Test rectifier if output is low or zero.

To check for a faulty rectifier diode, connect positive lead of an ohmmeter to alternator lead and negative ohmmeter lead to alternator stator. Ohmmeter should read infinity. With ohmmeter leads reversed, ohmmeter should indicate approximately 700 ohms.

Alternator and rectifier are available only as a unit assembly.

TORO

Model Series	Bore	Stroke	Displacement	Rated Power
GTS 150	2.56 in.	1.78 in.	9.2 cu.in.	5.5 hp
	(65.0 mm)	(45.2 mm)	(150 cc)	(4.1 kW)

ENGINE INFORMATION

The Toro GTS 150 engine model series is used on Toro walk-behind lawn mowers. The engine is an air-cooled, four-stroke, single-cylinder engine with a vertical crankshaft and an overhead valve system. Engine identification numbers are stamped in the rocker arm cover as shown in Fig. TO100. Engine models 97772 and 97777 are covered in this section.

MAINTENANCE

LUBRICATION. The engine is lubricated by oil supplied by a rotor type oil pump located in the bottom of the crankcase, as well as by a slinger driven by the camshaft gear.

Change oil after first two hours of operation and after every 25 hours of operation or at least once each operating season. Change oil weekly or after every 25 hours of operation if equipment undergoes severe usage.

Engine oil level should be maintained at full mark on dipstick. Engine oil should meet or exceed latest API service classification. Manufacturer recommends SAE 30 or SAE 10W-30 oil.

Crankcase capacity is 25 fl. oz. (0.74 L). Fill engine with oil so oil level reaches, but does not exceed, full mark on dipstick.

AIR CLEANER. The air cleaner consists of a canister and the foam filter element it contains.

The filter element should be cleaned after every 25 hours of operation, or more frequently if severe operating conditions are encountered. Clean filter in soapy water and squeeze until dry. Inspect filter for tears and holes or any other opening. Discard filter if it cannot be cleaned satisfactorily or if it is torn or otherwise damaged. Pour clean engine oil into the filter, then squeeze to remove excess oil and distribute oil throughout.

FUEL FILTER. A fuel filter is located in the fuel tank. The filter is molded in the tank. If filter is damaged or cleaning will not remove dirt or debris, the fuel tank must be renewed.

CRANKCASE BREATHER. The crankcase breather is built into the tappet chamber cover (C—Fig. TO101). A fiber disc acts as a one-way valve. Clearance between fiber disc valve and breather body should not exceed 0.045 inch (1.14 mm). If it is possible to insert a 0.045 inch (1.14 mm) wire gauge (W—Fig. TO102) between disc and breather body, renew breather assembly. Do not use excessive force when measuring gap. Disc should not stick or bind during operation. Renew if distorted or damaged. Inspect breather tube for leakage.

SPARK PLUG. Recommended spark plug is Champion RC12YC. Specified spark plug electrode gap is 0.030 inch (0.76 mm).

CARBURETOR. The engine is equipped with a Walbro LMS float type carburetor.

Adjustment. Initial setting of idle mixture screw (IM—Fig. TO103) is 1¼ turns out from a lightly seated position. Run engine until normal operating temperature is attained. Be sure choke is open. Run engine with speed control in slow position and adjust idle speed screw (IS) so engine speed is 1500 rpm. Turn idle mixture screw clockwise until engine begins to stumble and note screw position. Turn idle mixture screw counterclockwise until engine begins to stumble and note screw position. Turn the idle mixture screw clockwise to a position that is midway from clockwise (lean) and counterclockwise (rich) positions. With engine running at idle, rapidly move speed control to full throttle position. If engine stumbles or hesitates, slightly turn idle mixture screw counterclockwise and repeat test. Re-

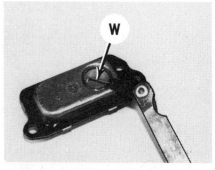

Fig. TO102—Clearance between fiber disc valve and crankcase breather housing must be less than 0.045 inch (1.15 mm). A spark plug wire gauge (W) may be used to check clearance as shown, but do not apply pressure against disc valve.

Fig. TO100—Engine identification numbers are stamped in rocker arm cover.

Fig. TO101—The crankcase breather is located in valve tappet cover (C).

Fig. TO103—View showing location of idle mixture screw (IM), idle speed screw (IS) and maximum governed speed screw (M).

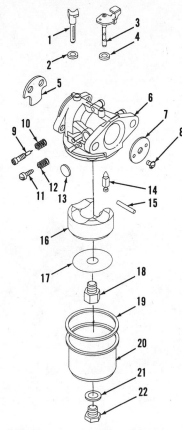

Fig. TO104—Exploded view of carburetor.

1. Choke shaft	12. Spring
2. Washer	13. Welch plug
3. Throttle shaft	14. Fuel inlet valve
4. Washer	15. Pin
5. Choke plate	16. Float
6. Body	17. Float stop plate
7. Throttle plate	18. Main jet
8. Screw	19. Gasket
9. Idle mixture screw	20. Fuel bowl
10. Spring	21. Gasket
11. Idle speed screw	22. Bowl screw

check idle speed and, if necessary, readjust idle speed screw.

High speed mixture is controlled by a fixed main jet (18—Fig. TO104) that is not adjustable.

R&R And Overhaul. Before removing carburetor from the engine, carefully note the position of all control linkage and springs to insure correct reassembly. Remove air cleaner assembly from the carburetor. Remove fuel from tank and disconnect fuel line from carburetor. Disconnect choke spring. Remove two screws retaining carburetor, pull carburetor away from engine and disconnect governor link rod.

Disassembly of carburetor is evident after inspection and referral to Fig. TO104. Note that choke plate (5) is retained in a slot in choke shaft with a slight interference fit. The choke plate can be removed by pulling it out of the slot. The throttle plate (7) is retained with a screw. To remove the Welch plug (13), pierce the plug with a small chisel

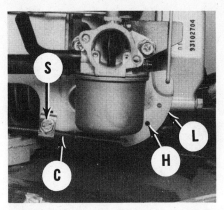

Fig. TO105—When speed control is in "FAST" position, holes (H) in throttle lever (L) and bracket should be aligned. Loosen clamp screw (S) to reposition cable housing (C).

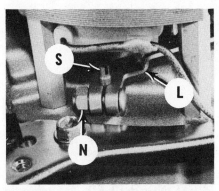

Fig. TO106—View of governor lever (L), shaft (S) and clamp nut (N). Refer to text for adjustment.

or other sharp pointed tool, then pry out the plug, but do not damage underlying metal. The fuel inlet valve seat can be removed by inserting a hooked wire through opening in the seat and pulling the seat out of the carburetor body.

Metal parts may be cleaned in carburetor cleaner. Direct compressed air through orifices and passageways in the opposite direction of normal fuel flow. Do not use wires or drill bits to clean orifices as enlargement of the orifice could affect the calibration of the carburetor. Inspect carburetor and renew any damaged or excessively worn components. The body must be replaced if there is excessive throttle shaft or choke shaft play as bushings are not available.

Install Welch plug using a punch that is larger in diameter than the Welch plug. Be careful not to indent the plug; the plug should be flat after installation. After plug is installed, seal outside edge of plug with fingernail polish or other suitable sealant. Use a 1/8 inch (3 mm) diameter rod to install the fuel inlet valve seat. The groove on the seat must be down (towards carburetor bore). Push in the seat until it bottoms. Install the choke plate so the numbers are visible when the choke is closed.

Install the throttle plate so the numbers are visible and towards the idle mixture screw when the throttle is closed. Apply Loctite 271 thread locking compound on throttle plate retaining screw. The float level is not adjustable. If the float is not approximately parallel with the body when the carburetor is inverted, then the float, fuel inlet valve and/or valve seat must be replaced. Float stop plate (17) is secured by main jet (18).

To install carburetor, reverse the removal procedure. Make certain that the gasket is in place between the air cleaner housing and the carburetor and that the breather vent hose on the back of the air cleaner housing mates with the breather vent tube.

SPEED CONTROL CABLE. When the speed control is in the "FAST" position, holes (H—Fig. TO105) in throttle lever (L) and bracket should be aligned. Unscrew throttle cable clamp screw (S) and relocate cable to align holes, then retighten clamp screw.

GOVERNOR. To adjust governor linkage, loosen governor lever clamp nut (N—Fig. TO106). Move throttle lever (L—Fig. TO105) to fast position and insert a 1/8 inch rod into holes (H) in the lever and bracket. Rotate governor shaft (S—Fig. TO106) counterclockwise as far as possible, hold shaft and tighten clamp nut.

Maximum governed speed is adjusted by turning screw (M—Fig. TO103). Maximum governed speed should be 3000 rpm.

IGNITION. The engine is equipped with a breakerless ignition system. All components are located outside the flywheel. Armature air gap should be 0.006-0.012 inch (0.15-0.30 mm).

VALVE ADJUSTMENT. Remove rocker arm cover. Remove spark plug. Rotate crankshaft so piston is at top dead center on compression stroke. Insert a suitable measuring device through spark plug hole, then rotate crankshaft clockwise as viewed at flywheel end so piston is 1/4 inch (6.4 mm) below TDC to prevent interference by the compression release mechanism with the exhaust valve. Clearance between rocker arm (R—Fig. TO107) and valve stem cap (V) should be 0.005-0.007 inch (0.12-0.18 mm) for intake and exhaust. Check clearance using feeler gauges. Loosen lock screw (S) and turn rocker arm pivot nut (N) to obtain desired clearance. Hold pivot nut and retighten lock screw to 45 in.-lbs. (5.1 N•m) torque. Install rocker arm cover

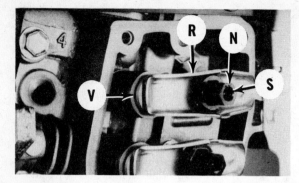

Fig. TO107—Loosen lock screw (S) and rotate adjusting nut (N) to adjust valve clearance. Clearance between rocker arm (R) and valve stem cap (V) should be 0.005-0.007 inch (0.12-0.18 mm) for intake and exhaust.

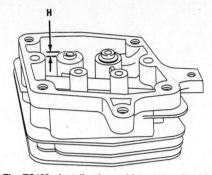

Fig. TO109—Install valve guide so top of guide protrudes (H) above boss 0.120-0.150 inch (3.05-3.81 mm).

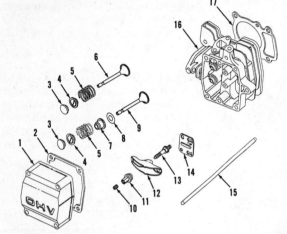

Fig. TO108—Exploded view of cylinder head assembly.

1. Rocker cover
2. Gasket
3. Valve cap
4. Valve retainer
5. Valve spring
6. Exhaust valve
7. Valve seal
8. Gasket
9. Intake valve
10. Lock screw
11. Adjusting nut
12. Rocker arm
13. Stud
14. Push rod guide
15. Push rod
16. Cylinder head
17. Head gasket

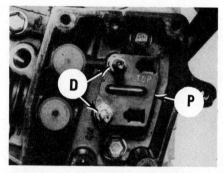

Fig. TO110—Install push rod guide plate (P) so "TOP" is toward flywheel side of head.

and tighten cover screws in a diagonal sequence to 45 in.-lbs. (5.1 N•m) torque.

NOTE: Pivot nut is 10 mm and lock screw is Torx design.

CYLINDER HEAD. The cylinder head should be removed periodically and cleaned of deposits.

REPAIRS

TIGHTENING TORQUES. Recommended tightening torque specifications are as follows:

Carburetor mounting screws 60 in.-lbs.
(6.8 N•m)
Connecting rod . . . . 100 in.-lbs.
(11.3 N•m)
Cylinder head . . . . 160 in.-lbs.
(18.1 N•m)
Flywheel nut . . . . 60 ft.-lbs.
(81.6 N•m)
Muffler . . . . 85 in.-lbs.
(9.6 N•m)
Oil pan . . . . 85 in.-lbs.
(9.6 N•m)
Rocker arm cover . . . . 45 in.-lbs.
(5.1 N•m)
Rocker arm studs . . . . 110 in.-lbs.
(12.4 N•m)
Spark plug . . . . 170 in.-lbs.
(19.2 N•m)

CYLINDER HEAD. To remove the cylinder head, first remove fuel tank, air cleaner, blower housing, carburetor, speed control bracket, muffler, and rocker arm cover. Loosen lock screws (10—Fig. TO108), and remove rocker arm pivot nuts (11), rocker arms (12) and push rods (15); mark all parts so they can be returned to original location. Unscrew cylinder head screws and remove cylinder head (16). Clean cylinder head thoroughly, then check for cracks, distortion or other damage.

When reinstalling cylinder head, do not apply sealer to cylinder head gasket. Be sure the cylinder head alignment pins are installed in cylinder block. Install the push rods making sure that they are in the valve tappets, then install the cylinder head over the push rods. Tighten cylinder head bolts in three steps following sequence of numbers embossed on cylinder head. Final torque reading should be 160 in.-lbs. (18.1 N•m).

VALVE SYSTEM. Valves are actuated by rocker arms mounted on studs threaded into the cylinder head. To remove valves from cylinder head, remove valve wear caps (3—Fig. TO108). Depress valve springs until slot in valve spring retainer (4) can be aligned with end of valve stem. Release spring pres-

sure and remove retainer, spring and valve from cylinder head.

Valve face and seat angles are 45° for intake and exhaust. Standard seat width is 0.060 inch (1.5 mm).

The cylinder head is equipped with renewable valve guides for both valves. Renew valve guide if inside diameter is 0.240 inch (6.10 mm) or more. Guides may be installed either way up. Top of guide should protrude 0.120-0.150 inch (3.05-3.81 mm) as shown in Fig. TO109.

Install push rod guide plate (P—Fig. TO110) so "TOP" mark is toward flywheel side of cylinder head. Rocker arm studs (D) are screwed into cylinder head. Hardening sealant should be applied to threads contacting cylinder head.

Install gasket (8—Fig. TO108) and valve seal (7) on intake valve. Do not lubricate wear caps (3).

R&R ENGINE. The following procedure applies to engines equipped with a blade brake clutch. The procedure for engines with a flywheel brake (zone start) is similar, although the blade brake components are absent.

To remove engine, disconnect and properly ground spark plug lead. Disconnect any electrical wires to engine. Remove blade mounting assembly and bottom cover (C—Fig. TO111). Unscrew nut (N—Fig. TO112), push back belt idler (I), and remove bearing block (B),

Fig. TO111—Remove cover (C) for access to mower drive components.

Fig. TO113—Pegs (G) on drive pulley (P) must engage slots (S—Fig. TO114) in flywheel.

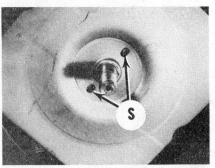

Fig. TO114—Be sure pegs (G—Fig. TO113) on drive pulley engage slots (S) in flywheel.

brake drum (D) and flywheel (F). Separate transmission drive belt from drive pulley (P—Fig. TO113) and remove drive pulley. If desired, drain engine oil. Unscrew engine mounting screws and remove engine while disconnecting throttle cable.

Reverse removal procedure to reinstall engine while noting the following: Pegs (G—Fig. TO113) on drive pulley must be toward bottom of crankshaft. Note that center of flywheel is tapered to fit crankshaft. Install flywheel so pegs on drive pulley fit in slots (S—Fig. TO114) on flywheel.

CAUTION: Drive pulley will be damaged if pegs on pulley do not fit properly in flywheel slots during assembly. Before tightening bearing block nut, rotate flywheel slightly. If pegs are in slots, flywheel rotation will be stopped by pegs.

Stepped side of bearing block (B—Fig. TO112) fits in center of brake drum. Note that flat side on head of blade mounting bolts must be toward center of brake drum. Tighten bearing block retaining nut to 58 ft.-lbs. (79 N•m).

When tightening nut, be sure blade drive belt is not trapped between drum and flywheel. Tighten blade retaining nuts to 27 ft.-lbs. (37 N•m).

CAMSHAFT. To remove camshaft (26—Fig. TO115) proceed as follows: Remove engine as previously outlined. Drain engine oil. Clean pto end of crankshaft and remove any burrs or rust. Remove rocker arm cover, rocker arms and push rods; mark all parts so they can be returned to original position. Unscrew fasteners and remove oil pan. Remove governor assembly (G—Fig. TO116). Rotate crankshaft so timing

marks (M—Fig. TO117) on crankshaft and camshaft gears are aligned (this will position valve tappets out of way). Withdraw camshaft and remove tappets. Mark the tappets so they can be reinstalled in their original position.

Renew camshaft if either camshaft bearing journal diameter is 0.615 inch (15.62 mm) or less. Renew camshaft if lobes are excessively worn or scored.

Lubricate valve tappets with SAE 30 engine oil prior to installation. Lubricate camshaft with SAE30 engine oil, then install camshaft while aligning timing marks (M—Fig. TO117) on crankshaft and camshaft gears. Install governor assembly on camshaft. Make certain that the governor plunger is positioned against the governor shaft arm and that the flyweights move freely. Note that roll pin (P) in end of camshaft must engage oil pump drive shaft in oil pan during assembly. Install oil pan and apply nonhardening sealant such as Permatex 2 to screw (4—Fig. TO118). Tighten cover screws to 85 in.-lbs. (9.6 N•m) in sequence shown in Fig. TO118. Do not force mating of oil pan with crankcase. Reassemble remainder of components.

PISTON, PIN AND RINGS. To remove piston and rod assembly, remove engine as previously outlined. Remove cylinder head and camshaft as previously outlined. Unscrew connecting rod screws and remove piston and rod.

Maximum allowable piston ring end gap is 0.030 inch (0.76 mm) for compression rings and 0.060 inch (1.52 mm) for oil ring rails. Renew piston if ring side clearance exceeds 0.005 inch (0.12 mm) with a new piston ring installed in groove. Oversize as well as standard size piston and rings are available.

Piston pin is a slip fit in piston and rod. Renew piston if piston pin bore diameter is 0.552 inch (14.02 mm) or greater. Renew piston pin if diameter is 0.551 inch (14.00 mm) or less.

Top piston ring may be installed with either side up. Second piston ring must be installed with notch (S—Fig. TO119) toward piston skirt.

When assembling piston and rod, note that arrow on piston crown and "MAG" on connecting rod must be on same side. See Fig. TO120. Install piston and rod assembly in engine with arrow on piston crown toward flywheel as shown in Fig. TO121. Install rod cap so arrow on cap (R—Fig. TO122) points in same direction as arrow (A) on rod. Tighten rod screws to 100 in.-lbs. (11.3 N•m).

Install camshaft and cylinder head as previously outlined.

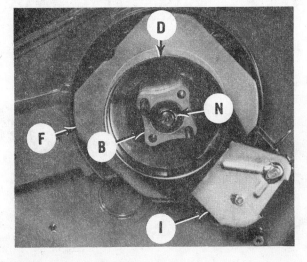

Fig. TO112—View of bearing block (B), brake drum (D), brake idler (I), flywheel (F) and retaining nut (N).

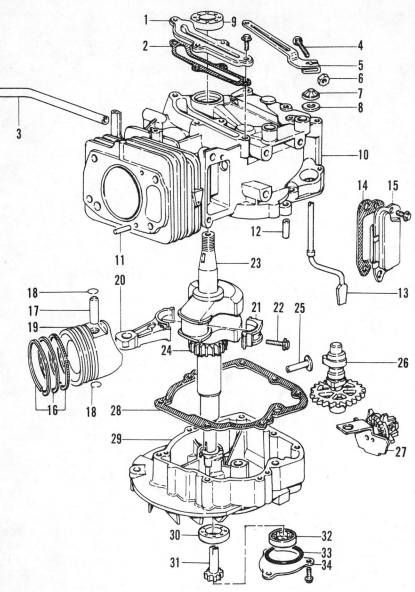

Fig. TO117—Install camshaft so timing marks (M) on camshaft and crankshaft gears are aligned. Pin (P) in camshaft end must engage slot in oil pump drive shaft when installing crankcase cover.

CONNECTING ROD. The connecting rod rides directly on crankpin. Connecting rod and piston are removed as an assembly as outlined in previous section.

Connecting rod reject size for crankpin hole is 1.127 inch (28.63 mm). Renew connecting rod if piston pin bore diameter is 0.5525 inch (14.03 mm) or greater. A connecting rod with 0.020 inch (0.51 mm) undersize big end diameter is available to accommodate a worn crankpin (machining instructions are included with new rod).

GOVERNOR. The engine is equipped with a mechanical governor (Fig. TO123) that is driven by the camshaft gear as shown in Fig. TO116. The flyweight assembly is mounted on end of camshaft along with oil slinger. The oil slinger and flyweight assembly are only available as a unit assembly. Inspect flyweight assembly for broken components. When installing governor, make certain that the governor plunger is positioned against the governor shaft arm and that the flyweights move freely.

CRANKSHAFT AND MAIN BEARINGS. The crankshaft rides di-

Fig. TO115—Exploded view of engine.

1. Cover	10. Crankcase	18. Retaining ring
2. Gasket	11. Dowel pin	19. Piston
3. Breather tube	12. Dowel pin	20. Connecting rod
4. Clamp bolt	13. Governor shaft	21. Rod cap
5. Governor lever	14. Gasket	22. Screw
6. Nut	15. Tappet cover	23. Crankshaft
7. Push nut	16. Piston rings	24. Gear
8. Washer	17. Piston pin	25. Tappet
9. Oil seal		

26. Camshaft
27. Governor
28. Gasket
29. Oil pan
30. Oil seal
31. Oil pump inner rotor
32. Outer rotor
33. "O" ring
34. Cover

Fig. TO116—The governor assembly (G) is mounted on end of camshaft.

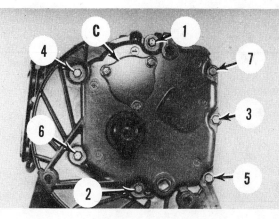

Fig. TO118—Tighten oil pan retaining screws in sequence shown. Apply sealant to screw (4).

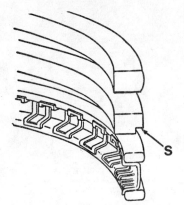

Fig. TO119—Install second compression piston ring so notch (S) is toward piston skirt.

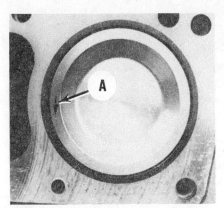

Fig. TO121—Install piston so arrow (A) on piston crown points toward flywheel side of engine.

Fig. TO123—View of governor and oil slinger assembly.

rectly in the crankcase and oil pan bearing bores. Rejection sizes for crankshaft are: pto-end bearing journal 1.060 inch (26.92 mm); flywheel-end bearing journal 0.873 inch (22.17 mm); crankpin 1.122 inch (28.50 mm). A connecting rod with 0.020 inch (0.51 mm) undersize big end diameter is available to accommodate a worn crankpin.

The crankcase main bearing bore rejection size is 0.878 inch (22.30 mm). The oil pan main bearing bore rejection size is 1.065 inch (27.05 mm).

Install oil seals so lip is toward inside of crankcase.

Tighten oil pan screws to 85 in.-lbs. (9.6 N•m) in sequence shown in Fig. TO118.

CYLINDER. If cylinder bore wear is 0.003 inch (0.76 mm) or more, or out-of-round is 0.0015 inch (0.038 mm) or greater, cylinder must be rebored to next oversize.

Standard cylinder bore diameter is 2.5615-2.5625 inch (65.06-65.09 mm).

OIL PUMP. A rotor type oil pump (Fig. TO124) driven by the camshaft is located in the bottom of the oil pan.

Remove engine from equipment for access to oil pump cover (C—Fig. TO118). Remove cover and extract

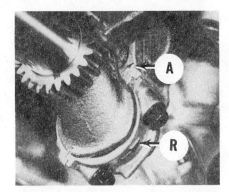

Fig. TO122—Install rod cap so arrow on cap (R) points in same direction as arrow (A) on rod.

pump rotors (31 and 32—Fig. TO115 or TO124). Mark rotors so they can be reinstalled in their original position. Renew any components which are damaged or excessively worn.

FLYWHEEL BRAKE. Some engines are equipped with a pad type flywheel brake. The brake should stop the engine within three seconds when the operator releases mower safety control and the speed control is in high speed position.

To service flywheel brake, remove fuel tank, dipstick and oil fill tube, blower housing and rewind starter. Disconnect brake spring (Fig. TO125). Disconnect stop switch wire. Remove two

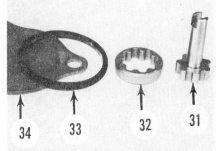

Fig. TO124—View of oil pump components.

31. Oil pump inner rotor   33. "O" ring
32. Outer rotor   34. Cover

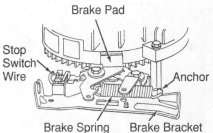

Fig. TO125—A flywheel brake is used on some engines.

retaining screws from brake pad arm and brake bracket.

The brake pad is available only as part of the bracket assembly. Minimum allowable brake pad thickness is 0.090 inch (2.3 mm). When installing brake assembly, tighten retaining screws to 35 in.-lbs. (4.0 N•m) torque.

REWIND STARTER. Refer to Fig. TO126 for exploded view of rewind starter.

To disassemble starter, remove rope handle and allow rope to wind slowly into starter. Remove plastic sleeve (11) and centering pin (10), then unscrew retainer screw (9). Remove retainer (8), brake spring (7), pawls (6) and springs (5). Wear appropriate safety eyewear and gloves before disengaging pulley

Fig. TO120—Assemble piston and rod so arrow (A) on piston crown and "MAG" on side of rod are positioned as shown.

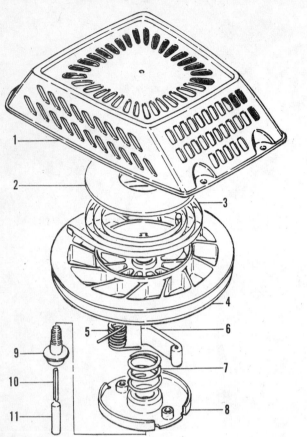

Fig. TO126—Exploded view of rewind starter.

1. Starter housing
2. Spring cover
3. Rewind spring
4. Pulley
5. Spring
6. Pawl
7. Brake spring
8. Retainer
9. Screw
10. Roll pin
11. Sleeve

Fig. TO129—Insert centering pin (P) in retainer screw until bottomed, then install sleeve (L) on pin.

installing pulley, be sure inner end of rewind spring engages spring retainer adjacent to housing center post. Install the pawl and spring so the spring end (E—Fig. TO127) forces the pawl toward the center of the pulley. Place the retainer on the pulley hub so bosses (B—Fig. TO128) are inside ends of pawls.

With starter assembled, except for rope, install rope as follows: Rotate pulley counterclockwise until tight, then allow to unwind so hole in pulley aligns with rope outlet. Insert rope through starter housing and pulley hole, tie a knot in rope end, allow rope to wind onto pulley and install rope handle.

Install centering pin (P—Fig. TO129) and sleeve (L) in retainer screw.

ELECTRIC STARTER MOTOR. Refer to Fig. TO130 for an exploded

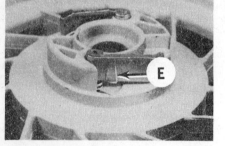

Fig. TO127—Spring end (E) should force pawl toward center of pulley.

Inspect components for damage and excessive wear. Reverse disassembly procedure to install components. Rewind spring coils wind in counterclockwise direction from outer end. When

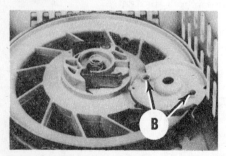

Fig. TO128—Bosses (B) on retainer must fit inside pawls during assembly.

from starter as rewind spring may uncoil uncontrolled. Place shop towel around pulley and lift pulley out of housing; spring should remain with pulley.

Fig. TO130—Exploded view of electric starter motor.

1. "E" ring
2. Helix
3. Nylon gear
4. Cover
5. Gasket
6. Felt
7. Clutch gear
8. Pinion gear
9. Drive end bracket
10. Plastic washer
11. Spring washer
12. Metal washer
13. Armature
14. Spacer
15. Metal washer
16. Plastic washer
17. Frame
18. Through-bolt
19. Brush end cap

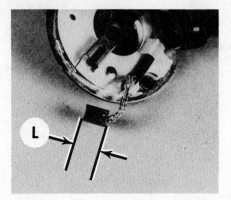

Fig. TO131—Minimum brush length (L) is 5/64 inch (2.0 mm).

view of 12-volt electric starter motor used on some models. Do not run starter for more than five seconds when testing. Starter pinion and helix must move without binding. Do not apply oil to helix (2) or nylon gear (3). Renew brushes if length (L—Fig. TO131) is 5/64 inch (2.0 mm) or less.

When assembling starter, note location of washers as shown in Fig. TO130. Be sure notches in end cap, frame and end bracket are aligned. Apply approximately 3/4 ounce of gear lubricant under clutch (7). Tap end cap (19) to seat bearings.

ALTERNATOR. Engines equipped with an electric starter motor are equipped with an alternator adjacent to the flywheel to charge the battery.

To test alternator output, disconnect black stator lead at white connector. Connect red lead of a DC ammeter to stator lead and black ammeter lead to engine ground. With engine running, alternator output should be at least 0.5 amps DC at 2800 rpm. If alternator output is zero or low and stator air gap is correct, replace stator.

The air gap between the stator and the flywheel magnets should be 0.010 inch (0.25 mm).

WISCONSIN ROBIN

TELEDYNE TOTAL POWER
3409 Democrat Road
Memphis, Tennessee 38118

2-STROKE

Model	Bore	Stroke	Displacement	Rated Power
WT1-125V	2.2 in.	1.97 in.	7.49 cu. in.	4.0 hp
	(56 mm)	(50 mm)	(123 cc)	(3.0 kW)

ENGINE INFORMATION

Model WT1-125V is a two-stroke, single-cylinder, air-cooled vertical crankshaft engine. Engine model and specification numbers are located on the name plate. Engine serial number is stamped either on the crankcase or the engine's identification tag. The model, specification and serial numbers must be furnished when ordering parts.

MAINTENANCE

SPARK PLUG. Recommended spark plug is a Champion CJ8 or equivalent. Manufacturer recommends removing and cleaning spark plug and adjusting electrode gap after every 50 hours of normal operation. Specified spark plug electrode gap is 0.024 inch (0.6 mm).

CARBURETOR. Model WT1-125V engine is equipped with a float type carburetor (Fig. W25) with fixed low speed and main fuel jets. Plastic float (15) is nonadjustable. Refer to Fig. W25 for exploded view of carburetor during disassembly and reassembly.

GOVERNOR. Model WT1-125V is equipped with a centrifugal flyweight type governor. The governor plate, governor sleeve and governor yoke are located in the crankcase and lubricated by the oil mixed with the fuel. To adjust the external governor linkage, first make certain all linkage operates freely. Inspect linkage for worn or broken parts. Governor spring (1—Fig. W26) is normally installed in the center hole of governor lever (9) as shown. Loosen adjusting nut (7) and adjusting plate screw (5). Push adjusting plate (6) downward and tighten screw (5). Place speed control lever (2) in high speed position and tighten nut (7). Adjust throttle stop screw (22) to limit maxi-

mum engine speed to rpm specified by equipment manufacturer.

IGNITION SYSTEM. Model WT1-125V engine is equipped with a solid-state electronic ignition system. Specified air gap between ignition coil

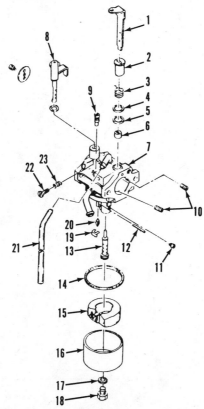

Fig. W25—Exploded view of the fixed jet float type carburetor used on Model WT1-125V engine.

1. Choke shaft	
2. Ring	
3. Spring	13. Nozzle
4. Cap	14. Gasket
5. Seal	15. Float
6. Ring	16. Float bowl
7. Body	17. Gasket
8. Throttle shaft	18. Bolt
9. Pilot jet	19. Clip
10. Air jets	20. Fuel inlet needle
11. Main jet	21. Tube
12. Pin	22. Throttle stop screw
	23. Spring

and flywheel is 0.016-0.020 inch (0.4-0.5 mm). To adjust, loosen retaining screws and move coil to obtain correct clearance. Tighten coil retaining screws to secure adjustment.

To check ignition system, inspect all components for broken, frayed, loose or disconnected ignition wires. Make certain spark plug is clean and in good condition. Use a standard test plug to check for ignition spark. Ignition coil can be checked using a good ohmmeter. Connect one ohmmeter lead to the high tension wire (spark plug) and connect remaining ohmmeter lead to the iron coil laminations. Ohmmeter reading should be 13,000 ohms. If resistance reading is infinite, open winding in ignition unit, loose or broken spark plug connector or faulty high tension lead is indicated. If resistance reading is low, secondary coil is shorted.

LUBRICATION. The engine is lubricated by mixing oil with the fuel. Manufacturer recommends mixing a good quality two-stroke, air-cooled en-

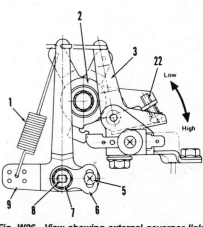

Fig. W26—View showing external governor linkage used on Model WT1-125V.

1. Governor spring	6. Adjusting plate
2. Speed control lever	7. Nut
3. Choke lever	8. Governor shaft
5. Adjusting plate	9. Governor lever
screw	22. Throttle stop screw

gine oil with gasoline at a 32:1 ratio. Use a separate container when mixing fuel and oil.

CARBON. Manufacturer recommends removing cylinder head and cleaning carbon and combustion deposits from exhaust port, muffler and piston after every 500 hours of normal operation. Use a wooden scraper to remove combustion deposits to prevent damaging piston or cylinder.

GENERAL MAINTENANCE. Check and tighten all loose bolts, nuts or clamps prior to each day of operation. Check for fuel leakage and repair if necessary.

Clean dust, dirt, grease or any foreign material from cylinder head and cylinder block cooling fins daily. Inspect fins for damage and repair if necessary.

Clean spark plug and air cleaner after every 50 hours of normal operation. Clean fuel strainer and fuel tank after every 200 hours of normal operation.

REPAIRS

TIGHTENING TORQUES. Recommended tightening torques are as follows:

Spark plug	18-22 ft.-lbs.
	(24-30 N·m)
Flywheel nut	28-30 ft.-lbs.
	(38-40 N·m)
Cylinder nuts	13-16 ft.-lbs.
	(18-22 N·m)
Crankcase bolts	80-88 in.-lbs.
	(9-10 N·m)

CYLINDER AND CYLINDER HEAD. To remove cylinder and cylinder head, remove spark plug and the four head bolts (1 and 2—Fig. W27). Carefully remove cylinder head (3) from cylinder. Remove cylinder head gasket (4). Remove the four nuts (6), lockwashers (7) and flat washer (8). Carefully slide cylinder off of piston and ring assembly. Remove gasket (9).

Clean carbon and combustion deposits from cylinder head and combustion chamber prior to reassembly. Inspect all parts for wear or damage. If cylinder head is warped 0.004 inch (0.1 mm) or more, renew cylinder head. Standard cylinder bore diameter is 2.0900-2.0907 inches (56.000-56.019 mm). If cylinder bore diameter is 2.0959 inches (56.15 mm) or more, cylinder should be resized for the next size piston and ring set which is available. Oversize piston and ring sets are available in 2.215 inches (56.25 mm) and 2.224 inches (56.50 mm). If cylinder is 0.0004 inch (0.01 mm) or more out-of-round, or if cylinder

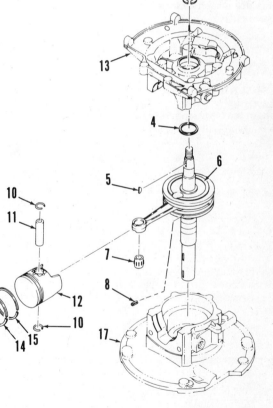

Fig. W27—Exploded view of WT1-125V engine crankcase, cylinder and cylinder head assembly.

1. Head bolt
2. Head bolt
3. Cylinder head
4. Gasket
5. Cylinder
6. Nut
7. Lockwasher
8. Flat wsher
9. Gasket
10. Cylinder stud
11. Cylinder stud
12. Seal
13. Crankcase half
14. Main bearing
15. Main bearing
16. Dowel pin
17. Crankcase half
18. Governor seal
19. Seal

Fig. W28—Exploded view of connecting rod and crankshaft assembly.

1. Nut
2. Washer
4. Shim
5. Key
6. Connecting rod & crankshaft assy.
7. Needle bearing
8. Spring pin
10. Retaining rings
11. Piston pin
12. Piston
13. Crankcase half
14. Piston ring
15. Piston ring
17. Crankcase half

taper exceeds 0.0006 inch (0.015 mm), cylinder must be resized.

When reassembling, install a new gasket (9) and make certain piston ring ends are around the locating pins in piston grooves. Carefully work cylinder down over piston. Install a new head gasket (4) with folded edge facing toward cylinder head. Install cylinder head and tighten head bolts following a

crisscross pattern to torque specified in TIGHTENING TORQUES.

PISTON, PIN AND RINGS. To remove piston, refer to CYLINDER AND CYLINDER HEAD paragraphs to remove cylinder head and cylinder. Remove the two retaining rings (10—Fig. W28) and slide piston pin (11) out of piston pin bore and connecting rod pin

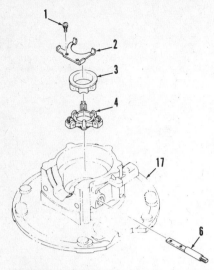

Fig. W29—Exploded view of governor assembly used on Model WT1-125V engine.

1. Screw	4. Governor flyweight
2. Governor yoke	plate
3. Governor sleeve	6. Governor shaft
	17. Crankcase half

bore. Remove needle bearing (7) as necessary. Remove piston rings (14 and 15).

Clean and inspect piston. Top piston ring is a chrome color and second ring is a darker color. Standard piston ring groove width is 0.0787 inch (2.0 mm). If ring groove width exceeds 0.0846 inch (2.15 mm), renew piston. Standard clearance between ring and ring groove is 0.0020-0.0035 inch (0.05-0.09 mm). If

clearance exceeds 0.0059 inch (0.015 mm), renew piston and/or rings. Standard piston ring end gap is 0.004-0.010 inch (0.10-0.25 mm). If ring end gap exceeds 0.070 inch (1.52 mm), renew rings and/or recondition cylinder bore. Standard piston ring width is 0.0925 inch (2.35 mm). If ring width is 0.0921 inch (2.34 mm) or less, renew ring.

Standard piston pin bore diameter in piston is 0.4724 inch (12 mm). If piston pin bore diameter exceeds 0.4738 inch (12.035 mm), renew piston. Standard piston pin diameter is 0.4724 inch (12 mm). If piston pin diameter is 0.4712 inch (11.97 mm) or less, renew piston pin.

Standard piston diameter is 2.203 inches (55.96 mm). If piston diameter is 2.199 inches (55.85 mm) or less, renew piston. Standard clearance between piston and cylinder bore is 0.0016-0.0029 inch (0.041-0.074 mm). If clearance exceeds 0.0109 inch (0.277 mm), renew piston and/or recondition cylinder bore.

When installing piston on connecting rod, the "F" mark on top of piston should face toward flywheel side of engine after assembly. Always install new piston pin retainers and make certain piston ring ends are around the locating pins in piston grooves.

CRANKCASE, CRANKSHAFT AND CONNECTING ROD. To remove crankshaft and connecting rod assem-

bly, remove cylinder head and cylinder as outlined in CYLINDER AND CYLINDER HEAD paragraphs. Remove piston and ring assembly. Remove the five bolts retaining upper and lower crankcase halves. Gently tap crankcase halves to separate. Remove crankshaft and connecting rod assembly. Crankshaft and connecting rod are serviced as an assembly only. Do not separate crankshaft and connecting rod. Remove retaining screws (1—Fig. W29), governor yoke (2), governor sleeve (3) and governor flyweight plate (4). It may be necessary to heat crankcase halves slightly to remove ball bearing type main bearings (14 and 15—Fig. W27).

Seals (12, 18 and 19) should be pressed into seal bores until seated. It may be necessary to heat crankcase slightly to install main bearings (14 and 15). Standard diameter for crankshaft main bearing journal is 0.79 inch (20.066 mm). If diameter is 0.7884 inch (20.025 mm) or less, renew crankshaft. Standard thrust clearance is 0.0000-0.0079 inch (0.0-0.2 mm). If thrust clearance exceeds 0.04 inch (1.0 mm), vary thickness of shim (4—Fig. W28). Shims are available in 0.0039 inch (0.1 mm) and 0.0118 inch (0.3 mm) thicknesses. Standard crankshaft runout is 0.002 inch (0.05 mm). If crankshaft runout exceeds 0.005 inch (0.13 mm), renew crankshaft.

Illustrations courtesy Teledyne Total Power

WISCONSIN ROBIN

4-STROKE OHV ENGINES

Model	Bore	Stroke	Displacement	Rated Power
WO1-115	2.28 in.	1.69 in.	6.93 cu.in.	3.3 hp
	(58 mm)	(43 mm)	(113 cc)	(2.5 kW)
WO1-150	2.52 in.	1.81 in.	9.02 cu.in.	4.8 hp
	(64 mm)	(46 mm)	(147 cc)	(3.6 kW)
WO1-210	2.83 in.	2.05 in.	12.92 cu.in.	7.0 hp
	(72 mm)	(52 mm)	(211 cc)	(5.2 kW)

ENGINE IDENTIFICATION

All models are air-cooled, four-stroke, single-cylinder engines. All models utilize an overhead valve system and are equipped with a horizontal crankshaft. Engine model and specification number are indicated on the name plate attached to the flywheel shroud. The engine serial number is stamped on the crankcase base.

MAINTENANCE

LUBRICATION. Engine oil level should be checked daily or after every 8 hours of operation. Oil should be changed after every 20 hours of operation.

Engine oil should meet or exceed latest API service classification. Use SAE 30 oil when temperature is above 40° F (4° C), SAE 20 oil when temperature is between 40° F (4° C) and 15° F (–9° C), and SAE 10W-30 oil when temperature is below 15° F (–9° C).

Crankcase capacity is 1.27 pints (0.6 L) for Models WO1-115 and WO1-150, and 1.37 pints (0.65 L) for Model WO1-210.

AIR FILTER. The air filter element should be cleaned weekly or after every 50 hours of operation, whichever occurs first. The foam element may be cleaned using warm water and soap. Dry filter then apply light oil. Squeeze excess oil from filter while noting if filter is completely coated with oil. Clean filter more frequently if extremely dirty or dusty conditions exist.

FUEL FILTER. On models so equipped, the fuel filter should be cleaned monthly or after every 150 hours of operation, whichever occurs first. The filter may be located in the bottom of fuel tank or in the fuel shut-off valve assembly.

SPARK PLUG. Recommended spark plug for all models is NGK BP6ES or Champion N9Y. Electrode gap should be 0.024-0.028 inch (0.6-0.7 mm). Tighten spark plug to 104-130 in.-lbs. (11.8-14.7 N•m).

CARBURETOR. Refer to Fig. WR101 or WR102 for an exploded view of carburetor. Both carburetors are equipped with an adjustable idle mixture screw (3) and a fixed main jet (21).

Initial setting of idle mixture screw (3) is one turn open from a lightly seated position for Models WO1-115 and WO1-150, or 1½ turns open for Model WO1-210. Make final adjustment of idle mixture screw with engine at normal operating temperature. Adjust idle mix-

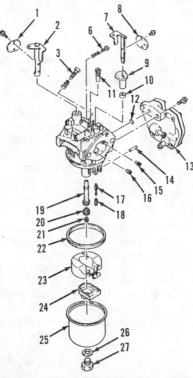

Fig. WR101—Exploded view of carburetor used on Models WO1-115 and WO1-150.

1. Throttle plate
2. Throttle shaft
3. Idle mixture screw
6. Idle speed screw
7. Choke shaft
8. Choke plate
9. Bushing
10. Seal
11. Idle mixture jet
12. "O" ring
13. Fuel pump
14. Pin
15. Air jet
16. Air jet
17. Fuel inlet valve
18. Spring
19. Main nozzle
20. Nozzle
21. Main jet
22. Gasket
23. Float
24. Cap
25. Fuel bowl
26. Gasket
27. Screw

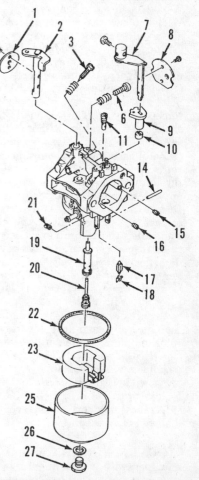

Fig. WR102—Exploded view of carburetor used on Model WO1-210. Refer to Fig. WR101 for parts identification except for: 18. Clip.

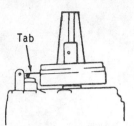

Fig. WR103—Float should be parallel to gasket surface of carburetor. Bend tab to adjust float level.

ture screw so engine idles smoothly and accelerates without hesitation. Adjust idle speed screw (6) so engine idle speed is 1200 rpm.

Before removing carburetor from the engine, carefully note the position of governor linkage and springs to insure correct reassembly. Remove air cleaner assembly from the carburetor. Disconnect fuel line from carburetor. Disconnect governor springs. Pull carburetor away from engine and disconnect governor link rod.

To disassemble carburetor, refer to Fig. WR101 or WR102 and remove float bowl (25). Remove float pin (14), float (23) and fuel inlet valve (17). On Model WO1-210, unscrew main jet (21—Fig. WR102) before removing nozzle (20). On all models, remove idle mixture screw (3) and idle jet (11). On Models WO1-115 and WO1-150, remove fuel pump mounting screws and separate pump (13) from carburetor body. Throttle plate (1) and choke plate (8) are retained by screws. Throttle shaft (2) and choke shaft (7) can be withdrawn after throttle and choke plates are removed.

Metal parts may be cleaned in carburetor cleaner. Direct compressed air through orifices and passageways in the opposite direction of normal fuel flow. Do not use wires or drill bits to clean orifices as enlargement of the orifice could affect the calibration of the carburetor. Inspect carburetor and renew any damaged or excessively worn components. Renew float if it is dented, contains fuel or if wear is evident in area of pivot pin or float tab.

To check or adjust float level, install fuel inlet valve and float. Invert carburetor and note position of float in relation to carburetor body. Float should be parallel to gasket surface of carburetor as shown in Fig. WR103. Bend float tab to adjust float level.

To reassemble carburetor, reverse the disassembly procedure.

GOVERNOR. All models are equipped with a centrifugal flyweight type governor. Governor assembly is located in the crankcase cover. To adjust governor linkage, loosen governor lever

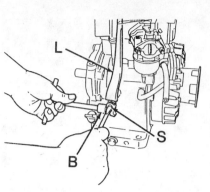

Fig. WR104—With governor lever (L) in wide open throttle position, turn governor shaft (S) clockwise as far as possible and tighten clamp bolt (B).

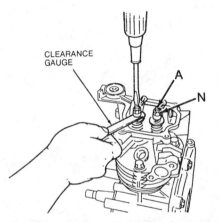

Fig. WR105—To adjust valve clearance, loosen nut (N) and turn adjusting screw (A). Clearance should be 0.003-0.005 inch (0.08-0.12 mm).

clamp bolt (B—Fig. WR104) so that governor lever (L) can be moved independently from governor shaft (S). Move governor lever (L) so carburetor throttle plate is fully open. Turn governor shaft (S) as far clockwise as possible and tighten clamp bolt.

IGNITION. A solid-state ignition system is used on all models. A one-piece ignition module/coil is mounted outside the flywheel. Air gap between ignition coil legs and flywheel should be 0.020 inch (0.5 mm). Adjust air gap by loosening ignition coil mounting screws and relocating coil. Ignition timing should be 21-25° BTDC on Models WO1-115 and WO1-210 or 26-30° BTDC on Model WO1-150. Ignition timing is not adjustable.

VALVE ADJUSTMENT. Engine must be cold when adjusting valve clearance. Rotate crankshaft so piston is at top dead center on compression stroke. Remove rocker arm cover and measure clearance between rocker arm and valve stem using a feeler gauge. Clearance between valve stem and rocker arm should be 0.003-0.005 inch

(0.08-0.12 mm) for intake and exhaust. To adjust clearance, loosen nut (N—Fig. WR105) and turn rocker arm adjusting screw (A) as necessary. Tighten nut and recheck clearance.

CYLINDER HEAD. Manufacturer recommends removal of carbon from cylinder head combustion chamber after 500-600 hours of operation.

REPAIRS

TIGHTENING TORQUES. Recommended tightening torques are as follows:

Connecting rod:
WO1-115 .... 78-100 in.-lbs. (8.8-11.3 N•m)
WO1-150, WO1-210 .. 148-173 in.-lbs. (16.7-19.6 N•m)
Crankcase cover:
WO1-115, WO1-150 .... 70-87 in.-lbs. (7.8-9.8 N•m)
WO1-210 .... 148-165 in.-lbs. (16.7-18.6 N•m)
Cylinder head .... 200-233 in.-lbs. (22.5-26.5 N•m)
Flywheel nut:
WO1-115, WO1-150 .... 44-46 ft.-lbs. (58.8-63.7 N•m)
WO1-210 .... 58-72 ft.-lbs. (78.4-98 N•m)
Spark plug .... 104-130 in.-lbs. (11.8-14.7 N•m)

FUEL PUMP. A diaphragm type fuel pump (13—Fig. WR101) is attached to the carburetor on Models WO1-115 and WO1-150. A diaphragm set is available for servicing the fuel pump.

CYLINDER HEAD. To remove cylinder head, remove interfering air baffles and shroud. Remove fuel tank on Model WO1-210. On all models, disconnect throttle linkage and fuel line and remove carburetor. Remove muffler. Remove spark plug and rotate crankshaft so piston is at top dead center on compression. Remove rocker arm cover. Mark push rods so they can be returned to original position. Remove cylinder head bolts and remove cylinder head and gasket.

Push rocker shaft (9—Fig. WR106) out of cylinder head and remove rocker arms (18). The rocker arms are identical, but they should be marked so they can be installed in original positions when reassembling. Compress valve springs and remove valve spring retainers (3). Remove valves from cylinder head.

Clean combustion deposits from combustion chamber. Check cylinder head for cracks or other damage. Maximum

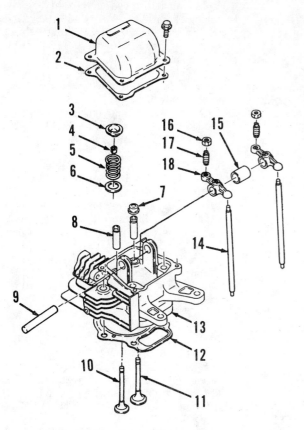

1. Rocker arm cover
2. Gasket
3. Valve spring retainer
4. Retainer lock
5. Valve spring
6. Washer
7. Seal
8. Valve guides
9. Rocker arm shaft
10. Exhaust valve
11. Intake valve
12. Head gasekt
13. Cylinder head
14. Push rod
15. Spacer
16. Nut
17. Adjuster screw
18. Rocker arm

allowable cylinder head warpage is 0.006 inch (0.15 mm).

Reverse disassembly procedure to install cylinder head. Tighten cylinder head bolts evenly in a criss-cross pattern to 200-233 in.-lbs. (22.5-26.5 N•m). Adjust valve clearance as previously outlined.

VALVE SYSTEM. Valve face and seat angles are 45° for intake and exhaust. Valve seat width should be 0.028-0.039 inch (0.7-1.0 mm); maximum allowable seat width is 0.080 inch (2.0 mm).

Specified valve stem diameter is 0.217 inch (5.50 mm) for Models WO1-115 and WO1-150, and 0.260 inch (6.60 mm) for Model WO1-210. Minimum allowable valve stem diameter is 0.211 inch (5.35 mm) for Models WO1-115 and WO1-150, and 0.254 inch (6.45 mm) for Model WO1-210.

Desired intake valve stem-to-guide clearance is 0.0008-0.0020 inch (0.020-0.050 mm) for Models WO1-115 and WO1-150, and 0.0010-0.0024 inch (0.025-0.062 mm) for Model WO1-210. Maximum allowable intake valve stem-to-guide clearance is 0.012 inch (0.3 mm) for all models.

Desired exhaust valve stem-to-guide clearance is 0.0022-0.0036 inch (0.056-0.092 mm) for Models WO1-115 and WO1-150, and 0.0022-0.0039 inch

(0.056-0.100 mm) for Model WO1-210. Maximum allowable exhaust valve stem-to-guide clearance is 0.012 inch (0.3 mm) for all models. Valve guides are renewable. A valve seal is used on the intake valve.

Standard valve spring free length is 1.201 inches (30.5 mm) for Models WO1-115 and WO1-150, and 1.319 inches (33.5 mm) for Model WO1-210. Minimum allowable valve spring length is 1.142 inches (29 mm) for Models WO1-115 and WO1-150, and 1.260 inches (32 mm) for Model WO1-210. Valve spring should be square within 0.039 inch (1.0 mm).

Rocker arm clearance on shaft should be 0.0006-0.0015 inch (0.016-0.039 mm) for Models WO1-115 and WO1-150, and 0.0002-0.0016 inch (0.006-0.015 mm) for Model WO1-210.

CAMSHAFT AND TAPPETS. To remove camshaft, proceed as follows: Remove rocker arm cover and disengage push rods from rocker arms. Drain engine oil. Remove crankcase cover. Position engine so tappets cannot fall out or hold tappets in bores. Withdraw camshaft. If tappets are to be removed, mark them so they can be returned to original bores.

Camshaft lobe height for both valves should be 1.142 inches (29.0 mm) for Models WO1-115 and WO1-150, and

1.213 inches (30.8 mm) for Model WO1-210. Minimum allowable lobe height is 1.132 inches (28.75 mm) for Models WO1-115 and WO1-150, and 1.203 inches (3.55 mm) for Model WO1-210. Specified camshaft bearing journal diameter on Models WO1-115 and WO1-150 is 0.591 inch (15.0 mm) for both ends, with a minimum allowable diameter of 0.589 inch (14.95 mm). Specified camshaft bearing journal on Model WO1-210 at gear end is 0.591 inch (15.0 mm) with a minimum allowable diameter of 0.589 inch (14.95 mm). Specified camshaft bearing journal on Model WO1-210 at flyweel end is 0.984 inch (25.0 mm) with a minimum allowable diameter of 0.982 inch (24.95 mm).

Specified outside diameter of tappets is 0.315 inch (8.0 mm). Tappet clearance in bore should be 0.0005-0.0015 inch (0.013-0.037 mm) on Models WO1-115 and WO1-150 and 0.0010-0.0024 inch (0.025-0.062 mm) on Model WO1-210.

The camshaft is equipped with a compression release device. Release lever (26—Fig. WR107) extends above the camshaft exhaust lobe at slow engine speed (starting) thereby holding the exhaust valve open and reducing compression pressure. At running speed, the weight on the release lever moves the lever below the camshaft lobe. Release lever should move freely without binding.

Install camshaft by reversing removal procedure. Lubricate camshaft and tappets with SAE 30 engine oil during assembly. Align timing marks (Fig. WR108) on crankshaft and camshaft gears. Tighten crankcase cover screws to 70-87 in.-lbs. (7.8-9.8 N•m) on Models WO1-115 and WO1-150, and to 148-165 in.-lbs. (16.7-18.6 N•m) on Model WO1-210.

PISTON, PIN, RINGS & ROD. To remove piston and rod, remove cylinder head and camshaft as previously outlined. Remove carbon deposits and ring ridge (if present) from top of cylinder. Detach rod cap (11—Fig. WR107) and remove piston (15) and rod (12). Remove retaining ring (13), then push piston pin (14) out of piston and connecting rod. Note that heating piston will make removal and installation of pin easier. A hair dryer or similar heat source can be used to heat the piston.

Measure piston diameter near bottom of skirt perpendicular to piston pin. Standard piston diameter is 2.2567 inches (57.32 mm) for Model WO1-115, 2.4929 inches (63.32 mm) for Model WO1-150, and 2.8132 inches (71.455 mm) for Model WO1-210. Piston clearance in cylinder should be 0.0009-0.0025 inch (0.024-0.063 mm) for

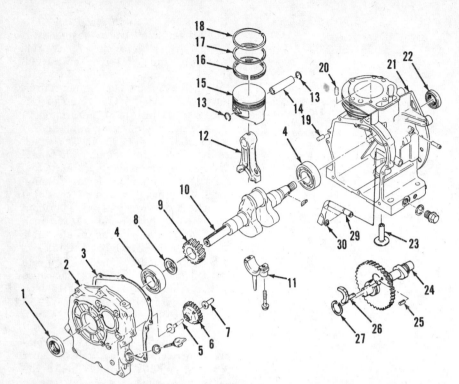

Fig. WR107—Exploded view of engine. Governor shaft (5) is attached to crankcase cover (2) and not available separately. A separate oil slinger plate is attached to rod cap (11) on Model WO1-210.

1. Seal		9. Gear	16. Oil ring	23. Tappet
2. Crankcase cover		10. Crankshaft	17. Second compression ring	24. Camshaft
3. Gasket		11. Rod cap	18. Top compression ring	25. Pin
4. Bearing		12. Connecting rod	19. Dowel pin	26. Compression release
5. Shaft		13. Retaining ring	20. Dowel pin	27. Retaining ring
6. Governor		14. Piston pin	21. Crankcase	29. Governor shaft
7. Sleeve		15. Piston	22. Seal	30. "E" ring
8. Shim				

Fig. WR108—View of timing marks on crankshaft and camshaft gears.

Models WO1-115 and WO1-150, and 0.0014-0.0029 inch (0.035-0.074 mm) for Model WO1-210. Pistons and rings are available in standard and oversizes of 0.010 inch (0.25 mm) and 0.020 inch (0.50 mm).

Specified piston ring end gap is 0.008-0.016 inch (0.2-0.4 mm) for compression rings on all models, with a maximum allowable end gap of 0.060 inch (1.5 mm). Specified piston ring end gap for oil ring is 0.008-0.016 inch (0.2-0.4 mm) for Models WO1-115 and WO1-150, and 0.006-0.014 inch (0.15-0.35 mm) for Model WO1-210.

Specified piston ring side clearance in piston ring groove for Models WO1-115 and WO1-150 is 0.0012-0.0030 inch (0.030-0.075 mm) for compression rings and 0.0008-0.0030 inch (0.020-0.075 mm) for oil ring; maximum side clearance for all rings is 0.006 inch (0.15 mm). Specified piston ring side clearance for Model WO1-210 is 0.0020-0.0037 inch (0.050-0.095 mm) for top compression ring, 0.0016-0.0033 inch (0.040-0.085 mm) for second compression ring, and 0.0004-0.0025 inch (0.010-0.063 mm) for oil ring; maximum allowable ring side clearance for all rings is 0.006 inch (0.15 mm). Renew piston if ring side clearance is excessive.

Specified piston pin bore in piston for Model WO1-115 is 0.5115-0.5119 inch (12.991-13.002 mm) with a maximum allowable bore of 0.5132 inch (13.035 mm). Specified piston pin diameter for Model WO1-115 is 0.5115-0.5118 inch (12.992-13.000 mm) with a minimum allowable diameter of 0.5102 inch (12.96 mm). Maximum allowable piston-to-pin clearance is 0.0023 inch (0.06 mm).

Specified connecting rod small end diameter for Model WO1-115 is 0.5122-0.5126 inch (13.010-13.021 mm) with a maximum diameter of 0.5150 inch (13.08 mm). Specified connecting rod-to-pin clearance is 0.0004-0.0011 inch (0.010-0.029 mm) and wear limit is 0.0047 inch (0.12 mm).

Specified piston pin bore in piston for Model WO1-150 is 0.6296-0.6300 inch (15.991-16.002 mm) with a maximum allowable bore of 0.6313 inch (16.035 mm). Specified piston pin diameter for Model WO1-150 is 0.6296-0.6299 inch (15.992-16.000 mm) with a minimum allowable diameter of 0.6283 inch (15.96 mm). Maximum allowable piston-to-pin clearance is 0.0023 inch (0.06 mm).

Specified connecting rod small end diameter for Model WO1-150 is 0.6303-0.6307 inch (16.010-16.021 mm) with a maximum diameter of 0.6331 inch (16.08 mm). Specified connecting rod-to-pin clearance is 0.0004-0.0011 inch (0.010-0.029 mm) and wear limit is 0.0047 inch (0.12 mm).

Specified piston pin bore in piston for Model WO1-210 is 0.6689-0.6694 inch (16.991-17.002 mm) with a maximum allowable bore of 0.6707 inch (17.035 mm). Specified piston pin diameter for Model WO1-210 is 0.6690-0.6693 inch (16.992-17.000 mm) with a minimum allowable diameter of 0.6677 inch (16.96 mm). Maximum allowable piston-to-pin clearance is 0.0023 inch (0.06 mm).

Specified connecting rod small end diameter for Model WO1-210 is 0.6695-0.6699 inch (17.005-17.016 mm) with a maximum diameter of 0.6725 inch (17.08 mm). Specified connecting rod-to-pin clearance is 0.0002-0.0009 inch (0.005-0.024 mm) and wear limit is 0.0047 inch (0.12 mm).

Specified connecting rod big end diameter is 1.0236-1.0241 inch (26.000-26.013 mm) for Model WO1-115, with a maximum allowable diameter of 1.0276 inch (26.1 mm). Rod bearing clearance for Model WO1-115 should be 0.0008-0.0018 inch (0.020-0.046 mm); maximum allowable clearance is 0.008 inch (0.2 mm).

Specified connecting rod big end diameter is 1.1811-1.1817 inch (30.000-30.016 mm) for Model WO1-150, with a maximum allowable diameter of 1.1850 inch (30.1 mm). Rod bearing clearance for Model WO1-150 should be 0.0008-0.0019 inch (0.020-0.049 mm); maximum allowable clearance is 0.008 inch (0.2 mm).

Specified connecting rod big end diameter is 1.3389-1.3396 inch (34.009-34.025 mm) for Model WO1-210, with a maximum allowable diameter of 1.3425

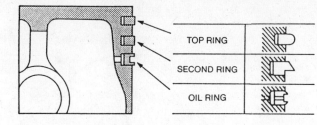

Fig. WR109—View of piston rings showing correct location.

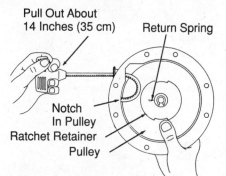

Fig. WR110—View showing method of releasing spring tension on rewind starter assembly.

inch (34.1 mm). Rod bearing clearance for Model WO1-210 should be 0.0013-0.0026 inch (0.034-0.066 mm); maximum allowable clearance is 0.008 inch (0.2 mm).

Refer to Fig. WR109 for piston ring identification and location. Lubricate piston pin, piston rings and connecting rod bearings with SAE 30 engine oil during assembly. Assemble piston and rod so "DF" or "BF" on piston crown is on same side as "MAG" on rod. Install piston and rod so piston and rod marks are towards flywheel. Install rod cap so marks align with rod. On Models WO1-115 and WO1-150, oil slinger on cap is towards governor. On Model WO1-210, install oil slinger plate so slinger is towards governor. Tighten connecting rod screws on Model WO1-115 to 78-100 in.-lbs. (8.8-11.3 N•m) and on Models WO1-150 and WO1-210 to 148-173 in.-lbs. (16.7-19.6 N•m).

GOVERNOR. The governor gear and flyweight assembly (6—Fig. WR107) rides on a stud (5) attached to the crankcase cover. The assembly is serviced only as a unit. Components must move freely without binding.

CRANKSHAFT. The crankshaft is supported by ball bearings located in the crankcase and crankcase cover. To remove crankshaft, remove starter and flywheel. Remove piston and connecting rod as previously outlined. Remove crankshaft. Remove ball bearings (4—Fig. WR107) and oil seals (1 and 22) from crankcase and crankcase cover as necessary.

Specified crankpin diameter is 1.0223-1.0228 inches (25.967-25.980 mm) for Model WO1-115; minimum allowable diameter is 1.018 inches (25.85 mm). Specified crankpin diameter is 1.1798-1.1803 inches (29.967-29.980 mm) for Model WO1-150; minimum allowable diameter is 1.175 inches (29.85 mm). Specified crankpin diameter is 1.3370-1.3376 inches (33.959-33.975 mm) for Model WO1-210; minimum allowable diameter is 1.333 inches (33.85 mm). Crankpin out-of-round and taper must be less than 0.0002 inch (0.005 mm).

Crankshaft main bearing journal diameter on Models WO1-115 and WO1-150 should be 0.9838-0.9841 inch (24.988-24.997 mm) with a minimum allowable diameter of 0.982 inch (24.95 mm). Crankshaft main bearing journal diameter on Model WO1-210 should be 1.1807-1.1811 inches (29.991-30.000 mm) with a minimum allowable diameter of 1.179 inches (29.95 mm).

Reverse disassembly procedure to reassemble. Align timing marks (Fig. WR108) on crankshaft and camshaft gears. Crankshaft end play should be 0.000-0.008 inch (0.0-0.2 mm) on all models. End play is adjusted using shim (8—Fig. WR107) which is available in thicknesses of 0.024 inch (0.6 mm), 0.031 inch (0.8 mm) and 0.039 inch (1.0 mm). Only one shim must be installed.

CRANKCASE AND CYLINDER. Standard cylinder bore diameter is 2.2835-2.2842 inches (58.000-58.019 mm) for Model WO1-115 with a maxi-

mum allowable diameter of 2.287 inches (58.1 mm). Standard cylinder bore diameter is 2.5197-2.5204 inches (64.000-64.019 mm) for Model WO1-150 with a maximum allowable diameter of 2.523 inches (64.1 mm). Standard cylinder bore diameter is 2.8346-2.8354 inches (72.000-72.019 mm) for Model WO1-210 with a maximum allowable diameter of 2.838 inches (72.1 mm). Maximum allowable cylinder out-of-round is 0.0004 inch (0.01 mm). Cylinder may be rebored for installation of oversize piston and rings.

REWIND STARTER. To disassemble the rewind starter, refer to Fig. WR110 and release spring tension by pulling rope handle until about 14 inches (35 cm) of rope extends from unit. Use thumb pressure to prevent pulley from rewinding and position rope in notch in outer rim of pulley. Release thumb pressure slightly and allow spring mechanism to slowly unwind. Remove clip or nut (2—Fig. WR111 or WR112), and remove components (3 through 7). Slowly remove pulley from support shaft in housing.

CAUTION: Take extreme care that rewind spring remains in recess of housing. Do not remove spring unless new spring is to be installed.

If rewind spring escapes from housing, form a wire ring of same circumference as recess in housing and twist ends together securely. Starting with outside loop, wind spring inside the ring in counterclockwise direction as shown in Fig. WR113.

NOTE: New rewind springs are secured in a similar wire ring for ease in assembly.

Place new spring assembly over recess in housing so hook in outer loop of spring is over the tension tab in the

Fig. WR111—Exploded view of rewind starter used on Models WO1-115 and WO1-150.

1. Starter cup
2. Nut
3. Friction plate
4. Return spring
5. Spacer
6. Thrust washer
7. Ratchet
8. Pivot pin
9. Pulley
10. Rewind spring
11. Housing

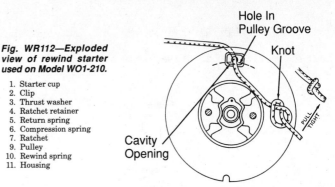

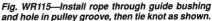

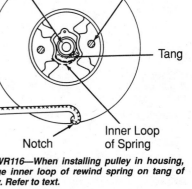

Fig. WR112—Exploded view of rewind starter used on Model WO1-210.

1. Starter cup
2. Clip
3. Thrust washer
4. Ratchet retainer
5. Return spring
6. Compression spring
7. Ratchet
9. Pulley
10. Rewind spring
11. Housing

Fig. WR115—Install rope through guide bushing and hole in pulley groove, then tie knot as shown.

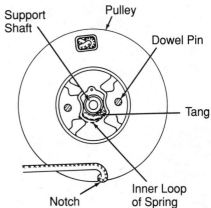

Fig. WR116—When installing pulley in housing, engage inner loop of rewind spring on tang of pulley. Refer to text.

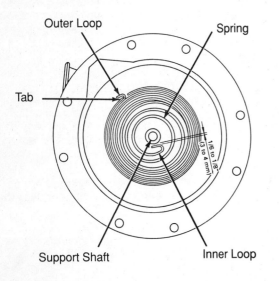

Fig. WR113—Illustration showing fabricated spring holder used to hold rewind spring.

Fig. WR114—Pulley must be installed with hook in outer loop of rewind spring engaged on tension tab and inner loop of spring spaced as shown from support shaft.

inch (25 mm) from end. Pull knot into top of handle. Install other end of rope through guide bushing of housing and through hole in pulley groove. Pull rope out through cavity opening and tie a knot about 1 inch (25 mm) from end. Tie knot as illustrated in Fig. WR115 and stuff knot into cavity opening. Wind rope 2½ turns clockwise on pulley, then lock rope in notch of pulley. Install pulley on support shaft and rotate pulley counterclockwise until tang on pulley engages hook on inner loop of rewind

housing. Carefully press spring from wire ring and into recess of housing. See Fig. WR114.

Using a new rope of same length and diameter as original, place rope in handle and tie a figure eight knot about 1

spring (see Fig. WR116). Place outer flange of housing in a vise and use finger pressure to keep pulley in housing. Hook a loop of rope into pulley notch and preload rewind spring by turning pulley four full turns counterclockwise. Re-

move rope from notch and allow pulley to slowly turn clockwise as rope winds on pulley and handle returns to guide bushing on housing. On Models WO1-115 and WO1-150, install pivot pin (8—Fig. WR111), if removed.

Refer to Figs. WR111 and WR112 and install components (2 through 7). On Model WO1-210, install return spring (5—Fig. WR112) with bent end hooked into hole of pulley hub and looped end toward outside, then mount ratchet retainer so loop end of return spring extends through slot. Rotate retainer slightly clockwise until ends of slots just begin to engage the ratchets. Press down on retainer, install washer and clip.

Illustrations courtesy Teledyne Total Power

WISCONSIN ROBIN

4-STROKE ENGINES

Model	Bore	Stroke	Displacement	Rated Power
W1-080	2.01 in.	1.50 in.	4.74 cu.in.	1.8 hp
	(51 mm)	(38 mm)	(77.6 cc)	(1.3 kW)
W1-145	2.48 in.	1.81 in.	8.73 cu.in.	3.4 hp
	(63 mm)	(46 mm)	(143 cc)	(2.4 kW)
W1-145V	2.48 in.	1.81 in.	8.73 cu.in.	4.0 hp
	(63 mm)	(46 mm)	(143 cc)	(3.0 kW)
W1-150	2.48 in.	1.81 in.	8.73 cu.in.	3.5 hp
	(63 mm)	(46 mm)	(143 cc)	(2.6 kW)
W1-185	2.64 in.	2.05 in.	11.2 cu.in.	4.6 hp
	(67 mm)	(52 mm)	(183 cc)	(3.4 kW)
W1-185V	2.64 in.	2.05 in.	11.2 cu.in.	5.0 hp
	(67 mm)	(52 mm)	(183 cc)	(3.7 kW)
W1-230	2.68 in.	2.44 in.	13.7 cu.in.	5.5 hp
	(68 mm)	(62 mm)	(225 cc)	(4.1 kW)

ENGINE IDENTIFICATION

All models are four-stroke, air-cooled, single-cylinder gasoline engines. Models W1-145V and W1-185V are vertical crankshaft engines. All other models are horizontal crankshaft engines. On all models, the engine model and specification numbers are located on the name plate on flywheel shroud. The serial number is stamped on the crankcase base. Always furnish engine model, specification and serial numbers when ordering parts.

MAINTENANCE

LUBRICATION. Check engine oil level daily and maintain oil level at full mark on dipstick or at lower edge of filler plug.

Engine oil should meet or exceed latest API service classification. Use SAE 30 oil when temperature is above 40° F (4° C), SAE 20 oil when temperature is between 15° F (–9° C) and 40° F (4° C), and SAE 10W-30 oil when temperature is below 15° F (–9° C).

Oil should be changed after every 50 hours of operation. Crankcase capacity is 0.85 pints (0.4 L) for Model W1-080, 1.25 pints (0.6 L) for Models W1-145, W1-145V, W1-150, W1-185 and W1-185V, and 1.5 pints (0.7 L) for Model W1-230.

SPARK PLUG. Recommended spark plug for Models W1-080, W1-145V and W1-185V is a NGK BM4A or equivalent. Recommended spark plug for Models W1-145, W1-150 and W1-185 is a NGK B6HS or equivalent. Recom-

mended spark plug for Model W1-230 is a NGK BP4HS or equivalent. Specified spark plug electrode gap for all models is 0.025 inch (0.6 mm).

CARBURETOR. All models are equipped with a Mikuni float type carburetor. Refer to Fig. WR25, WR26 or WR27 for exploded view of carburetor. On all models except W1-230, carburetor has fixed low speed and high speed jets.

Model W1-230 is equipped with carburetor shown in Fig. WR27. Carburetor has an adjustable low speed mixture screw (24) and a fixed main fuel (high speed) jet (15). Initial adjustment of low speed mixture screw should be 1⅜ turns open from a lightly seated position. Make final carburetor adjustment on Model W1-230 with engine at normal operating temperature and running. Adjust low speed mixture screw to obtain smoothest idle operation and acceleration.

On all models, adjust idle speed screw to obtain an idle speed of 1250 rpm at normal operating temperature.

Before removing carburetor from the engine, carefully note the position of governor linkage and springs to insure correct reassembly. Be careful not to stretch or otherwise damage governor springs or linkage as governor setting would be affected. Remove air cleaner assembly from the carburetor. Disconnect fuel line from carburetor. Disconnect governor springs. Pull carburetor away from engine and disconnect governor link rod.

To disassemble carburetor, refer to Fig. WR25, WR26 or WR27 and remove

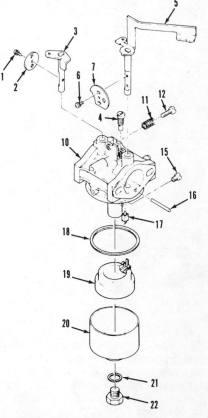

Fig. WR25—Exploded view of carburetor used on all models except W1-150 and W1-230.

1. Screw	12. Idle speed screw
2. Throttle plate	15. Main jet
3. Throttle shaft	16. Pin
4. Idle mixture jet	17. Fuel inlet valve
5. Choke shaft	18. Gasket
6. Screw	19. Float
7. Choke plate	20. Fuel bowl
10. Carburetor body	21. Gasket
11. Spring	22. Plug

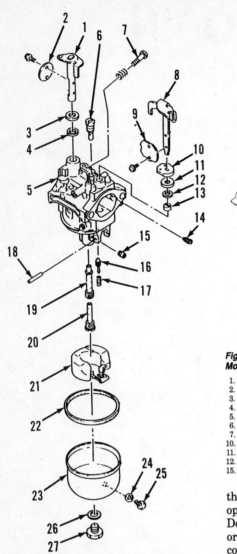

Fig. WR26—Exploded view of carburetor used on Model W1-150.

1. Throttle shaft
2. Throttle plate
3. Washer
4. Seal
5. Carburetor body
6. Idle mixture jet
7. Idle speed screw
8. Choke shaft
9. Choke plate
10. Detent plate
11. Washer
12. Seal
13. Bushing
14. Idle air jet
15. Main jet
16. Fuel inlet valve
17. Spring
18. Pin
19. Main nozzle
20. Nozzle
21. Float
22. Gasket
23. Fuel bowl
24. Gasket
25. Drain screw
26. Gasket
27. Screw

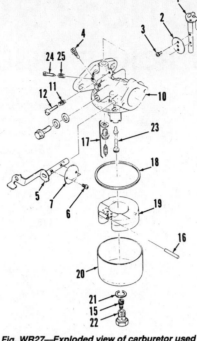

Fig. WR27—Exploded view of carburetor used on Model W1-230.

1. Throttle shaft
2. Throttle plate
3. Screw
4. Idle mixture jet
5. Choke shaft
6. Screw
7. Choke plate
10. Carburetor body
11. Spring
12. Idle speed screw
15. Main jet
16. Pin
17. Fuel inlet valve
18. Gasket
19. Float
20. Fuel bowl
21. Gasket
22. Plug
23. Nozzle
24. Idle mixture screw
25. Spring

through orifices and passageways in the opposite direction of normal fuel flow. Do not use wires or drill bits to clean orifices as enlargement of the orifice could affect the calibration of the carburetor. Inspect carburetor and renew any damaged or excessively worn components. Renew float if it is dented, contains fuel or if wear is evident in area of pivot pin or float tab.

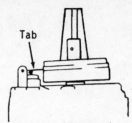

Fig. WR28—On Models W1-080, W1-150 and W1-230, float should be parallel with float bowl mating surface.

To check or adjust float level, remove fuel bowl and place carburetor body on end (on manifold flange) so float pin is in vertical position. Move float to close fuel inlet valve.

NOTE: Needle valve is spring loaded. Float tab should just contact needle valve pin but should not compress spring.

On Models W1-080, W1-150 and W1-230, float should be parallel with fuel bowl mating surface (Fig. WR28). On Model W1-145, W1-185, W1-145V and W1-185V, measure float setting as shown in Fig. WR29. Dimension "A" should be 0.492-0.571 inch (12.5-14.5 mm). On all models, carefully bend tab on float lever to obtain correct setting.

GOVERNOR. All models are equipped with a centrifugal flyweight type governor. To adjust external governor linkage on all models, loosen clamp screw on governor lever (see Fig. WR30, WR31, WR32 or WR33) so that governor shaft can be turned independently of governor lever. Move speed control lever so throttle valve in carburetor is opened fully. Hold lever in this position and rotate governor shaft clockwise as far as possible. Tighten clamp screw. Changing governor spring location in holes will vary engine speed.

float bowl. Remove float pin, float and fuel inlet valve. On Model WO1-210, unscrew main jet (15—Fig. WR26) before removing nozzle (19 and 20). On all models, remove idle mixture jet. Throttle plate and choke plate are retained by screws. Throttle shaft and choke shaft can be withdrawn after throttle and choke plates are removed. When installing choke plate and throttle plate, apply threadlocking compound such as Loctite 271 to threads of retaining screws.

Metal parts may be cleaned in carburetor cleaner. Direct compressed air

Fig. WR29—On Models W1-145, W1-145V, W1-185 and W1-185V, refer to text for float dimension "A."

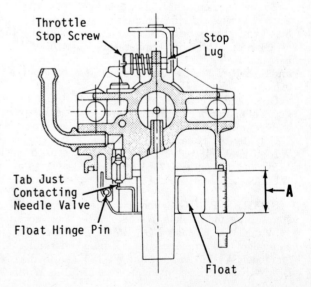

Illustrations courtesy Teledyne Total Power

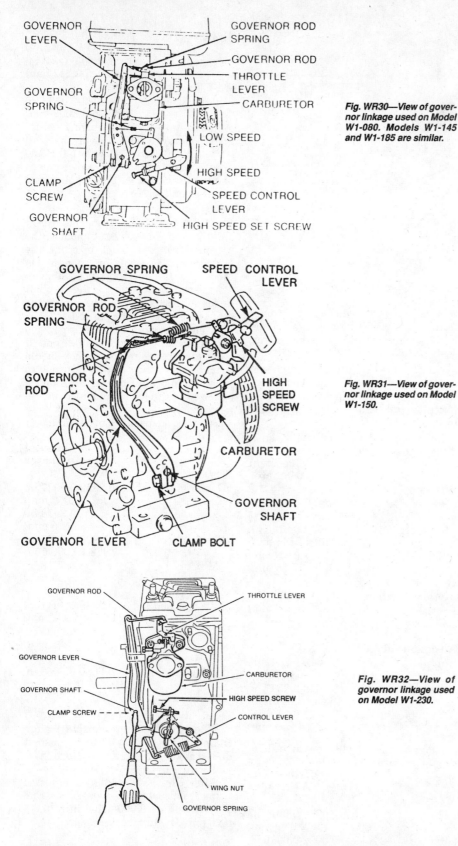

Fig. WR30—View of governor linkage used on Model W1-080. Models W1-145 and W1-185 are similar.

Fig. WR31—View of governor linkage used on Model W1-150.

Fig. WR32—View of governor linkage used on Model W1-230.

tween ignition coil and flywheel is 0.020 inch (0.5 mm).

To check and adjust engine timing, remove flywheel shroud. Disconnect lead wire from shut-off switch. Connect test lead from a continuity light to lead wire and ground remaining test lead to engine. Slowly rotate flywheel in normal operating direction until light goes out. Immediately stop turning flywheel and check location of timing marks. Timing mark "M" on crankcase and "P" mark on flywheel should align. If timing mark on flywheel is below timing mark on crankcase, breaker-point gap is too small. Carefully measure the distance necessary to align the two timing marks, then remove flywheel and breaker-point cover. Changing point gap 0.001 inch (0.03 mm) will change timing mark on flywheel approximately 1/8 inch (3.18 mm). Reassemble and tighten flywheel retaining nut to torque specified under TIGHTENING TORQUES in REPAIRS section.

Solid-State Ignition System. The electronic type solid-state ignition system does not have breaker points. There is no scheduled maintenance. Specified air gap between ignition coil and flywheel is 0.020 inch (0.5 mm).

VALVE ADJUSTMENT. Valve tappet gap (cold) should be 0.003-0.005 inch (0.08-0.12 mm) for all models except Model W1-230. Valve tappet gap (cold) for Model W1-230 should be 0.006-0.008 inch (0.16-0.20 mm). To check clearance, remove tappet cover/breather assembly. Position piston at top dead center on compression stroke and measure clearance between tappet and valve stem end using a feeler gauge.

Valves must be removed to adjust valve clearance. To increase clearance, grind off end of valve stem. To reduce clearance, grind valve seat deeper or renew valve and/or valve tappet.

REPAIRS

TIGHTENING TORQUES. Recommended tightening torque specifications are as follows:

Connecting rod:
W1-080 . . . . 53-71 in.-lbs.
(6-8 Nm)
W1-145, W1-145V,
W1-150 . . . . 80-97 in.-lbs.
(9-11 Nm)
W1-185, W1-185V,
W1-230 . . . . 150-168 in.-lbs.
(17-19 Nm)
Crankcase cover/oil pan:
W1-080, W1-145,
W1-150, W1-185 . . . . . 71-88 in.-lbs.
(8-10 Nm)

IGNITION SYSTEM. Early models are equipped with a breaker-point type ignition system. Late models are equipped with a solid-state ignition system. Refer to the appropriate paragraphs for ignition type being serviced.

Breaker-Point Ignition System. Breaker-points and condenser are located behind flywheel and the ignition coil is located outside flywheel. Initial breaker-point gap for all models is 0.014 inch (0.36 mm). Specified air gap be-

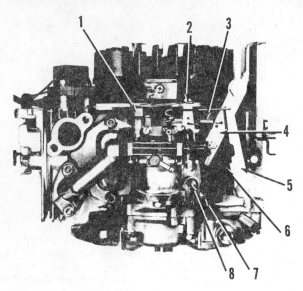

Fig. WR33—View of governor linkage used on Models W1-145V and W1-185V.

1. Throttle shaft
2. Governor lever
3. Governor spring
4. Speed control lever
5. Choke lever
6. Adjusting screw
7. Clamp bolt
8. Governor shaft

W1-145V, W1-185V . . . 71-80 in.-lbs.
(8-9 Nm)
W1-230. . . . . 150-160 in.-lbs.
(17-18 Nm)

Cylinder head bolts/nuts:
W1-080. . . . . 80-97 in.-lbs.
(9-11 Nm)
W1-145, W1-145V,
W1-185, W1-185V. . . . 14-17 ft.-lbs.
(19-23 Nm)
W1-150. . . . . 16-19 ft.-lbs.
(22-26 Nm)
W1-230. . . . . 24-26 ft.-lbs.
(33-35 Nm)

Flywheel nut:
All models . . . . 44-47 ft.-lbs.
(59-64 Nm)

Spark plug:
W1-080, W1-150,
W1-230 . . . . 17-19 ft.-lbs.
(23-26 Nm)
W1-145, W1-145V,
W1-185, W1-185V. . . . . 9-11 ft.-lbs.
(12-14 Nm)

CYLINDER HEAD. To remove cylinder head, first remove fuel tank and blower housing.

Clean carbon from cylinder head and piston. Check cylinder head mounting surface for distortion using a straight-edge and feeler gauge. On all models, renew cylinder head if warpage exceeds 0.006 inch (0.15 mm). Always use a new head gasket when installing cylinder head. Tighten cylinder head bolts or nuts evenly and in stages until torque specified in TIGHTENING TORQUES is obtained.

VALVE SYSTEM. To remove valves, first remove cylinder head and tappet cover/breather assembly. Position valve spring retainer so that notch in outer diameter of retainer faces outward. Compress valve spring and pull spring retainer outward to disengage retainer from valve stem. Remove valve and clean combustion deposits from valves, seats and ports.

On all models, valve face and seat angles should be 45°. Standard seat width is 0.047-0.059 inch (1.2-1.5 mm). Maximum allowable seat width is 0.098 inch (2.5 mm).

Refer to following list for valve stem dimensions:

Valve Stem Diameter

W1-080:
Intake . . . . 0.2152-0.2157 in.
(5.468-5.480 mm)
Exhaust. . . . . 0.2129-0.2143 in.
(5.408-5.444 mm)
Min. dia. (in. & ex.). . . . . . . 0.2105 in.
(5.35 mm)

W1-230:
Intake & exhaust . . 0.2740-0.2731 in.
(6.96-6.938 mm)
Min. dia. . . . . 0.2697 in.
(6.85 mm)

All other models:
Intake . . . . 0.2543-0.2549 in.
(6.460-6.475 mm)
Exhaust. . . . . 0.2528-0.2537 in.
(6.422-6.444 mm)
Min. dia. (in. & ex.) . . . . . . 0.2500 in.
(6.35 mm)

Refer to following list for valve stem-to-guide clearance:

Stem-to-Guide Clearance

W1-080:
Intake . . . . 0.0008-0.0019 in.
(0.02-0.05 mm)
Exhaust. . . . . 0.0022-0.0036 in.
(0.056-0.092 mm)
W1-145, W1-145V, W1-185, W1-185V:
Intake . . . . 0.0010-0.0024 in.
(0.025-0.062 mm)
Exhaust. . . . . 0.0022-0.0039 in.
(0.056-0.100 mm)

W1-150:
Intake. . . . . 0.0010-0.0024 in.
(0.025-0.062 mm)
Exhaust . . . . 0.0030-0.0046 in.
(0.075-0.117 mm)
W1-230:
Intake & exhaust. . 0.0022-0.0038 in.
(0.056-0.098 mm)

Maximum allowable valve stem-to-guide clearance for both valves on all models is 0.012 inch (0.3 mm). If clearance is excessive, renew valve guides using suitable valve guide puller and installation tool.

CAMSHAFT. On all models except W1-150 and some W1-230 models, the camshaft rides in bores in crankcase and crankcase cover or oil pan. On Model W1-150 and some W1-230 models, the camshaft serves as the pto shaft and is supported by a ball bearing in the crankcase cover at the pto end.

When removing camshaft, position engine so tappets will not fall out. If valve tappets are removed, they should be marked so they can be returned to their original positions. If camshaft is renewed, tappets should also be renewed.

Specified camshaft journal diameter is listed below:

Journal Diameter

W1-080:
Gear end. . . . . 0.3932-0.3937 in.
(9.987-10.000 mm)
Min. dia. . . . . 0.3918 in.
(9.95 mm)
Flywheel end . . . . . 0.3926-0.3937 in.
(9.972-10.000 mm)
Min. dia. . . . . 0.3918 in.
(9.95 mm)

W1-150:
Flywheel end . . . . . 0.9838-0.9841 in.
(24.988-24.997 mm)
Min. dia. . . . . 0.9823 in.
(24.95 mm)

W1-230:
Both journals . . . . . 0.6680-0.6688 in.
(16.967-16.977 mm)
Min. dia. . . . . 0.6670 in.
(16.942 mm)

When reinstalling camshaft, align timing marks (Fig. WR36) on camshaft and crankshaft gears. On Model W1-230 with camshaft pto and all Model W1-150, camshaft end play is adjusted using shim spacer (7—Fig. WR34). End play should be 0.000-0.008 inch (0.00-0.20 mm). Install only one shim.

GOVERNOR. The governor assembly is located in the crankcase cover on Models W1-080, W1-145, W1-150, W1-185 and W1-230 (Fig. WR34) or in the oil pan on Models W1-145V and W1-185V (Fig. WR35). The governor gear is

Illustrations courtesy Teledyne Total Power

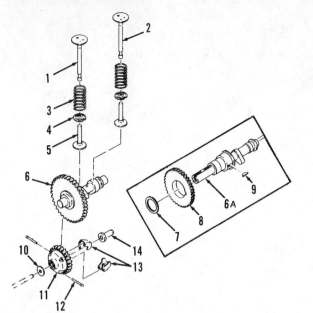

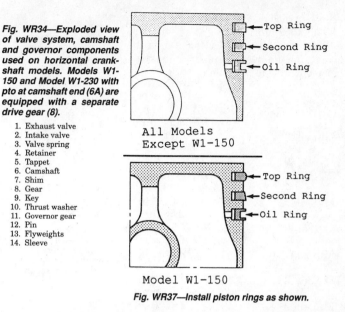

Fig. WR34—Exploded view of valve system, camshaft and governor components used on horizontal crankshaft models. Models W1-150 and Model W1-230 with pto at camshaft end (6A) are equipped with a separate drive gear (8).

1. Exhaust valve
2. Intake valve
3. Valve spring
4. Retainer
5. Tappet
6. Camshaft
7. Shim
8. Gear
9. Key
10. Thrust washer
11. Governor gear
12. Pin
13. Flyweights
14. Sleeve

Fig. WR37—Install piston rings as shown.

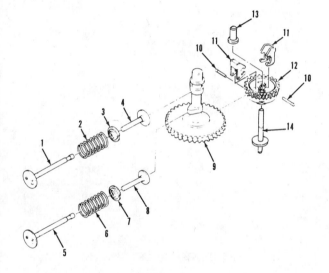

Fig. WR35—Exploded view of governor flyweight assembly used on vertical crankshaft models.

1. Exhaust valve
2. Spring
3. Retainer
4. Tappet
5. Intake valve
6. Spring
7. Retainer
8. Tappet
9. Camshaft & gear
10. Pins
11. Flyweights
12. Governor gear
13. Sleeve
14. Stem

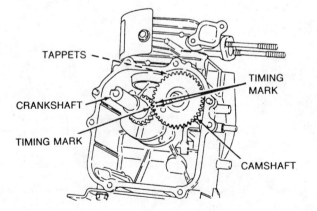

Fig. WR36—Align timing marks on crankshaft and camshaft gears when installing camshaft.

mm); maximum allowable gap is 0.040 inch (1.0 mm). Ring end gap for top two rings on all other models should be 0.008-0.016 inch (0.20-0.40 mm); maximum allowable gap is 0.059 inch (1.5 mm).

Specified piston ring side clearance in piston ring groove is listed below:

Ring Side Clearance

W1-080, W1-145, W1-145V:
Top ring	0.0035-0.0053 in. (0.090-0.135 mm)
Second ring	0.0023-0.0041 in. (0.060-0.105 mm)
Oil ring	0.0004-0.0025 in. (0.010-0.065 mm)

W1-150:
Top ring	0.0020-0.0037 in. (0.050-0.095 mm)
Second ring	0.0016-0.0033 in. (0.040-0.085 mm)

W1-185, W1-185V:
Top ring	0.0019-0.0037 in. (0.050-0.095 mm)
Second ring	0.0004-0.0021 in. (0.010-0.055 mm)
Oil ring	0.0004-0.0025 in. (0.010-0.065 mm)

W1-230:
Top ring	0.0020-0.0037 in. (0.050-0.095 mm)
Second ring	0.0016-0.0033 in. (0.040-0.085 mm)
Oil ring	0.0004-0.0025 in. (0.010-0.065 mm)

On all models, renew piston if ring side clearance exceeds 0.006 inch (0.15 mm).

Specified piston-to-cylinder clearance is 0.0003-0.0018 inch (0.008-0.047 mm) for Model W1-080 and 0.0012-0.0027 inch (0.030-0.069 mm) for Model W1-230. Specified piston clearance for

driven by the camshaft gear on all models. The governor assembly is serviced only as a unit. Components must move freely without binding.

PISTON, PIN AND RINGS. All models are equipped with one compression ring, one scraper ring and one oil control ring. Install rings as shown in Fig. WR37. Stagger ring end gaps at 90° intervals around piston.

Piston ring end gap for the top two rings on Models W1-185V and W1-230 should be 0.002-0.010 inch (0.05-0.25

all other models is 0.0008-0.0023 inch (0.020-0.059 mm).

Refer to following list for standard size piston diameter (measured at skirt perpendicular to pin hole):

	Piston Diameter
W1-080	2.0067-2.0075 in.
	(50.97-50.99 mm)
Min. dia.	2.0035 in.
	(50.89 mm)
W1-145, W1-145V,	
W1-150	2.4787-2.4795 in.
	(62.96-62.98 mm)
Min. dia.	2.4756 in.
	(62.88 mm)
W1-185, W1-185V . .	2.6362-2.6370 in.
	(66.96-66.98 mm)
Min. dia.	2.6331 in.
	(66.88 mm)
W1-230	2.6752-2.6760 in.
	(67.95-67.97 mm)
Min. dia.	2.6720 in.
	(67.87 mm)

Specified piston pin diameter for Model W1-080 is 0.4327-0.4330 inch (10.990-11.000 mm); minimum allowable diameter is 0.4325 inch (10.99 mm). Specified piston pin diameter for Model W1-150 is 0.5115-0.5118 inch (12.992-13.000 mm); minimum allowable diameter is 0.5102 inch (12.96 mm). Specified piston pin diameter for all other models is 0.5509-0.5512 inch (13.992-14.000 mm); minimum allowable diameter is 0.5496 inch (13.96 mm).

On all models, piston pin should be 0.00035 inch (0.009 mm) interference fit to a 0.00039 inch (0.010 mm) loose fit in piston pin bore. Piston pin looseness must not exceed 0.0023 inch (0.06 mm). Heating the piston will make removal and installation of piston pin easier.

Standard piston pin clearance in connecting rod is 0.0004-0.0012 inch (0.010-0.029 mm). If piston pin clearance exceeds 0.0047 inch (0.12 mm), renew piston pin and/or connecting rod.

Piston and rings are available in standard and oversizes.

CONNECTING ROD. To remove piston and connecting rod assembly, remove all cooling shrouds. Remove cylinder head and crankcase cover or oil pan. Remove connecting rod retaining bolts and remove connecting rod and piston assembly from cylinder bore.

Connecting rod-to-crankpin clearance for all models should be 0.0015-0.0025 inch (0.037-0.063 mm). If clearance exceeds 0.008 inch (0.20 mm), renew rod and/or crankshaft.

Connecting rod side clearance on crankshaft for all models should be 0.004-0.012 inch (0.1-0.3 mm). If side clearance exceeds 0.039 inch (1.0 mm),

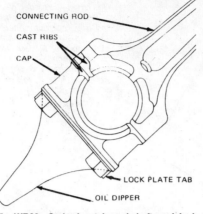

Fig. WR38—On horizontal crankshaft models, install rod cap on rod so ribs or aligning marks are adjacent. On Model W1-150, oil dipper is a part of rod cap and not separate.

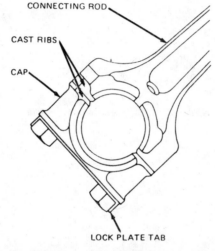

Fig. WR39—On vertical crankshaft models, install rod cap on rod so ribs are adjacent.

renew connecting rod and/or crankshaft.

Specified connecting rod big end diameter is 0.7874-0.7879 inch (20.000-20.013 mm) for Model W1-080, 1.0236-1.0241 inch (26.000-26.013 mm) for Models W1-185, W1-185V and W1-230, and 0.9449-0.9454 inch (24.000-24.013 mm) for all other models.

Connecting rod-to-piston pin clearance for all models should be 0.0004-0.0011 inch (0.010-0.029 mm). If clearance exceeds 0.005 inch (0.12 mm), renew piston pin and/or connecting rod.

When installing connecting rod and piston assembly, make certain match marks (cast ribs) on connecting rod and cap are adjacent to each other as shown in Figs. WR38 and WR39. On horizontal crankshaft engines equipped with a detachable dipper, install dipper with offset facing towards pto end of engine if engine is to be operated on a tilt towards pto side. Mount dipper with offset facing towards flywheel end if operated on

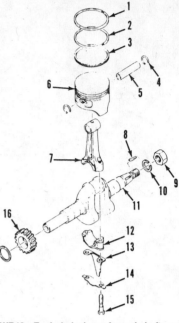

Fig. WR40—Exploded view of crankshaft, connecting rod and piston assembly on Model W1-185. Other models are similar.

1. Top ring	10. Lockwasher
2. Second ring	11. Crankshaft
3. Oil control ring	12. Rod cap
4. Retainer	13. Oil dipper
5. Piston pin	14. Lockplate
6. Piston	15. Cap screw
7. Connecting rod	16. Gear
8. Key	17. Shim
9. Nut	

level or tilted towards flywheel side. On all models, use a new lockplate and tighten connecting rod cap screws to torque specified in TIGHTENING TORQUES.

CRANKSHAFT. The crankshaft on all models is supported by a ball bearing at both ends. Renew bearings if rough, noisy, excessively worn or otherwise damaged.

Crankshaft end play should be 0.000-0.008 inch (0.00-0.20 mm) for all models. End play is controlled by an adjusting collar or shim (17—Fig. WR40) located between crankshaft gear and main bearing. Several thicknesses are available, but only one shim is installed.

To determine the correct thickness of adjusting collar or shim with crankcase cover or oil pan removed, measure distance (A—Fig. WR41) between machined surface of crankcase face and end of crankshaft gear. Measure distance (B) between machined surface of crankcase cover or oil pan and end of main bearing. The compressed thickness of gasket (C) is 0.009 inch (0.23 mm). Select shim or adjusting collar that is 0.001-009 inch (0.03-0.23 mm) less than the total of A, B and C. After

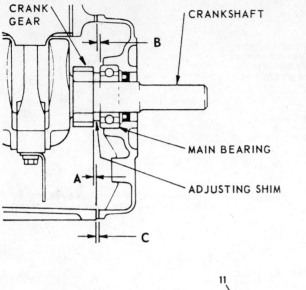

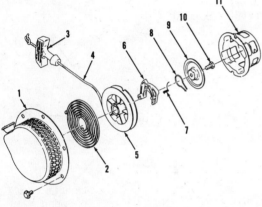

Fig. WR41—View showing location of measuring points to determine correct shim thickness for obtaining correct crankshaft end play. Refer to text.

Fig. WR42—Exploded view of rewind starter used on Model W1-080.

1. Starter housing
2. Rewind spring
3. Rope handle
4. Rope
5. Rope pulley
6. Ratchet
7. Return spring
8. Friction spring
9. Friction plate
10. Screw
11. Starter cup

reassembly, crankshaft end play can be checked with a dial indicator.

Refer to following list for crankpin diameter:

	Crankpin Diameter
W1-080	0.7854-0.7859 in. (19.950-19.963 mm)
Min. dia.	0.781 in. (19.84 mm)
W1-145, W1-145V, W1-150	0.9429-0.9434 in. (23.950-23.963 mm)
Min. dia.	0.9390 in. (23.85 mm)
W1-185, W1-185V, W1-230	1.0216-1.0222 in. (25.950-25.963 mm)
Min. dia.	1.0177 in. (25.85 mm)

Connecting rod-to-crankpin clearance for all models should be 0.0015-0.0025 inch (0.037-0.063 mm). If clearance exceeds 0.008 inch (0.20 mm), renew rod and/or crankshaft. Maximum allowable crankpin taper and out-of-round is 0.0002 inch (0.005 mm).

When reassembling engine, make certain timing marks (Fig. WR36) on crankshaft and camshaft gears are aligned. Install crankshaft seals with lips toward ball bearings.

CYLINDER. If cylinder bore is scored, or out-of-round more than 0.0004 inch (0.01 mm), the cylinder should be bored to next larger oversize for which piston and rings are available.

Refer to following list for standard bore size:

W1-080	2.0079-2.0086 in. (51.000-51.019 mm)
Max. dia.	2.0138 in. (51.15 mm)
W1-145, W1-145V, W1-150	2.4803-2.4811 in. (63.000-63.019 mm)
Max. dia.	2.4862 in. (63.15 mm)
W1-185, W1-185V	2.6378-2.6385 in. (67.000-67.019 mm)
Max. dia.	2.6437 in. (67.15 mm)
W1-230	2.6772-2.6779 in. (68.000-68.019 mm)
Max. dia.	2.6831 in. (68.15 mm)

REWIND STARTER. Model W1-080. To disassemble starter, remove rope handle and allow rope to wind into starter. Unscrew center retaining screw (10—Fig. WR42). Wear appropriate safety eyewear and gloves before disengaging pulley from starter as spring may uncoil uncontrolled. Place shop towel around pulley and lift pulley out of housing; spring should remain in starter housing. If spring must be removed from housing, position housing so spring side is down and against floor, then tap housing to dislodge spring.

Install rewind spring so outer end engages starter housing notch and coils are in counterclockwise direction from outer end. Wrap rope around pulley in counterclockwise direction viewed from flywheel side of pulley. Hook outer end of ratchet spring (7) under ratchet. Apply a light coat of grease to friction spring (8) and install spring in groove of friction plate (9). Install friction plate so the two ends of friction spring (8) fit around lug on ratchet (6). Install center screw. Pretension rewind spring by turning pulley two turns counterclockwise before passing rope through rope outlet of housing and attaching handle.

All Other Models. To disassemble the rewind starter, refer to Fig. WR43 and release spring tension by pulling rope handle until about 14 inches (35 cm) of rope extends from unit. Use thumb pressure to prevent pulley from rewinding and position rope in notch in outer rim of pulley. Release thumb pressure slightly and allow spring mechanism to slowly unwind. Remove clip or nut (2—Fig. WR44 or WR45), and remove components (3 through 7). Slowly remove pulley from support shaft in housing.

CAUTION: Take extreme care that rewind spring remains in recess of housing. Do not remove spring unless new spring is to be installed.

If rewind spring escapes from housing, form a wire ring of same circumference as recess in housing and twist ends

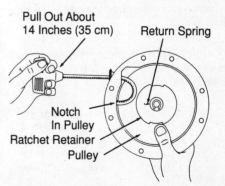

Fig. WR43—View showing method of releasing spring tension on rewind starter assembly.

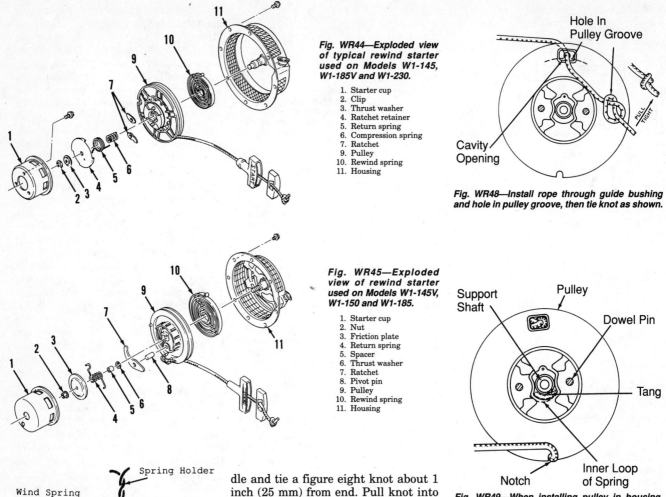

Fig. WR44—Exploded view of typical rewind starter used on Models W1-145, W1-185V and W1-230.

1. Starter cup
2. Clip
3. Thrust washer
4. Ratchet retainer
5. Return spring
6. Compression spring
7. Ratchet
8. ~~~
9. Pulley
10. Rewind spring
11. Housing

Fig. WR45—Exploded view of rewind starter used on Models W1-145V, W1-150 and W1-185.

1. Starter cup
2. Nut
3. Friction plate
4. Return spring
5. Spacer
6. Thrust washer
7. Ratchet
8. Pivot pin
9. Pulley
10. Rewind spring
11. Housing

Fig. WR48—Install rope through guide bushing and hole in pulley groove, then tie knot as shown.

Fig. WR49—When installing pulley in housing, engage inner loop of rewind spring on tang of pulley. Refer to text.

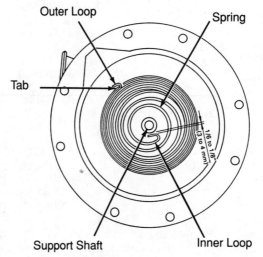

Fig. WR46—Illustration showing fabricated spring holder used to hold rewind spring.

dle and tie a figure eight knot about 1 inch (25 mm) from end. Pull knot into top of handle. Install other end of rope through guide bushing of housing and through hole in pulley groove. Pull rope out through cavity opening and tie a knot about 1 inch (25 mm) from end. Tie knot as illustrated in Fig. WR48 and stuff knot into cavity opening. Wind rope 2½ turns clockwise on pulley, then lock rope in notch of pulley. Install pul-

ley on support shaft and rotate pulley counterclockwise until tang on pulley engages hook on inner loop of rewind spring (see Fig. WR49). Place outer flange of housing in a vise and use finger pressure to keep pulley in housing.

together securely. Starting with outside loop, wind spring inside the ring in counterclockwise direction as shown in Fig. WR46.

NOTE: New rewind springs are secured in a similar wire ring for ease in assembly.

Place new spring assembly over recess in housing so hook in outer loop of spring is over the tension tab in the housing. Carefully press spring from wire ring and into recess of housing. See Fig. WR47.

Using a new rope of same length and diameter as original, place rope in han-

Fig. WR47—Pulley must be installed with hook in outer loop of rewind spring engaged on tension tab and inner loop of spring spaced as shown from support shaft.

Illustrations courtesy Teledyne Total Power

Hook a loop of rope into pulley notch and preload rewind spring by turning pulley four full turns counterclockwise. Remove rope from notch and allow pulley to slowly turn clockwise as rope winds on pulley and handle returns to guide bushing on housing. On Models W1-145V, W1-150 and W1-185, install pivot pin (8—Fig. WR45), if removed.

Refer to Figs. WR44 and WR45 and install components (2 through 7). On Models W1-145, W1-185V and W1-230, install return spring (5—Fig. WR44) with bent end hooked into hole of pulley hub and looped end toward outside, then mount ratchet retainer so loop end of return spring extends through slot. Rotate retainer slightly clockwise until ends of slots just begin to engage the ratchets. Press down on retainer, install washer and clip.

WISCONSIN CENTRAL PARTS DISTRIBUTORS

(Arranged Alphabetically by States)
**These firms carry extensive stocks of repair parts. Contact them
for the name of dealers in their area who will have replacement
parts.**

Joe H. Brady & Assoc.
Phone: (205) 252-8124
308 South 31st Street
Birmingham, Alabama 35233

Parts Service Company, Inc.
Phone: (205) 264-3464
301 Columbus Street
Montgomery, Alabama 36104

Keeling Company
Phone: (501) 945-4511
P.O. Box 15310
North Little Rock, Arkansas 72231

Southwest Products Corp.
Phone: (602) 269-3581
2949 N. 30th Avenue
Phoenix, Arizona 85017

Lanco Engine Services, Inc.
Phone: (619) 429-1313
664 Marsat Court, Suite D
Chula Vista, California 92011

E.E. Richter & Son
Phone: (415) 658-1100
P.O. Box 8186
Emeryville, California 94662

Lanco Engine Services, Inc.
Phone: (213) 329-9300
13610 Gramercy Avenue
Gardena, California 90249

Central Equipment Company
Phone: (303) 388-3696
4477 Garfield Street
Denver, Colorado 80216

Highway Equipment & Supply Co.
Phone: (904) 783-1630
5366 Highway Avenue
Jacksonville, Florida 32205

P.H. Neff & Sons, Inc.
Phone: (305) 592-5240
5295 N.W. 79th Avenue
Miami, Florida 33166

Highway Equipment & Supply Co.
Phone: (305) 843-6310
1016 West Church Street
Orlando, Florida 32805

Highway Equipment & Supply Co.
Phone: (813) 621-9634
6015 U.S. Highway 301 North
Tampa, Florida 33610

Georgia Engine Sales & Service
Phone: (404) 446-1100
5715 B Oakbrook Parkway
Norcross, Georgia 30093

Lanco Engine Services Hawaii, Inc.
Phone: (808) 836-1188
3219 Ualena Street
Honolulu, Hawaii 96819

Teledyne Total Power
Phone: (800) 851-7552
1230 North Skyline Drive
Idaho Falls, Idaho 83402

Wilson Engine Power, Inc.
Phone: (515) 266-1443
5727 N.E. 16th Street
Des Moines, Iowa 50313

Harley Industries, Inc.
Phone: (316) 262-5156
P.O. Box 336
Wichita, Kansas 67201

Wilder Motor & Equipment Co., Inc.
Phone: (502) 966-5141
4022 Produce Road
Louisville, Kentucky 40218

William F. Surgi Equipment Co.
Phone: (504) 733-0101
221 Laitram Lane
Harahan, Louisiana 70123

Diesel Engine Sales & Service
Phone: (617) 341-1760
199 Turnpike Street
Stroughton, Massachusetts 02072

Engine Supply of Novi, Inc.
Phone: (313) 349-9330
44455 Grand River
Novi, Michigan 48050

Teledyne Total Power
Phone: (612) 424-8234
1401 85th Avenue North
Brooklyn Park, Minnesota 55443

Northern Engine & Supply Inc.
Phone: (218) 741-2980
Hoover Road
Virginia, Minnesota 55792

Allied Construction Equipment Co.
Phone: (314) 371-1818
4015 Forest Park Avenue
St. Louis, Missouri 63108

King-McIver Sales, Inc.
Phone: (919) 294-4600
6375 New Burnt Poplar Road
Greensboro, North Carolina 27419

Northern Engine & Supply Co.
Phone: (701) 232-3284
2710 3rd Avenue, North
Fargo, North Dakota 58102

John Reiner Co.
Phone: (201) 460-9444
145 Commerce Road
Carlstadt, New Jersey 07072

Central Motive Power
Phone: (505) 884-2525
3740 Princeton Drive, N.E.
Albuquerque, New Mexico 87107

John Reiner & Co., Inc.
Phone: (315) 454-4490
6681 Moore Road
Syracuse, New York 13211

Cincinnati Engine & Parts Co., Inc.
Phone: (513) 221-3525
2863 Stanton Avenue
Cincinnati, Ohio 45206

Cincinnati Engine Parts Co.
Phone: (419) 589-4775
1051 Lucas Road
Mansfield, Ohio 44905

Harley Industries, Inc.
Phone: (918) 627-9220
5408 South 103rd East Avenue
Tulsa, Oklahoma 74147

Lucky Distributing
Phone: (503) 252-1249
8111 NE Columbia Blvd.
Portland, Oregon 97218

Joseph L. Pinto, Inc.
Phone: (215) 747-3877
719 East Baltimore Pike
East Lansdowne, Pennsylvania 19050

Teledyne Total Power
Phone: (412) 863-8874
1061 Main Street
N. Huntington, Pennsylvania 15642

Wilder Motor & Equipment
Phone: (803) 799-1220
1219 Rosewood Drive
Columbia, South Carolina 29201

Southern Supply Company
Phone: (901) 424-1900
366 North Royal Street
Jackson, Tennessee 38301

RCH Distributors, Inc.
Phone: (901) 345-2200
3140 Carrier St.
Memphis, Tennessee 38101

Wilder Motor & Equipment Co.
Phone: (615) 329-2365
301 15th Avenue North
Nashville, Tennessee 37203

Harley Industries, Inc.
Phone: (214) 638-4504
9226 Premier Row
Dallas, Texas 75247

Harley Industries, Inc.
Phone: (713) 466-8999
12227 L FM529 (Spencer)
Houston, Texas 77041

Teledyne Total Power
Phone: (703) 752-9397
1127 International Parkway
Fredericksburg, Virginia 22405

Engine Sales & Service Co.
Phone: (304) 342-2131
601 Ohio Avenue
Charleston, West Virginia 25302

Teledyne Total Power
Phone: (414) 786-1600
2244 West Bluemound Road
Waukesha, Wisconsin 53187

CANADIAN DISTRIBUTORS

Mandem Inc.
Phone: (403) 465-0244
5925 83rd Street
Edmonton, Alberta T6E 4

Pacific Engines & Equipment
Phone: (604) 254-0804
1391 William Street
Vancouver, British Columbia V5L 2

Mandem Inc.
Phone: (204) 885-9790
21 Murray Park Road
Winnipeg, Manitoba R3J 3

Mandem Inc.
Phone: (506) 854-0982
481 Edinburgh Drive
Moncton, New Brunswick E1E 4

Mandem Inc.
Phone: (416) 255-8158
3 Bestobell Road
Toronto, Ontario M8W 4

Mandem Inc.
Phone: (514) 342-9233
8550 Delmeade Road
Montreal, Quebec H4T 1

United Continental Engines
Phone: (514) 739-2751
8550 Delmeade Road
Montreal, Quebec H4T 1

Mandem Inc.
Phone: (306) 352-2631
1250 St. John Street
Regina, Saskatchewan S4R 1

YAMAHA

YAMAHA MOTOR CORPORATION
6555 Katella Avenue
Cypress, California 90630

Model	Bore	Stroke	Displacement
EF600	48 mm	39 mm	70.6 cc
	(1.89 in.)	(1.54 in.)	(4.3 cu.in.)

NOTE: Metric fasteners are used throughout engine.

ENGINE IDENTIFICATION

This Yamaha engine is used on the Yamaha Model EF600 generator. The engine is a four-stroke, single-cylinder, air-cooled, horizontal crankshaft engine.

MAINTENANCE

LUBRICATION. The engine is splash lubricated by an oil slinger mounted on bottom of connecting rod cap. Engine oil level should be checked prior to each operating interval. Maintain oil level at lower edge of fill plug opening.

Engine oil should be changed after the first 20 hours of operation and every 150 hours of operation or three months thereafter. Crankcase capacity is 0.35 L (0.37 qt.).

Engine oil should meet or exceed latest API service classification. Normally, SAE 30 oil is recommended for use when temperatures are above 60° F (15° C). SAE 20 oil is recommended when temperatures are between 60° F (15° C) and 32° F (0° C). SAE 10 oil is recommended when temperatures are below 32° F (0° C).

AIR FILTER. The engine is equipped with a foam type air filter element. The air filter element should be removed and cleaned after every 150 hours of operation, or more often if operating in extremely dusty conditions. To clean element, wash in a nonflammable cleaning solvent and gently squeeze element dry. Oil element with clean engine oil and gently squeeze out the excess oil. Apply grease to outer edge of element to seal edge. Install element so flocked side is out.

FUEL FILTER. A fuel filter is located in the fuel tank. The fuel filter should be inspected after the first 20 hours of operation and every 150 hours thereafter.

A fuel filter is also located in the fuel pump. See FUEL PUMP in REPAIRS section.

CRANKCASE BREATHER. The crankcase breather is located in the valve cover on the side of the engine. Regular maintenance is not required.

SPARK PLUG. Recommended spark plug is NGK BPMR6A or equivalent. Spark plug should be removed, checked and cleaned after every 50 hours of operation. Renew spark plug if electrode is corroded or damaged. Spark plug electrode gap should be set at 0.6-0.7 mm (0.024-0.028 in.).

CARBURETOR. Adjustment. Idle mixture is controlled by idle jet (8—Fig. Y1)) and idle mixture screw (7). Initial setting of idle mixture screw is ¾ turn out. Make final adjustment of idle mixture with engine running at normal operating temperature. Adjust idle mixture screw so engine runs smoothly at idle speed and engine will accelerate without hesitation. Standard idle mixture jet size is #37.5. High speed operation is controlled by fixed main jet. Standard main jet size is #56.3.

Overhaul. To disassemble carburetor, remove fuel bowl retaining screw (18), gasket (17) and fuel bowl (16). Remove float pin (14), float (13) and fuel inlet valve (11). Remove throttle and choke shaft assemblies after unscrewing throttle and choke plate retaining screws. Remove idle mixture screw (7), idle mixture jet (8), main jet and main fuel nozzle (15).

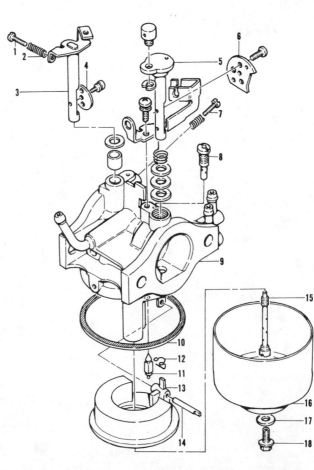

Fig. Y1—Exploded view of carburetor.

1. Idle speed screw
2. Spring
3. Throttle shaft
4. Throttle plate
5. Choke shaft
6. Choke plate
7. Idle mixture screw
8. Idle jet
9. Body
10. Gasket
11. Fuel inlet valve
12. Clip
13. Float
14. Float pin
15. Nozzle
16. Fuel bowl
17. Gasket
18. Screw

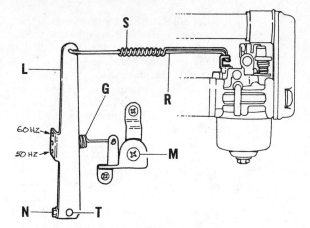

Fig. Y2—Drawing of governor linkage. Note governor spring attachment holes on governor lever and corresponding generator current frequency.

G. Governor spring
L. Governor lever
M. Maximum governed speed screw
N. Nut
R. Throttle rod
S. Spring
T. Governor shaft

Metal parts may be cleaned in carburetor cleaner. Direct compressed air through orifices and passageways in the opposite direction of normal fuel flow. Do not use wires or drill bits to clean orifices as enlargement of the orifice could affect the calibration of the carburetor. Inspect carburetor and renew any damaged or excessively worn components.

When assembling the carburetor note the following. With carburetor inverted, float height measured from fuel bowl mating surface (gasket removed) should be 17.0 mm (0.67 in.). Float height is not adjustable; replace any components which are damaged or excessively worn and adversely affect float position.

Refer also to FUEL PUMP in REPAIRS section for fuel system hose routing diagram.

GOVERNOR. The engine is equipped with a mechanical, flyweight type governor. To adjust governor linkage, proceed as follows: Loosen governor lever clamp nut (N—Fig. Y2), rotate governor lever (L) so throttle plate is fully open and hold lever in place. Turn governor shaft (T) clockwise as far as possible, then tighten nut (N).

Note attachment points in Fig. Y2 for governor spring on governor lever according to required generator current frequency.

Rotating maximum governed engine speed screw (M) will adjust generator current frequency. Engine speed (no-load) should be approximately 4350 rpm.

IGNITION SYSTEM. The engine is equipped with a capacitor discharge ignition system. The trigger coil and charge coil are attached to a stator plate behind the flywheel. The CDI module is attached to the side of the blower housing and the ignition coil is mounted on the inside of the blower housing. Ignition timing is not adjustable.

To check the trigger coil and charge coil, remove the flywheel. To check the trigger coil, connect an ohmmeter to white/red lead and black lead on trigger coil. Ohmmeter reading should be 33.3-40.7 ohms. To check the charge coil, connect an ohmmeter to brown lead and black lead on charge coil. Ohmmeter reading should be 315-385 ohms. When reinstalling flywheel, tighten flywheel nut to 32-36 N•m (24-26 ft.-lbs.).

To check ignition coil primary side, connect one ohmmeter lead to primary lead and touch iron coil laminations with remaining lead. Ohmmeter should register 1.4-1.8 ohms. To check ignition coil secondary, connect one ohmmeter lead to the spark plug lead wire and remaining lead to the iron core laminations. Ohmmeter should read 5.3k-7.9k ohms.

Test specifications are not available for CDI module. The CDI should be suspected if an ignition malfunction occurs and testing indicates other components are satisfactory.

VALVE ADJUSTMENT. Clearance between tappet and end of valve stem, with engine cold, should be 0.05-0.15 mm (0.002-0.006 in.) for intake and exhuast. To check clearance, position piston at top dead center on compression stroke. Remove tappet chamber cover/breather assembly from engine. Measure clearance between each tappet and valve stem end using a feeler gauge.

To adjust valve clearance, valves must be removed from engine. If clearance is too large, renew valve to reduce clearance. If clearance is too small, grind end of valve stem to increase clearance.

COMPRESSION PRESSURE. Specified compression pressure reading is 588 kPa (85 psi). Minimum compression pressure reading is 490 kPa (71 psi).

REPAIRS

TIGHTENING TORQUES. Recommended tightening torques specifications are as follows:

Generator rotor . . . . 18-23 N•m
(160-204 in.-lbs.)
Connecting rod. . . . . 5.4-6.4 N•m
(48-57 in.-lbs.)
Crankcase cover. . . . . 8.8-9.8 N•m
(78-87 in.-lbs.)
Cylinder head. . . . . 8.8-9.8 N•m
(78-87 in.-lbs.)
Flywheel. . . . . 32-36 N•m
(24-26 ft.-lbs.)

FUEL PUMP. A diaphragm type pump delivers fuel from the fuel tank to the carburetor. Refer to Fig. Y3 for a drawing of fuel system hose routing. A fuel valve (9) on the fuel pump is controlled by the speed control lever. A fuel filter in the fuel valve is accessible after removing sediment cup (9—Fig. Y4).

Refer to Fig. Y4 for an exploded view of fuel pump. A diaphragm kit is available for servcing the pump.

CYLINDER HEAD. The cylinder head (2—Fig. Y5) may be removed after removing air cleaner assembly, blower housing and cylinder air shrouds.

Clean combustion deposits from cylinder head, then inspect for cracks or other damage. Check cylinder head mounting surface for warpage using a straightedge and feeler gauge. Minor distortion of cylinder head surface can be trued by moving the cylinder head across 400 grit sandpaper placed on a surface plate.

Use a new head gasket when installing cylinder head. Tighten cylinder head screws in a crossing pattern to 8.8-9.8 N•m (78-87 in.-lbs.).

VALVE SYSTEM. The valves are accessible after removing cylinder head and valve cover/breather from engine (Fig. Y5). Identify parts as they are removed so they can be reinstalled in their original positions. Compress the valve springs and disengage the slotted spring retainers from valve stems. Remove valves from the cylinder block.

Clean carbon from valve stem and head. Inspect valves for excessive wear or damage and renew as necessary.

Valve face and seat angles are 45° for intake and exhaust. Recommended valve seating width is 0.7 mm (0.028 in.) with a maximum allowable seat width of 1.2 mm (0.047 in.). Valve should be

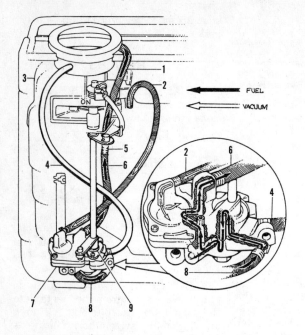

tappets fall away from cam lobes, then remove camshaft from crankcase. Identify tappets so they can be reinstalled in their original positions, then remove tappets from crankcase.

Specified camshaft journal diameter is 10.00 mm (0.394 in.). Minimum camshaft journal diameter is 9.90 mm (0.390 in.).

Make certain that camshaft lobes are smooth and free of scoring, pitting and other damage. Minimum lobe height for intake and exhaust lobes is 19.60 mm (0.772 in.). If camshaft is renewed, the tappets should also be renewed.

When installing camshaft, make certain that crankshaft gear and camshaft gear timing marks (T—Fig. Y7) are aligned.

PISTON, PIN AND RINGS. Piston (9—Fig. Y6) and connecting rod (12) are removed as an assembly after cylinder head and crankcase cover have been removed. Remove carbon deposits (if present) from upper end of cylinder before removing the piston. Remove connecting rod cap (19), then push piston and connecting rod assembly out top of cylinder. Remove retaining rings (11) and push pin (10) out of piston to separate piston from connecting rod.

Specified piston clearance in bore is 0.020-0.040 mm (0.0008-0.0016 in.). Measure piston diameter 10 mm (0.39

lapped to its seat if poor valve seating or sealing is evident.

Standard valve stem diameter for both valves is 5.50 mm (0.2165 in.). Minimum valve stem diameter is 5.40 mm (0.2126 in.) for intake valve and 5.38 mm (0.2118 in.) for exhaust valve.

Valve guides are integral with cylinder block. Cylinder block should be renewed if valve guides are worn excessively.

Specified valve spring standard free length is 23.00 mm (0.90 in.) and minimum free length is 20.0 mm (0.79 in.) for both valve springs.

CAMSHAFT. Camshaft (13—Fig. Y6) is supported directly in bores of the aluminum crankcase and crankcase cover. To remove camshaft, remove generator housing, generator and crankcase cover (24). Rotate crankshaft so piston is at top dead center on compression stroke. Turn engine upside down so

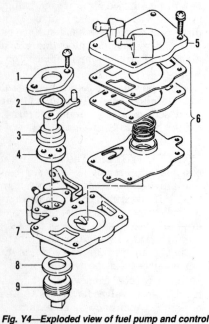

Fig. Y4—Exploded view of fuel pump and control valve.

1. Retainer plate
2. Spring washer
3. Control lever
4. Valve disc
5. Pump cover
6. Diaphragm & valve assy.
7. Pump housing
8. Gasket
9. Sediment bowl

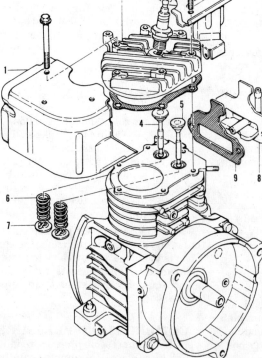

Fig. Y5—Exploded view of cylinder head and valves.

1. Air shroud
2. Cylinder head
3. Air shroud
4. Intake valve
5. Exhaust valve
6. Valve spring
7. Valve retainer
8. Valve cover & breather
9. Gasket

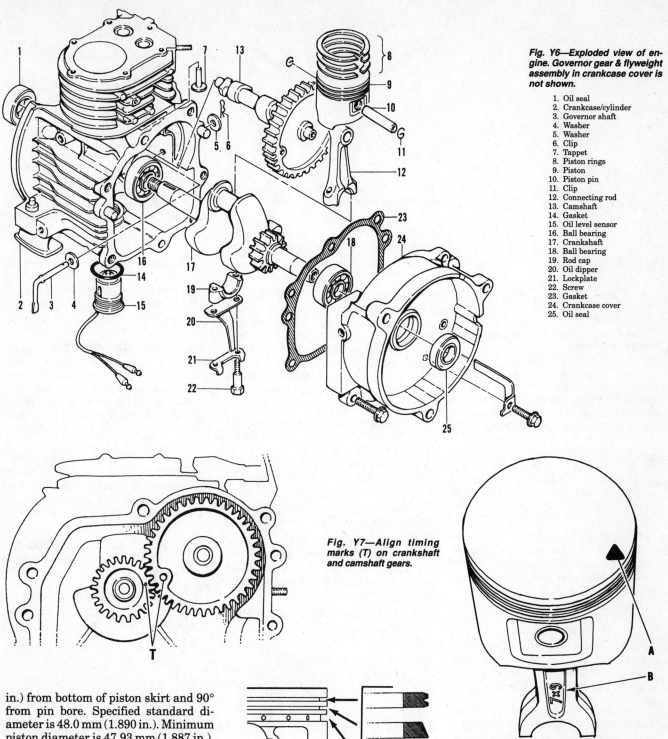

Fig. Y6—Exploded view of engine. Governor gear & flyweight assembly in crankcase cover is not shown.

1. Oil seal
2. Crankcase/cylinder
3. Governor shaft
4. Washer
5. Washer
6. Clip
7. Tappet
8. Piston rings
9. Piston
10. Piston pin
11. Clip
12. Connecting rod
13. Camshaft
14. Gasket
15. Oil level sensor
16. Ball bearing
17. Crankshaft
18. Ball bearing
19. Rod cap
20. Oil dipper
21. Lockplate
22. Screw
23. Gasket
24. Crankcase cover
25. Oil seal

Fig. Y7—Align timing marks (T) on crankshaft and camshaft gears.

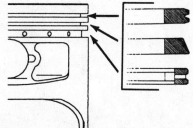

Fig. Y8—Install piston rings on piston in locations shown.

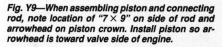

Fig. Y9—When assembling piston and connecting rod, note location of "7 × 9" on side of rod and arrowhead on piston crown. Install piston so arrowhead is toward valve side of engine.

in.) from bottom of piston skirt and 90° from pin bore. Specified standard diameter is 48.0 mm (1.890 in.). Minimum piston diameter is 47.93 mm (1.887 in.). Oversize piston and rings are available.

Specified ring side clearance is 0.05 mm (0.002 in.) for top compression ring and 0.04 mm (0.0016 in.) for second compression ring and oil control ring. Maximum piston ring side clearance for all rings is 0.15 mm (0.006 in.). Specified ring end gap is 0.3 mm (0.012 in.) for all rings. Maximum ring end gap is 1.0 mm (0.040 in.).

Piston pin should be a push fit in piston. Specified piston pin diameter is 10.00 mm (0.394 in.). Minimum allow-

able piston pin diameter is 9.95 mm (0.392 in.).

When installing rings on piston, make sure that manufacturer's mark on

face of ring is toward piston crown. Refer to Fig. Y8 for configuration and location of piston rings.

Install piston on connecting rod so side of connecting rod marked "7 × 9" and arrowhead on piston crown are positioned as shown in Fig. Y9. Lubricate cylinder, piston, piston pin, rings and

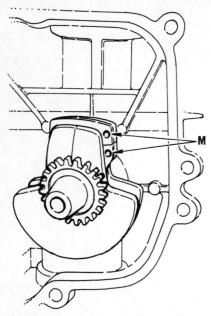

Fig. Y10—Align match marks (M) on rod and rod cap during assembly.

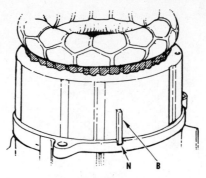

Fig. Y11—Align boss (B) on stator with notch (N) in crankcase cover during assembly.

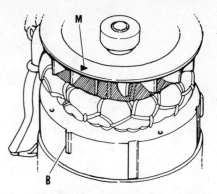

Fig. Y12—With piston at top dead center, boss (B) on stator and mark (M) on rotor fan must be aligned.

connecting rod bearing with SAE 30 engine oil prior to installation in engine. Install piston in engine so arrowhead (A) on piston crown points toward valve side of engine. Align match marks (M—Fig. Y10) on connecting rod and rod cap. Install oil dipper (20—Fig. Y6) so it points toward crankcase cover. Tighten connecting rod screws to 5.4-6.4 N·m (48-57 in.-lbs.).

CONNECTING ROD. Connecting rod rides directly on crankshaft journal. Connecting rod and piston are removed as an assembly as outlined in previous section.

Specified clearance between crankpin journal and connecting rod bore is 0.015-0.027 mm (0.0006-0.0011 in.). Specified connecting rod big end diameter is 18.00 mm (0.709 in.). Maximum allowable inside diameter for connecting rod big end is 18.050 mm (0.7106 in.). Specified connecting rod small end diameter is 10.00 mm (0.394 in.). Maximum allowable connecting rod small end diameter is 10.05 mm (0.3957 in.).

Standard connecting rod side clearance is 0.1-0.4 mm (0.004-0.016 in.). Maximum allowable side clearance is 1.0 mm (0.039 in.).

Install connecting rod as outlined in previous section.

GOVERNOR. The internal centrifugal flyweight governor is mounted on the crankcase cover. The governor is driven by the camshaft gear. Components must move freely without binding. The governor gear is retained by a

snap ring. Be sure snap ring is properly seated in groove on shaft during assembly.

CRANKSHAFT, MAIN BEARINGS AND OIL SEALS. The crankshaft is supported at each end by ball bearing type main bearings (16 and 18—Fig. Y6) located in the crankcase and crankcase cover. The crankshaft may be removed after removing piston and connecting rod as previously outlined.

Standard crankpin journal diameter is 18.00 mm (0.7087 in.). Minimum allowable crankpin diameter is 17.90 mm (0.705 in.). Maximum allowable crankshaft runout is 0.02 mm (0.001 in.).

When installing crankshaft, make certain crankshaft and camshaft gear timing marks (T—Fig. Y7) are aligned.

Inspect main bearings and renew if rough, loose or damaged. Install bearings by pressing against numbered side until bearing is seated against shoulder in crankcase or crankcase cover.

Install oil seals so lip is toward main bearing.

CYLINDER. If cylinder bore exceeds 48.15 mm (1.896 in.) or if out-of-round of bore exceeds 0.15 mm (0.006 in.), cylinder should be bored to the next oversize.

OIL WARNING SYSTEM. The engine may be equipped with a low-oil warning system that grounds the ignition system and lights a warning lamp if the oil level is low. The oil level sensor is located in the bottom of the crankcase.

To test oil level sensor, disconnect sensor leads and connect an ohmmeter or continuity tester to sensor leads. With sensor in normal position, there should be continuity. With sensor inverted, there should be no continuity.

GENERATOR STATOR AND ROTOR. Install stator in crankcase cover so boss (B—Fig. Y11) is aligned with notch (N) in crankcase cover. Install rotor so mark (M—Fig. Y12) on rotor fan is aligned with boss (B) on stator when piston is at top dead center. Tighten

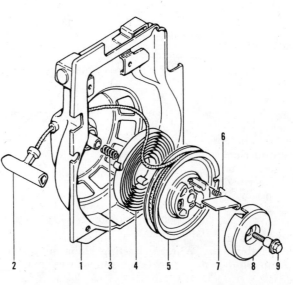

Fig. Y13—Exploded view of rewind starter.

1. Blower housing
2. Rope handle
3. Friction spring
4. Rewind spring
5. Pulley
6. Spring
7. Pawl
8. Retainer
9. Screw

rotor retaining screw to 18-23 N•m (160-204 in.-lbs.). Install rear cover so notch in bearing bore aligns with pin in rotor bearing outer race.

REWIND STARTER. The rewind starter is mounted on the blower housing. To disassemble starter, release preload tension of rewind spring by removing rope handle and allowing rope to wind slowly into starter. Remove retainer screw (9—Fig. Y13), retainer (8) and spring (3). Remove pawl (7) and spring (6). Remove pulley (5) with rewind spring (4). Wear appropriate safety eyewear and gloves before remov-ing pulley or rewind spring from pulley as spring may uncoil uncontrolled.

To reassemble, reverse the disassembly procedure. Spring (4) should be lightly greased. Install rewind spring on pulley so coil direction is counterclockwise from outer end. Wind rope around pulley in counterclockwise direction as viewed from pawl side of pulley. Position pulley in starter housing and turn pulley counterclockwise until spring tension is felt. Pass rope through rope outlet and install rope handle. Install pawl (7) and spring (6) so spring end forces pawl toward center of pulley.

Pawl should point in direction shown in Fig. Y13. Install spring (3), retainer (8) and screw (9). To apply spring tension to pulley, pull a loop of rope into notch in pulley and rotate pulley counterclockwise a couple of turns, then pull rope out of notch and allow rope to wind into starter housing. Rope handle should be snug against housing, if not, increase spring tension by rotating pulley another turn. Check for excessive spring tension by pulling rope to fully extended length. With rope fully extended it should be possible to rotate pulley at least $\frac{1}{2}$ turn counterclockwise, if not reduce spring tension one turn.

YAMAHA

Model	Bore	Stroke	Displacement
EF1000	50 mm	43 mm	84 cc
	(1.97 in.)	(1.69 in.)	(5.2 cu. in.)

NOTE: Metric fasteners are used throughout engine.

ENGINE INFORMATION

This Yamaha engine is used on the Yamaha Model EF1000 generator. The engine is a four-stroke, single-cylinder, air-cooled engine. The engine is equipped with a horizontal crankshaft and overhead valve system.

MAINTENANCE

LUBRICATION. The engine is splash lubricated by an oil slinger mounted on bottom of connecting rod cap. Engine oil level should be checked prior to each operating interval. Maintain oil level at lower edge of fill plug opening.

Engine oil should be changed after the first 20 hours of operation and every 100 hours of operation or six months thereafter. Crankcase capacity is 0.43 L (0.45 qt.).

Engine oil should meet or exceed latest API service classification. Normally, SAE 30 oil is recommended for use when temperatures are above 60° F (15° C). SAE 20 oil is recommended when temperatures are between 60° F (15° C) and 32° F (0° C). SAE 10 oil is recommended when temperatures are below 32° F (0° C).

The engine may be equipped with a low-oil warning system that grounds the ignition system and lights a warning lamp if the oil level is low. See OIL WARNING SYSTEM in REPAIRS section.

AIR FILTER. The engine is equipped with a foam type air filter element. The air filter element should be removed and cleaned after every 50 hours of operation, or more often if operating in extremely dusty conditions. To clean element, wash in a nonflammable cleaning solvent and gently squeeze element dry. Oil element with clean engine oil and gently squeeze out the excess oil. Apply grease to outer edge of element to seal edge. Install element so flocked side is out.

FUEL FILTER. A fuel filter screen is located on the pickup tube of the fuel valve attached to the bottom of the fuel tank.

CRANKCASE BREATHER. The crankcase breather is located in the side of the crankcase. Regular maintenance is not required.

SPARK PLUG. Recommended spark plug is NGK BPR6HS or equivalent. Renew spark plug if electrode is corroded or damaged. Spark plug electrode gap should be set at 0.6-0.7 mm (0.024-0.028 in.). Tighten spark plug to 14.7-19.6 N·m (130-173 in.-lbs.).

CARBURETOR. Adjustment. Idle mixture is controlled by idle jet (6—Fig. Y101) and is not adjustable. Standard idle mixture jet size is #50. High speed operation is controlled by fixed main jet (8). Standard main jet size is #58.8.

The choke control cable should be adjusted so choke plate opening and closing is synchronized with choke control lever position.

Overhaul. To disassemble carburetor, remove fuel bowl retaining screw (17), gasket (16) and fuel bowl (15). Remove float pin (13), float (12) and fuel inlet valve (9). Remove throttle and choke shaft assemblies after unscrewing throttle and choke plate retaining screws. Remove idle mixture jet (6), main jet (8) and main fuel nozzle (14).

Metal parts may be cleaned in carburetor cleaner. Direct compressed air through orifices and passageways in the opposite direction of normal fuel flow. Do not use wires or drill bits to clean orifices as enlargement of the orifice

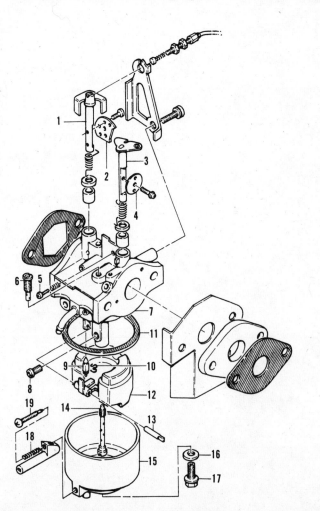

Fig. Y101—Exploded view of carburetor.

1. Choke shaft
2. Choke plate
3. Throttle shaft
4. Throttle plate
5. Idle speed screw
6. Idle jet
7. Body
8. Main jet
9. Fuel inlet valve
10. Clip
11. Gasket
12. Float
13. Float pin
14. Nozzle
15. Fuel bowl
16. Gasket
17. Screw
18. Spring
19. Drain screw

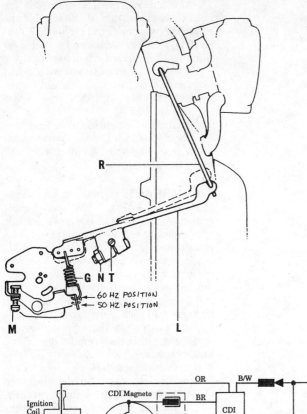

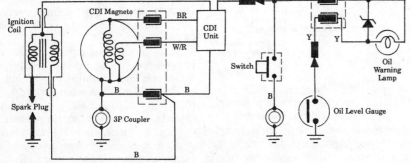

Fig. Y102—Drawing of governor linkage. Note governor spring attachment holes on governor lever and corresponding generator current frequency.

G. Governor spring
L. Governor lever
M. Maximum governed speed screw
N. Nut
R. Throttle rod
T. Governor shaft

Fig. Y103—Wiring schematic for ignition system and oil sensor circuit.

B. Black	Or. Orange	W. White
Br. Brown	R. Red	Y. Yellow

could affect the calibration of the carburetor. Inspect carburetor and renew any damaged or excessively worn components.

When assembling the carburetor note the following. With carburetor inverted, float height measured from fuel bowl mating surface (gasket removed) should be 17.0 mm (0.67 in.). Replace any components which are damaged or excessively worn and adversely affect float position. Install fuel bowl so drain screw (19) is located just to right of fuel inlet tube.

GOVERNOR. The engine is equipped with a mechanical, flyweight type governor. To adjust governor linkage, proceed as follows: Loosen governor lever clamp nut (N—Fig. Y102), rotate governor lever (L) so throttle plate is fully open and hold lever in place. Turn governor shaft (T) counterclockwise as far as possible, then tighten nut (N).

Note attachment points in Fig. Y102 for governor spring on governor lever according to required generator current frequency.

Rotating maximum governed engine speed screw (M) will adjust generator current frequency.

IGNITION SYSTEM. The engine is equipped with a capacitor discharge ignition system. The trigger coil and charge coil are attached to a stator plate behind the flywheel. The CDI module is attached to the side of the blower housing and the ignition coil is mounted on the inside of the blower housing. Ignition timing is not adjustable. Refer to Fig. Y103 for wiring schematic.

To check the trigger coil and charge coil, remove the flywheel. To check the trigger coil, connect an ohmmeter to white/red lead and black lead on trigger coil. Ohmmeter reading should be 34.2-41.8 ohms. To check the charge coil, connect an ohmmeter to brown lead and black lead on charge coil. Ohmmeter reading should be 315-385 ohms. When reinstalling flywheel, tighten flywheel nut to 35-45 N·m (26-33 ft.-lbs.).

To check ignition coil primary side, connect ohmmeter leads to wire terminals on ignition coil. Ohmmeter should register 0.21-0.25 ohms. To check ignition coil secondary, connect one ohmmeter lead to the spark plug lead wire and remaining lead to the large wire terminal. Ohmmeter should read 4.49k-6.73k ohms.

Test specifications are not available for CDI module. The CDI should be suspected if an ignition malfunction occurs and testing indicates other components are satisfactory.

The engine may be equipped with a low-oil warning system that grounds the ignition system and lights a warning lamp if the oil level is low. See OIL WARNING SYSTEM in REPAIRS section.

VALVE ADJUSTMENT. Valve-to-rocker arm clearance should be checked and adjusted after every 300 hours of operation. Engine must be cold when checking valve clearance. Specified clearance between valve stem and rocker arm is 0.1 mm (0.004 in.) for both valves.

To adjust valve clearance, remove rocker arm cover. Rotate crankshaft so piston is at top dead center (TDC) on compression stroke. Insert a 0.1 (0.004 in.) feeler gauge between rocker arm (3—Fig. Y104) and end of valve stem (4). Loosen rocker arm jam nut (1) and turn adjusting nut (2) to obtain desired clearance (a slight drag should be felt when withdrawing the feeler gauge). Tighten jam nut and recheck clearance. Install rocker arm cover.

COMPRESSION PRESSURE. Compression pressure reading may be checked with compression release

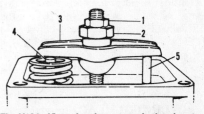

Fig. Y104—View of rocker arm and related parts.

1. Jam nut
2. Adjustment nut
3. Rocker arm
4. Valve stem clearance
5. Push rod

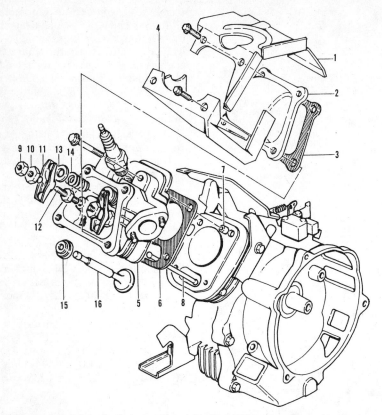

Fig. Y105—Exploded view of cylinder head assembly. Seal (15) is not used on exhaust valve.

1. Air shroud	5. Cylinder head	9. Jam nut	13. Valve retainer
2. Rocker arm cover	6. Head gasket	10. Adjustment nut	14. Valve spring
3. Gasket	7. Dowel pin	11. Rocker arm	15. Seal
4. Air shroud	8. Push rod	12. Stud	16. Intake valve

mechanism operational or disengaged. With compression release operating, compression reading should be 193-490 kPa (28-71 psi). With compression release disengaged (remove exhaust push rod), compression reading should be 1076-1276 kPa (156-185 psi).

REPAIRS

TIGHTENING TORQUES. Recommended tightening torques specifications are as follows:

Generator rotor . . . . 18-24 N•m
(160-212 in.-lbs.)
Connecting rod . . . . 7.8-9.8 N•m
(69-87 in.-lbs.)
Crankcase cover . . . . . . 8.8-10.8 N•m
(78-96 in.-lbs.)
Cylinder head . . . . 9.8-11.8 N•m
(87-104 in.-lbs.)
Flywheel . . . . 35-45 N•m
(26-33 ft.-lbs.)
Spark plug . . . . 14.7-19.6 N•m
(130-173 in.-lbs.)

CYLINDER HEAD. To remove cylinder head (5—Fig. Y105), remove carburetor and muffler assemblies. Remove blower shrouds and rocker arm

cover. Unscrew cylinder head mounting bolts and remove cylinder head.

Clean combustion deposits from cylinder head, then check for cracks or other damage. Check cylinder head mounting surface for warpage using a feeler gauge and straightedge. Minor distortion of cylinder head surface can be trued by moving cylinder head across 400 grit sandpaper placed on a surface plate.

Use a new head gasket when installing cylinder head. Tighten cylinder head bolts evenly in a crossing pattern to 9.8-11.8 N•m (87-104 in.-lbs.). Adjust valves as outlined in VALVE ADJUSTMENT paragraph.

VALVE SYSTEM. To remove valves, remove rocker arms. Compress valve springs and remove valve retainers. Remove valves and springs. Note that intake valve is equipped with a seal.

Clean carbon from valve stem and head. Inspect valves for excessive wear or damage and renew as necessary.

Valve face and seat angles are 45°. Recommended valve seating width is 0.7 mm (0.028 in.) with a maximum allowable seat width of 1.2 mm (0.047 in.). Valve should be lapped to its seat if poor valve seating or sealing is evident.

Standard valve stem diameter for both valves is 5.0 mm (0.1968 in.). Minimum valve stem diameter is 4.87 mm (0.1917 in.) for intake valve and 4.89 mm (0.1925 in.) for exhaust valve.

Valve guides are integral with cylinder head. Cylinder head should be renewed if valve guides are worn excessively.

Specified valve spring standard free length is 27.4 mm (1.08 in.) and minimum free length is 24.4 mm (0.96 in.) for both valve springs.

CAMSHAFT. Camshaft and camshaft gear (19—Fig. Y106) are an integral casting which is equipped with a compression release mechanism. The compression release pin (P—Fig. Y107) extends at cranking speed to hold the exhaust valve open slightly thereby reducing compression pressure.

To remove camshaft, remove blower shrouds and rocker arm cover and disengage push rods from rocker arms. Drain engine oil, then remove generator and crankcase cover (23—Fig. Y106) and withdraw camshaft.

Specified camshaft journal diameter is 12.00 mm (0.472 in.). Minimum camshaft journal diameter is 11.90 mm (0.4685 in.).

Make certain that camshaft lobes are smooth and free of scoring, pitting and other damage. Minimum lobe height for intake and exhaust lobes is 20.25 mm (0.797 in.). If camshaft is renewed, the tappets (12) should also be renewed.

Inspect compression release mechanism and check for proper operation. Compression release spring (S—Fig. Y107) should hold weight (W) against stops. With weight against stops, pin (P—Fig. Y108) projection (H) above cam lobe (L) should be 0.7-1.0 mm (0.028-0.039 in.).

When installing camshaft, make certain that crankshaft gear and camshaft gear timing marks (T—Fig. Y106) are aligned. Tighten crankcase cover screws in a crossing pattern to 8.8-10.8 N•m (78-96 in.-lbs.).

PISTON, PIN AND RINGS. Piston and connecting rod are removed as an assembly after cylinder head, crankcase cover and camshaft have been removed. Remove carbon deposits (if present) from upper part of cylinder before removing piston. Remove connecting rod cap, then push piston and connecting rod assembly out top of cylinder. Remove retaining rings and push pin out of piston to separate piston from connecting rod.

Specified piston clearance in bore is 0.020-0.035 mm (0.0008-0.0014 in.). Measure piston diameter 10 mm (0.39

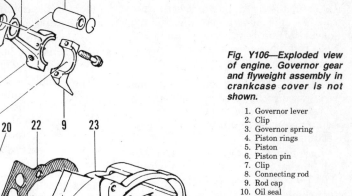

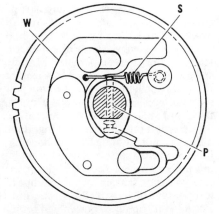

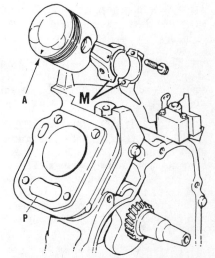

Fig. Y106—Exploded view of engine. Governor gear and flyweight assembly in crankcase cover is not shown.

1. Governor lever
2. Clip
3. Governor spring
4. Piston rings
5. Piston
6. Piston pin
7. Clip
8. Connecting rod
9. Rod cap
10. Oil seal
11. Crankcase/cylinder
12. Tappet
13. Gasket
14. Oil level sensor
15. Washer
16. Governor shaft
17. Ball bearing
18. Crankcase breather baffle
19. Camshaft
20. Crankshaft
21. Ball bearing
22. Gasket
23. Crankcase cover
24. Oil seal

Fig. Y107—Drawing of compression release mechanism on camshaft gear. When engine is stopped, spring (S) pulls weight (W) against stops on gear. Pin (P) is forced up ramp (R) so the pin extends above the exhaust camshaft lobe (see Fig. Y108), thereby opening the exhaust valve slightly.

Fig. Y108—With weight against stops, pin (P) projection (H) above cam lobe (L) should be 0.7-1.0 mm (0.028-0.039 in.).

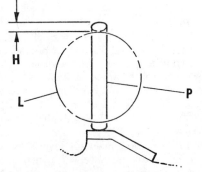

Fig. Y109—Install piston rings on piston in locations shown.

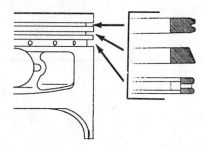

Fig. Y110—When assembling piston and rod note location of alignment mark on side of rod and arrowhead (A) on piston crown. Install piston so arrowhead is toward push rod side (P) of engine. Align match marks (M) on rod and rod cap during assembly.

in.) from bottom of piston skirt and 90° from pin bore. Specified standard diameter is 50.0 mm (1.968 in.). Minimum piston diameter is 49.93 mm (1.966 in.). Oversize piston and rings are available.

Specified ring side clearance is 0.06 mm (0.0024 in.) for top compression ring and 0.04 mm (0.0016 in.) for second compression ring and oil control ring. Maximum piston ring side clearance for all rings is 0.15 mm (0.006 in.). Specified ring end gap is 0.2 mm (0.008 in.) for all rings. Maximum ring end gap is 0.9 mm (0.035 in.).

Piston pin should be a push fit in piston. Specified piston pin diameter is 12.00 mm (0.472 in.). Minimum allowable piston pin diameter is 11.95 mm (0.470 in.).

When installing rings on piston, make sure that manufacturer's mark on face of ring is toward piston crown. Refer to Fig. Y109 for configuration and location of piston rings.

Install piston on connecting rod so arrowhead (A—Fig. Y110) on piston crown is on side of connecting rod with alignment marks (M). Lubricate cylinder, piston, piston rings, pin and connecting rod bearing with SAE 30 engine oil prior to installing piston in engine. Install piston in engine so arrowhead (A) on piston crown points toward push rod (P) side of engine. Align match marks (M) on connecting rod and rod cap. Tighten connecting rod screws to 7.8-9.8 N·m (69-87 in.-lbs.).

CONNECTING ROD. Connecting rod rides directly on crankshaft journal. Connecting rod and piston are removed as an assembly as outlined in previous section.

Specified clearance between crankpin journal and connecting rod bore is 0.016-0.034 mm (0.0006-0.0013 in.). Specified connecting rod big end diameter is 22.00 mm (0.866 in.). Maximum allowable inside diameter for connecting rod big end is 22.05 mm (0.868 in.). Specified connecting rod small end diameter is 12.00 mm (0.472 in.). Maximum allowable connecting rod small end diameter is 12.05 mm (0.474 in.).

Standard connecting rod side clearance is 0.2-0.6 mm (0.008-0.024 in.). Maximum allowable side clearance is 1.0 mm (0.039 in.).

Install connecting rod as outlined in previous section.

GOVERNOR. The internal centrifugal flyweight governor is mounted on the crankcase. The governor is driven by a gear on the crankshaft. Components must move freely without binding. The governor gear is retained by a snap ring. Be sure snap ring is properly seated in groove on shaft during assembly.

CRANKSHAFT, MAIN BEARINGS AND OIL SEALS. The crankshaft is supported at each end by ball bearing type main bearings (17 and 21—Fig. Y106) located in the crankcase and crankcase cover. The crankshaft may be removed after removing piston and connecting rod as previously outlined.

Standard crankpin journal diameter is 22.00 mm (0.866 in.). Minimum allowable crankpin diameter is 21.90 mm (0.862 in.). Maximum allowable crankshaft runout is 0.02 mm (0.001 in.).

When installing crankshaft, make certain crankshaft and camshaft gear timing marks are aligned.

Inspect main bearings and renew if rough, loose or damaged. Install bearings by pressing against numbered side until bearing is seated against shoulder in crankcase or crankcase cover.

Install oil seals so lip is toward main bearing.

CYLINDER. If cylinder bore exceeds 50.15 mm (1.974 in.) or if out-of-round of bore exceeds 0.15 mm (0.006 in.), cylinder should be bored to the next oversize.

OIL WARNING SYSTEM. The engine may be equipped with a low-oil warning system that grounds the ignition system and lights a warning lamp if the oil level is low. Refer to Fig. Y103 for wiring schematic. The oil level sensor (14—Fig. Y106) is located in the bottom of the crankcase. To test circuit, run engine then disconnect yellow sensor wire. Grounding the yellow wire to the engine should cause the warning light to flash and the engine should stop.

To check oil level sensor while installed in engine, disconnect oil sensor lead and connect an ohmmeter or continuity tester between switch lead and engine ground. With oil level correct, tester should indicate no continuity. With no oil in crankcase, tester should indicate continuity.

To test a removed oil level sensor, connect an ohmmeter or continuity tester to sensor lead and sensor mounting plate. With switch in normal position, there should be continuity. With switch inverted, there should be no continuity.

GENERATOR STATOR AND ROTOR. Install stator in crankcase cover so boss on stator is aligned with notch in crankcase cover. Install rotor so arrowhead on rotor fan aligns with cross-hatched area on stator when piston is at top dead center. Tighten rotor retaining screw to 18-24 N·m (160-212 in.-lbs.).

REWIND STARTER. The rewind starter is mounted on the blower housing. To disassemble starter, release preload tension of rewind spring by removing rope handle and allowing rope to wind slowly into starter. Remove retainer screw, retainer and spring. Remove pawl and pawl spring. Remove pulley with spring. Wear appropriate safety eyewear and gloves before removing pulley or rewind spring from pulley as spring may uncoil uncontrolled.

To reassemble, reverse the disassembly procedure. Rewind spring should be lightly greased. Install rewind spring on pulley so coil direction is counterclockwise from outer end. Wind rope around pulley in counterclockwise direction as viewed from pawl side of pulley. Position pulley in starter housing and turn pulley counterclockwise until spring tension is felt. Pass rope through rope outlet and install rope handle. Install pawl and pawl spring so spring end forces pawl toward center of pulley. Install remainder of components. To apply spring tension to pulley, pull a loop of rope into notch in pulley and rotate pulley counterclockwise a couple of turns, then pull rope out of notch and allow rope to wind into starter housing. Rope handle should be snug against housing, if not, increase spring tension by rotating pulley another turn. Check for excessive spring tension by pulling rope to fully extended length. With rope fully extended it should be possible to rotate pulley at least ½ turn counterclockwise, if not, reduce spring tension one turn.

YAMAHA

Model	Bore	Stroke	Displacement
EF1600	60 mm	43 mm	121 cc
	(2.36 in.)	(1.69 in.)	(7.4 cu. in.)
EF2500	67 mm	48 mm	169 cc
	(2.64 in.)	(1.89 in.)	(10.3 cu. in.)
EF3800	75 mm	56 mm	247 cc
	(2.95 in.)	(2.20 in.)	(15.1 cu. in.)

NOTE: Metric fasteners are used throughout engine.

ENGINE INFORMATION

These Yamaha engines are used on the Yamaha Model EF1600, EF2500 and EF3800 generators. Model EF3800E is equipped with an electric starter and, unless noted, service specifications are same as for Model EF3800. Each engine is a four-stroke, single-cylinder, air-cooled engine with a horizontal crankshaft and overhead valve system.

MAINTENANCE

LUBRICATION. The engine is splash lubricated by an oil slinger mounted on bottom of connecting rod cap. Engine oil level should be checked prior to each operating interval. Maintain oil level at lower edge of fill plug opening.

Engine oil should be changed after the first 20 hours of operation and every 100 hours of operation or six months thereafter. Crankcase capacity is 0.6 L (0.63 qt.) on Models EF1600 and EF2500, and 1.1 L (1.2 qt.) on Model EF3800.

Engine oil should meet or exceed latest API service classification. Normally, SAE 30 oil is recommended for use when temperatures are above 60° F (15° C). SAE 20 oil is recommended when temperatures are between 60° F (15° C) and 32° F (0° C). SAE 10 oil is recommended when temperatures are below 32° F (0° C).

The engine may be equipped with a low-oil warning system that grounds the ignition system and lights a warning lamp if the oil level is low. See OIL WARNING SYSTEM in REPAIRS section.

AIR FILTER. The engine is equipped with a foam type air filter element. The air filter element should be removed and cleaned after every 50 hours of operation, or more often if operating in extremely dusty conditions. To clean element, wash in a nonflammable cleaning solvent and gently squeeze element dry. Oil element with clean engine oil and gently squeeze out the excess oil.

FUEL FILTER. A fuel filter screen is located on the pickup tube of the fuel valve attached to the bottom of the fuel tank.

CRANKCASE BREATHER. The crankcase breather is located in the side of the cylinder head. Regular maintenance is not required.

SPARK PLUG. Recommended spark plug is NGK BPR6HS or equivalent. Renew spark plug if electrode is corroded or damaged. Spark plug electrode gap should be set at 0.6-0.7 mm (0.024-0.028 in.). Tighten spark plug to 17 N•m (150 in.-lbs.).

CARBURETOR. Adjustment. Idle mixture is controlled by idle jet (3—Fig. Y201) and idle mixture screw (1). Initial setting of idle mixture screw is ¼ turn out on Model EF1600, and 1¼ turns out on Models EF2500 and EF3800. Adjust idle mixture screw so engine runs smoothly at idle speed and engine will accelerate without hesitation. Standard idle mixture jet size is #37.5 on Model EF1600, #41.3 on Model EF2500, and #67.5 on Model EF3800. High speed operation is controlled by fixed main jet (9). Standard main jet size is #85 on Model EF1600, #101.3 on Model EF2500, and #113.8 on Model EF3800.

Overhaul. To disassemble carburetor, remove fuel bowl retaining screw (16—Fig. Y201), gasket (15) and fuel bowl (14). Remove float pin (12), float (11) and fuel inlet valve (7). Remove throttle and choke shaft assemblies after unscrewing throttle and choke plate retaining screws. Remove idle mixture screw (1), idle mixture jet (3), main jet (9), and if so equipped, main fuel nozzle (10).

Metal parts may be cleaned in carburetor cleaner. Direct compressed air through orifices and passageways in the opposite direction of normal fuel flow. Do not use wires or drill bits to clean orifices as enlargement of the orifice could affect the calibration of the carburetor. Inspect carburetor and renew any damaged or excessively worn components.

When assembling the carburetor note the following. With carburetor in-

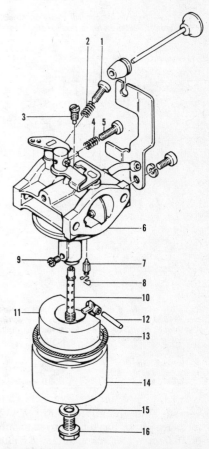

Fig. Y201—Exploded view of carburetor used on Model EF3800. Other models are similar.

1. Idle mixture screw
2. Spring
3. Idle jet
4. Spring
5. Idle speed screw
6. Body
7. Fuel inlet valve
8. Clip
9. Main jet
10. Nozzle
11. Float
12. Float pin
13. Gasket
14. Fuel bowl
15. Gasket
16. Screw

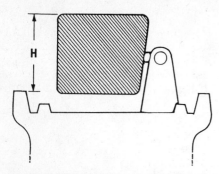

Fig. Y202—Float height (H) measured with gasket removed should be 16.0 mm (0.63 in.) on Models EF1600 and EF2500, and 15.9 mm (0.63 in.) on Model EF3800. Float height is not adjustable.

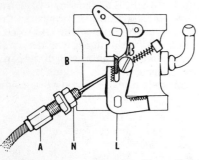

Fig. Y203—Drawing of control cable on Models EF1600 and EF2500. Refer to text for adjustment.

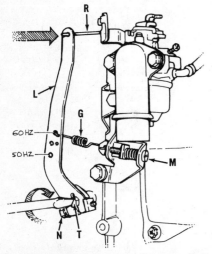

Fig. Y204—Drawing of governor linkage on Models EF1600 and EF2500. Note governor spring attachment holes on governor lever and corresponding generator current frequency.

G. Governor spring
L. Governor lever
M. Maximum governed
   speed screw
N. Nut
R. Throttle rod
T. Governor shaft

verted, float height (H—Fig. Y202) measured with gasket removed should be 16.0 mm (0.63 in.) on Models EF1600 and EF2500, and 15.9 mm (0.63 in.) on Model EF3800. Replace any components which are damaged or excessively worn and adversely affect float position.

On Models EF1600 and EF2500, adjust control cable as follows: Loosen locknut (N—Fig. Y203), then rotate ca-

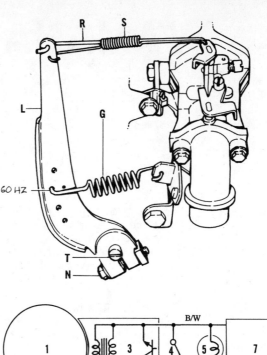

Fig. Y205—Drawing of governor linkage on Model EF3800. Note governor spring attachment hole on governor lever for 60 Hz generator operation. Maximum governed speed screw is located as shown in Fig. Y204.

G. Governor spring
L. Governor lever
N. Nut
R. Throttle rod
S. Spring
T. Governor shaft

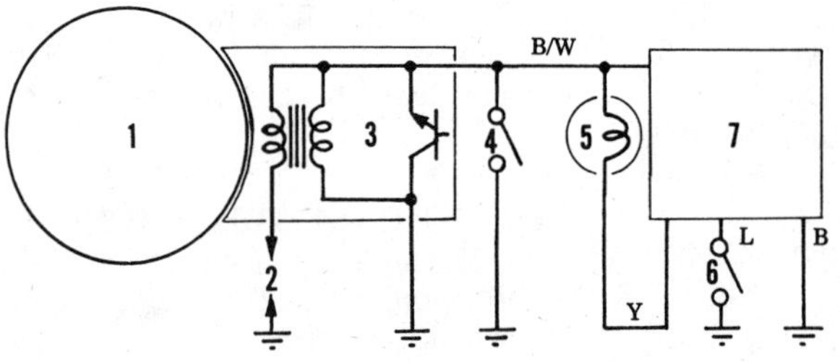

Fig. Y206—Wiring schematic for engines except Model EF3800E.

B. Black
L. Blue
W. White
Y. Yellow
1. Flywheel magneto
2. Spark plug
3. TCI unit
4. Stop switch
5. Oil warning lamp
6. Oil warning switch
7. Oil warning unit

ble adjuster (A) so choke lever (L) contacts carburetor boss (B) when control panel lever is in "ON" position (choke plate open). Tighten locknut (N). When control panel lever is in choke position, carburetor choke plate should be closed.

Some engines are equipped with an economy idle system. When the economy idle switch is actuated, an electrically controlled vacuum solenoid routes vacuum to a diaphragm that controls the carburetor throttle. Engine speed at economy idle should be approximately 2300 rpm.

Some engines are equipped with an automatic choke. A vacuum diaphragm is connected to the choke lever. Vacuum opens the choke half way when the engine starts. After the engine starts, current through a bimetallic strip moves the choke to the full-open position after approximately two minutes of running.

GOVERNOR. The engine is equipped with a mechanical, flyweight type governor. To adjust governor linkage, proceed as follows: Loosen governor lever clamp nut (N—Fig. Y204 or Y205), rotate governor lever (L) so throttle plate is fully open and hold lever in place. Turn governor shaft (T) clockwise as far as possible, then tighten nut (N).

The governor spring (G) must be connected to the proper hole in the governor lever according to required generator

current frequency. See Fig. Y204 or Y205.

Rotating maximum governed engine speed screw (M—Fig. Y204) will adjust generator current frequency. Maximum governed no-load engine speed should be 3180-3210 rpm for 50 Hz operation or 3780-3810 rpm for 60 Hz operation.

IGNITION SYSTEM. The breakerless ignition system requires no regular maintenance. The ignition coil unit is mounted outside the flywheel. Air gap between flywheel and ignition coil legs should be 0.5 mm (0.020 in.). Ignition timing is not adjustable. Refer to Fig. Y206 or Y207 for wiring schematic.

The engine may be equipped with a low-oil warning system that grounds the ignition system and lights a warning lamp if the oil level is low. See OIL WARNING SYSTEM in REPAIRS section.

VALVE ADJUSTMENT. Valve-to-rocker arm clearance should be checked and adjusted after every 300 hours of operation. Engine must be cold when checking valve clearance. Specified clearance is 0.03-0.08 mm (0.001-0.003 in.) for both valves.

To adjust valve clearance, remove rocker arm cover. Rotate crankshaft so piston is at top dead center (TDC) on compression stroke. Insert proper

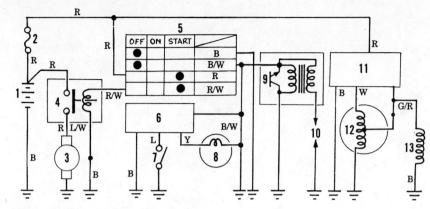

Fig. Y207—Wiring schematic for Model EF3800E.

B. Black
G. Green
L. Blue
R. Red
W. White

Y. Yellow
1. Battery
2. Fuse
3. Starter
4. Starter relay

5. Main switch
6. Oil warning unit
7. Oil warning switch
8. Oil warning lamp
9. TCI unit

10. Spark plug
11. Regulator
12. Flywheel magnets
13. Auto choke coil

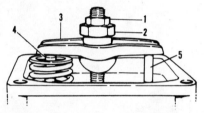

Fig. Y208—View of rocker arm and related parts.

1. Jam nut
2. Adjustment nut
3. Rocker arm

4. Valve stem clearance
5. Push rod

thickness feeler gauge between rocker arm (3—Fig. Y208) and end of valve stem (4). Loosen rocker arm jam nut (1) and turn adjusting nut (2) to obtain desired clearance (a slight drag should be felt when withdrawing the feeler gauge). Tighten jam nut and recheck clearance. Install rocker arm cover.

COMPRESSION PRESSURE. Compression pressure reading should be 392-490 kPa (57-71 psi).

REPAIRS

TIGHTENING TORQUES. Recommended tightening torques specifications are as follows:

Generator rotor:
EF1600, EF2500 . . . . 22 N·m
(195 in.-lbs.)
EF3800 . . . . 42 N·m
(31 ft.-lbs.)

Connecting rod:
EF1600, EF2500 . . . . 17 N·m
(150 in.-lbs.)
EF3800 . . . . 27 N·m
(19.8 ft.-lbs.)

Crankcase cover:
EF1600 . . . . 11 N·m
(97 in.-lbs.)
EF2500, EF3800 . . . . 20 N·m
(177 in.-lbs.)

Cylinder head . . . . 29 N·m
(21.3 ft.-lbs.)

Flywheel:
EF1600 . . . . 35 N·m
(25.7 ft.-lbs.)
EF2500 . . . . 65 N·m
(47.8 ft.-lbs.)
EF3800 . . . . 70 N·m
(51.5 ft.-lbs.)
Spark plug . . . . 17 N·m
(150 in.-lbs.)

CYLINDER HEAD. To remove cylinder head (11—Fig. Y209), remove air cleaner, carburetor and muffler. Disconnect breather tube, then remove blower shroud and rocker arm cover. Unscrew cylinder head bolts and remove cylinder head.

Clean combustion deposits from cylinder head, then check for cracks or other damage. Check cylinder head mounting surface for warpage using a feeler gauge and straightedge. Maximum allowable cylinder head warpage is 0.030 mm (0.0012 in.). Minor distortion of cylinder head surface can be trued by moving cylinder head across 400 grit sandpaper placed on a surface plate.

When installing cylinder head, tighten head screws evenly in a crossing pattern to a final torque of 29 N·m (21 ft.-lbs.). Adjust valve clearance as outlined in VALVE ADJUSTMENT paragraph.

VALVE SYSTEM. Remove rocker arms, compress valve springs and remove valve retainers. Remove valves, springs and valve spring retainers. Note that intake valve is equipped with a seal.

Clean carbon from valve stem and head. Inspect valves for excessive wear or damage and renew as necessary.

Valve face and seat angles are 45° for both intake and exhaust. Recommended valve seating width is 1.1 mm (0.043 in.) with a maximum allowable seat width of 1.7 mm (0.067 in.). Valve should be lapped to its seat if poor valve seating or sealing is evident.

Refer to following table for valve stem dimensions:

Valve Stem Diameter

EF1600, EF2500:
Intake . . . . 5.468-5.480 mm
(0.2152-0.2157 in.)
Exhaust . . . . 5.440-5.460 mm
(0.2142-0.2150 in.)

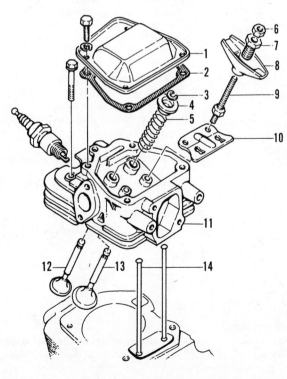

Fig. Y209—Exploded view of cylinder head assembly. Seal used on intake valve is not shown.

1. Rocker arm cover
2. Gasket
3. Valve key
4. Valve spring retainer
5. Valve spring
6. Jam nut
7. Adjustment nut
8. Rocker arm
9. Stud
10. Guide plate
11. Cylinder head
12. Exhaust valve
13. Intake valve
14. Push rod

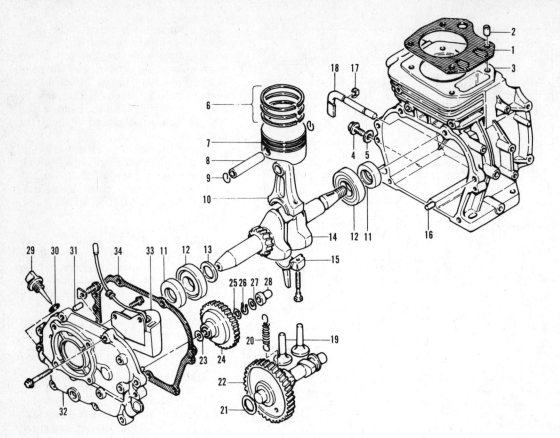

Fig. Y210—Exploded view of engine.

1. Head gasket
2. Dowel pins
3. Crankcase/cylinder
4. Drain screw
5. Gasket
6. Piston rings
7. Piston
8. Piston pin
9. Clip
10. Connecting rod
11. Oil seal
12. Ball bearing
13. Thrust washer
14. Crankshaft
15. Rod cap
16. Dowel pin
17. Snap ring
18. Governor shaft
19. Tappet
20. Compression release spring
21. Thrust washer
22. Camshaft
23. Washer
24. Governor
25. Washer
26. Snap ring
27. Washer
28. Sleeve
29. Dipstick
30. "O" ring
31. Dowel pin
32. Crankcase cover
33. Oil level sensor
34. Gasket

EF3800:
Intake . . . . 6.460-6.475 mm
          (0.2543-0.2549 in.)
Exhaust . . . . 6.440-6.460 mm
          (0.2535-0.2543 in.)

Refer to following table for valve stem-to-guide clearance:

Stem-to-Guide Clearance

EF1600, EF2500:
Intake . . . . 0.020-0.044 mm
          (0.0008-0.0017 in.)
Exhaust . . . . 0.040-0.072 mm
          (0.0016-0.0028 in.)
EF3800:
Intake . . . . 0.025-0.055 mm
          (0.0010-0.0022 in.)
Exhaust . . . . 0.040-0.075 mm
          (0.0016-0.0030 in.)

Valve guides are integral with cylinder head. Cylinder head should be renewed if valve guides are worn excessively.

Specified valve spring minimum free length is 33.2 mm (1.31 in.) for both valve springs on Models EF1600 and EF2500, and 33.0 mm (1.30 in.) on Model EF3800. Valve spring pressure for both valves on Models EF1600 and EF2500 should be 5.9 kg (13.0 lb.) at compressed height of 22.5 mm (0.88 in.). Valve spring pressure for both valves on

Model EF3800 should be 6.4 kg (14.0 lb.) at compressed height of 27.0 mm (1.06 in.).

Maximum push rod runout is 0.2 mm (0.008 in.).

CAMSHAFT. Camshaft and camshaft gear (22—Fig. Y210) are an integral casting which is equipped with a compression release mechanism. The compression release pin extends at cranking speed to hold the exhaust valve open slightly thereby reducing compression pressure.

To remove camshaft, remove blower shrouds and rocker arm cover and disengage push rods from rocker arms. Drain engine oil, then remove generator and crankcase cover (32) and withdraw camshaft. Mark tappets so they can be installed in original position.

Inspect camshaft journal and refer to following table for specified camshaft journal diameter:

Camshaft Journal Diameter

EF1600 . . . . 13.966-13.984 mm
          (0.5498-0.5506 in.)
EF2500 . . . . 14.966-14.984 mm
          (0.5892-0.5899 in.)
EF3800 . . . . 17.966-17.984 mm
          (0.7073-0.7080 in.)

Inspect camshaft lobes for excessive wear and damage. Specified camshaft

lobe height for both lobes is 24.65 mm (0.970 in.) for Model EF1600, 26.1 mm (1.028 in.) for Model EF2500, and 32.44 mm (1.277 in.) for Model EF3800.

Inspect compression release mechanism and check for proper operation. Compression release spring should hold weight against stops.

When installing camshaft, make certain that crankshaft gear and camshaft gear timing marks are aligned. Tighten crankcase cover screws following proper tightening sequence shown in Fig. Y211A or Y211B to 11 N·m (97 in.-lbs.) on Model EF1600 or to 20 N·m (177 in.-lbs.) on Models EF2500 and EF3800.

PISTON, PIN AND RINGS. Piston and connecting rod are removed as an assembly after cylinder head, crankcase cover and camshaft have been removed. Remove carbon deposits (if present) from upper end of cylinder before removing the piston. Remove connecting rod cap (15—Fig. Y210), then push piston and connecting rod assembly out top of cylinder. Remove retaining rings (9), use a suitable puller to remove piston pin (8), then separate piston (7) from connecting rod (10).

Specified piston clearance in bore is 0.020-0.055 mm (0.0008-0.0022 in.) for Model EF1600 and 0.030-0.070 mm

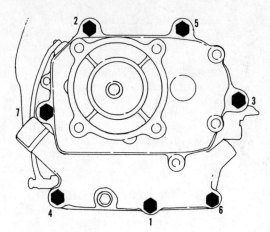

Fig. Y211A—Crankcase cover screw tightening sequence for EF1600 and EF2500 engines.

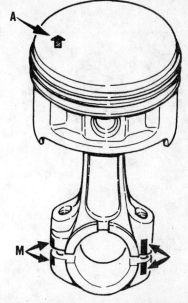

Fig. Y213—Arrow (A) on piston crown must point toward flywheel side of engine. Marks (M) on rod and cap must be aligned and toward crankcase cover side of engine. Marks (M) may appear on either side of rod and cap depending on engine model, but must be aligned during assembly.

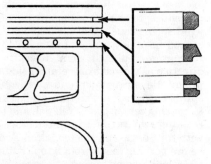

Fig. Y211B—Crankcase cover screw tightening sequence for EF3800 engine.

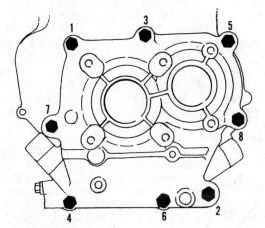

Fig. Y212—Install piston rings on piston in locations shown.

(0.0012-0.0027 in.) for Models EF2500 and EF3800.

Measure piston diameter 10 mm (0.39 in.) from bottom of piston skirt and 90° from pin bore. Refer to following table for standard piston diameter:

Piston Diameter	
EF1600	59.965-59.980 mm
	(2.3608-2.3614 in.)
Min. dia.	59.87 mm
	(2.357 in.)
EF2500	66.955-66.970 mm
	(2.6360-2.6366 in.)
Min. dia.	66.87 mm
	(2.633 in.)

EF3800	74.950-74.970 mm
	(2.9508-2.9516 in.)
Min. dia.	74.87 mm
	(2.948 in.)

Specified ring side clearance is 0.02-0.06 mm (0.0008-0.0024 in.) for compression rings. Maximum compression ring side clearance is 0.10 mm (0.004 in.). Specified ring end gap is 0.2-0.4 mm (0.008-0.016 in.) for all rings. Maximum ring end gap is 0.9 mm (0.035 in.).

Inspect piston pin for excessive wear and damage. Refer to following table for specified piston pin diameter:

Piston Pin Diameter	
EF1600	13.000-13.005 mm
	(0.5118-0.5120 in.)
Min. dia.	12.950 mm
	(0.5098 in.)
EF2500	15.000-15.005 mm
	(0.5906-0.5907 in.)
Min. dia.	14.950 mm
	(0.5886 in.)
EF3800	18.000-18.005 mm
	(0.7087-0.7088 in.)
Min. dia.	17.950 mm
	(0.7067 in.)

When installing rings on piston, make sure that manufacturer's mark on face of ring is toward piston crown. Refer to Fig. Y212 for configuration and location of piston rings.

Install piston on connecting rod so arrow (A—Fig. Y213) on piston crown will point toward flywheel and alignments marks (M) on side of connecting rod will point toward crankcase cover. Note that alignment marks may be on left or right side of rod depending on engine model. Heat piston to approximately 180° F (80° C) before installing piston pin. Lubricate cylinder, piston, piston pin, rings and connecting rod bearing before installing piston in engine. Install piston in engine so arrow (A) on piston crown points toward flywheel side of engine. Align match marks (M) on connecting rod and rod cap. Tighten connecting rod screws to 17 N·m (150 in.-lbs.) on Models EF1600 and EF2500 or to 27 N·m (19.8 ft.-lbs.) on Model EF3800.

CONNECTING ROD. Connecting rod rides directly on crankshaft journal. Connecting rod and piston are removed as an assembly as outlined in previous section.

Specified clearance between crankpin journal and connecting rod bore is 0.015-0.040 mm (0.0006-0.0016 in.). Refer to following table for specified connecting rod big end diameter:

Rod Big End Diameter	
EF1600	25.500-25.515 mm
	(1.0039-1.0045 in.)

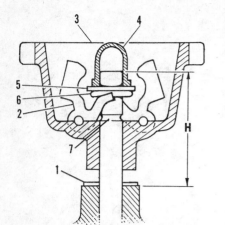

Fig. Y214—Governor shaft height (H) should be 36 mm (1.42 in.).

1. Washer
2. Flyweight
3. Governor gear
4. Sleeve
5. Washer
6. Snap ring
7. Shaft groove

EF2500.......... 30.000-30.015 mm
(1.1811-1.1817 in.)

EF2500.......... 33.500-33.515 mm
(1.3189-1.3195 in.)

Refer to following table for specified connecting rod small end diameter:

Rod Small End Diameter

EF1600.......... 13.015-13.025 mm
(0.5124-0.5128 in.)

EF2500.......... 15.015-15.025 mm
(0.5911-0.5915 in.)

EF3800.......... 18.015-18.025 mm
(0.7092-0.7096 in.)

Standard connecting rod side clearance is 0.4-1.1 mm (0.016-0.043 in.). Maximum allowable side clearance is 1.3 mm (0.051 in.).

Install connecting rod as outlined in previous section.

GOVERNOR. The internal centrifugal flyweight governor (24—Fig. Y210) is mounted on the crankcase cover. The governor is driven by the camshaft gear. Components must move freely without binding. The governor gear is retained on governor shaft by a snap ring.

Governor shaft height (H—Fig. Y214) should be 36 mm (1.42 in.). Refer to Fig. Y214 for governor assembly order. Be sure snap ring (6) is properly seated in groove (7) on shaft during assembly.

CRANKSHAFT, MAIN BEARINGS AND OIL SEALS. The crankshaft is supported at each end by ball bearing type main bearings (12—Fig. Y210) located in the crankcase and crankcase cover. The crankshaft may be

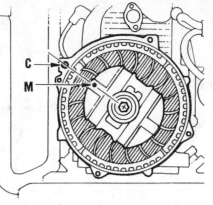

Fig. Y215—Install rotor so mark (M) on rotor fan aligns with crosshatched area (C) on generator case when piston is at top dead center.

removed after removing piston and connecting rod as previously outlined.

Refer to following table for specified crankpin diameter:

Crankpin Diameter

EF1600.......... 25.475-25.485 mm
(1.0029-1.0033 in.)

EF2500.......... 29.975-29.985 mm
(1.1801-1.1805 in.)

EF2500.......... 33.475-33.485 mm
(1.3179-1.3183 in.)

Maximum allowable crankshaft runout is 0.04 mm (0.0016 in.).

Inspect main bearings and renew if rough, loose or damaged. Install bearings by pressing against numbered side until bearing is seated against shoulder in crankcase or crankcase cover.

Install oil seals so lip is toward main bearing. When installing crankshaft, make certain crankshaft and camshaft gear timing marks are aligned.

CYLINDER. If cylinder bore exceeds following diameter or if out-of-round of bore exceeds 0.05 mm (0.002 in.), cylinder should be bored to the next oversize.

Cylinder Diameter

EF1600.......... 60.000-60.020 mm
(2.3622-2.3630 in.)

Min. dia.............. 60.12 mm
(2.367 in.)

EF2500.......... 67.000-67.020 mm
(2.6378-2.6386 in.)

Min. dia.............. 67.12 mm
(2.642 in.)

EF3800.......... 75.000-75.020 mm
(2.9528-2.9535 in.)

Min. dia.............. 75.12 mm
(2.957 in.)

OIL WARNING SYSTEM. The engine may be equipped with a low-oil warning system that grounds the igni-

tion system and lights a warning lamp if the oil level is low. The oil level sensor (33—Fig. Y210) is attached to the inside of the crankcase cover. See Fig. Y206 or Y207 for wiring schematic.

To test oil level sensor, disconnect sensor leads and connect an ohmmeter or continuity tester to sensor lead and sensor base. With sensor in normal position, there should be continuity. With sensor inverted, there should be no continuity.

GENERATOR STATOR AND ROTOR. Install rotor so mark (M—Fig. Y215) on rotor fan aligns with crosshatched area (C) on generator case when piston is at top dead center. Tighten rotor retaining screw to 22 N·m (195 in.-lbs.) on Models EF1600 and EF2500 or 42 N·m (31 ft.-lbs.) on Model EF3800.

REWIND STARTER. Models EF1600 and EF2500. To disassemble starter, remove rope handle and allow rope to wind slowly into starter. Unscrew center screw and remove retainer, brake spring, pawls and pawl springs. Wear appropriate safety eyewear and gloves before disengaging pulley from starter as spring may uncoil uncontrolled. Place shop towel around pulley and lift pulley out of housing; spring should remain with pulley.

Inspect components for damage and excessive wear. Reverse disassembly procedure to install components. Rewind spring coils wind in clockwise direction in pulley from outer end. Wind rope around pulley in counterclockwise direction as viewed from retainer side of pulley. Be sure inner end of rewind spring engages spring retainer adjacent to housing center post. Pulley should be rotated two turns counterclockwise with rope engaged in notch in pulley or before passing rope through rope outlet to preload rewind spring.

Model EF3800. To disassemble starter, release preload tension of rewind spring by removing rope handle and allowing rope to wind slowly into starter. Remove retainer nut, retainer, friction plate, pawl and pawl spring. Remove pulley. Wear appropriate safety eyewear and gloves before removing pulley as rewind spring in housing may uncoil uncontrolled.

To reassemble, reverse the disassembly procedure. Rewind spring should be lightly greased. Install rewind spring in starter housing so coil direction is clockwise from outer end. Wind rope around pulley in counterclockwise direction as viewed from pawl side of pulley. Position pulley in starter housing and turn pul-

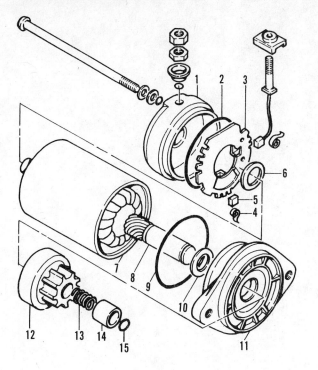

Fig. Y216—Exploded view of electric starter motor used on Model EF3800E.

1. End cap
2. "O" ring
3. Brush plate
4. Brush spring
5. Brush
6. Washer
7. Frame
8. Armature
9. "O" ring
10. Washer
11. Drive plate
12. Drive gear
13. Spring
14. Stop
15. Snap ring

ley counterclockwise until spring tension is felt. Wind rope around pulley in counterclockwise direction as viewed from pawl side of pulley. Position pulley in starter housing and turn pulley counterclockwise until spring tension is felt. Install pawl and pawl spring so spring end forces pawl toward center of pulley. Install remainder of components. Rotate pulley four turns counterclockwise, pass rope through rope outlet and install rope handle.

ELECTRIC STARTER. Model EF3800E is equipped with an electric starter motor and starter relay. Refer to Fig. Y207 for wiring schematic.

Refer to Fig. Y216 for an exploded view of electric starter motor. Minimum brush length is 9.3 mm (0.36 in.). Minimum commutator diameter is 26.4 mm (1.039 in.). Note alignment bosses and notches on end cap, frame and drive plate during assembly.

NOTES

NOTES

NOTES

NOTES

NOTES

NOTES

NOTES